FRONTIERS IN MODELING AND CONTROL OF BREATHING

INTEGRATION AT MOLECULAR, CELLULAR, AND SYSTEMS LEVELS

ADVANCES IN EXPERIMENTAL MEDICINE AND BIOLOGY

Recent Volumes in this Series

Volume 490
MECHANISMS OF LYMPHOCYTE ACTIVATION AND IMMUNE REGULATION VIII: Autoimmunity 2000 and Beyond
Edited by Sudhir Gupta

Volume 491
THE MOLECULAR IMMUNOLOGY OF COMPLEX CARBOHYDRATES—2
Edited by Albert M. Wu

Volume 492
NUTRITION AND CANCER PREVENTION: New Insights into the Role of Phytochemicals
Edited under the auspices of the American Institute for Cancer Research

Volume 493
NEUROIMMUNE CIRCUITS, DRUGS OF ABUSE, AND INFECTIOUS DISEASES
Edited by Herman Friedman, Thomas W. Klein, and John J. Madden

Volume 494
THE NIDOVIRUSES (CORONAVIRUSES AND ARTERIVIRUSES)
Edited by Ehud Lavi, Susan R. Weiss, and Susan T. Hingley

Volume 495
PROGRESS IN BASIC AND CLINICAL IMMUNOLOGY
Edited by Andreij Mackiewicz, Jan Zeromski, and Macief Kurpisz

Volume 496
NONINVASIVE ASSESSMENT OF TRABECULAR BONE ARCHITECTURE AND THE COMPETENCE OF BONE
Edited by Sharmila Majumdar, Ph.D., and Brian K. Bay, Ph.D.

Volume 497
INTRACTABLE SEIZURES: Diagnosis, Treatment, and Prevention
Edited by W. McIntyre Burnham, Peter L. Carlen, and Paul A. Hwang

Volume 498
DIABETES AND CARDIOVASCULAR DISEASE: Etiology, Treatment, and Outcomes
Edited by Aubie Angel, Naranjan Dhalla, Grant Pierce, and Pawan Singal

Volume 499
FRONTIERS IN MODELING AND CONTROL OF BREATHING
Edited by Chi-Sang Poon and Homayoun Kazemi

A Continuation Order Plan is available for this series. A continuation order will bring delivery of each new volume immediately upon publication. Volumes are billed only upon actual shipment. For further information please contact the publisher.

FRONTIERS IN MODELING AND CONTROL OF BREATHING

INTEGRATION AT MOLECULAR, CELLULAR, AND SYSTEMS LEVELS

Edited by

Chi-Sang Poon
Massachusetts Institute of Technology
Cambridge, Massachusetts

and

Homayoun Kazemi
Massachusetts General Hospital
and Harvard Medical School
Boston, Massachusetts

Kluwer Academic / Plenum Publishers
New York, Boston, Dordrecht, London, Moscow

Library of Congress Cataloging-in-Publication Data

Frontiers in modeling and control of breathing: integration at molecular, cellular, and systems levels/edited by Chi-Sang Poon and Homayoun Kazemi.
p. cm. — (Advances in experimental medicine and biology; v. 499)
"The VIII Oxford Conference convened on October 11–15, 2000 at ... North Falmouth on Cape Cod in Massachusetts, USA"—Pref.
Includes bibliographical references and index.
ISBN 0-306-46647-3
1. Respiration—Regulation—Congresses. I. Poon, Chi-Sang, 1952- II. Kazemi, Homayoun, 1934- III. Oxford Conference (8th: 2000: North Falmouth, Mass.) IV. Series.

QP123 .F766 2001
612.2—dc21

2001038337

Proceedings of the VIII Oxford Conference, held October 11–15, 2000, in North Falmouth, Massachusetts.

ISBN 0-306-46647-3

233 Spring Street, New York, N.Y. 10013

http://www.wkap.nl/

10 9 8 7 6 5 4 3 2 1

A C.I.P. record for this book is available from the Library of Congress

Printed in the United States of America

Conference Proceedings

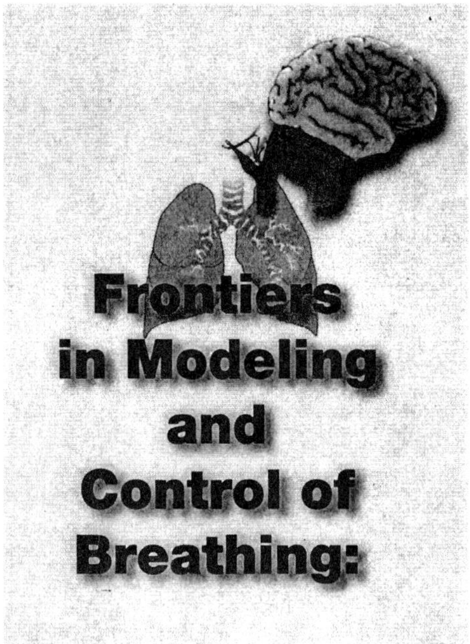

Integration at Molecular, Cellular, and Systems Levels

October 11 –15, 2000
N. Falmouth (Cape Cod), Massachusetts, USA

Conference Co-Chairs: Chi-Sang Poon, (Cambridge, MA)
Homayoun Kazemi, (Boston, MA)

International Program Committee
M. Aoki (Japan)
E.N. Bruce (USA)
J. Champagnat (France)
J. Duffin (Canada)
A.Guz (UK)
M.C.K. Khoo (USA)
J. Lipski (New Zealand)
E.E. Nattie (USA)
N. Prabhakar (USA)
J.E. Remmers (Canada)
D.W. Richter (Germany)
G. Song (China)
S.A. Ward (UK)

International Organizing Committee
G. Benchetrit (France)
Y. Honda (Japan)
R.L. Hughson (Canada)
C.-S. Poon (USA)
P.A. Robbins (UK)
J.W. Severinghaus (USA)
S.A. Ward (UK)
B.J. Whipp (UK)

Local Organizing Committee
S.A. Shea (Boston, MA)
M. Siniaia (Cambridge, MA)
D. Systrom (Boston, MA)
D. Ward (Rochester, NY)

The Organizing Committee gratefully acknowledges support from the following groups:

Sponsored by

The American Physiological Society

The Physiological Society (UK)

The Whitaker Foundation

Respiration Physiology

GlaxoSmithKline Inc.

Harvard Apparatus, Inc.

Endorsed by

The American Thoracic Society

Never mind the troubles before you
They all will fade away
You can count on me to stand and to hold you
'Cause I am here to stay
If only for today
I am unafraid

— *Anthem of VIII Oxford Conference*

VIII Oxford Conference Participants

PREFACE

Over the past two decades the international *Oxford Conference* series has cultivated an acclaimed tradition of periodically bringing together scientists in diverse disciplines of neuroscience, physiology, and clinical medicine as well as mathematics and engineering for the exchange of ideas and concepts, and presentation of new and exciting data concerning respiratory control, in a placid setting conducive to rigorous yet genial scientific interactions. The history of this triennial event was eloquently recounted at this most recent conference by Dr. Brian Whipp, whose elegant narrative forms the opening chapter in this book.

Building on this rich legacy, the VIII Oxford Conference convened on October 11-15, 2000, at the Sea Crest Conference Center in North Falmouth on Cape Cod in Massachusetts. With the spectacular Old Silver Beach on the shores of Buzzard's Bay as a backdrop, the conference presented a superb forum for formal and informal discussions among the participants on the latest discoveries and observations related to modeling and control of breathing. The environment itself and the surprise blessing of unseasonably balmy weather at the time of the meeting allowed a most inspiring and enjoyable experience for all. The camaraderie amongst the conferees—and the unique opportunity to get to know one another up close and personal in a pleasant environment—added to the luster of the occasion. All told, more than 140 scientists, students and their companions from 12 different countries on 4 continents around the globe attended the conference.

The scientific sessions were the highlights of the meeting. We were fortunate to have three outstanding plenary lectures: the first one by Dr. John Severinghaus on the history of respiratory control and modeling, the second by Dr. Peter Scheid on central chemosensitivity, and the third by Dr. Jean Champagnat on genetic and developmental models for neural control of breathing. Apart from these invited lectures, the scientific program was devoted mainly to investigator contributed presentations, both in oral sessions and in poster sessions. The over one hundred papers presented at the conference covered the wide spectrum of modeling and experimental studies of respiratory control ranging from genetics and ion channels to respiratory diseases. In addition to these formal sessions the scientific program was complemented by five modeling-related workshops, each of which was led by prominent scientists in the field with extensive discussions among the audience. Four of these workshops addressed some emerging controversial issues regarding the complex and multiplex mechanisms of the respiratory central pattern generator, peripheral and central chemoreceptors, chemoreflex pathways, and respiratory instability and variability. A fifth workshop reviewed the accomplishments and state of the art of biomedical modeling, the impending opportunities and challenges in this field, and the recommended long-range plans of interest to the Oxford Conference community in the future. An executive summary of this biomedical modeling workshop—which was drafted by the workshop chair, Dr. Neil Cherniack, and other panelists with extensive feedback from all participants—is presented in the second chapter of this book. Beginning with Chapter 3, the bulk of the book is comprised of papers based on the lectures, presentations of original work, and summaries of workshops of the VIII Oxford Conference. Altogether, this book represents an excellent compendium of the current state of knowledge in the area of modeling and control of breathing.

One of the other highlights of the meeting was a research competition for pre- and postdoctoral trainees, in which some 19 abstracts were received and reviewed by an international committee. Six finalists made oral presentations in a dedicated session at the conference, and a panel of senior scientist attendees chose three as the recipients of the special cash awards and certificates presented at the final banquet: Christopher Del Negro from the U.S. National Institutes of Health, David Kline from Case Western Reserve University, and Daniel Young from Massachusetts Institute of Technology. The selection committee stressed the difficulty of selecting these three as award winners in view of the high quality of science and presentation demonstrated by all the other trainee participants, whose works fill many of the chapters in this book.

We are particularly grateful to all the colleagues who contributed to the conference and, in a timely manner, provided us with the first-rate chapters that constitute this book. We also appreciate the help of our International Program Committee and our International and Local Organizing Committees in sustaining the high standards of science and a modus operandi worthy of the prestige of the *Oxford Conference* series. Special thanks are also due to Denise Small, whose continuing efforts through the past months in helping to organize and manage the infrastructure of the meeting made this a successful conference and, by all expectations, a valuable book. We also acknowledge with particular thanks all the sponsors whose generous financial and material support made this event possible.

For ourselves, the privilege of being able to organize the VIII Oxford Conference has been a most humbling and breathtaking experience in the past three years. The enormous advances in this field and the overwhelming enthusiasm of all the participants are reflected in this proceedings—a volume that should usher in the new millennium as an important reference for all investigators interested in the modeling and control of breathing for many years to come.

Chi-Sang Poon
Cambridge, Massachusetts

Homayoun Kazemi
Boston, Massachusetts

December 2000

CONTENTS

Oxford Conference: The Past, Present and Future

Central and Peripheral Chemoreceptors

Respiratory Neurons and Neural Networks

Development and Plasticity of Respiratory Control

Chemoreflex and Exercise Control of Breathing

Respiratory Sensation and Behavioral Control

Stability and Variability of Breathing

Oxford Conference: The Past, Present and Future

REMEMBRANCE OF "OXFORD" CONFERENCES PAST

Brian J. Whipp*

The seeds of what subsequently flowered into the triennial "Oxford Conferences" on modelling and the control of breathing were sown at the conference dinner of a scientific meeting at the Polish Academy of Sciences in Warsaw in 1971 at which Dan Cunningham, a respiratory physiologist at the University of Oxford and Richard Herczynski a mathematician and fluid dynamicist at the Polish Academy of Sciences, happened to be seated beside each other. As recounted by Cunningham in the interview that he gave to the Physiological Society (of the United Kingdom) in 1994 as part of its Oral History project:

> *"My paper at the Warsaw meeting was about the pattern of breathing ...... I was trying to express the thing quantitatively and he drew it all out on a paper table napkin for me ...... we thought at that time, supposing people like him and people like myself got together and talked this kind of thing out in public at a meeting, it might be to the benefit of both sides"*

The actual conference which brought together "people like Cunningham" and "people like Herczynski" did not take place until September 1978. It was held at the University Laboratory of Physiology in Oxford under the joint organisational aegis of the Institute of Management and Control and the Physiological Society. The symposium was entitled "Modelling of a Biological Control System: the Regulation of Breathing ". It attracted 90 participants, from 10 different countries who presented papers which utilised a wide variety of experimental and theoretical approaches to respiratory control: in vivo, in vitro and "in silico". In addition to the scientific sessions themselves, the participants were able to enjoy the attendant attractions of the Oxford academic milieu, "dreaming spires" and all: the participants were housed and had meals at The Queen's College, the conference banquet was held in the dining hall of New College and with a recital arranged at the Holywell Music Room. Richard Herczynski, sadly, was not able to attend owing to the political climate of the times. The meeting itself was highly successful and

* Department of Physiology, St. George's Hospital Medical School, Cranmer Terrace, London, SW17 0RE, UK

Frontiers in Modeling and Control of Breathing, edited by Poon and Kazemi.
Kluwer Academic/Plenum Publishers, New York, 2001.

is still recalled with pleasure by those who attended it. There were, however, no published proceedings of this conference: rather the participants received a strikingly-yellow 212 page soft-backed pre-print that contained prolegomena, running to 164 pages. This was co-authored by: E. R. Carson, of the City University, London; D. J. C. Cunningham; R. Herczynski; D. J. Murray-Smith, of the University of Glasgow and E. S. (Ebbe) Petersen of Oxford University. The entire prolegomenon attempted to harness the resources of the physiological and mathematical approaches to the regulation of breathing, including considerations of the conceptual justification for the procedures used in addition to the details of the procedures themselves. Although more than 20 years old much in the prolegomenon remains useful. The remainder of the pre-print is taken up with the abstracts of the presentations, totalling 43. Sadly however, not everyone who presented provided an abstract and so those not in attendance can only surmise what, for example, Fred Grodins might have said on the tantalising topic of "Exercise Hyperpnea - the Ultra-Secret".

There were, at that time however, no apparent plans for a second such conference. But the success of the Oxford meeting and the importance of the issues being addressed, along with the range of techniques being offered to address them, seemed to justify a follow-up. To this effect Karl Wasserman, Susan Ward and I, representing the physiological perspective, and J. Weldon Bellville, Fred Grodins, George Swanson and Don. Wiberg, representing that of the control system engineers, met at the Harbor-UCLA Medical Center in Torrance, California, to plan a follow-up conference . This took place in September 1982 at the University of California Conference Center at Lake Arrowhead under the sponsorship of the American Physiological Society and the IEEE Control Systems and Engineering in Medicine and Biology Societies. This lakeside, forested retreat in the San Bernadino mountains of Southern California was chosen to provide what the organisers termed:

> "*a sufficiently placid and contemplative milieu to be conducive to discussions which would extend beyond the formal sessions*"

And to maximise the opportunity for such informal discussions the scientific sessions were scheduled for the mornings and evenings - the afternoons being free. The conference Proceedings were published[1] with a yellow hard-back cover in recognition of the original Oxford yellow book. It contains 49 chapters of the presentations including introductory and concluding chapters by Cunningham and William Yamamoto respectively. But if the Arrowhead conference has any importance in the documentary history of the Oxford Conferences it will likely be because a session was convened to consider whether a continuing cycle of conferences on the topic was justified. There was overwhelming, and to my recollection unanimous, support for the idea. The only concern being the "when" and the "where" of the next conference. It was the general view that the four years that had elapsed between the Oxford and Arrowhead conferences was, perhaps, too long and that a three year period might be more appropriate: and so it came to pass. But where? All, I recall, were delighted when Gila Benchetrit and her colleagues, P. Baconnier and J. Demongeot, of the University of Grenoble offered to host the next conference in France.

They did so in September 1985. And in so doing, outdid all other conferences in the series with respect to "a quiet, contemplative milieu" by convening the meeting at the medieval Abbey of Solignac in the Limoges area of France. Being waited on at table by monks produced an air of calm and even meditative serenity that remains unsurpassed in the series. Among the scientific attractions of the conference, for which we remain grateful to Benchetrit and her colleagues, was the opportunity to hear the great French mathematician Rene Thom, the founder and chief exponent of Catastrophe Theory, address issues relating to respiratory control from his perspective. In his chapter "Towards a Typology of Regulation" in the subsequent Proceedings[2] Thom provides (page 93) what must be the most seductive sentence in the entire canon of respiratory control:

> "*among the lineaments of these lips, only one curve would be of experimental importance*"

The published proceedings of the Solignac conference, contained 37 chapters, and retained the yellow binding of the previous proceedings. The volume is also meritorious in another sense, that being the published discussions of each of the presentations. This adds greatly to what, in retrospect, seems to have been the flavour of the conference. This is the only volume in the series, however, that has provided this additional feature. A highlight of the conference, undoubtedly, was the excursion to the prehistoric cave paintings of Lascaux in the Dordogne. Even though the originals are no longer available for public viewing the facsimiles in the duplicate parallel caves provide a profound and intellectually stimulating experience.

The torch was then passed to George Swanson who, along with Fred Grodins, Richard Hughson and Denham Ward planned the 4th conference in the series. This was held in September of 1988 at Grand (also known as Spirit) Lake in the Rocky Mountains of Colorado, chosen according to Swanson's preface to the subsequently published proceedings[3]:

> "*to continue the meditative atmosphere of the previous meetings and to put the conference on a new higher plane (8,500 ft)*".

The rusticity of the accommodation detracted in no way from the enjoyment of the setting. Several features of this conference are memorable, not least of which was that Richard Herczynski was now able to attend, presenting a paper co-authored with Peter Robbins entitled "3 D Theory of Respiration: the Steady State Case". And so with Dan Cunningham also a providing the introductory address we had, for the first time, both progenitors of the series making presentations. The organisers continued the recent and welcome tradition of scheduling an excursion into the program: in this case horse-back riding (western style) in the mountains. The published proceedings[3] provided 49 chapters of the presentations with a cover now no longer yellow but white with a yellowish tint to the golden-brown lettering – which, at least in my copy, seems to become yellower with age. However, a major detail in Swanson's preface to this volume was that he referred to the meeting as the "Oxford Conference". For that also the series owes George Swanson a debt of gratitude.

The 5th Conference in the series was held in September 1991, at the Fuji Institute of Training in Japan. It was organised by Professor Yoshiyuki Honda and his associates

Yoshimi Miyamoto, Kimio Konno and John Widdecombe from London: this being the first in the series to be officially titled the "Oxford Conference". It proved to be the largest Conference in the series, both in terms of the 198 participants, more than a third of whom came from outside Japan, and the 80 chapters in the published proceedings[4]. The cover now eschewed yellow altogether, being white with red lettering. The proceedings included a commemorative lecture by Stan Yamashiro for Fred Grodins, a pioneer in the application of control system theory to respiratory control issues and a sustaining supporter of the "Oxford" enterprise. The site for the conference had been carefully chosen a by the planning committee: the foot of majestic Mount Fuji. Sadly, the weather proved not to be as kind and generous as our Japanese hosts: the mountain being mostly shrouded from view by fog, cloud or rain. The weather also hastened our departure at the end of the conference as a powerful typhoon headed towards the region at a disquietingly rapid rate. A highlight the conference was assuredly the spectacular (and whatever its aural equivalent is) evening concert by the rightly highly-acclaimed Japanese drum troupe: the power of the drumming and the subtlety of its shifting rhythms was matched only by the physical demands of providing it. The musicians were generous enough to allow the participants to use, or rather pound, their instruments after the performance - not "such sweet thunder" you may be assured.

Steve Semple, along with Lewis Adams, Abe Guz and I, served as the organising committee for the 6th Conference in the series. This was held at the Royal Holloway College, considered to be the "country campus" of the University of London, with its fine Victorian buildings set in 100 acres of wooded parkland. The College is located some 30 miles west of central London. - far enough, it was thought, to deter too-ready access to the cultural, and other, attractions of the city during the meeting. The conference banquet was held in the College's highly-acclaimed, and security-guarded, picture gallery. Both Dan Cunningham and Richard Herczynski, the " founding fathers " of the conference series, each made after-dinner comments, which were greatly appreciated by the diners. The conference proceedings[5] yielded 68 chapters, the cover reverting predominantly to the original yellow.

Richard Hughson, along with David Cunningham and James Duffin, took responsibility for the 7th Conference in the series. This was held at the appropriately named, Grandview Inn in Huntsville, Northern Ontario, overlooking Fairy Lake: this managed to meld the luxurious with the arcadian. In addition to the scientific content of the conference good use was made of the lakeside facilities, with reports of Yoshiuki Honda and John Severinghaus, two giants in the field of respiratory control, canoeing across (or at least on) the Lake - an indelible recollection for those at the meeting and image for those not. The death of Dan Cunningham was sadly noted at the Conference ; Peter Robbins movingly recalled Dan and his contributions to respiratory control in an after-dinner presentation. Another ingredient adding to the distinctive flavour of the Oxford Conferences was the decision to provide student awards for the "best" papers in the categories of "Control of Breathing" and "Modelling of Breathing". These were, awarded to Judith Thornton of Oxford University, for her paper "Cardiorespiratory responses to the imagination of exercise and altered perception of exercise load", and to Ravi Mohan of the University of Toronto, for his paper "Measurement of chemoreflex model parameters", respectively. Congratulations to the students who won these inaugural awards (and, of course, the other worthy contributors) and also to the organisers for establishing, what one suspects will be, a continuing feature of future

Conferences. The proceedings[6], containing 37 chapters, were published in 1998 with a cover the dominant colour of which is difficult to categorise precisely, the closest being that of expensive cocoa.

In the collected Conference Proceedings one can find evidence of the genesis, evolution, transience and even last gasp of proposed control mechanisms. As such, they provide an important historical perspective on the growth and development of our understanding of respiratory control. The contributions of this "Cape Cod" conference properly awaits a future perspective. What is clear however is that the Conferences themselves continue and continue to evolve, with the expanding interest in organising subsequent Conferences now requiring difficult choices by the International Organising Committee from among the worthy and attractive proposals. They are likely to continue to attract scientists working, at what some term, the "cutting edge" of the disciplines that contribute to furthering our understanding of respiratory control.

These "remembrances" naturally provide only one perspective on the Conferences; while others might have recalled different particulars one aspect, I have no doubt, would be common to any other's recollection: that being the congeniality and lack of rancour in the discussions. The participants were obviously well aware of the adage that:

> "*people quarrel because they've never learned to argue*."

REFERENCES

1. *Modelling and Control of Breathing,* edited by Whipp, B.J. and Wiberg, D.M. (Elsevier Biomedical, New York, 1983)
2. *Concepts and Formalizations in the Control of Breathing,* edited by Benchetrit, G., Baconnier, P. and Demongeot, J., (Manchester University Press, Manchester, UK, 1987)
3. *Respiratory Control: A Modeling Perspective,* edited by Swanson, G., Grodins, F.S. and Hughson, R.H., (Plenum Press, New York, 1989)
4. *Control of Breathing and its Modeling Perspective,* edited by Honda, Y., Miyamoto, Y., Konno, K., and Widdicombe, J.G., (Plenum Press, New York, 1992)
5. *Modeling and Control of Ventilation,* edited by Semple, S. J. G., Adams, L., and Whipp, B. J., (Advances in Experimental Medicine and Biology, vol. 393, Plenum Press, New York, 1995
6. *Advances in Modeling and Control of Ventilation,* edited by Hughson, R.L., Cunningham, D.A. and Duffin, J., (Advances in Experimental Medicine and Biology, vol. 450, Plenum Press, New York, 1998)

WORKSHOP ON MODELING IN THE 21ST CENTURY

An executive summary

VIII Oxford Conference Panel on Biomedical Modeling*

1. INTRODUCTION

The first Oxford Conference was organized in 1978 by the late Dr. D.J.C. Cunningham (Oxford) and Dr. R. Hercynski (Warsaw)—among others—who saw the need to bring physiologists and mathematicians together to address critical issues in understanding the control of breathing. Four years later, a group at UCLA led by Drs. B.J. Whipp, S.A. Ward (both respiratory and exercise physiologists), J.W. Bellville (an anesthesiologist) and D.M. Wiberg (a systems engineer) organized a sequel meeting at Lake Arrowhead, CA, which proved to be as much a success as the first one. Since then, the 'Oxford Conference' has been continued every 3 years in various parts of the world with this eighth meeting being held in N. Falmouth, Massachusetts, USA.

Since its inception in the late '70s the Oxford Conference series has served a unique role in bringing mathematical modelers and systems physiologists together to address the control of breathing. In the two decades of the Conference's illustrious history the field of the respiratory control has advanced tremendously. The methodologies of modeling and physiology have also become increasingly sophisticated and varied, as reflected in the rich and diverse contributions to the Oxford 2000 meeting. As such, the Year 2000 Oxford Conference was an important occasion in which researchers interested in the control of breathing sought to understand its mechanisms through multidisciplinary and integrative research efforts. A central theme of the Conference was the bringing together of experimental and modeling approaches to the study of respiratory control at the molecular, cellular and systems levels. Both experimentalists and modelers in respiratory control shared a common forum for presenting the current state of the art in their disciplines.

* Panelists: NS Cherniack (Chair), UMDNJ-New Jersey Medical School, Newark, NJ 07103; R Butera, Georgia Institute of Technology, Atlanta, GA; J Champagnat, Centre National de La Recherche Scientifique, France; J Duffin, University of Toronto, Canada; MCK Khoo, University of Southern California, Los Angeles, CA; NR Prabhakar, Case Western Reserve University, Cleveland, OH; C-S Poon, Massachusetts Institute of Technology, Cambridge, MA; PA Robbins, Oxford University, UK; JC Smith, National Institute of Neurological Disorders and Stroke, Bethesda, MD; KP Strohl, Case Western Reserve University, Cleveland, OH.

Frontiers in Modeling and Control of Breathing, edited by Poon and Kazemi.
Kluwer Academic/Plenum Publishers, New York, 2001.

2. CONTRIBUTIONS OF MODELS TO THE UNDERSTANDING OF BIOLOGICAL SYSTEMS

From a historical perspective, modeling has always been integral to biological research (even though it often may not be recognized as such). Perhaps the most famous example of this is the double-helix model of Watson and Crick, which provided a detailed account of the structure of the DNA molecule. In this historic discovery, the reductionist's approach (using chromatographic and X-ray crystallographic methods) produced an array of interesting (but fragmented) pieces of information about the DNA molecule. Yet it was the ingenious modeling effort of Watson and Crick (in a heated race with other modelers including L. Pauling) that ultimately managed to fit all the pieces of the puzzle together, making sense of them.

One can think of models as ranging from ones that simply simulate known data based on certain first principles or axioms (bottom-up models) to those that are capable of integrating a wide variety of data to reveal certain novel underlying mechanisms or general principles (top-down models). Examples of the latter are the Mendelian law of heredity and the Darwinian theory of evolution. Most models are neither totally bottom-up nor top-down but lie somewhere in between. Some models could be very simple and specific to a small set of data whereas others could be highly complex and large-scale, involving huge data and multiple known or hypothetical mechanisms. Simple models may be limited in its scope and range of prediction while complex models may run the risk of becoming intractable and overreaching. A model is of great value if it can integrate a large amount of information to yield a satisfactory explanation of their interrelationships (or a simple and testable hypothesis about them) without invoking an excessive number of assumptions or presumptions.

Models have served several functions:

1. Archival; organizing existing data in an orderly, readily understood fashion.
2. Analytical; characterizing, predicting, emulating and displaying new data based on available data and first principles or assumptions.
3. Hypothesis generating; examining the feasibility of new ideas quantitatively with the prediction of novel outcomes.
4. Hypothesis testing; analyzing proposed ideas of the interrelationship among biological elements to see if they can account for known observations.
5. Study design; streamlining the experimental protocol and statistical and data analyses to yield the maximum information from an experiment.

In the area of respiratory control, the landmark contribution of Grodins and coworkers in the 1960's on the respiratory chemostat model[1] ushered in an era of dynamic systems modeling that has prevailed in the past three decades. The extensive modeling effort has led to many new concepts and methodologies in respiratory control, including (just to name a few):

- the dynamics of the chemical control of breathing[2]
- the rebreathing technique for CO_2 response measurement[3]
- systems analysis of respiratory instability[4]
- systems analysis of the exercise hyperpnea[5]
- optimization of ventilatory pattern[6, 7]
- variational analysis of chemosensory neurotransmission[8]
- nonlinear analysis of breathing pattern variability[9, 10]

The advances in the modeling and control of breathing in the past decades culminated in a recent journal special issue[11] and in the present volume.

3. NEW OPPORTUNITIES

Until recently, research in respiratory control and modeling has focused on mechanisms at the systems level, in step with the traditional trend in systems physiology. In the past decade, however, important advances in the study of respiratory control at the cellular and molecular levels have opened new opportunities for respiratory modeling. For example, microarrays of chips allow information to be rapidly accumulated on the genetic composition of cells and protein production not only in tissues removed from the whole animal but also in reduced preparations. The explosion of information regarding the structures and properties of the respiratory central pattern generator and peripheral sensors, their corresponding modes of neural transmission, and the increasing knowledge of the genetic influence on the development of respiratory neurons and neural circuitry, etc., call for new dimensions in modeling the respiratory control system. At the same time, the advent of modern complex systems theory in the past two decades — which has led to a wide spectrum of new and elaborate concepts such as chaotic dynamics, multifractals, neural networks, wavelets, intelligent (adaptive optimal) control — and the emergence of homologous modeling techniques such as neuromorphic and biomimetic (e.g., silicon-based very-large-scale-integrated circuits or microelectromechanical systems) models of brain systems, or those that act as an interface between computers and real neural systems, offer innovative and promising approaches to the modeling of the respiratory control system. Further, the birth of the information age has greatly accelerated the use of computers and the Internet as indispensable tools in biomedical research. The confluence of such revolutionary developments in the fields of experimental biology and systems/computational theory at the turn of the millennium has created many exciting opportunities for respiratory system modeling that were almost unthinkable just a few years ago.

4. NEW CHALLENGES

On the other hand, further progress in the mutually beneficial interchange between respiratory control and modeling research in a sense is complicated by an increasing exigency to go beyond disciplinary boundaries in order to capitalize on emerging opportunities. Ironically, the rapid developments in experimental biology and systems/computational theory have not only created exciting research avenues but may have also inadvertently deepened the cultural divide between experimental physiologists and modelers, making it all-the-more difficult for them to collaborate or even communicate with one another. The tension between these select corps of scientists is subtle but acute; some of the issues confronting them are discussed in a recent commentary[12].

However, neither of these two groups itself is monolithic, and each consists of individuals with diverse expertise and interests. Collaborations between the generic modeler and the generic physiologist are probably inadequate and need to be larger and broader including for example experimentalists from varying disciplines (such as physiology, biochemistry, psychology and genetics) and modelers who have worked with cellular automata, genetic algorithms and evolutionary computation, data mining and data fusion, etc.

Another challenge facing the new biology is a dire need to decipher exorbitant amounts of data which—thanks to the efficiency of the reductionist approach—are accumulating at a rate much faster than could be assimilated readily by human comprehension. At one extreme, the phenomenal proliferation of information about the human genome has greatly outpaced the complex processes involved in the mapping of the genetic code to biologic structures and functions ranging from proteins to the whole organism. Increasingly, computational approaches such as computer-aided bioinformatics are becoming a staple of biomedical research in dealing with these complex issues. However, the question now looming of whether functional and structural genomics problems may represent a class of intractable (NP-hard) computational problems sends a grim reminder of the intrinsic limitations of brute-force computation as a means of problem solving. A fundamental challenge to the modeler and theoretician at present is the discovery of any universal laws or principles or design motifs (top-down models) underlying such genotypic-phenotypic relationships—and for that matter any integrative physiology relationships—that would help to greatly simplify the solutions to these complex problems.

Although the new biology stresses the subcellular and molecular biology, genomics, proteomics, nanobiology, and nanotechnology, there is increasing acceptance of the idea that biological events needed to be studied across space and time as with other complex systems, with particular reference to such intricate properties as emergent and adaptive behavior, adaptation, plasticity, acclimatization, development and senescence, degeneration and regeneration, apoptosis and self-organization, etc. An important goal for the next decade might be to merge models of overall systems and whole-body behaviors with models of neuronal and genetic networks to develop a kind of more global "super model" of the respiratory system that transcends disciplinary and organization boundaries. On the way toward such a goal new areas of modeling could include, for example:

1. Models and hypotheses that attempt to integrate multiple sub-processes in order to explain their emergent behavior, such as the apparent homeostatic regulation of arterial blood gases and pH in some but not all physiologic states or the generation of finely tuned respiratory pattern in neuronal assemblies driven by multiple afferent inputs.
2. Modeling methodologies that allow us to incorporate knowledge about regional (spatial) inhomogeneities in structure and function. This issue has received considerable attention in the pulmonary gas exchange and respiratory mechanics literature but so far, has not been addressed seriously by the respiratory control community. Potential applications of such modeling methodologies include a comprehensive muscle mechanics model of upper airway dynamics or models of the population dynamics of neuronal assemblies.
3. Modeling methodologies that allow us to quantitate processes that cover a wide range of time scales (milliseconds to days or years), e.g., time-varying (non-stationary) behaviors resulting from physiologic adaptation, neural plasticity or learning and memory during development or pathogenesis, or under changes in environment or physiologic states.

4. Models and methods that describe the stochastic and deterministic (mechanistic) components of respiratory variability and help sort out what is physiologic and what is noise.
5. Models that employ Bayesian (or other) statistical inference techniques to "customize" population-based models for the generation of predictions in individual subjects.
6. Modeling methodologies that can incorporate imprecise knowledge of certain aspects of the physiology of a given process – i.e. linking of "traditional" mathematical models with fuzzy models.
7. Models that incorporate behavioral and volitional influences (sleep/waking cycles; emotions, cortical commands, respiratory sensations, etc.).
8. Modeling methodologies that employ high-speed computers, silicon chips, microelectromechanical or nanotechnological devices to emulate the behaviors of the respiratory central pattern generator, respiratory neurons or their ion channels, not only for the purpose of simulation but also as a means to interact with or even control them.
9. And in addition, models that take into account variations or abnormalities in respiratory control that are the result of aberrant genetic transmission.

5. WHAT WE CAN DO TO SEIZE THE NEW OPPORTUNITIES

The Oxford Conference has been highly successful in its goal of bringing together the modeler and the experimentalist and any changes in the workings of future meetings should be made only after careful deliberation.

The Conference could have as one of its aims to articulate, discuss, and refine, new approaches to developing the infrastructure and scientific framework that will ultimately allow the merging of models that deal with different time and size scales and help relate reductionist approaches to the big questions still unsolved of "whole" animal behavior. New programs at the U.S. National Institutes of Health, National Science Foundation and at some universities, as well as similar programs in other countries, indicate that this need for new and improved restructuring is attracting national and international attention albeit slowly. Of particular interest are new funding mechanisms that allow for long-range, high-risk cross-cutting integrative modeling approaches with potential high payoffs in terms of fundamental paradigm shifts. New criteria for evaluation of such model-driven research need to be established independently of the conventional criteria that are geared toward low-risk, methods-driven research approaches with predictable short-term payoffs. An area of strategic importance at the national and institutional levels is the need to initiate new training and career development programs that foster combined experimental-modeling approaches to biomedical research. Cross-cutting training programs of this nature are vital in order to nurture the next generation of biomedical scientists who are well equipped with such multidisciplinary research skills. We need to consider whether the Oxford Conference can help accelerate this process.

One of the attractive features of the Conference has been its relatively small size allowing participants ample time to meet informally and exchange ideas outside of formal presentations. Nonetheless, two ideas for change come immediately to mind. It might be possible in future conferences to bring to the meeting people working in systems other than respiration, or with expertise in new experimental or modeling techniques, without sacrificing the meeting's intimacy. The Oxford Conference could be a place where modelers/experimentalists who work with cellular and genetic systems meet and exchange ideas with those who work at the level of the whole organism and with chronic illness and disease in humans. One important new direction for the Oxford Conference that has emerged at this meeting is the promotion of trainee participation and interest. This is the first meeting in which trainees had a major and prominent presence and many of them seemed very enthused with the meeting. Even greater participation in future conferences should be seriously considered.

Another way to enhance the breadth and size of the group collaborating is through the establishment of an electronic forum (website and a listserve) that would allow the exchange of hypotheses, models and simulation data to a larger audience than usually attends the Conference. A virtual Oxford Conference Forum could take place more often than the every third year meetings. Many of the participants at the Oxford Conference already maintain similar websites of their own and these could be linked by a permanent Oxford Conference site. On a larger level the website could be used to generate new interest in modeling as a way of looking at complex systems and complex behavior as well as showcasing successful "models" at universities and research institutes of collaborative efforts between theoreticians and experimentalists. Sufficient material might be gathered in this manner to generate a comprehensive synopsis of the advances in modeling in the form of a perspective paper that could be presented at major scientific meetings or published in a broad-based biomedical journal.

Much more could be done but it may be better to begin with smaller, then larger steps evaluating their impact as we go. We can't afford though to let opportunities pass us by.

6. REFERENCES

1. F. S. Grodins, S. J. Buell, and A. J. Bart, Mathematical analysis and digital simulation of the respiratory control system, *J. Appl. Physiol.* **22**:260, 1967.
2. D. J. C. Cunningham, P. A. Robbins, and C. B. Wolff, Integration of respiratory responses to changes in alveolar partial pressures of CO_2 and O_2 and in arterial pH, in: *Handbook of Physiology*, sec. 3, the respiratory System, vol 2: Control of Breathing, part 2, edited by N. S. Cherniack, and J. G. Widdicombe, (Washington, DC, American Physiological Society, 1986), pp. 475-528.
3. D. J. C. Read, A clinical method for assessing the ventilatory response to carbon dioxide, *Australasian Ann. Med.* **16**:20, 1967.
4. M. C. K. Khoo, R. E. Kronauer, K. P. Strohl, et al, Factors inducing periodic breathing in humans: A general model, *J. Appl. Physiol.* **53**:644, 1982.
5. B. J. Whipp, Control of the exercise hyperpnea, in: *Regulation of breathing, II*, edited by T. F. Hornbein, (Dekker, New York), pp 1069-1139, 1981.
6. C-S. Poon, Ventilatory control in hypercapnia and exercise: Optimization hypothesis, *J. Appl. Physiol.* **62**:2447, 1987.
7. C-S. Poon, S-L. Lin, and O. B. Knudson, Optimization character of inspiratory neural drive, *J. Appl. Physiol.* **72**:2005-2017, 1992.
8. W. S. Yamamoto, Transmission of information by the arterial blood stream with particular reference to carbon dioxide, *Biophys. J.* **2**:143-159, 1962.
9. E. N. Bruce, Temporal variations in the pattern of breathing, *J. Appl. Physiol.* **80**:1079.
10. B. Hoop, M. D. Burton, and H. Kazemi, Fractal noise in breathing, in: *Bioengineering Approaches to Pulmonary Physiology and Medicine*, edited by M.C.K. Khoo, (New York, Plenum, 1996), pp. 161-173.
11. C-S. Poon, and H. Kazemi, Special issue: Modeling and Control of Breathing, *Respirat. Physiol.* **122** (2-3): 81-290, 2000.
12. J. M. Bower, C. Koch, Experimentalists and modelers: can we all just get along?, *Trends Neurosci.* 15:458-61, (1992).

Central and Peripheral Chemoreceptors

CENTRAL RESPIRATORY CHEMOSENSITIVITY: CELLULAR AND NETWORK MECHANISMS

David Ballantyne and Peter Scheid*

1. INTRODUCTION

In this chapter we assess recent experimental evidence concerning the mechanisms by which CO_2/H^+ stimulates certain classes of brainstem neurone, and then ask whether this qualifies the identification of these neurons as respiratory chemosensors. We begin by considering what the concept of a "respiratory chemosensor" is intended to explain.

2. THE CONCEPT OF A BRAINSTEM RESPIRATORY CHEMOSENSOR

In air breathing vertebrates an increase in the CO_2 or H^+ concentration of the blood perfusing the brain exerts a powerful stimulating action on ventilation. The starting point for modern work is that the receptors mediating this response are located within the substance of the lower brainstem, and so are represented at cellular level by neurones or glia (or both). From this position it is possible to distinguish between two general ideas as to the nature of such respiratory chemosensors. The first is to suppose that the brainstem is equipped with a class of cell – analogous perhaps to glomus cells of the carotid body – which is uniquely specialized to detect a change in CO_2 concentration or pH and to signal this change to the neurones controlling ventilation. This, of course, is an old idea, and in the very general form in which it is stated here does not depend on any particular theory as to exactly where the receptors are located in the brainstem nor on any particular proposal as to their mechanism of action. A quite different idea is to suppose that the task of detecting and signalling a change in CO_2/pH does not require a specialized chemosensory cell, but may be performed by cells which are also engaged in a variety of other functions. On this view the property of chemosensitivity might be shared generally by brainstem networks involved in cardiorespiratory control, i.e. the chemosensors would be represented not by specialized cells dedicated to this one function, but by molecular or morphological specializations of otherwise quite "ordinary" neurones.

---

* Institut für Physiologie, Ruhr-Universität Bochum, 44780 Bochum, Germany

Frontiers in Modeling and Control of Breathing, edited by Poon and Kazemi.
Kluwer Academic/Plenum Publishers, New York, 2001.

The most important obstacle in the way of experimentally differentiating between these two possibilities is that, for mammals, there is no wholly suitable experimental model of central respiratory chemosensitivity. This would require a preparation which shows a normal respiratory response to acidosis, i.e. the behaviour which it is the object of this work to explain, and at the same time allows cells to be tested for evidence of chemosensitivity appropriate to such a response. The key issue here is contained in the word "appropriate". The term itself implies that some cells may exhibit a response to acidosis which in principle might qualify them for a role in chemoreception but which is nevertheless irrelevant to the control of respiration. There might be several reasons for this, but most obviously because the response occurs over an inappropriate range of pH, or because it does not possess a signal character. Even among neurones which appear to meet these two particular criteria there might be other grounds, for example their location in "non-respiratory" structures or, less compellingly, the small size of the response, which make it unlikely that they function as respiratory chemosensors.[1] The absence of readily identified respiratory activity, i.e. the usual situation in most conventional brainstem slices and invariably so in tissue culture, is not itself a bar to identifying neurones with appropriate pH sensitivity, but with few exceptions – notably neurones of the medullary ventral raphé,[2] pontine locus coeruleus[3] and hypoglossal nucleus[4] – little or nothing is known about the natural activity of these neurones or their response to acidosis in a brain with intact blood supply. Conversely, in those *in vitro* preparations which do show respiratory rhythm so that respiratory neurones can be unequivocally identified[5] it is not immediately evident that the way these neurones respond to acidosis is linked to the global respiratory response. (For a discussion of this point see ref 6). This last illustrates a difficulty which is, however, common to all of this work, and that is the problem of correctly identifying the cellular chemosensory signal.

3. IDENTIFYING THE CELLULAR CHEMOSENSORY SIGNAL

3.1. Excitation as Signal

The simplest measure of signalling in neurones is provided by a change in spike frequency. It follows that the magnitude and time course of a spike frequency change elicited by acidosis can be regarded as constituting a chemosensory signal. In principle, this signal could be excitatory or inhibitory in character, but inhibitory responses have in general attracted little interest in relation to respiratory chemosensitivity because within the CNS as a whole they are very much more common than excitatory responses. Nevertheless, there may be instances where inhibition is relevant. For example, in the isolated brainstem-spinal cord preparation of the neonatal rat, medullary neurones showing an expiratory-phased pattern of discharge exhibit a hyperpolarizing displacement of membrane potential and discharge is inhibited during exposure to hypercapnic acidosis (2 to 8% CO_2).[5] Inasmuch as the hyperpolarizing response is retained after suppression of respiratory rhythm in a low Ca^{2+}-high Mg^{2+} medium, it appears to be intrinsic to these neurones and clearly conforms to the particular definition of a chemosensory signal given above. The response also appears to be significant from a functional point of view in this particular preparation since, following an initial period of activation, expiratory intercostal motoneurones are inhibited by such acidosis.[7]

The usual assumption nevertheless is that the signal generated by brainstem respiratory chemosensors is likely to be excitatory, and when this criterion is adopted it appears that a significantly large number of lower brainstem neurones may be capable of functioning as chemosensors. These include neurones of the solitary complex,[8-10] ventral medullary raphé,[2,11] ventrolateral medulla,[2, 5, 12] locus coeruleus[3, 13-14] and the hypoglossal nucleus.[4] More significantly, when compared under identical conditions (in tissue culture) and in response to the same stimulus (3 to 9% CO_2) it turns out that the incidence of such CO_2-excited neurones is substantially greater (22%) in the ventral raphé, a structure known to be involved in the respiratory response to CO_2 *in vivo*[15] than in "non-respiratory" structures (total of 5% in hippocampus, neocortex and cerebellum) where the response is also, on average, weaker.[1] Finally, it is important to note that at least some of these putative chemosensors also form part of the synaptically connected network of brainstem neurones showing respiratory rhythm: this is true of some ventral medullary neurones,[5] of a large majority of locus coeruleus neurones,[13] and of at least some hypoglossal motoneurones.[16]

It is evident that, while convenient, the term "CO_2-excited" is an inadequate basis on which to gather all these neurones together under a single heading. In some cases the response may consist of a "pure" increase in spike frequency with no change in the prevailing baseline potential; this is true, for example, of endogeneously rhythmic raphé neurones.[2] In others it may involve a combination of depolarization and increased spike frequency, and this appears to be the usual form of the response in neurones of the solitary complex,[8-10] hypoglossal nucleus,[4] and neonatal locus coeruleus.[13] Finally, it may be restricted to a depolarizing displacement of membrane potential accompanied by an increase in input resistance, but with no evident effect on the timing of discharge; this last is typical of many synaptically-driven neonatal inspiratory neurones of the ventrolateral medulla.[5] At present it is unknown whether these differences are of any functional significance since there is little information available as to how the CO_2-induced response is integrated with the normal activity of these neurones under conditions of intact chemical synaptic transmission. From an experimental point of view, however, these differences are significant because they are evidence of non-uniformities in the nature or cellular distribution of pH-sensitive conductances. This in turn prompts the question whether these various neurones exhibit a chemosensory signal which is independent of the particular conductances with which they are equipped, and therefore independent of the detailed form of their membrane potential response to CO_2.

3.2. Intracellular pH as Signal

One obvious candidate for such a signal would be intracellular pH (pHi)[17-19] so that one approach towards answering this question would be to compare the pattern of changes in pHi elicited by acidosis in chemosensitive neurones with that elicited in non-chemosensitive neurones. Two issues are involved. Do these neurones behave differently with respect to pHi? And if so, is this difference of a kind which would justify referring to pHi as a chemosensory signal? The comparative data published by Ritucci et al.[17] are indirect in that measurements of pHi were not made on electrophysiologically identified chemosensitive neurones, but on neurones from known chemosensitive regions (solitary complex and ventrolateral medulla) and then compared with neurones from regions presumed to be non-chemosensitive and therefore regarded as a control. What this comparison shows is that while both sets of neurones responded to CO_2-induced acidosis with a significant decrease in pHi, they differed markedly in their subsequent response, with "non-chemosensitive" neurones showing a brisk recovery and "chemosensitive" neurones showing a sustained acidification. Since in both groups recovery of pHi from an acid load depends on the activation of a sodium-proton exchanger – which in the ventrolateral medulla has been identified as the NHE3 isoform[20] – it would appear that differences in the sensitivity of this exchanger to extracellular acidification are responsible for the observed differences in the behaviour of pHi, i.e., inhibition of proton transport in "chemosensitive" neurones requires a smaller acid shift in extracellular pH than in "non-chemosensitive" neurones, the measured difference being > 0.3 units at room temperature.[18]

If, as this evidence suggests, intracellular acidification constitutes the primary chemosensory signal, and so determines the time course of the membrane potential response to a chemostimulus, then clearly the sensitivity of the sodium-proton exchanger to extracellular pH becomes a major factor regulating the duration of the intracellular signal (see also ref 20). This means that in a situation where extracellular protons are instantaneously buffered chemosensitive neurones should tend to behave like non-chemosensitive neurones in respect of the kinetics of pHi, and this is roughly what is seen when the former are exposed to isohydric hypercapnia (external pH held constant) instead of hypercapnic acidosis.[17]

These results are clearly very encouraging for the idea that brainstem chemosensitive neurones, for all the diversity in their anatomical location and functional relations, might work in a similar way, utilizing intracellular pH as the chemosensory signal. This, however, depends on the assumption, not yet proven, that the neurones examined by Ritucci et al.[17] really do fall into two groups in respect of their membrane potential response to intracellular acidification. If this assumption turns out to be justified, then clearly a good case can be made for identifying pHi as the relevant signal. The crucial issue, however, is whether an acid change in pHi is required for the chemosensory response of the membrane potential. The behaviour of ventrolateral medullary neurones in organotypically cultured slices would suggest that this may be true of these neurones,[19] but it remains unclear whether intracellular acidification will account for all aspects of the time course of the membrane potential response. One reason why there is a pressing need for simultaneous measurements of pHi and membrane potential in identified chemosensitive neurones is that there is now a large and growing catalogue of pH-sensitive ion channels, and sorting out which of these channels function as physiological chemosensors would be greatly simplified if it were known that the chemostimulus acts at an intracellular site. However, even if this is true of some chemosensitive neurones, the variability in the exact form of the membrane potential response suggests that protons may "make use" of whatever pH-sensitive channels are available in a given neurone. In some cases then, the relevant signal may be intracellular pH and in others extracellular pH.

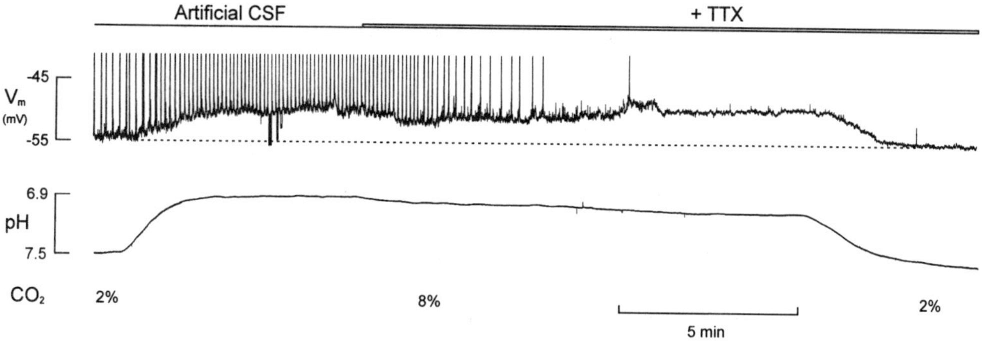

Figure 1. pH-sensitivity of the membrane potential of an inspiratory neurone in the ventrolateral medulla in the *in-vitro* brainstem-spinal cord of the neonatal rat. Switching from 2 to 8% CO_2 resulted in a depolarizing displacement of membrane potential (Vm) and increase in inspiratory burst frequency. Subsequent addition of TTX (bar) eliminated the respiratory rhythm, but Vm remained depolarized. On return to 2% CO_2 the membrane potential recovered. (A. Kawai, D. Ballantyne, K. Mückenhoff and P. Scheid, unpublished)

4. pH-SENSITIVE CONDUCTANCES IN BRAINSTEM CHEMOSENSITIVE NEURONES

Figure 1 shows a recording from an inspiratory neurone of the ventrolateral medulla in the isolated brainstem-spinal cord preparation of the neonatal rat. In this experiment (some details of which we have previously illustrated on a fast time base as Figure 4 of ref. 6) the preparation was initially exposed to artificial csf equilibrated with 2% CO_2 to establish control conditions and then the solution was switched to one equilibrated with 8% CO_2. This reduced the bath pH, measured close to the surface of the medulla, from 7.5 to 6.9. The response consisted of a depolarizing shift in the baseline membrane potential and an increase in the frequency of a rhythmically repeated, synaptically mediated burst of action potentials; this last is typical of the respiratory response to hypercapnic acidosis in the isolated neonatal brainstem. At the point shown TTX was added, at unchanged CO_2 concentration, to suppress voltage-gated Na^+ conductances. This eliminated the synaptically driven respiratory rhythm but left the baseline membrane potential essentially unchanged. Finally, on return to 2% CO_2 in the continued presence of TTX the membrane potential repolarized to its original level. Thus, in inspiratory neurones of this kind, and also in some tonically discharging neurones of the ventrolateral medulla, the baseline membrane potential is sensitive to hypercapnic acidosis.[5]

The simplest explanation for such depolarization is an acid-induced inhibition of one or more K^+ conductances. Identifying the relevant conductances is important because K^+ channels setting baseline or "resting" membrane potential are a frequent target of neurotransmitters, whose effectiveness in displacing the membrane potential, or otherwise modifying neuronal excitability, may therefore depend on the prevailing pH. The conductance mechanism underlying the depolarization of ventrolateral medullary inspirat-ory neurones is unknown. However, the membrane potential of hypoglossal motoneurones has recently also been shown to be sensitive to pH, and for these neurones there is direct evidence that acidification inhibits (pK of 7.4) a K^+ current through a "leak" channel, i.e. one showing a linear current-voltage characteristic, identified as of TASK-1 type.[4]

Importantly, the pH sensitivity of this channel depends on extracellular acidification. Why this is important is because the hypoglossal nucleus was one of the regions selected by Ritucci et al.[17] as a control "non-chemosensitive" region for comparison with pHi regulation in neurones from "chemosensentive" regions (see section 3.2); i.e. it formed part of the basis for their hypothesis that intracellular acidification constitutes a chemosensory signal. It now appears that the effect of extracellular acidification is not indirect, i.e. confined to its action on sodium-proton exchange, but that in hypoglossal neurones (and probably also in many other neurones, see below) such acidification functions directly as a chemostimulus.

One of the striking facts to emerge from the work of Talley et al.[4] is that TASK-1 mRNA is expressed at high levels in many cranial and spinal motoneurones, and although this includes the nucleus ambiguus it appears not to include more ventrolaterally located medullary neurones, i.e. neurones traditionally associated with respiratory chemo-sensitivity. Nor does it include neurones of the ventral raphé, the serotonergic component of which is quite clearly chemosensitive.[2, 11] In contrast, another monoaminergic chemosensitive group, the locus coeruleus,[3, 13] shows a high level of TASK-1 labelling.

It is clear from this patchy distribution that the presence of a pH-sensitive K^+ leak channel is not the defining characteristic of brainstem chemosensitive neurons, and although the pH-sensitivity of hypoglossal motoneurones may be adequately accounted for by the properties of TASK-1, in other neurones quite different channels may be used. Strong candidates are represented by some members of the inwardly rectifying family of K^+ channels,[21] at least one of which (Kir 2.3) is sensitive to external pH in the physiological range[22] though the single channel properties of Kir 2.3 expressed in oocytes reveal both internal and external pH sensors with somewhat different pK values.[23] Finally, in some neurones the full expression of pH sensitivity may depend on a combination of different kinds of channel, and very probably also on how these channels are distributed between cell soma and dendrites. This leads to the topic of the integration of pH-sensitive mechanisms in a network of neurones, which is considered next.

5. pH-SENSITIVE SIGNALLING WITHIN A BRAINSTEM NETWORK

Many of the points which we have been reviewing in general terms come together when we consider the chemosensitivity of the locus coeruleus (LC). Our reason for selecting this nucleus for more detailed consideration is that it has been the object of much of our own work in recent years, and because there is a very substantial background of information, unrelated to the problem of chemosensitivity, on its morphology, electrophysiology and pharmacology.

5.1. pH-Sensitive Rhythms in the LC Network

One fact which makes this nucleus of interest in any discussion of respiratory chemosensitivity is that it is composed of neurones which, at least in the neonate, receive a respiratory innervation.[13] Figure 2A illustrates an example of this in a simultaneous whole cell recording from an LC neurone and suction electrode recording from a phrenic nerve root in the *in vitro* brainstem-spinal cord preparation. The whole cell recording

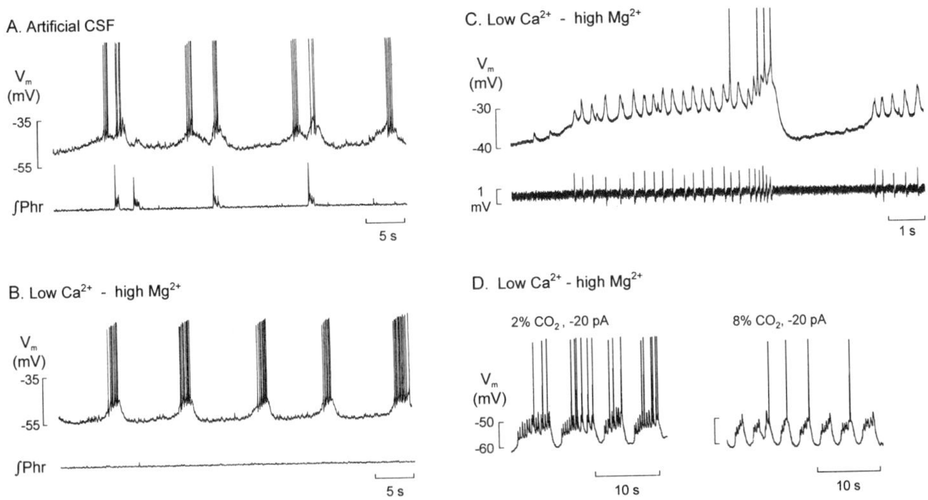

Figure 2. pH-sensitivity of membrane potential oscillations in LC neurones in the *in vitro* brainstem-spinal cord of the neonatal rat. A and B, show rhythmic changes in membrane potential (Vm) of the same neurone under conditions of intact chemical synaptic transmission in artificial csf (A) and after suppression of phrenic discharge (Phr) in low Ca^{2+} (0.2 mM)-high Mg^{2+} (5.0 mM) (B). C, shows a simultaneous whole cell (Vm) recording from one LC neurone and an extracellular recording from a second LC neurone in low Ca^{2+} -high Mg^{2+}; spike discharge in the whole cell recorded neurone was partly suppressed by injection of hyperpolarizing current from the pipette. D, shows in another LC neurone in low Ca^{2+} -high Mg^{2+} that switching from 2 to 8% bath CO2 concentration resulted in an increase in frequency of the slow rhythm; Records A-C: M. Andrzejewski, D. Ballantyne, K. Mückenhoff and P. Scheid, unpublished; Record D, modified after 14.

reveals a sequence of rhythmic depolarizing waves, each of which triggers a burst of action potentials. Some of these waves are synchronized to the inspiratory burst recorded on the phrenic root, while the others occur independently. Figure 2B shows that in the presence of a low Ca^{2+}-high Mg^{2+} medium, phrenic discharge is suppressed so that the respiratory-phased component of the membrane potential rhythm is eliminated. What then remains is a rhythm which does not depend on intact chemical synaptic transmission, but which is nevertheless synchronized throughout the nucleus.

This last is illustrated in Fig. 2C which shows a paired whole cell recording from one neurone and an extracellular recording from a second neurone. It is clear that these two neurones are synchronized both on the time scale (seconds) of the slow oscillation which underlies burst discharge and on the much faster time scale (tens of milliseconds) of events underlying the occurrence of spikes within a burst. Inspection of the whole cell recording shows that these faster events consist of a train of small subthreshold oscillations of membrane potential, and that almost every small oscillation is accompanied by a spike in the extracellularly recorded neurone. Multi-unit extracellular recordings confirm that such synchronization of discharge is a quite general feature of LC activity in low Ca^{2+}-high Mg^{2+} [14] while paired whole cell recordings confirm that this feature is attributable to the simultaneous occurrence of small subthreshold oscillations in each member of the pair (M. Andrzejewski, D. Ballantyne, K. Mückenhoff, P. Scheid; unpublished). This last is consistent with earlier work on slice preparations showing that neonatal LC neurones are organized into a low-pass electrically coupled network.[24, 25]

Figure 2D shows that the frequency of the slow oscillation is sensitive to CO_2 concentration, in this particular example approximately doubling following the switch from 2 to 8% CO_2.

What these findings show is that in the isolated neonatal brainstem, LC neurones exhibit two rhythms: one related to respiration, which is of course sensitive to CO_2, and presumably originates within the medulla; and one which is not dependent on chemical input, but which occurs throughout the LC network and which is also sensitive to CO_2 (Fig. 2D). The question which is of immediate interest here, is how exactly does CO_2 modify this intrinsic rhythm of the LC network, and how is this accomplished?

5.2. pH-Sensitive Oscillations and Electrical Coupling in the LC Network

Although in Fig. 2D the most prominent feature of the response to increased CO_2 was an increase in frequency of the slow rhythm, more generally the response also includes a modest (a few mV) depolarizing displacement of baseline membrane potential.[13] These two components can be separated by exposure to the gap junction uncoupler carbenoxolone which suppresses all oscillatory activity, and so desynchronizes LC discharge, but leaves the depolarizing response to CO_2 intact.[14] Given the high level of TASK-1 mRNA expression in (adult) LC neurones,[4] this depolarization might be attributable to the effect of reduced external pH and inhibition of the TASK-1 K^+ leak current. Alternatively, or in addition, it might be due to the effect of intracellular acidification and inhibition of a Kir current, for which there is also evidence in adult LC neurones.[3] In either case, the anticipated effect of closure of these channels would be to reduce net outward current across the membrane and so should tend to promote oscillatory activity driven by a source of inward (Na^+ and/or Ca^{2+}) current. It is of course entirely possible that the inward current driving the depolarizing phase of the oscillation is also subject to control by pH-sensitive channels with a reversal potential close to the prevailing baseline potential. This might account for the fact, as shown in Fig. 2D, that in some neurones an acid-induced increase in oscillation frequency is unaccompanied by a shift in baseline potential. Evidence suggestive of the involvement of other conductances is the fact that, in the presence of TTX and when exposed to Ba^{2+} at concentrations (up to 2 mM) which might be expected to inhibit TASK-1[4] and which should certainly inhibit the LC pH-sensitive Kir current,[3] LC neurones oscillate at a frequency which is clearly sensitive to pH (Andrzejewski, Ballantyne, Mückenhoff, Scheid; unpublished).

An intriguing aspect of the behaviour of LC neurones is that when they are uncoupled by carbenoxolone (or very high concentrations of CO_2) all spontaneous oscillatory activity ceases. Indeed this forms part of the evidence attributing such oscillation to gap junction coupling.[14, 25] One explanation for this loss of oscillation would be to suppose that the LC nucleus contains only a small number of endogenously oscillatory neurones, but paired recordings from various sites in slice preparations reveal no evidence of latency shifts of a magnitude consistent with the notion of a discreetly located pacemaker.[26] Moreover, this kind of behaviour, i.e. the ability of gap junction-coupled cell clusters, but not single cells, to oscillate, has well known precedents in the islet-cells of the pancreas (for references see 27) and in neurones of the inferior olive.[28] Simple resistive coupling between iso-potential cells will not itself generate a sustained oscillation, so that one requirement for oscillation in a gap junction network is a population of non-identical cells. This "non-identity" means only that there should be a non-zero coupling current between neurones; this current might result from quite modest differences in their channel densities. The important result which emerges from modelling such coupling (in a two-cell network of the inferior olive) is that a sustained oscillation does not require either member of the coupled pair to be an endogenous oscillator.[28] In other words, gap junction coupling imparts dynamic properties to a network, manifest as oscillatory activity, which may not be present in the individual (uncoupled) members of this network.

Currently available evidence suggests that gap junction coupling within the LC nucleus is likely to be dendro-dendritic.[24, 26] If membrane potential oscillations of LC neurones are indeed a consequence of gap junction coupling and therefore arise within the dendritic network, then the sensitivity of oscillation frequency to pH suggests that the property of chemosensitivity may also be a dendritic phenomenon; i.e. inserting pH-sensitive channels into dendritic membrane would seem to "make sense" if it is the function of these channels to drive the oscillation at higher frequency with no loss of synchronization. Both of these features are typical of the behaviour of LC neurones exposed to acidosis,[14] while the retention of coupling during such exposure is true also of chemosensitive neurones of the solitary complex.[10]

6. ARE ANY OF THESE NEURONES RESPIRATORY CHEMSENSORS?

At the outset we asked whether the kind of evidence reviewed here would qualify the currently known list of brainstem chemosensitive neurones as respiratory chemosensors. It is obvious that however detailed the information about pH-sensitive cellular mechanisms may become in the next few years, this will not of itself be evidence of such identification. There is, however, a good reason for pursuing such work. If it can be shown that these neurones are equipped with pH-sensitive mechanisms which modify their activity in a way which also occurs under natural conditions, then this would support the view that chemosensitivity is a generalized property of brainstem respiratory networks. The evidence available to date is certainly incomplete, but would tend to support this view.

7. ACKNOWLEDGMENT

Work reported from the authors' laboratory is supported by the Deutsche Forschungsgemeinschaft.

8. REFERENCES

1. Wang, W. & Richerson, G.B. (2000). Chemosensitivity of non-respiratory rat CNS neurons in tissue culture. *Brain Research* 860, 119-129.
2. Richerson, G.W. (1995). Response to CO_2 of neurons in the rostral ventral medulla *in vitro*. *Journal of Neurophysiology* 73, 933-944.
3. Pineda, J. & Aghajanian, G. (1997). Carbon dioxide regulates the tonic activity of locus coeruleus neurons by modulating a proton and polyamine-sensitive inward rectifier potassium current. *Neuroscience* 77, 519-528.

4. Talley, E.M., Lei, Q., Sirois, J.E., & Bayliss, D.A. (2000). TASK-1, a two-pore domain K^+ channel, is modulated by multiple neurotransmitters in motoneurones. *Neuron* 25, 399-410.
5. Kawai, A., Ballantyne, D., Mückenhoff, K. & Scheid, P. (1996). Chemosensitive medullary neurons in the brainstem-spinal cord preparation of the neonatal rat. *Journal of Physiology* 492, 277-292.
6. Ballantyne, D. & Scheid, P. (2000). Mammalian brainstem chemosensitive neurones: linking them to respiration *in vitro*. *Journal of Physiology* 525, 567-577 .
7. Iizuka, M. (1999). Intercostal expiratory activity in an *in vitro* brainstem-spinal cord-rib preparation from the neonatal rat. *Journal of Physiology* 520, 293-302.
8. Dean, J.B., Lawing, W.L. & Millhorn, D.L. (1989). CO_2 decreases membrane conductance and depolarizes neurons in the nucleus tractus solitarius. *Experimental Brain Research* 76, 656-661 .
9. Dean, J.B., Bayliss, D.A., Erickson, J.T., Lawing, W.L. & Millhorn, D.L. (1990). Depolarization and stimulation of neurons in nucleus tractus solitarii by carbon dioxide does not require chemical synaptic input. *Neuroscience* 36, 207-216.
10. Huang, R.-Q., Erlichman, J.S. & Dean, J.B. (1997). Cell-cell coupling between CO_2-excited neurons in the dorsal medulla. *Neuroscience* 80, 41-57.
11. Wang, W., Pizzonia, J.H. & Richerson, G.B. (1998). Chemosensitivity of medullary raphe neurones in primary tissue culture. *Journal of Physiology* 511, 433-450.
12. Jarolimek, W., Misgeld, U. & Lux, H.D. (1990). Neurons sensitive to pH in slices of the rat ventral medulla. *Pflügers Archiv* 416, 247-253.
13. Oyamada, Y., Ballantyne, D., Mückenhoff, K. & Scheid, P. (1998). Respiration-modulated membrane potential and chemosensitivity of locus coeruleus neurones in the *in vitro* brainstem-spinal cord of the neonatal rat. *Journal of Physiology* 513, 381-398.
14. Oyamada, Y., Andrzejewski, M., Mückenhoff, K., Scheid, P. & Ballantyne, D. (1999). Locus coeruleus neurons *in vitro*: pH-sensitive oscillations of membrane potential in an electrically coupled network. *Respiration Physiology* 118, 131-147.
15. Bernard,D.G., Li, A. & Nattie, E.E. (1996). Evidence for central chemoreception in the medullary raphe *Journal of Applied Physiology* 52, 131-140.
16. Withington-Wray, D.J.S., Mifflin, S. & Spyer, K.M. (1988). Intracellular analysis of respiratory-modulated hypoglossal motoneurones in the cat. *Neuroscience*, 25, 1041-1051.
17. Ritucci, N., Dean, J.B. & Putnam, R.W. (1997). Intracellular pH response to hypercapnia in neurons from chemosensitive areas of the medulla. *American Journal of Physiology* 273, R433-441 .
18. Ritucci, N., Chambers-Kersh, L., Dean, J.B. & Putnam, R.W. (1998) Intracellular pH regulation in neurons from chemosensitive and nonchemosensitive areas of the medulla. *Am. J. Physiol.* 275, R1152-1163.
19. Wiemann, M., Baker, R.E., Bonnet, U. & Bingmann, D. (1998). CO_2-sensitive medullary neurons: activation by intracellular acidification. *Neuroreport* 9, 167-170.
20. Wiemann, M., Schwark, J-R., Bonnet, U., Jansen, H.W., Grinstein, S., Baker, R.E., Lang, H-J., Wirth, K. & Bingmann, D. (1999). Selective inhibition of the Na^+/H^+ exchanger type 3 activates CO_2/H^+-sensitive medullary neurones. *Pflügers Archiv* 438, 255-252.
21. Nichols, C.G. & Lopatin, A.N. (1997). Inward rectifier potassium channels. *Annual Review of Physiology* 59, 171-191.
22. Coulter, K.L., Périer, F., Radeke, C.M., & Vandenberg, C.A. (1995). Identification and molecular localization of a pH-sensing domain for the inward rectifier potassium channel HIR. *Neuron* 15, 1157-1168.
23. Zhu, G., Chanchevalap, S., Cui, N. & Jiang, C. (1999). Effects of intra- and excitracellular acidifications on single channel Kir R2.3 currents. *Journal of Physiology* 516, 699-710.
24. Christie,M.J. & Jelinek, H.F. (1993). Dye-coupling among neurns of the rat locus coeruleus during postnatal development. *Neuroscience* 56, 129-137.
25. Travagli, R.A., Dunwiddie, T.V. & Williams, J.T. (1995). Opiod inhibition in locus coeruleus. *Journal of Neurophysiology* 74, 519-528.
26. Ishimatsu, M. & Williams, J.T. (1996). Synchronous activity in locus coeruleus results from dendritic interactions in pericoerulear regions. *Journal of Neuroscience* 16, 5196-5204.
27. Smolen, P., Rinzel, J. & Sherman, A. (1993). Why pancreatic islets burst but single BETA cells do not. The heterogeneits hypothesis. *Biophysical Journal* 64, 1668-1680.
28. Manor, Y., Rinzel, J., Segev, I., & Yarom, Y. (1997). Low amplitude oscillation in the inferior olive: a model based on electrical coupling of neurons with heterogeneous channel densities. *Journal of Neurophysiology* 77, 2736-2752.

CHEMORECEPTION AND TONIC DRIVE IN THE RETROTRAPEZOID NUCLEUS (RTN) REGION OF THE AWAKE RAT

Bicuculline and muscimol dialysis in the RTN

Eugene E. Nattie*

1. INTRODUCTION

In anesthetized animals, focal acidification of a portion of the rostral ventrolateral medulla (RVLM), the retrotrapezoid nucleus (RTN), increases respiratory output[3, 6]. We define the RTN region to include the RTN proper, described by retrograde tracing experiments as lying ventral to the facial nucleus[19], and adjacent neurons of the parapyramidal region and the juxta-facial portion of the nucleus paragigantocellularis lateralis. Focal disruption of the RTN region by cooling[2] or by neurotoxin injection[10,12,13] substantially reduces baseline respiratory output and the response to systemic hypercapnia. The RTN region can provide a tonic drive to breathe and is a site of central chemoreception.

The importance of these functions in the absence of anesthesia is less certain. Large bilateral RVLM lesions, including the RTN, result in hypoventilation and a markedly diminished response to hypercapnia[18]. In the goat[5, 16] cooling of the RVLM, including the RTN region, decreases ventilation and the response to hypercapnia in the waking state but the effects are smaller than observed under anesthesia.

In contrast, awake rats with RTN lesions produced by prior neurotoxin injection breathe normally at rest but have a 39% reduction in the response to systemic hypercapnia[1]. Focal acidification of the RTN of the rat using microdialysis of CO_2 laden artificial cerebrospinal fluid (aCSF) stimulates breathing only in wakefulness[7]. These data in awake rats suggest that only chemoreception is present in the RTN region.

In this chapter I summarize two previously published experiments[14, 15], performed in the unanesthetized rat, designed to examine whether both chemoreception and a tonic drive can be attributed to the RTN region. First, we focally, acutely, and reversibly altered the function of neurons in the RTN region by dialysis of muscimol. Second, we

* Department of Physiology, Dartmouth Medical School, Lebanon, N.H., 03756-0001

Frontiers in Modeling and Control of Breathing, edited by Poon and Kazemi.
Kluwer Academic/Plenum Publishers, New York, 2001.

evaluated whether neurons in the RTN region are under ongoing, endogenous GABAergic inhibition by dialysis of the $GABA_A$ receptor antagonist, bicuculline.

2. METHODS

The experiments were approved by the Institutional Animal Care and Use Committee of Dartmouth Medical School. The methods have been recently described in detail[14, 15]. In brief, in anesthetized Sprague-Dawley rats a guide tube was inserted and fixed such that the tip was just above the RTN region; for each experiment a dialysis probe was inserted. We measured ventilation and metabolism in a Pappenheimer-style plethysmograph[21] and body temperature. Dialysis of aCSF continued through the experimental protocol. In one control, aCSF dialyzed continuously through a 4 hr period showed no effects on breathing[14, 15]. In each test run, two to four sets of baseline measurements breathing air were repeated over 40 min then 1 mM muscimol or bicuculline was dialyzed for 30 min. In separate experiments, we examined the effect of muscimol or bicuculline dialysis on the response to 7% CO_2.

3. RESULTS

3.1. Unilateral RTN Dialysis with 1 mM Muscimol or Bicuculline; Effects on Ventilation During Rest

Figure 1 shows the main findings of the two experiments in respect to breathing. With muscimol dialysis, V_T was significantly decreased during and just after dialysis (one-way repeated measures ANOVA, $P < 0.05$ at 15 and 50 min; Dunnett's). Frequency was not significantly affected. There was an apparent initial decrease in frequency at 5 min of dialysis but during the dialysis frequency tended to increase. Ventilation is decreased ($P < 0.02$, one-way ANOVA; 5, and 15 mins; $P < 0.05$ Dunnett's).

With bicuculline dialysis, V_T was significantly increased during and just after dialysis ($P < 0.001$, one-way repeated measures ANOVA; $P < 0.05$ for all times during and after dialysis compared to control except 5, and 60 min, Dunnett's test). Frequency was not significantly affected. There was an apparent initial increase in frequency at 5 min of dialysis but during the dialysis frequency tended to decrease. Ventilation is increased ($P < 0.05$, one-way ANOVA; 5, 20, and 25 min points, $P < 0.05$ Dunnett's test). Metabolic rate and body temperature were not significantly affected by muscimol or bicuculline dialysis (data not shown).

3.2. Unilateral RTN Dialysis with 1 mM Muscimol or Bicuculline; Effects on Response to CO_2

In the 1 mM muscimol dialysis group (data not shown), V_E was significantly decreased by 20 % both during and post muscimol dialysis ($P < 0.002$; $P< 0.05$, Dunnett's). V_T was decreased both during (-10%)($P < 0.05$, Dunnett's) and after muscimol (-9%)($P < 0.05$, Dunnett's) and f was decreased ($P < 0.004$) both during (-16%) ($P < 0.05$) and after (-11%) ($P < 0.05$) muscimol dialysis.

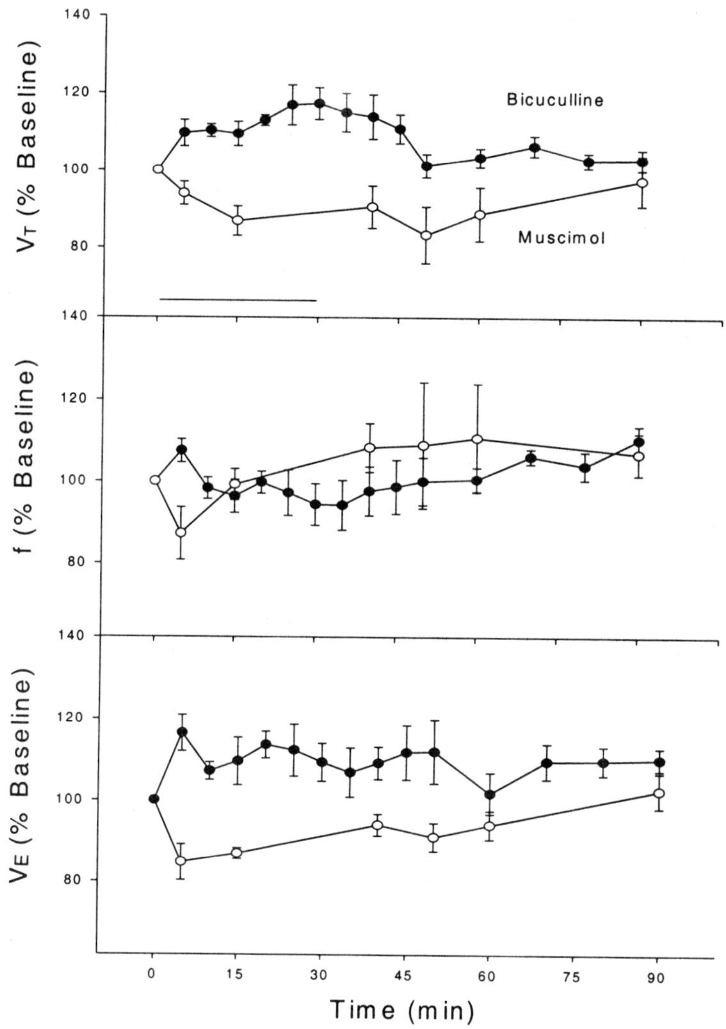

Figure 1. The change (% of baseline; mean +/- SEM) in tidal volume, breathing frequency, and ventilation during and after 30 min dialysis (horizontal line in top panel) with aCSF containing 1 mM muscimol (N=5 rats) or bicuculline (N=6 rats).

With 1 mM bicuculline treatment (data not shown), neither the absolute values of CO_2 stimulated V_E, V_T, and f nor the change in these values differed significantly among the three dialysis tests.

3.3. Anatomical Location of Dialysis Probes

Post mortem histological analysis showed that in each case, the probe tip was in the region of the RTN (data not shown).

4. DISCUSSION

4.1. Limitations of Technique

We have used microdialysis to deliver neuroactive substances (TRH, muscimol, bicuculline, CO_2) to specific regions of the medulla in the conscious rat[4, 7, 14, 15]. In respect to tissue spread, for CO_2, we have measured tissue pH at different distances from the dialysis probe under anesthesia[7, 11] and shown that the change in is confined to within ~600 um.[11] For TRH, muscimol, and bicuculline it is more difficult to measure the exact tissue region affected. Our use of a fluorescein marker with 1 mM concentration in the probe and the use of data from the literature[4] allow us to estimate that the tissue distribution is likely to be within 500-900 um of the probe tip and the concentration at the probe is likely to be 0.35 mM or less.

4.2. Effects of Muscimol and Bicuculline Dialysis in the RTN on Breathing at Rest

Dialysis of muscimol into the RTN region of the conscious rat decreases V_T and V_E; bicuculline increases V_T and V_E. Neither had effects on metabolic rate or body temperature. Frequency appears to increase within, and shortly after, the period of muscimol dialysis and decrease within and shortly after the period of bicuculline dialysis. The net effect on V_E shows initial changes, which are predominantly the result of the strong muscimol/bicuculline effects on V_T, followed by a return toward baseline reflecting the delayed changes in f.

Our interpretation depends on the presence of other chemoreceptor sites both centrally[3, 6, 7, 9, 10, 12] and in the periphery and the fact that our animal model has a 'closed-loop' control system. In either case, muscimol or bicuculline, the final level of V_E likely reflects a balance of inhibition/disinhibition in the RTN region and hyper/hypocapnic induced excitation/inhibition at other chemoreceptor locations.

4.3. Effects of Muscimol and Bicuculline Dialysis in the RTN on the Response to Hypercapnia

With 1 mM muscimol in the dialysate, the V_E response to hypercapnia was significantly decreased by 20% suggesting an effect on the central chemoreceptor system. Focal stimulation of the RTN region by microdialysis of CO_2 in the awake rat increases V_E by approximately 18%.[7] These separate results indicate that neurons in this region provide a portion of the central chemoreceptor response to increased CO_2 in the awake rat. In contrast, with 1 mM bicuculline in the dialysate, there was no effect on the response to breathing 7% CO_2.

4.4. What is the Quantitative Importance of the RTN in the Control of Breathing?

In conscious rats, neurotoxin induced lesions had no effect on resting breathing but decreased the response to 7 % CO_2 by 39% [1] while muscimol dialysis decreased V_E by 15% and the response to 7 % CO_2 by 20%. The RTN tissue volumes affected by neurotoxin lesion and muscimol dialysis are likely to be similar. The greatest difference in the two protocols is in the timing and reversibility of the RTN disruption. Muscimol has an effect in 5 min that is reversible within 60 min of cessation of dialysis. The lesions[1] were produced under anesthesia, are permanent and their effects not studied until 48 hrs after their production. It is possible that some type of neuronal plasticity or adaptation occurred in the two days of recovery allowed in that protocol.

4.5. Endogenous GABAergic Inhibition

Microinjection of bicuculline on some inspiratory and expiratory premotor neurons of the ventral respiratory group that project to the spinal cord of anesthetized dogs can amplify their discharge pattern[8]. This effect was termed 'gain modulation' and while the source of the endogenous GABAergic input is undefined the concept is that the respiratory system is under some type of ongoing endogenous GABAergic modulation in resting conditions.

Our results support this concept. We have previously proposed that the RTN neurons provide a source of endogenous drive to the respiratory system. RTN neurons also appear to be under endogenous GABAergic inhibition. We emphasize that our data are obtained in conscious rats adding to the potential importance of this GABAergic modulation.

That this effect does not involve the response to systemic hypercapnia is surprising. We suggest that many RTN neurons are susceptible to inhibition with exogenously applied muscimol. However, only some receive endogenous GABAergic input, those involved in providing an ongoing drive to the respiratory system.

4.6. Conclusions and Physiological Significance

The RTN provides an important excitatory input to neurons that determine the tidal volume and the overall level of ventilation in the awake resting state and the RTN is involved in the response to systemic hypercapnia during wakefulness. Further, RTN neurons are modulated by an ongoing, endogenous GABAergic influence of unknown anatomical origin.

5. ACKNOWLEDGMENTS

This research was supported by HL 28066.

6. REFERENCES

1. M. R. Akilesh, M. Kamper, A. Li, and E. E. Nattie. Effects of unilateral lesions of retrotrapezoid nucleus on breathing in awake rats, *J. Appl. Physiol.* **82**, 469-479 (1997).
2. K. Budzinska, C. von Euler, F. F. Kao, T. Panteleo, and Y. Yamamoto. Effects of graded focal cold block in the rostral areas of the medulla, *Acta. Physiol. Scand.* **124**, 329-340 (1985).
3. E. L. Coates, A. Li, and E. E. Nattie. Widespread sites of brainstem ventilatory chemoreceptors, *J. Appl. Physiol.* **75,** 5-14 (1993).
4. C. L. Cream, E. Nattie, and A. Li. TRH microdialysis into the RTN of the conscious rat increases breathing, metabolism, and temperature, *J. Appl. Physiol.* **87**, 673-682 (1999).

5. H. V. Forster, P. J. Ohtake, L. G. Pan, and T. F. Lowry. Effect on breathing of surface ventrolateral medullary cooling in awake, anesthetized and asleep goats, *Respir. Physiol.* **110,** 187-197 (1997).
6. A. Li and E. E. Nattie. Focal central chemoreceptor sensitivity in the RTN studied with a CO_2 diffusion pipette *in vivo*, *J. Appl. Physiol.* **83**, 420-428 (1997).
7. A. Li, M. Randall, and E. E. Nattie. CO_2 microdialysis in the retrotrapezoid nucleus of the rat increases breathing in wakefulness but not in sleep, *J. Appl. Physiol.* **87,** 910-919 (1999).
8. D. R. McCrimmon, Zuperku, E. J., Hayashi, F., Dogas, Z., Hinrichsen, C. F. L., Stuth, E. A., Tonkovic-Capin, M., Krolo, M., Hopp, F. A. Modulation of the synaptic drive to respiratory premotor and motor neurons, *Respir. Physiol.* **110**, 161-176 (1997).
9. E. E. Nattie. Central chemoreceptors, pH, and respiratory control, in: *PH and Brain Function*, edited by K. Kaila and B. R. Ransom (John Wiley and Sons, Inc., New York, 1998), pp 535-560.
10. E. E. Nattie. CO_2, brainstem chemoreceptors, and breathing. *Prog. Neurobiol.* **59**, 299-331 (1999).
11. E. E. Nattie and A. Li.Central chemoreception in the region of the ventral respiratory group of the rat, *J. Appl. Physiol.* **81**, 1987-1995 (1996).
12. E. E. Nattie and A. Li. Retrotrapezoid nucleus lesions decrease phrenic activity and CO_2 sensitivity in rats, *Respir. Physiol.* **97**, 63-77 (1994).
13. E. E. Nattie, A. Li, and W. M. St. John. Lesions in retrotrapezoid nucleus decrease ventilatory output in anesthetized or decerebrate cats, *J. Appl. Physiol.* **71**, 1364-1375 (1991).
14. E. E. Nattie, J. Shi, and A. Li. Bicuculline dialysis in the retrotrapezoid nucleus (RTN) region stimulates breathing in the awake rat, *J. Appl. Physiol.* (In press) (2001).
15. E. Nattie and A Li. Muscimol dialysis in the retrotrapezoid nucleus (RTN) region inhibits breathing in the awake rat, *J. Appl. Physiol.* **89**, 153-162 (2000).
16. P. J. Ohtake, H. V. Forster, L. G. Pan, T. F. Lowry, M. J. Korducki, E. A. Aaron, and E. M. Weiss. Ventilatory responses to cooling the ventrolateral medullary surface of awake and anesthetized goats, *J Appl. Physiol.* **78**, 247-257 (1995).
17. J. R. Pappenheimer. Sleep and respiration of rats during hypoxia, *J. Physiol.* **266**, 191-207 (1977).
18. M. E. Schläfke. Elimination of central chemosensitivity by coagulation of a bilateral area on the ventral medullary surface in awake cats, *Pflüg. Arch.* **379**, 231-241 (1979).
19. J. C. Smith, D. E. Morrison, H. H. Ellenberger, M. R. Otto, and J. Feldman. Brainstem projections to the major respiratory neuron populations in the medulla of the cat, *J. Comp. Neurol.* **281**, 69-96 (1989).

CHRONIC INTERMITTENT HYPOXIA ENHANCES CAROTID BODY CHEMORECEPTOR RESPONSE TO LOW OXYGEN

Y. Peng, D.D. Kline, T.E. Dick, and N.R. Prabhakar*

1. INTRODUCTION

Episodic or intermittent hypoxia occurs in many pathophysiological situations, including sleep apneas and recurrent apneas in premature infants. In addition, intermittent hypoxia is not all uncommon in humans who have lung diseases such as chronic obstructive pulmonary disease (COPD), asthma or pulmonary fibrosis. Thus mammals experience episodic hypoxia, often in life, even perhaps more so than sustained hypoxia that occurs in situations like high altitude.

Studies on humans[1] and in experimental animals have documented that chronic intermittent hypoxia (CIH) increases blood pressure[2] and augment ventilatory response to hypoxia[3]. Denervation of carotid bodies no longer cause cardio-respiratory alterations by CIH[2], indicating that peripheral chemoreceptors are critical for cardio-respiratory alterations by CIH. Despite their importance in inducing profound changes in cardio-respiratory system, there are no studies demonstrating the effects of CIH on carotid body function. Therefore, in the present study, using a rodent model, we examined the effects of CIH on carotid body chemoreceptor activity. Our results suggest that sensory response of the carotid bodies to acute hypoxia is significantly enhanced by CIH, and this effect appears to be direct consequence of CIH on the oxygen sensing process in the chemoreceptor tissue.

2. METHODS

Experiments were performed on male Sprague-Dawley rats weighing 250-350g. Awake unrestrained animals were exposed to intermittent hypoxia by placing them in a chamber, wherein the inflow of gases is controlled by timed solenoid valves. Intermittent

* Department of Physiology and Biophysics, Case Western Reserve University, Cleveland, OH 44106.
Tel: 617-368-8636, Fax: 216-368-1693, e-mail: nrp@po.cwru.edu

Frontiers in Modeling and Control of Breathing, edited by Poon and Kazemi.
Kluwer Academic/Plenum Publishers, New York, 2001.

hypoxia comprised of 15sec of 5% O_2 followed by 21% O_2 for 5min.Intermittent hypoxia was given 8h/day (9:00AM-5:00PM) for 10 consecutive days. Parallel experiments were performed on animals exposed to room air for 10 days or to 12% O_2 continuously for 4 hours. Following the exposure to gas challenges, animals were anesthetized with urethane (1.2g/kg; i.p.). After tracheal intubation, a femoral artery and vein were cannulated for measuring arterial blood pressure and for intravenous administration of fluids and drugs, respectively. The animals were paralyzed with pancuronium bromide (2.5mg/kg_h, i.v.) to prevent spontaneous breathing, and ventilated with a respirator (Harvard Apparatus). Arterial blood samples were collected for measurement of P_{O2}, P_{CO2}, and pH. The rectal temperature was maintained at 38±1°C by means of a heating pad.

2.1. Recording of Carotid Body Activity *In Vivo* and *In Vitro*

2.1.1. In-Vivo Carotid Body Activity

The carotid sinus nerve was dissected free of connective tissue, transected where it joins the glossopharengeal nerve. Afferent electrical activity was recorded using platinum-iridium electrodes. The nerve was immersed in a pool of mineral oil to prevent drying. Electrical activity was amplified by an AC-amplifier (Grass P511), with a band with of 100-3000 Hz, displayed on an oscilloscope (Tektronix 5B12N). The action potentials above the baseline noise were converted to standardized pulses using a window discriminator (Winston RAD II-A), and fed into a rate meter (CWE Inc. RIC-830) to display the magnitude of the sensory discharge. The signals from the rate meter as well as action potentials, along with the blood pressure signals, were recorded on a chart recorder (Astro-Med, Inc.) The following criteria were used to identify the chemorcceptor activity: a) prompt augmentation of sensory discharge by stopping the respirator (asphyxia) for 30 s; b) marked decrease in sensory discharge in response to ventilating the animal with 100% O_2.

2.1.2. In-Vitro Carotid Body Activity

The carotid bifurcation along with the sinus nerves were harvested from anesthetized animals. The carotid body and the sinus nerve was cleaned of connective tissue, placed in a recording chamber and superfused with Dubelco's minimum essential medium (DMEM) equilibrated with 100% O_2 at a rate of 1.5 ml/min at 36.7°C. Chemoreceptor activity was recorded with a glass suction electrode. The response to acute hypoxia was obtained by switching the perfusate from hyperoxia to medium equilibrated with 12% O_2. Hypoxic challenge was maintained for 5 min.

2.2. Experimental Protocols

In the experiments involving carotid bodies *in vivo*, baseline chemoreceptor activity was recorded while the animal was ventilated with 100% O_2. paCO_2 was kept close to 35mmHg. After obtaining the baseline activity for 5min, sensory response to 12% O_2 was recorded. Hypoxic challenge was maintained for 1min. Arterial blood samples were collected at the end of gas challenge for measurements of blood gases. The protocols were essentially the same in animals exposed to 4h of continuous hypoxia or with carotid bodies *in vitro*. pO_2 of the solution was monitored in *in vitro* experiments.

2.3. Data Analysis

The following variables were recorded *in vivo* experiments: a) mean arterial blood pressure (mmHg); chemoreceptor activity (impulses/s); arterial pO_2, pCO_2 and pH. In the experiments with carotid bodies *in vitro*, chemoreceptor activity (impulses/s) and superfusate pO_2 were analyzed. The data are expressed as mean ± S.E.M. One-way analysis of variance (ANOVA) with repeated measures was used to evaluate the statistical significance. P-values less than 0.05 were considered significant.

3. RESULTS

3.1. Chronic Intermittent Hypoxia (CIH) Enhances Carotid Body Sensory Response to Acute Hypoxia *in vivo*

Hypoxia (paO_2 =36±3 mmHg; $paCO_2$ = 36±1 mmHg) augmented carotid body sensory discharge in both controls as well as in rats exposed to CIH. However, the baseline sensory discharge as well as the magnitude of the hypoxic response were significantly greater in CIH animals than the controls. Average results are summarized in figure 1. Basal activity was 78% greater in rats conditioned with CIH (Normoxic controls 78±12 impulses/s *vs* CIH 139±26 impulses/s $P<0.05$; paO_2= 252±27 mmHg; $paCO_2$ 36±1 mmHg; n = 6 each group). The magnitude of the hypoxic response was 47% higher in CIH animals than the controls (Normoxic controls +174±25 impulses/s *vs* CIH +255±32 impulses/s; $P<0.05$).

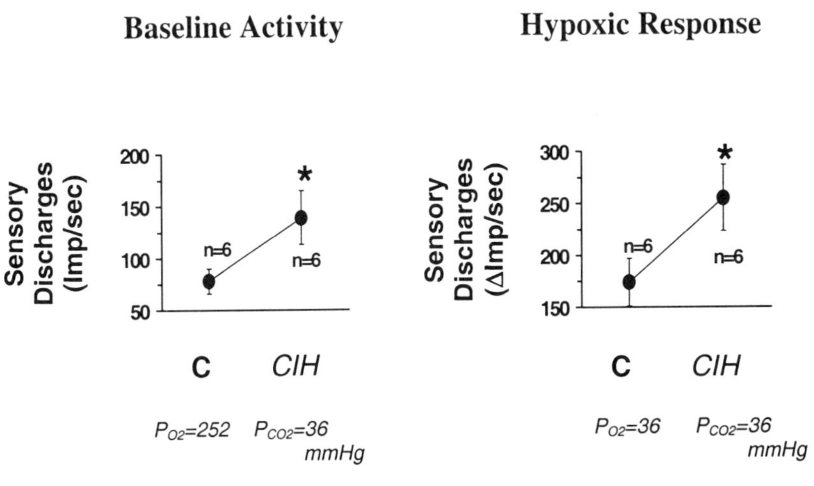

Figure 1. Changes of carotid body chemoreceptor activity following chronic intermittent hypoxia exposure in the intact rats. C, control rats, CIH, chronic intermittent hypoxia rats. * $P<0.05$, compared with the control rats.

The enhanced hypoxic response was associated with significant alterations in arterial blood pressure. Basal mean arterial blood pressure was significantly greater in animals exposed to CIH than controls (Normoxic control 113±6 mmHg *vs* CIH 122±4 mmHg $P<0.05$). However, arterial blood pressure decreased in both groups of rats during hypoxia, but the magnitude of decrease was not significantly different ($P > 0.05$).

3.2. Chemosensory Sensory Response to Acute Hypoxia is Attenuated by Comparable Duration of Sustained Hypoxia

Ten days of intermittent hypoxia (15 s hypoxia- 5 min normoxia) 8 h/day equals to 4h of continuous hypoxia (including the lag time for exchange of hypoxic gas). We tested whether 4h of continuous hypoxia enhance the carotid body response to hypoxia. To test this possibility, chemoreceptor activity was recorded in animals exposed to 4h of isobaric hypoxia(12% O_2, n =5) or 21% O_2 (controls, n =5). Unlike the CIH conditioned animals, baseline activity was comparable between both groups (Normoxic controls 78±12 impulses/sec *vs* continuous hypoxia 86±19 impulses/s, $P > 0.05$). Control animals responded with significant increase in chemosensory activity in response to hypoxia (paO_2 36±3 mmHg +96±6 impulses/s; p<0.01). By contrast, hypoxic response was markedly blunted in animals exposed to four hours of continuous hypoxia. Sensory discharge increased only by +19 impulses/s. (From 86±19 to 105±28 impulses/s, paO_2 =36±2 mmHg; $P > 0.05$).

Basal arterial pressure between both groups of animals was comparable between both groups of animals (Normoxic controls: 113±6 mmHg *vs* Continuous Hypoxia: 117±9 mmHg; $P > 0.05$).

3.3. Enhanced Hypoxic Sensory Response in CIH Animals is not Secondary to Alterations in Arterial Blood Pressure

The results described above demonstrate that CIH enhances carotid body sensitivity to hypoxia and this effect was associated with significantly elevated arterial blood pressure. To assess whether enhanced hypoxic response is secondary to blood pressure changes, carotid body activity was monitored *in vitro*, wherein the influence of arterial blood pressure is effectively absent. The results from these experiments are summarized in Figure 2. In the carotid bodies from control animals, hypoxia (pO_2 of the medium 32±4 mmHg) increased the sensory discharge from 45±10 impulses/s to 86±18 impulses/s (+41±12 impulses/s; n=6). As shown in figure 2, carotid bodies from CIH animals have significantly elevated baseline sensory activity. Furthermore, they also exhibited significantly greater hypoxic sensory response (pO_2 of the medium 36±4 mmHg). On average, sensory activity in CIH carotid bodies increased from 137±21 impulses/s to 331±55 impulses/s (+195±51 impulses/s, n=6).

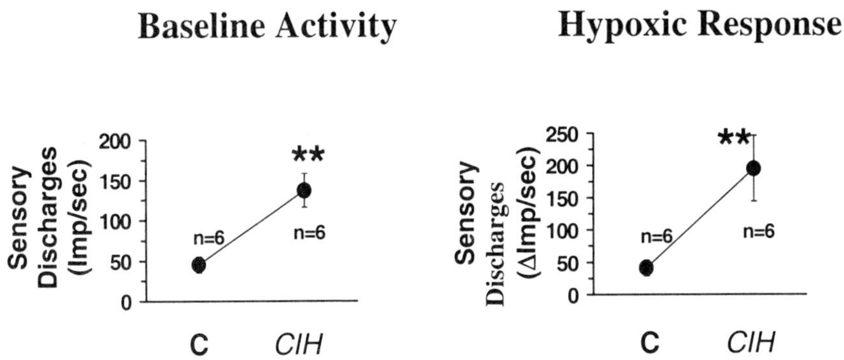

Figure 2. Changes of carotid body chemoreceptor activity following chronic intermittent hypoxia exposure in the *in vitro* CBs. C, control rats, CIH, chronic intermittent hypoxia rats. ** $P<0.01$, compared with control rats.

4. DISCUSSION

In the present study, using a rodent model, we examined the effects of CIH on carotid body sensory response to hypoxia. Our results demonstrate that CIH enhances carotid body sensory response to hypoxia. These observations provide, for the first time, direct evidence for the alterations in peripheral chemoreceptor function by CIH. Consistent with the previous studies[2], we also observed elevated blood pressures in CIH conditioned animals. Carotid chemoreceptor sensitivity to acute hypoxia is augmented in mild hypertension[4]. Consequently, it could be argued that the enhanced hypoxic sensitivity by CIH might be secondary to blood pressure changes and/or elevated levels of circulating catecholamines[5]. However, such a possibility seems unlikely, because enhanced sensory response to hypoxia was also seen in carotid bodies *in vitro*, wherein influence of arterial blood pressure are effectively absent. Therefore, we believe that the enhanced hypoxic sensitivity is due to direct effect of CIH on body oxygen sensing mechanisms in the carotid body. Oxygen sensing mechanisms in the carotid body involves two interdependent processes, i.e., transduction and transmission of the hypoxic signal to the afferent nerve ending[6]. Further studies are necessary to determine whether CIH affects transduction or transmission or both.

It is interesting that hypoxic sensitivity of the rat was blunted rather than enhanced with equal cumulative duration of continuous hypoxia. Furthermore, unlike CIH, continuous hypoxia did not elevate basal sensory activity or increased basal blood pressure. In view of the previous reports suggesting that sustained hypoxia augments hypoxic sensory response[7] the finding that continuous hypoxia blunted hypoxic sensitivity is unexpected. It is possible that the duration of continuous hypoxia employed in the present study (i.e., 4h) may not be adequate to enhance the carotid body sensitivity. None-the-less, these observations suggest that CIH but not comparable duration of continuous hypoxia enhances carotid body sensitivity to hypoxia.

5. CONCLUSION

In conclusion, the results of the present study demonstrate that conditioning with CIH enhances carotid body sensitivity to hypoxia, and this effect was associated with significant elevation in baseline sensory activity and arterial blood pressure. Since enhanced hypoxic sensitivity is also seen in an *in vitro* carotid body preparation, it is suggested that CIH affects oxygen sensing processes in the carotid body chemoreceptors.

6. ACKNOWLEDGMENTS

This work is supported by grants from National Institutes of Health, Heart, Lung and Blood Institute HL-25830 (N.R. Prabhakar); D.D. Kline is supported by training grant T32 HL- 07887.

7. REFERENCES

1. E.C. Fletcher, R.D.DeBehnke, M.S.Lovoi, and A.B.Gorin, Undiagnosed sleep apnea in patients with essential hypertension. *Ann. Intern. Med.* **103**:*190-195, 1985.*
2. E.C. Fletcher, G.Bao, and C.C.Miller, III, Effects of recurrent episodic hypocapnic, eucapnic, and hypercapnic hypoxia on systemic blood pressure. *J.Appl.Physiol.* **78**:1516-1521, 1995.
3. L.Ling, E.B.Olson, and G.S.Mitchell, Chronic intermittent hypoxia enhances the hypoxic ventilatory control system. *FASEB J.* **12**: A781, 1998.
4. J.Przybyski, A.Trzebski, T.Czyzewski, and J. Jodkowski, Responses to hyperoxia, hypoxia, hypercapnia and almitrine in spontaneously hypertensive rats. *Bull Eur. Physiopath. Respir.* 4: 145-154, 1980.
5. E.C. Fletcher, J.Miller, J.W.Schaaf, and J.G.Fletcher, Urinary cathecholamines before and after tracheostomy in patients with obstructive sleep apnea and hypertention. *Sleep.* **10**: 35-44, 1987.
6. N.R. Prabhakar, Oxygen sensing by the carotid body chemoreceptors. *J.Appl.Physiol.*. **88**: 2287-2295, 2000.
7. F.L. Powell, W.K. Milsom, and G.S.Mitchell, Time domine of the hypoxic ventilatory response. *Respir. Physiol.* **112**: 123-134, 1998.

NEUROTRANSMITTER RELEASE FROM THE RABBIT CAROTID BODY: DIFFERENTIAL EFFECTS OF HYPOXIA ON SUBSTANCE P AND ACETYLCHOLINE RELEASE

Dong-Kyu Kim[*], Beth A. Summers[†], Nanduri R. Prabhakar[†], and Ganesh K. Kumar[*]

1. INTRODUCTION

The mechanism(s) by which carotid body senses low oxygen remains to be an active area of investigation. It is currently believed that hypoxia causes membrane depolarization, activation of voltage-dependent Ca^{2+} channel(s), elevation in the cytosolic calcium, and release of neurotransmitter(s) leading to increase in the sensory discharge of the carotid body[6,7]. Multiple neurotransmitters are expressed in the carotid body including acetylcholine, catecholamines, substance P-like neuropeptides and gas molecules (see Ref. [7] for review). While acetylcholine (ACh) and substance P (SP) augment the sensory response of the carotid body, dopamine, enkephalins, and atrial natriuretic peptide inhibit the carotid body activity. Based on a variety of observations reported in the literature, it has been postulated that hypoxia facilitates the concurrent release of several neurotransmitters from the carotid body and potential interactions among these neurotransmitters are of importance in shaping the sensory response of the carotid body to hypoxia[7]. As a first step toward testing this hypothesis, in the present study, we have determined the release of SP and ACh from the carotid body in response to hypoxia and assessed whether the magnitude of their release is comparable or differs from each other.

* Department of Biochemistry, Case Western Reserve University, Cleveland, OH 44106. Tel: 216-368-3349; Fax: 216-368-3419; e-mail: kgk@po.cwru.edu
† Department of Physiology and Biophysics, Case Western Reserve University, Cleveland, OH 44106.

Frontiers in Modeling and Control of Breathing, edited by Poon and Kazemi.
Kluwer Academic/Plenum Publishers, New York, 2001.

2. MATERIALS AND METHODS

2.1. Isolation of Carotid Bodies

Experiments were performed in adult male New Zealand White rabbits (n = 42) weighing 2.0 ~ 2.5 kg. The animals were killed by CO_2, and the bifurcations at the common carotid arteries were rapidly excised following protocols approved by the Institutional Animal Care and Use Committee.

2.2. Substance P Release

For the analysis of SP release, freshly harvested carotid bodies were placed in a polypropylene tube fitted with a porous filter and superfused with Krebs Ringer bicarbonate (KRB) solution, pH 7.4 containing 0.1% bovine serum albumin and 1 mM of phenylmethylsulfonyl fluoride. The temperature of the superfusing medium was maintained at 37 $\pm$ 0.5°C. The carotid bodies were superfused at a rate of 0.3 ml/min in succession with medium saturated with 95 % O_2 + 5% CO_2 (hyperoxia), 1% O_2 + 5% CO_2 + N_2 (hypoxia) and 95 % O_2 + 5% CO_2 (hyperoxic recovery) for 15 min each. Superfusates were collected at 5 min intervals, acidified with glacial acetic acid and stored at –20°C for SP analysis. At the end of the experiment, carotid bodies were homogenized in acetic acid (2M) solution containing 1 mM EDTA at 4°C. The homogenate was centrifuged at 32,000*x* g for 15 min and SP content of the supernatant was determined by procedure outlined below.

Prior to SP analysis, both superfusates and peptide extracts from the carotid body were purified using Sep-Pak (C18) cartridge columns. The amount of SP was determined using an enzyme immunoassay (EIA) kit (Cayman Chemicals Inc., Ann Arbor, MI) and following the protocol described by the manufacturer. The sensitivity of the assay was ~ 17 picogram of SP/ml. The rate of SP released into the superfusion medium was expressed in femtomoles per min per carotid body.

2.2. Acetylcholine Release

The carotid bodies were incubated at 37°C with 100 µl of KRB solution, pH 7.4 containing 300 µM neostigmine bromide, an inhibitor of acetylcholinesterase. For the analysis of ACh release, the carotid bodies were incubated for 30 min successively with KRB solutions, pre-equilibrated with 95% O_2 + 5% CO_2 (hyperoxia), 1% O_2 + 5% CO_2 + N_2 (hypoxia) and 95% O_2 + 5% CO_2 (hyperoxic recovery) and the medium was constantly mixed during the incubation. Following each gas challenge, the incubation medium was collected and ACh content was determined using procedures described below. After the final incubation, the carotid bodies were homogenized in 0.1 M perchloric acid, neutralized with 0.1 M sodium triphosphate and analyzed by HPLC-ECD method.

Acetylcholine was measured using high performance liquid chromatography coupled with an immobilized enzyme reactor (IMER) column containing acetylcholinesterase and choline oxidase. ACh was separated from choline by reverse phase HPLC and then converted quantitatively into hydrogen peroxide by the IMER column. The amount of hydrogen peroxide, a measure of ACh, was determined using a LC-4C model BAS electrochemical detector and peroxidase-wired glassy carbon electrode. The release of ACh was quantified using a standard curve relating the concentrations of ACh and the corresponding peak areas and expressed in femtomoles per min per carotid body.

2.3. Data Analysis

All data are expressed as means ± SEM. Statistical significance was evaluated by a paired *t*-test or ANOVA for repeated measures, and, if a significant difference was indicated, the results were further compared using Tukey's test. *P* values < 0.05 were considered significant.

3. RESULTS

Basal levels of SP and ACh in a single rabbit carotid body were ~0.06 and 6.6 picomoles, respectively (Table 1). The results suggest that in the rabbit carotid body ACh is expressed at ~ 100-fold higher level than SP.

In the first series of experiments, SP release from the rabbit carotid body was determined. Basal release of SP during hyperoxia was 2.0 ± 0.1 fmol per min per carotid body and was equal to ~3% of the total SP in the carotid body (Figure 1). Hypoxia caused an increase in SP release. On average, SP release during hypoxia was 5 ± 0.2 fmol per min per carotid body. Thus, compared to the basal hyperoxic release hypoxia increased SP release by ~150% ($n = 6$; $p < 0.05$). Within 10 min after termination of the hypoxic challenge, SP levels in the medium tend to return to controls.

In the second series of experiments, ACh release from the rabbit carotid body was investigated. As shown in Figure 2, there was substantial basal release of ACh under hyperoxic condition. On average, the basal release of ACh was ~6 fmol per min per carotid body. The basal release represents about ~0.1% of the total ACh in the carotid body. In response to hypoxia, there was a marked decrease in ACh release (Figure 2). During hypoxia, there was ~50% decrease in ACh release as compared to the hyperoxic control ($n = 4$; $p < 0.05$). After terminating hypoxic challenge, there was a prompt return in ACh release towards controls. The results suggest that hypoxia inhibits ACh release from the carotid body and this effect is fully reversible. These results further suggest that hypoxia-induced decrease in ACh release is not secondary to deterioration of the CB preparation *per se*.

Table 1. Substance P, acetylcholine and choline contents of the rabbit carotid body[*]

Neurochemicals	Concentration (pmoles per carotid body)
[1]Substance P	0.058 ± 0.01
Acetylcholine	6.5 ± 0.8
Choline	121.1 ± 14.1

[*]The data represent mean ± SEM (n^1 = 28 rabbits for SP and 14 rabbits for ACh studies)

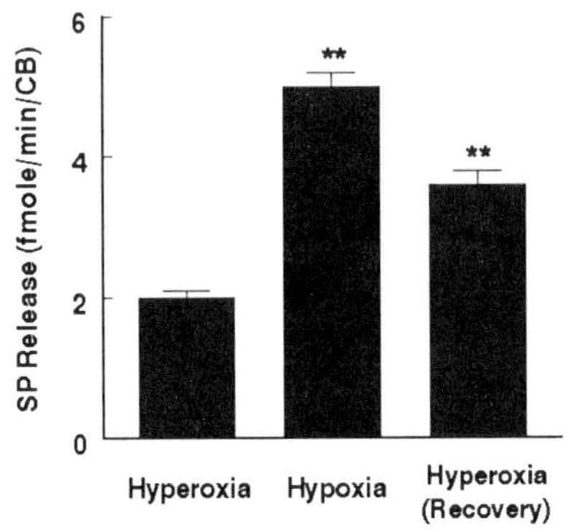

Figure 1. Hypoxia increases SP release from the rabbit carotid body (CB). Data represent mean ± SEM.

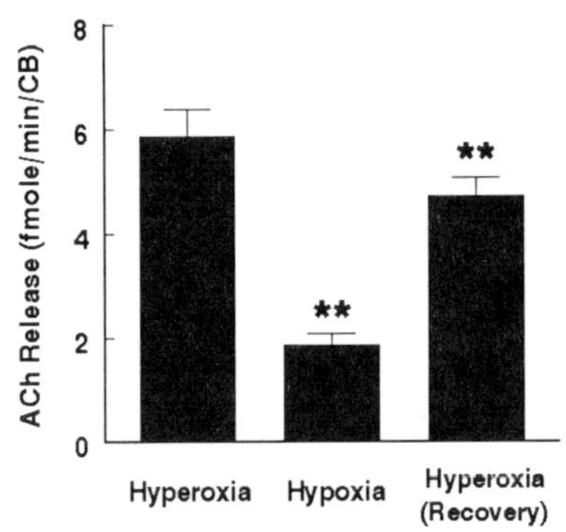

Figure 2. Inhibition of ACh release from the rabbit carotid body by hypoxia. Data represent mean ± SEM.

4. DISCUSSION

One of the major findings of the present study is that hypoxia enhances SP release in the carotid body. The increase in SP release observed during hypoxia was consistent with the previous study by Hansen et al.[4] who reported a decrease in tissue content of both SP and met-enkephalin in the rabbit carotid body in response to hypoxia *in vivo*. Our present finding that SP is released during hypoxia together with the previous reported dopamine release by low oxygen (see Ref. [7] for review) suggest that hypoxia stimulates multiple neurochemicals in the rabbit carotid body.

An intriguing finding of the present study is that hypoxia strongly inhibits ACh release from the rabbit carotid body. This is in contrast to a recent study on the cat carotid body wherein an increase in ACh release in response to normocapnic hypoxia has been reported[2]. Experimental conditions being identical, it is likely that the differential effect of hypoxia on ACh release may be due to differences in the species used in the two studies. However, it has been reported that in the central nervous system and peripheral tissues hypoxia either exerts no effect or inhibits the spontaneous release of ACh. For instance, hypoxia induced a gradual decline in the spontaneous release of ACh from the myenteric plexus of guinea pig ileum[5]. ACh release was unaffected by hypoxia in the striatal slices[8] whereas hypoxia inhibited ACh release in the whole brain and synaptosomal preparations[1,3,9]. These results taken together suggest that hypoxia predominantly exerts an inhibitory influence on ACh release. Our present finding that hypoxia inhibits ACh release from the rabbit carotid body is consistent with the reported inhibition of ACh release by low oxygen in the central and peripheral neural tissues.

It is interesting to note that in the rabbit carotid body the concentration of ACh was ~ 100-fold higher than the level of SP but their basal release was comparable to each other. It is likely that SP and ACh occur in two distinct pools i.e., releasable and non-releasable and that the size of the "releasable pool" is similar for the two neurotransmitters whereas the size of the non-releasable pool of ACh is considerably larger than that of SP.

In conclusion, our results demonstrated that despite marked differences in their content, spontaneous release of SP and ACh from the rabbit carotid body was of similar magnitude. More importantly, under identical experimental conditions, hypoxia augmented SP release whereas it inhibited ACh release. Studies are underway to determine the cellular mechanisms associated with this differential effects of hypoxia on SP and ACh release from the rabbit carotid body.

5. ACKNOWLEDGMENT

This study was supported by grants from National Heart, Lung, and Blood Institute HL-25830 (NRP) and HL-46462 (GKK).

6. REFERENCES

1. E. Chleide, and K. Ishikawa, Hypoxia-induced decrease of brain acetylcholine release detected by microdialysis, *Neuroreport* **1** (1990) 197-199.
2. R.S. Fitzgerald, M. Shirahata, and H.Y. Wang, Acetylcholine release from cat carotid bodies, *Brain Res.* **841** (1999) 53-61.
3. G. E. Gibson, and C. Peterson, Decreases in the release of acetylcholine in vitro with low oxygen, *Biochem. Pharmacol.* **31** (1982) 111-115.
4. G. Hanson, L. Jones, and S. Fidone, Physiological chemoreceptor stimulation decreases enkephalin and substance P in the carotid body, *Peptides.* **7** (1986) 767-769.
5. R. E. Larson, and H. R. Martins, Early effects of glucose and oxygen deprivation on the spontaneous acetylcholine release from the myenteric plexus of the guinea pig ileum, *Can J. Physiol. Pharmacol.* **59** (1981) 555-561.
6. J. Lopez-Barneo, Oxygen-sensing by ion channels and the regulation of cellular functions, *Trends. Neurosci.* **19** (1996) 435-440.
7. N.R. Prabhakar, Oxygen sensing by the carotid body chemoreceptors, *J. Appl. Physiol.* **88** (2000) 2287-2295.
8. K. Saijoh, H. Fujiwara, and C. Tanaka, Influence of hypoxia on release and up-take of neurotransmitters in guinea pig striatal slices: dopamine and acetylcholine, *Japan. J. Pharmacol.* **39** (1985) 529-539.
9. J. Sanchez-Prieto, S. A. K. Harvey, and J. B. Clark, Effects of in vitro anoxia and low pH on acetylcholine release by rat brain synaptosomes, *J. Neurochem.* **48** (1987) 1278-1284.

MUSCARINIC RECEPTORS INFLUENCE CATECHOLAMINE RELEASE FROM THE CAT CAROTID BODY DURING HYPOXIA

Hay-Yan J. Wang* and Robert S. Fitzgerald*†‡

1. INTRODUCTION

Many frontiers in the modeling of the ventilatory system and its control are at the cellular and molecular levels. The carotid body's response to hypoxia is no exception to this. A commonly accepted model of the carotid body's chemotransduction of hypoxia presents the transmitter-laden glomus cell being depolarized by the hypoxic challenge in as-yet-to-be-determined processes. This depolarization activates voltage-gated calcium channels with the subsequent entrance of extracellular calcium. This event, in turn, precipitates the release of several transmitters. Many investigators have reported the hypoxia-generated release of catecholamines from the carotid body. And recently we have reported the release of acetylcholine (ACh) from the cat carotid body during hypoxia[1]. These transmitters are presumed to bind to "postsynaptic" receptors located on the apposed sensory afferent fibers which have their cell bodies in the petrosal ganglion with their axons terminating in *nucleus tractus solitarii*. The transmitters are also presumed to bind to "presynaptic" autoreceptors on the glomus cells. Over the last decade our working hypothesis has been that ACh is *the* (or at least one) critically important excitatory transmitter. And in the mid 1990s we generated some neurophysiological recordings[2] which suggested the presence of both M1 and M2 muscarinic receptors in the carotid body. The data could not answer whether the receptors were located postsynaptically, presynaptically, or both. Since then we have reported preliminary results suggesting that they are at least located on the glomus cell.

The present study addresses the question: Do these glomus cell muscarinic autoreceptors influence the release of dopamine (DA) and norepinephrine (NE) during hypoxia? Interaction of catecholaminergic and cholinergic systems has previously been reported in the *corpus striatum* and *nucleus accumbens* of rats. We have reported earlier that the cholinergic biology of the carotid body seems to resemble certain characteristics found in other systems such as the neural control of airways smooth muscle. Specifically, prejunctional efferent neurons of the vagus exhibit M2 muscarinic receptors. ACh released from these neurons

---

Department of Environmental Health Sciences*, Physiology†, Medicine‡, The Johns Hopkins Medical Institutions, Baltimore, MD 21205.

Frontiers in Modeling and Control of Breathing, edited by Poon and Kazemi.
Kluwer Academic/Plenum Publishers, New York, 2001.

binds to these autoreceptors and reduces the subsequent release of ACh. When these receptors are blocked, the amount of ACh released increases. We have found this to be true of the carotid body challenged with hypoxia. When gallamine or methoctramine is included in the hypoxic Krebs Ringer bicarbonate incubation medium, the amount of ACh released from the carotid bodies is greater than in the blocker-free hypoxic medium. The question, again is: What is the effect of carotid body muscarinic receptors on the release of DA from a carotid body challenged with hypoxia?

2. METHODS

In every experiment a pair of cat carotid bodies was removed from the animal and rapidly cleaned of connective tissues in iced Krebs Ringer bicarbonate solution (KRB). Each carotid body was then placed in an Eppendorf tube containing 170 μL of 40 μM L-DOPA-containing KRB; one of the tubes also contained the appropriate concentration of a muscarinic receptor antagonist, either pirenzepine (M1) or methoctramine (M2). Both carotid bodies were then incubated at 37°C and bubbled with 95% O_2 / 5% CO_2 for 30 minutes.

Subsequently, the 170 μL solutions were discarded and the carotid bodies were incubated in 85 μL of fresh DOPA-containing KRB bubbled with 95% O_2 /5% CO_2 for 10-20 min, followed by exposure to 4% O_2 /5% CO_2 for a given duration depending on the experiment. These procedures were then repeated with alternating hyperoxic and hypoxic incubation periods. At the end of each hyperoxic or hypoxic period the incubation medium was collected and prepared for HPLC analysis, and a second 85 μL was immediately introduced into each incubation tube. After filter-centrifugation the collected samples were stored at -80° C for subsequent analysis.

HPLC analysis was carried out on a Bioanalytical Systems 480 instrument. Briefly, this consisted of a PM-80 solvent delivery pump, an analytical column, an amperometric detector using a glassy carbon working electrode with a standard Ag-AgCl reference electrode operating at +750 mV. The mobile phase used an acetate-citrate base with the pH adjusted to 3.1 with phosphoric acid. Data was acquired with the BAS INJECT software and analyzed with the BAS CHROMGRAPH software. Statistical significance ($P < 0.05$) was determined with the Sigma Stat software package.

3. RESULTS

Figure 1 summarizes several sets of experiments showing (mean ± SEM) that the release of DA increased when the hyperoxic KRB (95/5) was made hypoxic (4/5). This increase was *less* (by ~ 14%) when the hypoxic KRB contained 1 μM of pirenzepine (cross-hatched bar with + sign) than when the hypoxic KRB was blocker-free (open bar with - sign). This figure summarizes the three different durations of exposure to hypoxic KRB in 11 experiments. There was a significant difference with the 20' and 30' exposures but not with the 10' exposure (data not shown).

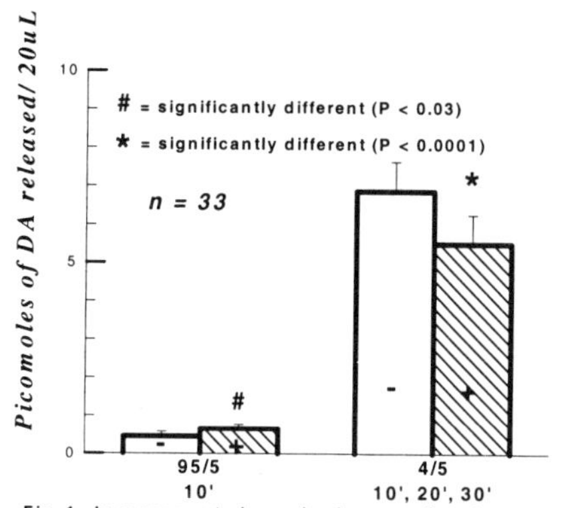

Fig. 1. In response to hypoxia glomus cells release *less* DA in the presence of 1 uM pirenzepine, an M1 muscarinic receptor antagonist.

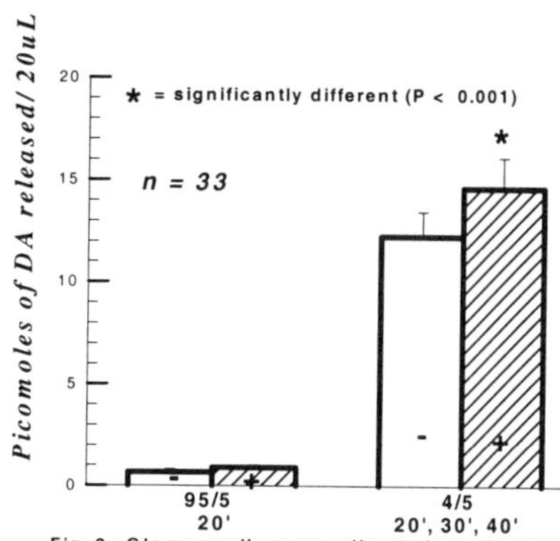

Fig. 2. Glomus cells responding to hypoxia release *more* DA if an M2 muscarinic receptor blocker, 1.37 uM methoctramine, is present.

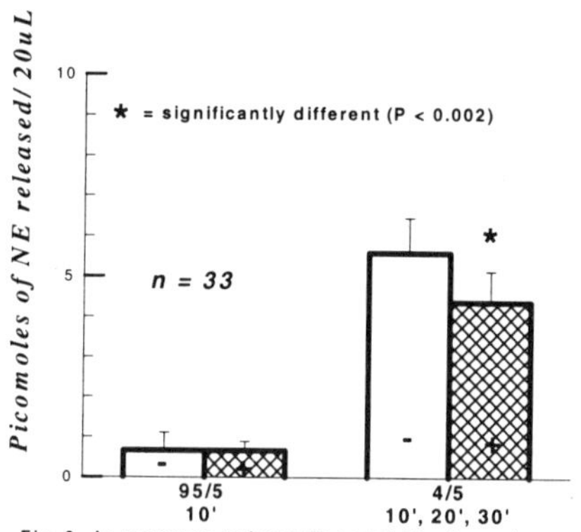

Fig. 3. In response to hypoxia carotid bodies release *less* NE if an M1 muscarinic receptor blocker like 1 uM pirenzepine is present.

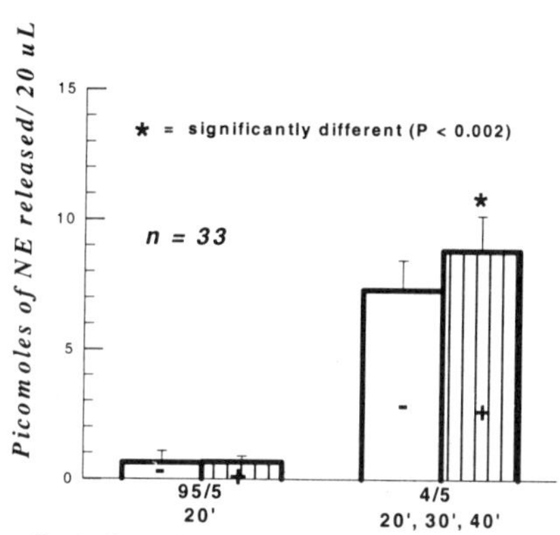

Fig. 4. More NE is released from carotid bodies incubated in a hypoxic medium if an M2 muscarinic receptor blocker, 1.37 uM methoctramine, is present.

Figure 2 shows that the release of DA was greater (by ~ 22%) when the hypoxic KRB contained methoctramine. As above this is a summary of three different durations of exposure to the hypoxic KRB in 11 experiments. There was no significant difference using the hyperoxic exposure nor with the 20-min exposure to hypoxic KRB; there were significant differences with the 30 and 40-min exposures (data not shown).

Figure 3 presents the effects of pirenzepine on the hypoxia-induced release of norepinephrine—a second catecholamine found in relative abundance in the cat carotid body. Though there was a decrease in the release at 10' and at 30', the only individual exposure that was statistically significant was that at 20'.

Figure 4 shows the effect of the M2 muscarinic receptor blocker on the release of norepinephrine from the carotid body challenged with hypoxia. Again there was only one individual exposure—the 40' exposure—which was significantly greater than without the blocker, though the 20' and 30' exposures showed a greater release in the presence of methoctramine.

4. DISCUSSION

We have shown that with an M1 muscarinic receptor blocker in the incubation medium a hypoxic challenge will provoke a smaller release of dopamine and norepinephrine than when the receptor blocker is not present. Contrariwise in the presence of an M2 receptor blocker a hypoxic challenge produces a larger release of dopamine and norepinephrine than in the absence of the blocker. We conclude from this that the hypoxic-induced release of catecholamines is modulated by the hypoxic-induced release of ACh acting on glomus cell muscarinic M1 and M2 receptors. M1 stimulated receptors enhance the release of DA and NE whereas M2 stimulated receptors attenuate this release.

Dinger and his colleagues[3,4] showed the presence of muscarinic receptors in the carotid body using ^{3}H-QNB. However, this agent is non-specific in that it binds to all muscarinic receptors. Hence, their data illustrate the *presence* of M1-M5 muscarinic receptors in the carotid body. Recently we have used RT-PCR techniques to determine if there was a *message* specifically for the M1 and the M2 muscarinic receptors. The results were positive.

Several investigators exploring the acetylcholine/catecholamine interaction in other systems have found similar results. In a microdialysis study of the rat striatum in 1993 DeKlippel and his colleagues[5] found that 75 μM of methoctramine produced a significant increase in DA release while 5 μM pirenzepine produced a significant decrease in DA release. In 1997 Smolders *et al.*[6] reported data which led them to conclude that M1 receptors might tonically facilitate striatal DA release and that M2 receptors might tonically inhibit striatal DA efflux. Finally, since 6 hydroxydopamine (which destroys dopamine terminals) also depletes the level of muscarinic receptors in *corpus striatum* and *nucleus accumbens*, de Belleroche *et al.*[7] concluded: "presynaptic muscarinic receptors capable of being depleted by 6-hydroxydopamine are likely to be mediators of the action of oxotremorine and acetylcholine on dopamine release from *nucleus accumbens*."
Our data support a role for muscarinic receptors acting to enhance or attenuate the release of DA from the glomus cells of the cat carotid bodies challenged with hypoxia.

5. REFERENCES

1. R.S. Fitzgerald, M. Shirahata, H-Y. Wang. Acetylcholine release from cat carotid bodies. *Brain Res.* **84**: 53-61 (1999).
2. R.S. Fitzherald, M. Shirahata, T. Ide. Further cholinergic aspects of carotid body chemotransduction of hypoxia in cats. J. Appl. Physiol. **82** : 819-827 (1997).
3. B. Dinger, T. Hirano, S.J. Fidone. Autoradiographic localization of muscarinic receptors in rabbit carotid body. *Brain Res.* **367**: 328-331 (1986).
4. T. Hirano, B. Dinger, K. Yoshizaki, C. Gonzalez, S. Fidone. Nicotinic versus muscarinic binding sites in cat and rabbit carotid bodies. *Biol. Signals* **1**: 143-149 (1992).
5. N. DeKlippel, S. Sarre, G. Ebinger, Y. Michotte. Effect of M1- and M2- muscarinic drugs on striatal dopamine release and metabolism: an in vivo microdialysis study comparing normal and 6-hydroxydopamine-lesioned rats. *Brain Res.* **630**: 57-64 (1993).
6. I. Smolders, L. Bogaert, G. Ebinger, Y. Michotte. Muscarinic modulation of striatal dopamine, glutamate, and GABA release, as measured with in vivo microdialysis. *J. Neurochem.* **68**: 1942-1948 (1997).
7. J.S. de Belleroche, I.M. Gardiner. Cholinergic action in the nucleus accumbens: modulation of dopamine and acetylcholine release. *Br. J. Pharmac.* **75**: 359-365 (1982).

PHARMACOLOGICAL AND IMMUNOCHEMICAL EVIDENCE OF THE DOPAMINE D_3 RECEPTOR IN THE GOAT CAROTID BODY

Zun-Yi Wang, Jay K. Herman, Ken D. O'Halloran, Ingegerd M. Keith, and Gerald E. Bisgard*

1. INTRODUCTION

Dopamine (DA) represents the predominant amine in the carotid body (CB) of several species. Over 3 decades, numerous studies of the effects of DA on carotid chemoreceptor function have produced a variety of results, whether conducted *in vivo or in vitro.* The reported effects of DA on carotid chemoreflex function are mostly inhibitory, although several investigators have reported excitatory effects, or both inhibitory and excitatory effects[1]. DA D_2 receptors have been found on the CB type I cells, petrosal ganglion and in afferent nerves[2,3,4]. It is possible that multiple DA receptor types may exist within the carotid chemoreflex pathway.

Recently, a DA D_3 receptor has been cloned, revealing a high degree of sequence homology with the D_2 receptor[5]. While the D_2 receptor is abundantly and widely expressed in all major dopaminoceptive brain areas, and peripherally, the D_3 receptor is localized in restricted regions of the brain such as the nucleus accumbens, the islands of Calleja and the archicerebellum[5,6]. D_3 receptors are of considerable interest, as they are a potential target of drugs employed in the treatment of neuropsychiatric disorders[5,7]. The present study examines the potential presence and functional role of D_3 receptors in the CB.

* Zun-Yi Wang, Jay K. Herman, Ingegerd M. Keith and Gerald E. Bisgard, Department of Comparative Biosciences, University of Wisconsin, Madison, WI 53706 USA. Ken D. O'Halloran, Department of Physiology, University College, Dublin, Dublin2, Ireland.
Correspondence: Z-Y Wang, Department of Comparative Biosciences, 2015 Linden Dr. West, University of Wisconsin, Madison, WI 53706, e-mail: zwang@svm.vetmed.wisc.edu, phone: (608) 263-8374, fax: (608) 263-3926

Frontiers in Modeling and Control of Breathing, edited by Poon and Kazemi.
Kluwer Academic/Plenum Publishers, New York, 2001.

2. METHODS

2.1. Immunocytochemical Receptor Localization

Normal, adult, female goats were anesthetizied with thiamylal sodium (10-15 mg/kg, i.v.) and maintained with α-chloralose (100-150 mg/kg/h, i.v.). The carotid arteries were perfused with heparinized saline, followed with phosphate-buffered 4 % paraformaldehyde (pH 7.4). The carotid bodies were then excised and immersed in Bouin's fixative overnight. After routine paraffin processing, serial sections were cut at 5-μm and mounted on poly-L-lysine coated slides. The antiserum against D_3 receptor was raised in rabbits (Research & Diagnostic Antibodies). This antiserum was demonstrated by the manufacturer to label specifically the plasma membranes of neurons in the glomeruli of the olfactory bulb; these neurons are known to express D_3 receptors. There was no cross-reactivity between the D_3 and the other identified DA receptors. De-paraffinized sections were treated with 0.3 % Triton-X-100 for 30 min and background was blocked with 10 % normal goat serum. Sections were then incubated overnight at 4° C with D_3 receptor antiserum, diluted at 1:1000. After rinsing in PBS, the sections were incubated with fluorescein isothiocyanate (FITC)-conjugated goat anti-rabbit IgG (1:160; Sigma) for 90 min. They were rinsed and coverslipped using a antifading solution (Molecular Probes). For negative controls, serum from a normal, non-immunized rabbit was used, to replace the specific D_3 receptor antiserum.

2.2. *In-Vivo* Experiments

Adult goats were anesthetized with pentothal sodium (10-15 mg/kg, i.v.) and maintained with α-chloralose by continuous i.v. infusion at the rate of 100 to 150 mg/kg/h. Body temperature was maintained at 38-39 °C with a circulating water heating pad. After catheterizing the left femoral artery and vein for injections, blood pressure monitoring and arterial sampling, the goat was paralyzed with pancuronium at a dose of 2-4 mg i.v., followed by 1-2 mg hourly. The goats were artificially ventilated and blood gases were measured frequently (Radiometer Model 500).

The carotid sinus nerve was exposed via a ventral midline incision in the neck of the goat. The nerve was cut at its junction with the glossopharyngeal nerve, desheathed and placed on a dissection platform for further dissection and recording. Small fibers were separated and placed on a platinum electrode for unipolar recording. Single or few-fiber units were accepted as chemoafferents by their typical firing pattern and positive response to NaCN infusion (50 μg/kg) and hypoxemia (10 % O_2 in N_2). Activity was assessed by measuring mean discharge frequency and the duration of actions induced by the tested agents. Nerve action potentials, ratemeter activity of neural frequency, systemic arterial blood pressure and end-tidal CO_2 were monitored and recorded.

The DA D_3 receptor agonist R(+)-7-Hydroxy-DPAT hydrobromide (7-OHDPAT, 1-10 µg/kg)[8] was administered by intracarotid injection via small catheter placed below the carotid bifurcation into the common carotid artery. The DA D_3 receptor antagonist UH232 (0.22-1 mg/kg)[8] was injected intravenously.

3. RESULTS

D_3 receptor-like immunoreactivity (LI) was noted on nerve fibers located between and surrounding type I cell clusters. Few positive nerve fibers were also observed within the cell clusters; they seemed to be unmyelinated and did not appear to form synaptic interactions with either type I or type II cells (Fig. 1A). In the nearby carotid sinus nerve, some fibers also displayed D_3 receptor-LI (Fig. 1B). No positive immunoreactivity was seen within the type I and type II cells. Using normal rabbit serum in place of the D_3 receptor antiserum completely abolished the immunoreactivity (not shown).

Close arterial bolus DA (10-50 µg/kg) injections induced a dose dependent inhibition of CB afferent activity (Fig. 2, left). Additionally, bolus injections of the D_3 receptor agonist, 7-OH DPAT (0.1 - 10 µg/kg) also induced a very potent and dose dependent inhibition of CB afferent activity (Fig. 2, right). Systemic administration of the DA D_3 receptor antagonist, UH232 (0.2-1.0 mg/kg i.v.) increased CB activity from a control value of 3.2 ± 1.2 (mean ± SE) Hz to 10.0 ± 2.3 Hz. The inhibition induced by exogenous DA (10 µg/kg) administration was attenuated following the administration of UH232; the inhibition was 75.1 ± 9.2 and 50 ± 7.6 s ($n = 7$, $p = 0.045$), before and after the application of UH232, respectively. The effect of 7-OH-DPAT (10 µg/kg) was only slightly reduced by UH232; the inhibition was 196.3 ± 26.5 and 165 ± 29.6 s ($n = 4$), before and after the application of UH232, respectively.

4. DISCUSSION

The D_3 and D_2 receptors share a high degree of sequence homology and, to some extent, also possess similar pharmacological properties and intracellular signalling systems[5,9]. The functional significance of this molecular diversity, maintained along evolution, is unclear. Interestingly, whereas the D_2 receptor is abundantly and widely expressed in all major dopaminergic pathways throughout the body, the distribution of D_3 receptors is very restricted[5]. Also, the distribution of these two receptor types does not seem to overlap[5,6,7]. Thus, the distinctive distribution patterns of the these two closely related receptors may reflect the different roles they play in the dopaminergic pathways.

We have here described presence of DA D_3 receptor-LI on the nerve fibers in the CB. The fact that selective D_3 receptor agonist, 7-OH DPAT[8], inhibited the carotid sinus nerve sensory chemoafferent activity suggests an inhibitory function for these receptors in the CB. This was further confirmed by the use of selective D_3 receptor antagonist UH232[8], which augmented sinus nerve activity. The present finding of inhibitory D_3 receptors in nerve fibers, together with evidence of D_2 receptors in the type I cells and post-synaptic regions of afferent nerve terminals, reflect that fact that more than one type of DA receptors may account for the observed actions of DA in CB.

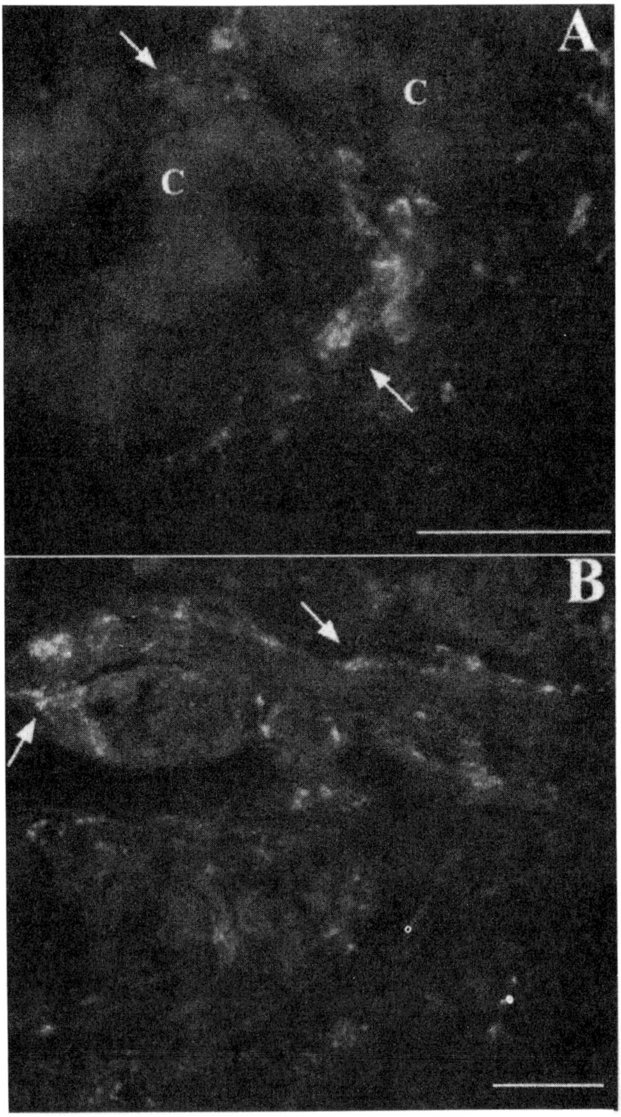

Figure 1. Representative photoimages of dopamine D_3 receptor-LI in the goat CB. Positive staining was noted on nerve fibers located between and surrounding type I cell clusters (A). In the nearby carotid sinus nerve, some fibers also displayed D_3 receptor-LI (B). C indicates type I cell clusters. Calibration bar = 50 μm.

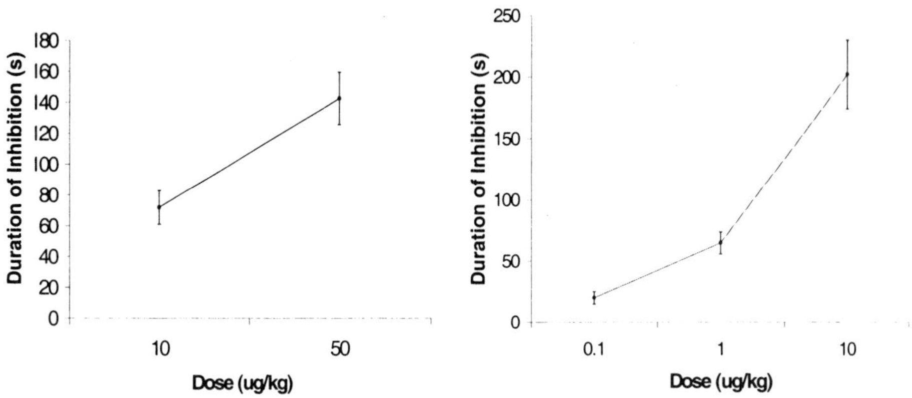

Figure 2. Close arterial bolus DA (left) or D_3 receptor agonist, 7-OH DPAT (right) injections induced a dose dependent inhibition of CB afferent activity. Data presented as mean ± sem.

5. ACKNOWLEDGMENTS

We would like to thank Gordon Johnson for his excellent technical assistance. (Supported by NIH Grants HL 15473 and HL10069).

6. REFERENCES

1. G. E. Bisgard, Carotid body mechanisms in acclimatization to hypoxia, *Respir. Physiol.* **121**, 237-246 (2000).
2. M. F. Czyzyk-Krzeska, Molecular aspects of oxygen sensing in physiological adaptation to hypoxia, *Respir. Physiol.* **110**, 99-111 (1997).
3. A. Bairam, C. Dauphin, F. Rousseau and E. W. Khandjian, Expression of dopamine D2-receptor mRNA isoforms at the peripheral chemoreflex afferent pathway in developing rabbits, *Am. J. Respir.Cell Mol. Biol.* **15**, 374-381 (1996).
4. E. B. Gauda, O. Bamford and C. R. Gerfen, Developmental expression of tyrosine hydroxylase, D2-dopamine receptor and substance P genes in the carotid body of the rat, *Neuroscience* **75**, 969-977 (1996).
5. R. A. Shafer and B. Levant, The D3 dopamine receptor in cellular and organismal function, *Psychopharmacol.* **135**, 1-16 (1998).
6. M.-L. Bouthenet, E. Souil, M.-P. Martres, P. Sokoloff, B. Giros and J.-C. Schwartz, Localization of dopamine D3 receptor mRNA in the rat brain using in situ hybridization histochemistry: comparison with dopamine D2 receptor mRNA, *Brain Res.* **564**, 203-219 (1991).
7. D. Levesque, Aminotetralin drugs and D3 receptor functions, *Biochem. Pharmacol.* **52**, 511-518 (1996).
8. C. W. Aretha, A. Sinha and M. P. Galloway, Dopamine D3-preferring ligands act at synthesis modulating autoreceptors, *J. Pharmacol. Exp. Ther.* **274**, 609-613 (1995).
9. L.-X. Liu, F. J. Monsma, Jr., D. R. Sibley and L. A. Chiodo, D2L, D2S, and D3 dopamine receptors stably transfected into NG108-15 cells couple to a voltage-dependent potassium current via distinct G protein mechanism, *Synapse* **24**, 156-164 (1996).

THE EXCITATORY EFFECT OF NITRIC OXIDE ON CAROTID BODY CHEMORECEPTION IS BLOCKED BY OLIGOMYCIN

Rodrigo Iturriaga and Matias Mosqueira*

1. INTRODUCTION

According to the current model of chemoreception in the carotid body (CB), glomus cells are the primary sites of hypoxic transduction. In response to hypoxia, glomus cells are expected to release one (or more) Ca^{2+}-dependent excitatory transmitter(s), which in turn increases the rate of discharge in nerve terminals of chemosensory petrosal neurons[8]. In addition to the excitatory transmitter(s), other molecules produced within the CB may act as modulators of the chemosensory process. It has been proposed that nitric oxide (NO) is a tonic inhibitory modulator of chemoreception to hypoxia[7, 18-21]. This proposition is based on the immunocytochemical localization of NO synthase (NOS) in autonomic neurons[10,20], and petrosal C sensory fibers[20], on the use of pharmacological tools (i.e. NO donors and NOS inhibitors[7,18-21]) and more recently on NO gas[13]. In the cat CB perfused *in vitro*, the administration of L-arginine and NO donors reduces the chemosensory response to hypoxia[7,21]. On the other hand, the inhibition of NOS increases the frequency of chemosensory discharges (f_x) and partly enhances the response to hypoxia in the cat CB *in situ*[12] and *in vitro*[21].

Recently, we studied the effects of the NO donor, sodium nitroprusside (SNP) on the carotid chemosensory responses to excitatory and inhibitory stimuli in paralyzed and artificially ventilated cats[12]. Surprisingly, we found that SNP increased basal f_x. On the contrary, we[1] and others[7,21] have found that SNP reduced or had no effect on basal f_x in the cat CB superfused or perfused *in vitro*. A major difference between *in vitro* and *in situ* CB preparations is the presence of large amounts of endothelium and vascular smooth muscle tissue in the whole cat, which is considered to be necessary to activate SNP for releasing NO[15]. Thus, large amounts of NO released from SNP *in situ* may account for the increased f_x. We tested this hypothesis studying the effects of medium and high NO concentrations produced by NO donors on chemosensory activity in normoxia and hypoxia. We found that, S-nitroso-N-acetylpenicillamine (SNAP) and 6-

* Laboratorio de Neurobiologia. Facultad de Ciencias Biológicas , P. Universidad Católica de Chile, Santiago. Chile.

(2-hydroxy-1-methyl-nitrosohydrazino)-N-methyl-1-hexanamine (NOC-9) reduced chemosensory responses to hypoxia[14]. However, in normoxia the same NO donors increased f_x in a dose-dependent manner, showing a dual effect of NO on carotid chemoreception depending on PO_2 levels[14]. It is likely that the excitatory effect of NO may be mediated by an impairment of mitochondrial chain electron transport and oxidative phosphorylation[4-6], conditions that are expected to increase f_x [16,17]. To test this hypothesis, we used oligomycin and carbonyl cyanide p-trifluoromethoxyphenyl-hydrazone (FCCP), which produce differential blockade of NaCN and hypoxic responses, leaving the response to nicotine less affected[16].

2. METHODS

Experiments were performed on cats anesthetized with sodium pentobarbitone (40mg/Kg i.p.). Cats were tracheotomized and one femoral vein was cannulated for additional anesthetic doses. The carotid bifurcation including the CB and the carotid nerve was perfused with a modified Tyrode as previously described[13, 14]. The carotid bifurcation was cannulated through the common carotid artery, excised and placed on a chamber. Each CB was perfused by gravity with Tyrode at pH 7.40, equilibrated with room air and 5% CO_2 and simultaneously superfused with Tyrode equilibrated with 95% N_2 and 5% CO_2. The temperature of the solution was maintained at 38.0 ±0.5 ºC with a regulated system. Chemosensory discharges were recorded from the carotid sinus nerve placed on a pair of platinium electrodes lifted into mineral oil. The neural signals were preamplified, amplified, filtered (10 Hz-1 KHz, notch filter 50Hz) and fed to an electronic amplitude discriminator. Selected chemosensory impulses were counted to assess f_x, expressed in Hz. The f_x signal was digitized with an analog-digital board DIGIDATA 1200A (Axon Instruments, USA) for later analyses. In some experiments, NO was measured through a chronamperometric system[14] (IVEC 10, Medical System Corp., USA). NO-sensitive electrodes consisting of single carbon fibers (30 μM diameter) covered with porphyrin and Nafion® (Quanteon, USA), were inserted into the CB. A potencial of 0.9V with respect to the Ag-AgCl reference electrode was applied for 100 ms at a rate of 5Hz. The oxidation current was displayed at a rate of 1Hz. Electrodes were calibrated with SNAP (100-600 μM) in Tyrode at pH 7.40. Hypoxic solution (95% N_2 and 5% CO_2, $PO_2 \approx 30$ Torr) was perfused for 120 s. Nicotine bitartrate and NaCN (1-100μg) and SNAP (200-900 μg) were injected as a boluses of 0.2-0.5 ml into the perfusate line. Oligomycin and FCCP were dissolved first in ethanol (90%) and then in Tyrode (1: 10,000). Blank solution consisting of ethanol diluted in Tyrode did not modify f_x. Results were expressed as means ± SEM´s. Statistical differences of dependent samples were assessed with Student´s paired test.

3. RESULTS

Figure 1A shows the chemosensory responses to hypoxic perfusion, nicotine (1-10 μg), SNAP (450 μg) and NaCN (1-10 μg) during control condition. Hypoxia produced a maintained increase in f_x , while injections of SNAP, nicotine and NaCN produced a transient increase in f_x. The first dose of oligomycin (12.5 μg) stimulated f_x as shown in

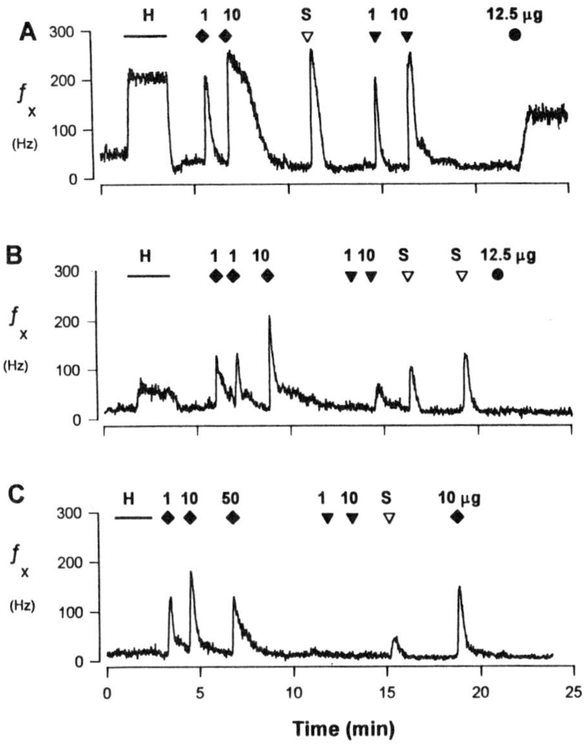

Figure 1. Effects of two doses of oligomycin (12.5 µg) on chemosensory responses induced by hypoxia, SNAP, nicotine and NaCN in one CB. **A**, control responses. **B**, after 30 min of the first dose of oligomycin. **C**, after 30 min of the second dose of oligomycin. **H**, hypoxia; ♦, nicotine; ∇, SNAP 450 µg; ▼, NaCN; •, oligomycin.

Fig 1A. The excitatory effect of the first dose of oligomycin persisted for about 20 min, after which it returned to basal values. Fig. 1B shows the effects of oligomycin on the chemosensory responses elicited by the excitatory stimuli 30 min after the injection of the first dose of oligomycin. Certainly, oligomycin strongly reduced chemosensory responses to hypoxia, SNAP and NaCN, leaving the response to nicotine less affected. Normally, the second dose of oligomycin did not increase f_x , but further reduced or even abolished the responses to hypoxia, NaCN and SNAP. Note that the responses to nicotine persisted, although they were reduced. The effect of oligomycin persisted for 2-3 hrs, and no recovery was observed. In preliminary experiments, we found that 5 µM FCCP perfused for 20 min also reduced the responses to hypoxia, SNAP and NaCN. The effect of FCCP was reversible upon its removal.

The effects of the initial dose of oligomycin 12.5 µg on the chemosensory responses to hypoxia, NaCN, SNAP and nicotine of 5 CBs are summarized in Fig 2.

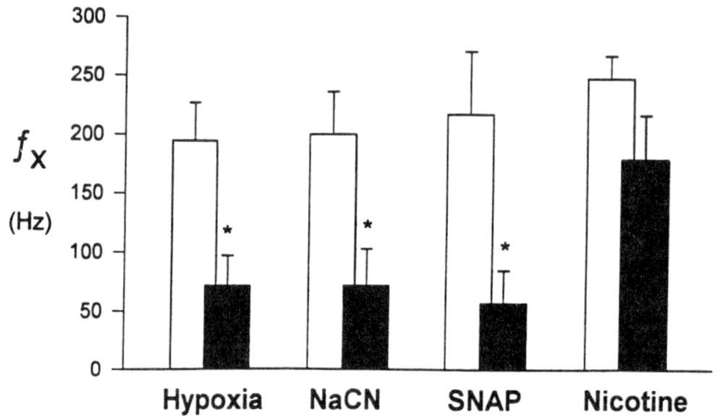

Figure 2. Summary of the effect of oligomycin (12.5 μg) on the amplitude of chemosensory responses to hypoxia, NaCN (10 μg), SNAP (250-900 μg), NaCN and nicotine (1-10 μg) in 5 CBs. * $p < 0.01$ (Student's paired Test). Control, empty box; oligomycin, black box.

Figure 3 shows that injections of 450 μg SNAP produced a similar NO peak before and after oligomycin 12.5 μg, while the chemosensory response was strongly reduced. Note that a high a dose of NaCN 100 μg was also detected by the NO sensitive electrode.

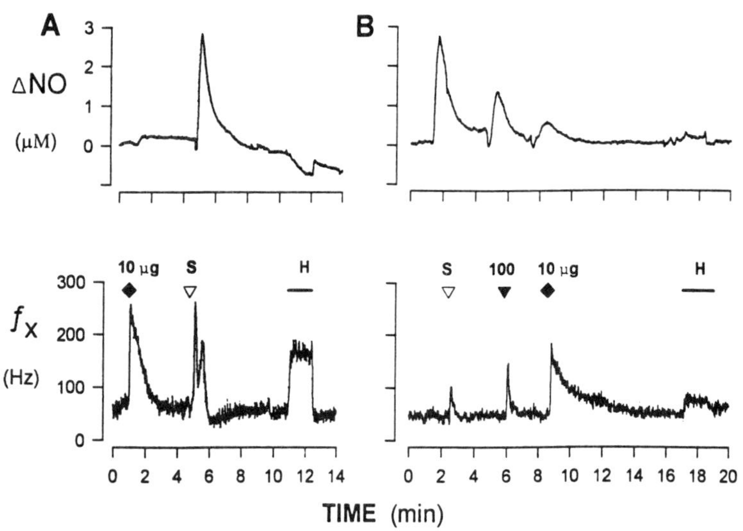

Figure 3. Simultaneous recordings of the effects of oligomycin on chemosensory response and NO signal. **A,** control responses. **B,** after oligomycin (12. 5 μg). **H**, hypoxia; ♦, nicotine; ▽, SNAP 450 μg; ▼, NaCN; Note that the NO signal decreased during the first hypoxic challenge due to movements of the preparation.

4. DISCUSSION

Present results show that oligomycin and FCCP reduced the carotid chemosensory excitation induced by SNAP in normoxia to the same degree as they reduced the chemosensory responses to hypoxia and NaCN, leaving the response to nicotine less affected. Our results suggest that NO may act on the respiratory chain of chemoreceptor cells, as hypoxia and NaCN do[16, 17] lending further support to the metabolic hypothesis of chemoreception[8,16]. The separation of the metabolic responses to cyanide and hypoxia to the response to nicotine and CO_2 indicates that there may be distinct pathways of chemoreception[16,17]. In isolated glomus cells[3], the administration of oligomycin increases $[Ca^{2+}]$, but the blockade of cytochrome oxidase by cyanide or anoxia failed to produce further increases in intracelular $[Ca^{2+}]$. Therefore, there is a parallelism between the results obtained by Duchen and Biscoe[3] in isolated glomus cells and the observations by Mulligan et al [16,17] in the whole CB.

The potential targets for the action of NO in the CB are the soluble guanylyl cyclase and the cytochrome oxidase aa_3 because these enzymes are highly sensitive to NO in the nM - μM range[5]. However, the activation of guanylyl cyclase in CB smooth muscle is expected to produce vasodilatation and a reduction in f_x [2, 7, 21], but any impairment of cytochrome oxidase redox activity is expected to increase f_x[16,17]. In recent times, the effects of NO on the respiratory chain and oxidative phosphorylation have received great attention[6]. It is known that NO inhibits mitochondrial respiration at different levels[4], reducing oxygen consumption in endothelial[5] and in the whole CB[2]. In fact, NO at low and medium concentration (50 nM - 5 μM) specifically and reversibly inhibits cytochrome oxidase competing with oxygen[6]. However, at higher concentrations NO, or its metabolite peroxynitrite may also inhibit other complexes of the respiratory chain in isolated mitochondria[4-6]. Several mechanisms, such as nitrosylation, oxidizing of protein thiols groups and removal of iron from iron-sulphur centers have been proposed to explain the effects of NO on the respiratory chain[6]. At higher levels, NO may also interact with other non-respiratory molecules such as free radicals, O_2, superoxide anion, and protein thiol groups.

In addition to the excitatory effect of NO observed during normoxia, NO reduced the response to hypoxia[7,13,14,18,20,21]. It has been proposed that the inhibitory effects of NO on hypoxic chemoreception were in part due vasodilatation[2,21] inhibition of glomus cells excitability by cGMP[21] or inhibition of thiol groups in Ca^{2+} channels[19]. Therefore, it is clear that NO may influence oxygen sensing in the CB by several mechanisms. In addition, a novel possibility is that the ratio NO/O_2 may be crucial for regulation of the glomus cells respiratory rate, playing a physiological role in oxygen sensing. Moreover, the recent finding that the mitochondria produces significant amounts of NO to regulate their own respiration[9,11], suggests that NO may be important for the physiological regulation of energy metabolism and rate of oxygen consumption[2] in the CB.

In summary, present results show that the excitatory effect of NO on carotid chemosensory activity during normoxia is reduced by oligomycin and FCCP as well as the responses to hypoxia and NaCN, providing new evidence to support the metabolic hypothesis of CB chemoreception.

5. ACKNOWLEDGMENTS

This work was supported by FONDECYT grant 198-0965. We would like to thank Mrs. Carolina Larraín for her assistance in the preparation of this manuscript.

6. REFERENCES

1. J. Alcayaga, R. Iturriaga, J. Ramirez, R. Readi, C. Quezada, and S. Salinas, Cat carotid body chemosensory responses to non-hypoxic stimuli are inhibited by sodium nitroprusside both *in situ* and *in vitro*, *Brain Res.* 767, 384-387 (1997).
2. D. G. Buerk, and S. Lahiri, Evidence that nitric oxide plays a role in O_2 sensing from tissue NO and PO_2 measurements in cat carotid body, *Adv. Exp. Med. Biol.* 475, 337-347 (2000).
3. M. R. Duchen and T. J. Biscoe, Relative mitochondrial membrane potential, and Ca^{2+} in type I cells isolated from rabbit arterial chemoreceptors, *J. Physiol. London.* 450, 33-61 (1992).
4. B. Borutaite, and C. G. Brown, Rapid reduction of nitric oxide by mitochondria, and reversible inhibition of mitochondrial respiration by nitric oxide, *Biochem. J.* 315, 295-299 (1996).
5. E. Clementi, C. G. Brown, N. Foxwell, and S . Moncada, On the mechanism by which vascular endothelial cells regulate their oxygen consumption. *Proc. Natl. Acad. Sci. USA.* 96, 1559-1562 (1999).
6. C. G. Brown, Nitric Oxide and mitochondrial respiration, *Biochim. Biophys. Acta* 1411, 351-369 (1999)
7. D. K. Chugh, M, Katayama, A. Mokashi, D. E. Debout, D. K. Ray, S. Lahiri, Nitric oxide-related inhibition of carotid chemosensory activity in the cat, *Resp. Physiol.* 97, 147-152 (1994).
8. C. Gonzalez, L. Almaraz, A. Obeso, and R. Rigual, Carotid body chemoreceptors: from natural stimuli to sensory discharges, *Physiol. Rev.* 74, 829-898 (1994).
9. P. Ghafourifar, and C. Richter. Nitric oxide synthase activity in mitochondria, *Febs Letters.* 418, 291-296 (1997).
10. P. A Grimes, S. Lahiri, R. Stone, A. Mokashi, and D Chugh, Nitric oxide synthase occurs in neurons and nerve fibres of the carotid body, *Adv. Exp. Med. Biol.* 360: 221-224 (1994).
11. C. Giulivi, Functional implications of nitric oxide produced by mitochondria in mitochondrial metabolism, *Biochem. J.* 332, 673-679 (1998).
12. R. Iturriaga, J. Alcayaga, S. Rey, Sodium nitroprusside blocks the cat carotid chemosensory inhibition induced by dopamine, but not that by hyperoxia, *Brain Res.* 799, 26-34 (1998)..
13. R. Iturriaga, M. Mosqueira and S Villanueva, Effects of nitric oxide gas on carotid body chemosensory response to hypoxia, *Brain Res*, 855,282-286 (2000).
14. R .Iturriaga, S. Villanueva, and M. Mosqueira, Dual effects of nitric oxide on cat carotid body chemoreception, *J. Appl. Physiol.* 89, 1005-1012 (2000).
15. E. Kowaluk, E. Seth and H. Fung, Metabolic activation of sodium nitroprusside to nitric oxide in vascular smooth muscle, *J. Pharm. Exp. Ther.* 262, 916-922 (1992).
16. E. Mulligan, S. Lahiri, and B.T. Storey. Carotid body O_2 chemoreception and mitochondrial oxidative phosphorylation, *J. Appl. Physiol.* 51, 438-466 (1981).
17. E. Mulligan, and S. Lahiri. Separation of carotid body chemoreceptor responses to O_2 and CO_2 by oligomycin and by actinomycin A, *Am. J. Physiol. Cell. Physiol.* 242, C200-C206 (1982).
18. N. R. Prabhakar, G. K. Kumar, C. H. Chang, F. H. Agani, and M. A. Haxhiu, Nitric oxide in the sensory function of the carotid body, *Brain Res.* 625, 16-22 (1993).
19. B. A Summers, J. L Overholt and N. R. Prabhakar, Nitric oxide inhibits L-type Ca^{2+} current in glomus cells of the rabbit carotid body via a cGMP-independent mechanism, *J. Neurophysiol.* 81, 1449-1457 (1999).
20. Z.-Z. Wang, L. J. Stensaas, D. S. Bredt, B. G. Dinger, and S. J. Fidone, Localization and actions of nitric oxide inthe cat carotid body, *Neuroscience* 60, 275-286 (1994).
21. Z.-Z. Wang, L.J. Stensaas, B.G. Dinger, and S.J. Fidone, Nitric oxide mediates chemoreceptor inhibition in the cat carotid body, *Neuroscience* 65, 217-229 (1995).

CO_2/HCO_3^- MODULATES K^+ AND Ca^{2+} CURRENTS IN GLOMUS CELLS OF THE CAROTID BODY

Jeffrey L. Overholt, Beth A. Summers, Eckhard Ficker, and Nanduri R. Prabhakar*

1. INTRODUCTION

Arterial CO_2 and O_2 are natural stimuli at the carotid body, the primary peripheral chemoreceptor. CO_2 acts synergistically with low oxygen (hypoxia) to stimulate carotid body activity and ultimately ventilation. Carotid body glomus cells are believed to be the primary site for chemotransduction. The cellular mechanism(s) of O_2 chemoreception by glomus cells are being delineated. Currently it is believed that chemoreception involves depolarization induced by inhibition of K^+ channels that are active at the resting membrane potential in glomus cells. Depolarization initiates Ca^{2+} influx via voltage-gated Ca^{2+} channels causing an increase in intracellular Ca^{2+} that leads to neurotransmitter release. On the other hand, relatively little is known about the mechanism of CO_2 chemoreception. As a natural stimulus, CO_2 could have direct cellular effects or indirect effects by changing pH (another natural stimulus to the carotid body). This complicates investigation of CO_2 effects since the effects could depend on the experimental conditions i.e., whether a physiological buffer is used. In fact, a number of studies have found that using bicarbonate-buffered solutions significantly improves the response of the *in-vitro* carotid body to hypoxia.[1-3]

One possible site for modulation of carotid body activity by CO_2 is to regulate glomus cell Ca^{2+} current, which should affect neurotransmitter release. Additionally, modulation of K^+ current should affect carotid body activity by altering the excitability of the cell. One important K^+ current is a HERG-like K^+ current that is active at the resting membrane potential in glomus cells [4]. Hypoxia has been shown to have effects on both Ca^{2+} [5] and K^+ current (for references see Prabhakar and Overholt [6]). In addition, pH can affect both Ca^{2+} [7] and HERG K^+ [8] current. However, it is not known whether CO_2 can directly affect Ca^{2+} or K^+ current or whether it may affect these currents indirectly through effects on pH. Therefore, we investigated the effects of CO_2 on HERG K^+ and Ca^{2+} current under standard conditions for whole-cell patch clamp (HEPES buffered media) and under more physiological conditions (bicarbonate buffered media). Our results show that CO_2 can have effects on both HERG K^+ and Ca^{2+} current. Moreover, the effects of CO_2 on both Ca^{2+} and HERG K^+ current

* Department of Physiology and Biophysics, Case Western Reserve University, Cleveland, OH 44106-4970.

Frontiers in Modeling and Control of Breathing, edited by Poon and Kazemi.
Kluwer Academic/Plenum Publishers, New York, 2001.

are greatly affected by the buffer system used (i.e., the presence of bicarbonate) for the studies. These effects on ion channels could account for the cellular effects of CO_2 at the carotid body.

2. METHODS

2.1. Isolation of Glomus Cells and Maintenance of HEK Cells

Carotid body cells were acutely isolated from adult male rabbits as described previously. Dissociated cells were maintained at 37°C in a CO_2 incubator in medium composed of a 50/50 mixture of Dulbecco's Modified Eagle Medium (DMEM) and HAM F12 (GIBCO) supplemented with antibiotics (penicillin and streptomycin), 10% fetal bovine serum (FBS), and insulin, transferrin and selenium (ITS, Sigma). Cells were used within 4-36 hours. HEK cells stably transfected with the HERG gene were grown at 37°C in a CO_2 incubator. The medium contained DMEM with 10% FBS and was supplemented with 0.005 mg/ml G418 and penicillin and streptomycin. Experiments were performed at RT (21°C).

2.2. Membrane Current Recording

K^+ and Ca^{2+} currents were measured in the whole cell configuration of the patch clamp technique [9] using an Axopatch 200 amplifier (Axon Instruments). Patch pipettes had resistances of 2-5 MΩ. For isolation of HERG currents the pipette solution contained (in mM): K aspartate (120), KCl (20), $CaCl_2$ (1), EGTA (5), HEPES (5), TrisGTP (0.2), MgATP (5) and glucose (10), pH 7.2 with KOH. The HEPES-buffered extracellular solution contained: NaCl (140), KCl (5.4), $MgCl_2$ (0.5), $CaCl_2$ (2.5), HEPES (5.5) and glucose (11), pH 7.4 with NaOH. The bicarbonate-buffered extracellular solution contained: NaCl (125), KCl (5), $MgCl_2$ (2), $CaCl_2$ (1.8), $NaHCO_3$ (25), NaH_2PO_4 (1.2) and glucose (11), pH 7.4 under continuos bubbling with 5% CO_2. Ca^{2+} current was isolated by using K^+- and Na^+-free intra and extracellular solutions. The intracellular solution contained (in mM): CsCl (115), TEA-Cl (20), MgATP (5), TrisGTP (0.2), phosphocreatine (10), EGTA (5) and HEPES (5), pH 7.2 with CsOH. The HEPES-buffered extracellular solution contained: NMGCl (140), CsCl (5.4), $BaCl_2$ (10), HEPES (10) and glucose (11), pH 7.4 with CsOH. The bicarbonate-buffered extracellular solution contained: NMGCl (120), CsCl (4.8), $BaCl_2$ (10), $NaHCO_3$ (25), NaH_2PO_4 (1.2) and glucose (11), pH 7.4 under continuous bubbling with 5% CO_2. Current recordings were not corrected for leak. PClamp programs (Axon Instruments) were used for data acquisition and analysis. The extracellular solution was changed using a fast-flow device [10]. Where appropriate data are expressed as mean ± SEM.

3. RESULTS

3.1. Effect of CO_2 on HERG Current

To begin to understand the mechanism of CO_2 action at the carotid body we examined the effects of CO_2 and HCO_3^- on HERG K^+ current, which is expressed in glomus cells. However, HERG K^+ currents in glomus cells are relatively small and very difficult to isolate from other large K^+ currents. Therefore, we chose to examine the effects of CO_2 on HERG channels heterologously expressed in human HEK 293 cells where the HERG current can be studied in isolation.

Figure 1A shows typical, slowly activating outward HERG K^+ currents at +10 mV followed by larger inward tail currents at -120 mV owing to the removal of a fast inactivation process that limits outward current at depolarized potentials. Figure 1B shows a time course for the effect of 5% CO_2 alone in a HEPES buffered external solution followed by 5% CO2 in a bicarbonate buffered solution. HERG current was elicited every 15 seconds by the protocol indicated in A. It can be seen that 5% CO_2 in HEPES inhibits both the inward tail and outward currents, probably due to a secondary effect of CO_2 on pH (pH of the HEPES solution bubbled with CO_2 was 6.53). In contrast, the time course in figure 1B and the current trace in figure 1A show that exposure to 5% CO_2 in a *bicarbonate* buffered extracellular solution *augments* the HERG current. Figures 1C and 1D further characterize the effects of 5% CO_2 in bicarbonate on the HERG current over a range of membrane potentials. These figures show that CO_2/bicarbonate shifts the activation curve for the HERG current to more negative potentials. These results suggest that CO_2 can affect HERG current, but the effect is critically dependent on the buffer used.

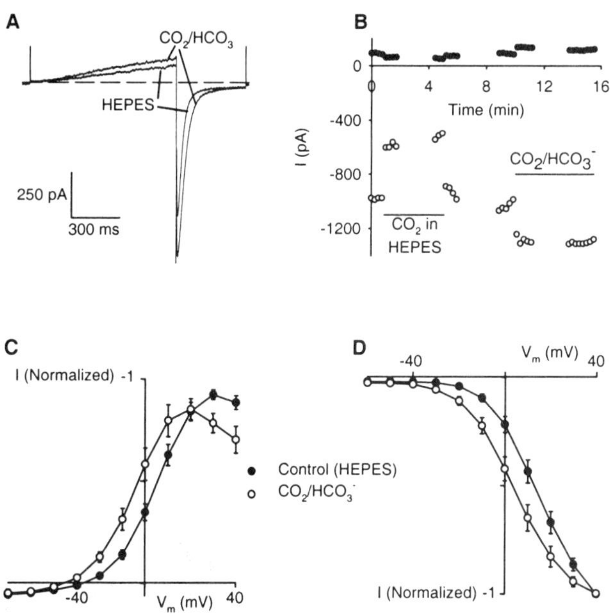

Figure 1. CO_2 alone inhibits, but CO_2/HCO_3^- augments HERG K^+ currents heterologously expressed in HEK 293 cells. A. Current elicited by a 900 ms voltage step to +10 mV followed by a voltage step to -120 mV for 325 ms (HP = -90). Cells were exposed to external solutions buffered with HEPES or bicarbonate (CO_2/HCO_3^-) bubbled with 5% CO_2/95% O_2. Dashed line indicates zero current level. B. Time course for the effects of CO_2 applied in a HEPES or bicarbonate buffered external solution (5% CO_2/ 95% O_2) on the HERG current elicited at +10 and -120 mV by the protocol indicated in A repeated every 15 seconds. C. and D. Average (n=13) current-voltage relations for current elicited under the conditions indicated in panel A. Outward (C) and tail (D) currents elicited by the protocol indicated in A, except the 900 ms step was varied from -60 to +40 mV in 10 mV increments.

3.2. Effect of CO_2 on Ca^{2+} Current

Regulation of Ca^{2+} channel activity in response to depolarization should affect neurotransmitter release from glomus cells and therefore activity in the carotid sinus nerve. To determine whether CO_2 has any effect on this phase of the transduction process, we investigated the effects of CO_2 on Ca^{2+} current in isolated, rabbit glomus cells. Figures 2A, 2B and 2C show current traces elicited by a step to 0 mV from a holding potential of -85 mV. Figure 2A shows that exposing glomus cells to a hypercapnic (10% CO_2), bicarbonate buffered external solution augments the Ca^{2+} current. Similarly, figure 2B shows that exposing glomus cells to a hypoxic, bicarbonate buffered external solution reversibly augments the Ca^{2+} current. This effect of hypoxia is seen in a bicarbonate, but not HEPES, buffered external solution at physiological pH [5].

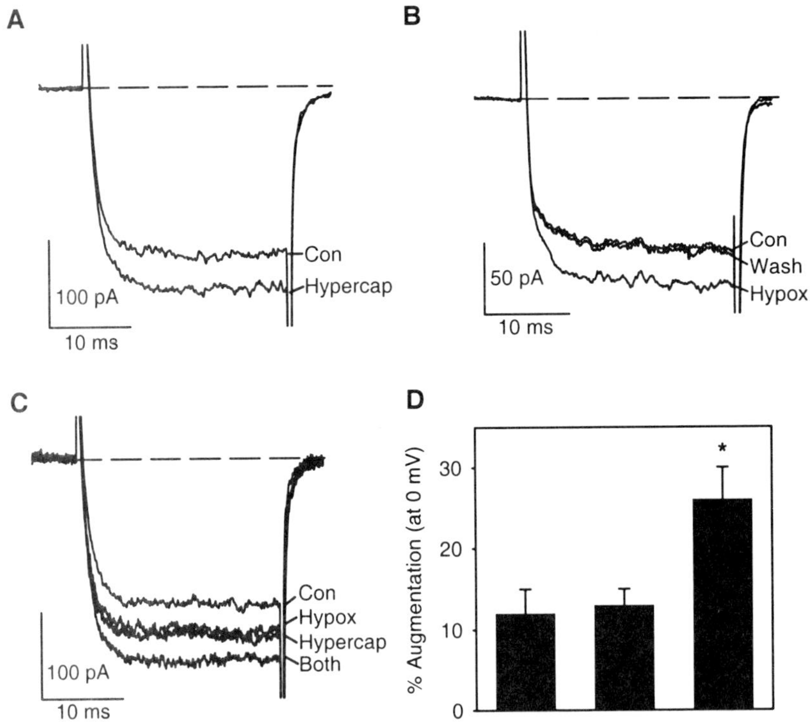

Figure 2. Hypoxia and hypercapnia have synergistic effects on Ca^{2+} current in glomus cells. Ca^{2+} current elicited by a 25 ms step to 0 mV in a bicarbonate buffered external solution (HP = -85 mV). A. Currents elicited in an external solution bubbled with 5% (Con) or 10% (Hypercap) CO_2 (21% O_2). B. Currents elicited before (Con), during (Hypox) and after (Wash) exposing cell to an external solution bubbled with a gas containing 1% O_2/5% CO_2. C. Currents elicited before (Con, 21% O_2/5% CO_2) and while exposing cell to an external solution bubbled with either 1% O_2/5% CO_2 (Hypox), 21% O_2/10% CO_2 (Hypercap) or 1% O_2/10% CO_2 (Both). All gas mixtures balanced with N_2. Dashed line indicates zero current level. D. Average (n=7) percent augmentation of the Ca^{2+} current at 0 mV elicited under the conditions described in C. * Indicates statistical significance, $p<0.05$.

3.3. Synergistic Effects of CO_2 and O_2 on Ca^{2+} Current

Hypoxia and CO_2 have synergistic effects on carotid body activity. Therefore, it is possible that the effects of hypoxia and hypercapnia converge at the level of the Ca^{2+} channel. To test this possibility, we examined the effects of hypoxia and hypercapnia alone and together in single cells. Figure 2C shows that both hypoxia and hypercapnia alone augment the Ca^{2+} current. It can also be seen that the effects of hypoxia and hypercapnia on the Ca^{2+} current are additive. This is shown more clearly in figure 2D, which compares the average percent augmentation of the Ca^{2+} current by hypoxia and hypercapnia alone and together. These results suggest that CO_2, like O_2, affects Ca^{2+} current in glomus cells. In addition, the effects of O_2 and CO_2 appear to converge at the level of the Ca^{2+} channel.

4. SUMMARY AND DISCUSSION

Our results show that CO_2 affects both HERG K^+ and Ca^{2+} current. In a HEPES buffered solution CO_2 inhibits, whereas CO_2 stimulates HERG current in a bicarbonate buffered solution. Further, hypercapnia augments Ca^{2+} current in glomus cells, and the effects of CO_2 and O_2 on Ca^{2+} current in glomus cells is additive.

The results presented in figure 1 clearly show that the external buffer used to study ionic currents can affect the behavior of the channels. In the case of HERG channels, the most obvious effect is a shift in the activation curve to more negative potentials in bicarbonate buffer. This shift could affect our interpretation of the role of the HERG channels in cell excitability. The shift to more negative potentials in bicarbonate suggests that more channels would be available at negative membrane potentials, and therefore give HERG channels a greater role in the control of the resting membrane potential. In glomus cells, the HERG-like current may actually have a half-activation potential of -54 mV, a shift of 10 mV (as in figure 1D) from our previous measurement in HEPES of -44 mV.[4] It should be noted that the effect may be more complicated as tail currents are also somewhat attenuated at more depolarized potentials in CO_2/bicarbonate (not shown). Therefore, to fully understand the effects, a more thorough study of the effects of CO_2/bicarbonate on other kinetic parameters of the HERG current is needed. In addition, further studies on the effect of hypercapnia on the glomus cell HERG current will be required to more fully understand the effects of CO_2 on carotid body activity.

The results presented in figure 2 show that hypercapnia can affect Ca^{2+} current in glomus cells. Moreover, the effects of CO_2 and O_2 on Ca^{2+} current are additive, suggesting that the synergistic effects of CO_2 and O_2 on carotid body activity may converge at the Ca^{2+} channel. We have previously shown that O_2 activates Ca^{2+} current via a PKC-dependent pathway [5]. The fact that the effects of CO_2 and O_2 are additive, suggests that they may activate Ca^{2+} current via different pathways. Future studies will investigate this possibility. Nonetheless, the present results indicate that one of the pathways by which CO_2 could enhance carotid body activity is by augmenting Ca^{2+} current in glomus cells.

Taken together, our data shows that use of a CO_2/bicarbonate buffered solution can affect ion channel activity when compared to HEPES. This is especially important since it suggests that the buffer used can affect the activity of excitable cells, since it is dependent on the activity of ion channels. The effects of CO_2 may be related to the interdependent relationships between CO_2, bicarbonate (HCO_3^-) and pH under physiological conditions.

However, our results also show that CO_2 can affect ionic current, and these effects are independent of pH.

5. ACKNOWLEDGMENTS

This work was supported by a National Institute of Health grant HL-25830 to Nanduri R. Prabhakar, and Jeffrey L. Overholt is a Parker B. Francis Fellow in Pulmonary Research.

6. REFERENCES

1. R. Iturriaga and S. Lahiri, Carotid body chemoreception in the absence and presence of CO2-HCO3-, *Brain Res.* **568**, 253-260 (1991).
2. M. Shirahata and R. S. Fitzgerald, The presence of CO_2/HCO_3^- is essential for hypoxic chemotransduction in the in vivo perfused carotid body, *Brain Res.* **545**, 297-300 (1991).
3. J. M. Panisello and D. F. Donnelly, Catecholamine secretion from glomus cells is dependent on extracellular bicarbonate, *Adv.Exp.Med.Biol.* **410**, 267-273 (1996).
4. J. L. Overholt et al., A HERG-Like potassium current regulates the resting membrane potential in glomus cells of the rabbit carotid body, *J.Neurophysiol.* **83**, 1150-1157 (2000).
5. B. A. Summers, J. L. Overholt, N. R. Prabhakar, Augmentation of L-type calcium current by hypoxia in rabbit carotid body glomus cells: evidence for a PKC-sensitive pathway, *J.Neurophysiol.* **84**, 1636-1644 (2000).
6. N. R. Prabhakar and J. L. Overholt, Cellular mechanisms of oxygen sensing at the carotid body: heme proteins and ion channels, *Respir.Physiol* **122**, 209-221 (2000).
7. W. Zhou and S. W. Jones, The effects of external pH on calcium channel currents in bullfrog sympathetic neurons, *Biophys.J.* **70**, 1326-1334 (1996).
8. J. Berube, M. Chahine, P. Daleau, Modulation of HERG potassium channel properties by external pH, *Pflugers Arch.* **438**, 419-422 (1999).
9. O. P. Hamill, A. Marty, E. Neher, B. Sakmann, F. J. Sigworth, Improved patch-clamp techniques for high-resolution current recording from cells and cell-free membrane patches, *Pflugers Arch.* **391**, 85-100 (1981).
10. J. L. Overholt and N. R. Prabhakar, Ca^{2+} current in rabbit carotid body glomus cells is conducted by multiple types of high voltage-activated Ca^{2+} channels, *J.Neurophysiol.* **78**, 2467-2474 (1997).

INHIBITION OR LACK OF NAD(P)H OXIDASE SUBUNITS DO NOT ALTER THE NORMAL CYTOSOLIC CALCIUM, SENSORY AND RESPIRATORY RESPONSE OF CHEMORECEPTORS

C. Rozanov*, A. Roy, A. Mokashi, P. Daudu, C. Ross†, and S. Lahiri

1. INTRODUCTION

Although investigated extensively[1, 2] at the present time the mechanism of O_2 sensing in the carotid body (CB) is still not known. One of the hypotheses states that NAD(P)H oxidase mediated production of reactive oxygen species (ROS) is decreased during hypoxia. Accordingly, hypoxia decreases superoxide anion production (O_2^-) and with it H_2O_2. It has been suggested that the reduction of H_2O_2 suppresses K^+ conductance of glomus cell membrane, leading to increase in cytosolic calcium ($[Ca^{2+}]_c$), neurotransmitter secretion and carotid sinus nerve (CSN) discharge. In this way, to counter hypoxia, the frequency and depth of ventilation is increased via the central respiratory center.

The NAD(P)H oxidases are a group of plasma membrane associated enzymes which during normal cellular metabolism catalyze the production of ROS using NADPH as the electron donor. The core enzyme is composed of five components: three of them exist in cytosol as a complex ($p40^{phox}$, $p47^{phox}$ and $p67^{phox}$) and two are located in the membrane ($p22^{phox}$ and $gp91^{phox}$) as a heterodimeric flavoprotein known as cytochrome b_{558}. Associated is also a membrane bound 22 kDa GTP binding protein (rac2). Defect in any of the components can lead to a dysfunctional NAD(P)H oxidase.

To test the hypothesis that NAD(P)H oxidase may serve as the O_2 sensor in CB glomus cells, we compared respiratory responses to normoxia (21% O_2), hyperoxia (100% O_2) and hypoxia (8% O_2) in p47*phox* and gp91*phox* knock-out mice with wild type mice.

We also investigated the effect of diphenyliodonium (DPI), an inhibitor of flavoprotein and PR-39, a proline-arginine rich antibacterial protein which inhibits NAD(P)H oxidase, on CSN discharge rates. In addition, since Ca^{2+} is a key element in

* C. Rozanov, A. Roy, A. Mokashi, P. Daudu and S. Lahiri, University of Pennsylvania School of Medicine, Philadephia, PA. †C. Ross, Kansas State University, Manhattan, KS.

Frontiers in Modeling and Control of Breathing, edited by Poon and Kazemi.
Kluwer Academic/Plenum Publishers, New York, 2001.

the cascade of chemotransduction, we compared changes in $[Ca^{2+}]_c$ in rats under normoxia and hypoxia with DPI and PR-39.

The responses to normoxia and hypoxia were normal in both mutant and wild type mice. In rats the inhibitors of NAD(P)H oxidase did not alter either CSN or calcium response under normoxia and hypoxia. Thus, we conclude either that NAD(P)H oxidase is not required for acute hypoxic responses in the CB or that there are other factors that are essential to work in conjunction with for chemotransduction to occur.

2. METHODS

2.1. Cytosolic Calcium

Changes in $[Ca^{2+}]_c$ is detected by fluorescence studies using Indo-1AM as the fluorescence indicator. During superfusion studies glomus cells were excited by a xenon light at 340 nm and the ratio of the emitted light at 405 and 495 nm were used to calculate the increase in $[Ca^{2+}]_c$ by the Grynkiewicz equation.[3]

2.2. Carotid Sinus Nerve Stimulation

The CB along with the intact CSN was surgically isolated from an anesthetized rat and transferred to a perfusion chamber maintained at 37^0C. Here the common carotid artery was connected to a cannula through which the various test solutions were introduced. A small cut under the tied internal carotid artery served as an exit port for the perfusate. The nerve was lifted and connected to an electrode from which neural discharge was detected. This is an in vitro experiment.[4]

2.3. Ventilation

These experiments were conducted in situ. The base of a T-tube was introduced into the trachea of an anesthetized mouse. Gas of different O_2 composition was introduced through one arm of the T-tube and the other arm served as the exit port for the exhaled gas. Positive pressure from the port of entry of gases ensured that the animal was inhaling the gas composition desired. A pressure transducer connected to the base of the T-tube was used to monitor the respiratory rate and tidal volume of breaths. Details of this procedure were described earlier.[5]

3. RESULTS

3.1. Changes in Cytosolic Calcium

Increase in glomus cell $[Ca^{2+}]_c$ is postulated to be a hallmark in the sequence of events leading to the final chemosensory response, ventilation. Thus, glomus cell calcium changes were monitored as the superfusate was switched from normoxic to hypoxic conditions in the presence of inhibitors of NAD(P)H oxidase in rat glomus cells. As seen in figure 1a, 10 μM DPI did not change $[Ca^{2+}]_c$ under normoxia. Under hypoxia,

$[Ca^{2+}]_c$ increased both in the presence and absence of DPI and the difference in the $[Ca^{2+}]_c$ levels were not significant. Similarly, 10 μM PR-39 did not change either normoxic or hypoxic $[Ca^{2+}]_c$ compared to control conditions (figure 1b).

3.1. Changes in Carotid Sinus Nerve Discharge

Another strategy used to test the importance of NAD(P)H oxidase in chemoreception was the use of NAD(P)H Oxidase inhibitors. In normoxia the flavoprotein inhibitor, DPI (10 μM, figure 2a) raised basal discharge slightly and the proline-arginine rich potent inhibitor of NAD(P)H oxidase, PR-39 (10 μM, figure 2b), did not change baseline CSN discharge. Neither compound influenced hypoxia-mediated neural discharge. Instead it was the severity of hypoxia that determined the relative increase in neural activity with both inhibitors.

3.3. Ventilation Response

To complete the sequence in chemotransduction to the final response, i.e. tidal volume and respiratory rate, ventilation experiments were conducted. Studies on respiratory rates and tidal volumes at 8, 21 and 100% O_2 levels were conducted on gp91*phox* knock-out mice and wild type mice of the same strain (figure 3a). There was no difference between gp91*phox* and wild type animals. Similarly, a comparison of respiratory rates and tidal volumes were made at the same Po_2 levels between p47*phox* knock-out mice and the corresponding wild type mice. Knocking out the p47 subunit from the NAD(P)H oxidase did not effect ventilation rates (figure 3b).

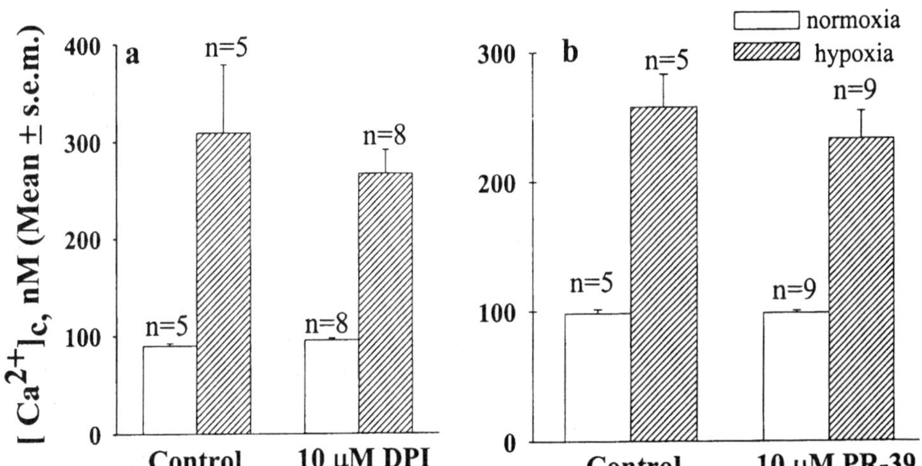

Figure 1. Superfusion studies of glomus cells showing cytosolic calcium changes with normoxia and hypoxia under the various conditions tested: (a) control versus DPI in the rat (b) control versus 10 μM PR-39 in the rat. Note differences in scale.

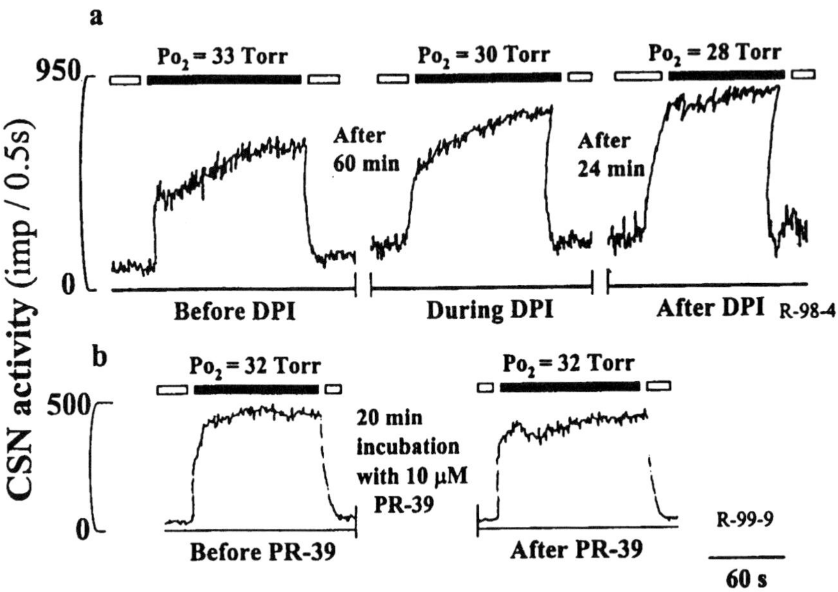

Figure 2. Carotid Sinus Nerve activity of the rat carotid body recorded under normoxia and various levels of hypoxia (a) before, during and after perfusion with 10 μM DPI (b) before and after perfusion with 10 μM PR-39.

4. DISCUSSION AND CONCLUSION

Hypoxia has been shown to decrease ROS which is generated by NAD(P)H oxidase.[6] If low ROS levels did play an important role in CB oxygen sensing, NAD(P)H oxidase may serve as a good candidate to be the sensor for low oxygen levels. Lack of one of the subunits required to mobilize the fully functional enzyme should impair its ability to produce constitutive levels of ROS. Under this condition a constitutive hypoxic response by default would replace normal normoxic base level signals and the normal normoxic-hypoxic response relationship should be nonfunctional.

Although we have not checked out every possible approach that NAD(P)H oxidase could be tested for its role as the oxygen sensor, we have addressed the question in two different ways. One of them was by using two different inhibitors of NAD(P)H oxidase and the other by testing the ventilation response of mice with either one of two subunits of NAD(P)H oxidase missing. By doing this we hoped to have covered a lot of ground in testing NAD(P)H oxidase as the oxygen sensor in the CB.

In our experiments, none of the inhibitors of NAD(P)H oxidase appeared to change either normoxic or hypoxic levels of $[Ca^{2+}]_c$ in rat CB glomus cells from what is seen in the absence of these inhibitors. Neither did the glomus cells of knock-out mice show any difference in $[Ca^{2+}]_c$ changes in both normoxia and hypoxia compared to those of their wild type mates (not shown). Similarly, hypoxic response of the CSN was unchanged in the presence of inhibitors of NAD(P)H oxidase compared to controls.

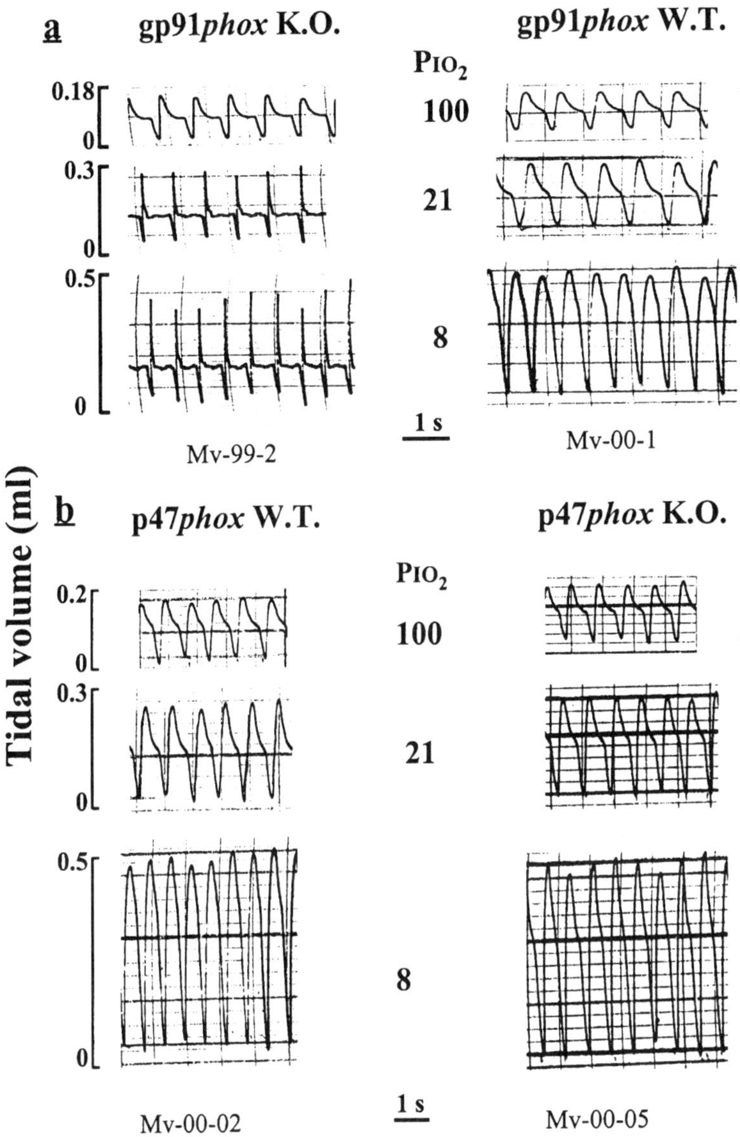

Figure 3. Ventilation experiments demonstrating the change in tidal volume under various percentages of oxygen in wild type (W.T.) versus knock-out (K.O.) mice (a) comparison between wild type and gp91*phox* K.O. mice and (b) comparison between wild type and p47*phox* K.O. mice.

Finally, in the absence of either the gp91*phox* subunit or the p47*phox* subunit of NAD(P)H oxidase, tidal volume and the respiratory rates were fully functional. Our results corroborate those of Archer et al.[7] who reported that in spite of lowered levels of ROS, gp91*phox* K. O. mice demonstrated normal hypoxic vasoconstriction. In addition, Obeso et al.[8] also reported that inhibition of NAD(P)H oxidase did not hamper hypoxia mediated catecholamine release from rat and rabbit CB glomus cells. Therefore, we conclude that NAD(P)H oxidase does not play a major role as the O_2 sensor in the CB.

5. ACKNOWLEDGMENT

Supported in part by grants R-37-HL-43413, RO1-HL-50180 and T-32-HL-07027.*

6. REFERENCES

1. S. Lahiri, Historical perspectives of cellular oxygen sensing and responses to hypoxia, *J. Appl. Physiol.* **88**, 1467-1473 (2000).
2. H. Acker and D. Xue, Mechanisms of O_2 sensing in the carotid body in comparison with other O_2-sensing cells, *News Physiol. Sci.* **10**, 211-216 (1995).
3. A. Mokashi, A. Roy, C. Rozanov, S. Osanai, B.T. Storey and S. Lahiri, High Pco does not alter pH_i, but raises $[Ca^{2+}]_i$ in rat carotid body glomus cells in the absence and presence of $CdCl_2$, *Brain Res.* **803**, 194-197 (1998).
4. A. Roy, C. Rozanov, D.G. Buerk, A. Mokashi and S. Lahiri, Suppression of glomus cell K^+ conductance by 4-aminopyridine is not related to $[Ca^{2+}]_i$, dopamine release and chemosensory discharge, *Brain Res.* **785**, 228-235 (1998).
5. A. Roy, C. Rozanov, A. Mokashi, P. Daudu, A. B. Al-Mehdi, H. Shams and S. Lahiri, Mice lacking gp91*phox* subunit of NAD(P)H oxidase showed glomus cell $[Ca^{2+}]_i$ and respiratory responses to hypoxia, *Brain Res.* **872**, 188-193 (2000).
6. S.L. Archer, J. Huang, T. Henry, D. Peterson and E.K. Weir, A redox-based O_2 sensor in rat pulmonary vasculature, *Cir. Res.* **73**, 1100-1112 (1993).
7. S.L. Archer, H.L. Reeve, E. Michelakis, L. Puttagunta, R. Waite, D.P. Nelson, M.C. Dinauer and E.K. Weir, O_2 sensing is preserved in mice lacking the gp91*phox* subunit to NADPH oxidase, *Proc. Natl. Acad. Sci. USA* **96**, 7944-7949 (1999).
8. A. Obeso, A. Gomez-Nino and C. Gonzalez, NADPH oxidase inhibition does not interfere with low Po_2 transduction in rat and rabbit CB chemoreceptor cells, *Am. J. Physiol.* **276**, C593-C601 (1999).

PERIPHERAL AND CENTRAL CHEMOSENSITIVITY: MULTIPLE MECHANISMS, MULTIPLE SITES?

A workshop summary

Workshop Chairs: Eugene E. Nattie* and Nanduri R. Prabhakar*

1. PERIPHERAL CHEMORECEPTION

1.1. Why Multiple Oxygen Sensors and Multiple Transmitters in the Carotid Body? (N.R. Prabhakar*)

The importance of peripheral chemoreceptors, especially the carotid bodies in control of breathing during hypoxia is being increasingly appreciated. Currently there are two views as to how hypoxia augments carotid body activity[1]. According to one view, a redox sensitive protein in the glomus cell is the oxygen sensor, and a variety of mitochondrial and non-mitochondrial redox-sensitive proteins have been proposed as potential O_2 sensors. The other view assumes that a K^+ channel in glomus cell is the primary O_2 sensor. The most challenging question is whether transduction involves a "single" or "multiple" O_2 sensors[1]. It is more than likely that multiple sensors are needed for oxygen sensing allowing the carotid body to respond to a wide range of arterial PO2's resulting in a curvilinear stimulus-response curve.

It is known from earlier studies by Lahiri and his-co-workers[2] that CO_2 has profound influence on the hypoxic sensory response of the carotid body. Also, the effects of hypoxia on glomus cell ionic conductances can be markedly influenced by CO_2(Overholt et al, this volume). These observations raise even a more fundamental question i.e., whether the O_2 sensing ability of the sensor depends on CO_2?

---

* Eugene E. Nattie, Department of Physiology, Dartmouth Medical School, Lebanon, N.H. 03756 (Co-Chair); Nanduri R. Prabhakar, Department of Physiology and Biophysics, Case Western Reserve University, Cleveland, OH 44106 (Co-Chair); R. Fitzgerald, Division of Physiology, Johns Hopkins University, Baltimore, MD 21205; Sukhamay Lahiri, Department of Physiology, University of Pennsylvania, Philadelphia, PA 19104; Peter A. Robbins, University Laboratory of Physiology, Oxford, Oxon OX1 3PT, UK; Curtis A. Smith, The Hohn Rankin Laboratory of Pulmonary Medicine, University of Wisconsin, Madison, WI 53706; Christina M. Spengler, Exercise Physiology, Swiss Federal Institute of Technology and University of Zurich, Zurich 8057, Switzerland.

Frontiers in Modeling and Control of Breathing, edited by Poon and Kazemi.
Kluwer Academic/Plenum Publishers, New York, 2001.

There is little doubt that neurotransmitters play essential role in the sensory transmission in the carotid body. Some of the transmitters excite, whereas others inhibit the sensory activity. It is possible that the excitatory and inhibitory transmitters work in concert with each other as a push-pull regulatory system[1]. Which one of them acts as an excitatory transmitter during hypoxia, however, remains unresolved. Dr. Fitzgerald presents his view on acetylcholine (ACh) as a potential excitatory transmitter for mediating the effects of hypoxia in the cat carotid body. However, release of ACh from rabbit carotid body is inhibited by hypoxia (see Kim et al; this volume). Given that hypoxia uniformly augments the sensory discharge in all species tested thus far, raises doubts as to the contribution of ACh to the sensory response to hypoxia in species other than cats. Another important question that need to be addressed is that are there separate sensory transmitters for hypoxia and hypercapnia? Presentations by Drs Lahiri and Fitzgerald address some issues relevant to these questions.

1.2. O_2-CO_2 Interactions in the Carotid Body and Glomus Cells (S. Lahiri*)

Po_2- Pco_2 interaction in ventilation was published in 1952/53 by Nielson and Smith. At that time, Dan Cunningham and Brian Lloyd in Oxford were working on this problem and came out with publications along with Loeschcke (Bochum). We subsequently measured the stimulus interaction (Po_2- Pco_2) in single fiber responses of cat carotid body, and showed resemblance with those in ventilation[2]. This strengthened the view that it mimicked the O_2-Hb dissociation curve. More recently we studied Po_2- Pco_2 interaction in Ca^{2+} in the rat glomus cells, and concluded that although changes in $[Ca^{2+}]$i. were greater with combined hypoxic-hypercapnia than either stimulus alone, but they may not entirely account for the sensory response elicited by both stimuli together[3]. We also found that ATP applied externally resulted in a huge $[Ca^{2+}]_i$ increase in glomus cells and this effect is coupled to activation surface ATP receptors to IP3 receptor system. ATP is generated metabolically and released during hypoxia. ATP then activated via the surface receptor the to release intracellular Ca^{2+} from the store. This is similar to CO_2 mediated response from the central chemoreceptors[4]. These are few examples that might provide some insights in to the possible influence of CO_2 on oxygen sensing.

1.3. Mechanisms of O_2-CO_2 Interactions in the Carotid Body: Role of Neurotransmitters (R. Fitzgerald*)

We have seen several times the impact of hypoxia on the CO_2 response curve where ventilation is the dependent variable. Hypoxia increases the slope of the ventilatory response to increasing hypercapnia. I would like to show you the response of the carotid body to hypoxia and CO_2. As CO_2 increases, neural activity from the carotid body rises and the slope of the rise is larger at lower values of pCO_2. Also important to note is that at the lower levels of PaO_2 the neural response to increasing CO_2 saturates. That is, as CO_2 increases there is no appreciable rise in neural activity. However, after a $paCO_2$ of about 50 mmHg, one observes a very brisk rise in the neural activity[5]. This suggests the possibility of two different mechanisms involved in the neuronal output from the carotid body, one responding to hypercapnia (and saturateable) and a second responding to hypoxia. This possibility receives support from the recent study of Dasso et. al.[6] in which they show response of $[Ca^{2+}]_i$ in a single glomus cell in response to hypoxia or to hypercapnia individually, along with a larger response when both are administered simultaneously. This suggests that the two different mechanisms are housed in one and the same glomus cell. The net effect of each is to increase $[Ca^{2+}]_i$, which may lead to release of neurotransmitter.

We have demonstrated that ACh fulfilled essential criteria needed for a synaptic transmitter[7]. These criteria have been more extensively supported by studies demonstrating with a variety of techniques that there are specific cholinergic receptors (a4 and a7 subunit-containing neuronal nicotinic, M2 and M1 muscarinic) present in the carotid body chemotransducing apparatus (Wang and Fitzgerald, this conference). Recently Zhang, et al 2000[8] found what seems to be an equally important role for ATP. It remains to be seen if ACh is the principal excitatory neurotransmitter during hypercapnia or hypoxia. Perhaps ACh plays the major role in hypoxia, but ATP plays the major role during hypercapnia. Or perhaps hypercapnia acidosis changes the sensitivity of the process to hypoxia, or hypoxia changes the sensitivity of the receptors to CO_2. Many tantalizing questions remain to be answered.

2. CENTRAL CHEMORECEPTION

2.1. Central Chemoreceptors Are Widespread. Why So Many? Why Have Any? (E.E. Nattie*)

Recent evidence obtained *in vivo* and *in vitro* indicate that central chemoreceptor sites have a widespread distribution within the brain[9,10,11,12,13]. Why do we need so many central chemoreceptor locations? There are a number of possibilities. Many sites allow redundancy, which provides reliability. Sites may differ in threshold and sensitivity, which provides stability. In evolution with new physiological requirements new sites may have emerged. Sites may vary in what they primarily 'sense', e.g., arterial blood vs. brain tissue or cell. Sites may vary in effectiveness depending on arousal state. There is support for this hypothesis. Focal acidification via microdialysis of CO_2 laden artificial cerebrospinal fluid in the region of the retrotrapezoid nucleus (RTN) increases ventilation only in the awake state via an increase in tidal volume[11]. Similar focal acidification of the medullary raphe increases ventilation only in sleep via an increase in breathing frequency[12].

An even more fundamental question, given the presence and sensitivity to CO_2 of the carotid body, is why do we need any central chemoreceptors? The first portion of this workshop focuses on this issue with brief presentations by Christine Spengler on patients with central alveolar hypoventilation who have no response to increased CO_2, by Peter Robbins on humans who have no carotid bodies, and by Curtis Smith on animal models with absent carotid bodies.

2.2. An Experiment of Nature: Congenital Central Hypoventilation Syndrome (C.M. Spengler*)

I would like to report on a very special group of patients suffering from the so called congenital central hypoventilation syndrome (CCHS). These patients breathe normally during the day but severely hypoventilate (or stop breathing) at night. The severity of the disease varies widely, i.e., some patients even need 24h ventilatory support. Most of these patients when challenged with hypercapnia (or hypoxia), they do not increase their breathing, even when awake. However, they do increase their breathing in response to other stimuli such as exercise.On the other hand, these subjects seem to arouse during sleep when challenged with increased exogenous CO_2.The question arises why these subjects don't increase their breathing to CO_2 but do seem to arouse from this same stimulus. It is possible that either central CO_2-receptors are the same for ventilatory and arousal responses but they are not functionally connected to the respiratory center thus only the arousal response 'works', or those chemoreceptors that mediate the arousal response are different from the 'respiratory' chemoreceptors and only the 'arousal' receptors work in these patients. Preliminary results from David Gozal and Ron Harper, imaging the brain (using fMRI) of CCHS patients and control subjects during hypercapnic challenges suggest indeed that brain areas responsive to CO2 are extensive (and include cerebellar, midbrain and frontal regions) and that CCHS patients are characterized by paucity of activation in multiple sites (especially at medial cerebellar and frontal regions). Looking at these results, we could speculate that CO_2-responsive areas are indeed different depending on their function (arousal or breathing). However, despite these interesting results, we can not yet conclusively tell whether the areas that were activated, are functional at all.

2.3. Central vs. Peripheral Chemoreception in Humans (P.A. Robbins*)

Results from a number of acute studies in humans[14], as well as in experimental animals[15], suggest that the contributions of the peripheral chemoreflex and central chemoreflex are independent and additive in respect to the ventilatory response. Given this, it is meaningful to consider how much of the ventilatory sensitivity to CO_2 arises from the peripheral chemoreceptors and how much from the central chemoreceptors.

One approach to this problem has been to determine the response to CO_2 under conditions of hyperoxia, when the peripheral chemoreceptors may be silenced[16]. However, a difficulty with this approach is that studies in anaesthetized animals in general show some peripheral chemoreceptor response to CO_2 under hyperoxic conditions[2]. An alternative approach has been to try to separate the peripheral and central chemoreflex components based upon the dynamics of their response, as the time constant for the central chemoreflex is approximately an order of magnitude greater than that for the peripheral chemoreflex. Recently, Pedersen et al[17] described a set of transitions between two different values for end-tidal PCO_2 which had been optimized for separating the peripheral and central components of the response based on their differing dynamics (a so-called multi-frequency binary sequence). Using this sequence and under conditions of hyperoxia, they demonstrated the presence of a fast component within the ventilatory response to CO_2 in six out of nine subjects. In this volume, Fatemian et al[18] report that no such fast component could be detected in any of four patients who had had bilateral carotid body resection. Taken together, these results suggest that a peripheral component to the ventilatory response to CO_2 remains under conditions of hyperoxia.

In patients who have undergone bilateral carotid body resection, not only is there a loss of the fast (peripheral) chemoreflex sensitivity, there is also a reduction in the magnitude of the slow (central) component of the CO_2 response[19,20]. These responses suggest that, while the acute sensitivities of the peripheral and central chemoreflexes may be independent and additive, this is not the case over longer time scales. Over longer time scales, it would appear that there is a loss of central chemoreflex sensitivity following a loss of peripheral input. This conclusion is reasonably consistent with denervation studies in experimental animals[20].

2.4. Chemoreception in Animals and Carotid Body (C.A. Smith*)

I would like to summarize studies of carotid body denervation performed in our laboratory and that of H.V. Forster. Although we came to the problem from different perspectives, we both asked essentially the same question, namely, is there redundancy/plasticity of the ventilatory control system following carotid body denervation? Forster's group[20] denervated adult goats and followed changes in ventilation and CO_2 sensitivity (steady-state method, normoxic background) for 15 days. They found an initial 60% decrement in the slope of the CO_2 response that was maintained for approximately 7 days, after which the slopes began to return to control values. By day 15 the slopes were virtually normal. In contrast, the CO_2 response slope (Read rebreathing method, hyperoxic) of our denervated adult dogs decreased by about 40% and remained at that level for the 21-day duration of the experiment (Rodman et al., submitted for publication). Accordingly, the Forster group concluded that there was central and/or peripheral plasticity for CO_2 sensitivity following removal of the carotid bodies in the goat, while we found no evidence of this in the dog. Why the discrepancy? We speculate that the goat, perhaps by virtue of its high altitude evolutionary history, may retain central and/or peripheral plasticity beyond the neonatal period, where plasticity has been well established (e.g., Refs. 21,22). Whether or not other commonly used laboratory species show time-dependent changes in CO_2 sensitivity following carotid body denervation remains to be determined.

3. DISCUSSION

Nattie: CCHS patients do not respond to CO_2 and they get into trouble when they fall sleep. Do patients with carotid body denervation have problems when they go to sleep?

Robbins: No.

H. Kazemi: The parasympathetic influence on heart rate is absent in CCHS patients as is the pupillary light reflex suggesting autonomic abnormalities. And, in some studies, they can respond appropriately in muscular exercise. How do these observations tie into the absent chemoreceptor response?

J.M. Ramirez: Maybe some of these differences are state related due to changes in the function of channels and the respiratory network that are state dependent. RET syndrome patients have problems in wakefulness; sudden infant death occurs in sleep; John Orem has shown state-dependent changes in apnea.

R. Shannon: In that CCHS patients don't breathe when they go to sleep, doesn't this suggest a problem in peripheral chemoreceptor processing?

Fitzgerald: Do they have a response to hypoxia in the daytime?

Spengler: No. Most of the studies have shown an absent response to hypoxia.

Fitzgerald: Have these patients been tested in respect to circulatory control.

Spengler: Not very much at all.

Ramirez: Why should not the central neurons also respond to hypoxia? The distinction that peripheral chemoreceptors are there primarily to sense hypoxia and the central chemoreceptors are there primarily to sense hypercapnia may be artificial. The blood vessels go right into the brain; why shouldn't neurons sense hypoxia as well.

Prabhakar: Irene Solomon is looking into central hypoxic sensitive neurons in the pre-Botzinger region. Don Reis has studied such neurons. And 10 years ago, Neil Cherniack and his group examined neurons in the ventrolateral medulla that were sensitive to hypoxia. The question is how much of this central hypoxic sensing is related to a change in ventilatory output.

Ramirez: In studies with clamping of carotid body input, as opposed to denervation of the carotid bodies, there is a central sensitivity to hypoxia. Denervation experiments can fool you as there is a loss of tonic input as well as hypoxic dependent increases in input.

Prabhakar: Maybe cardiovascular variables should be measured as a possible output related to central hypoxic detectors.

Lahiri: About 20 years ago, Neil Cherniack, Norm Edelman, and I tested the effects of dilute HCl on ourselves and a patient who had an absent ventilatory response to hypercapnia. All had a normal 'taste' response. The tongue of the patient with abnormal central chemosensitivity was perfectly all right.

Kazemi: What's the moral of the story?

Prabhakar: You can test the acid several ways. (Laughter.) But the question is are you breathing more.

N.S. Cherniack: I think there is a moral to the story. I think we should not get carried away with what we call 'receptors." There are lots of cells that respond to hypoxia in some way and there appear to be many cells that can respond to CO_2. We should be clear about what we mean by the term 'receptor'.

Fitzgerald: Well a receptor, if you are talking about a control mechanism, has also afferent pathways to a center, efferent pathways...

Cherniack: Wait a minute...Sukhamay just told us about a fellow with acid sensitive receptors on his tongue that had no relevance to respiratory control. One wonders how much some of this other stuff has any relevance to respiratory control.

Nattie: I want to get back to the question of why do we need central chemoreceptors?

Robbins: I think that the carotid body is altering, if you will, the magnitude of the response of the central chemoreceptors. In patients with carotid body resection what is being lost is a chunk of the central chemosensitivity.

Nattie: Do you think that this lost portion is that attributable to tonic afferent excitation from the carotid body perfused with normo-capnic and normoxic blood?

Robbins: How does the brain decide what is an appropriate output for what is happening at a receptor? I think of the carotid body as calibrating what is happening at the central chemoreceptor.

Nattie: So if there is tonic input from the carotid body then the central chemoreceptor response would be greater.

Robbins: Yes. For example in exposure to high altitude with increased stimulation of the carotid body, the CO_2 response is much greater.

Nattie: But again, why do we need central chemoreceptors? (Pause). I don't mean to put you on the spot, it's just something I like to think about. All your points are well taken and I agree with them. Curt (Smith) and I have had this discussion for some years now and agree on the importance of peripheral input on the central chemoreceptor response. But why have central chemoreceptors? Why so many?

Fitzgerald: Some years ago we did an experiment in dogs with perfused carotids. When the arterial PCO_2 was 25 to 30 mm Hg and we perfused the carotid with blood with PCO_2 5-7 mm Hg above that we did not get much of a ventilatory response. But the same amount perfused when the arterial PCO_2 was much higher produced a greater response. At even higher arterial values, say 55-60 mm Hg, perfusion of blood 5-7 mm Hg higher again produced a smaller response. Our interpretation was that after a certain point the central chemoreceptors took over for all levels of CO_2 input. The carotid body effect was minimal at low and at high PCO_2 levels but at normal levels they were quite effective.

Shannon: In response to the question, why have central chemoreceptors, in Comroe's textbook I read that central chemoreceptors are there to protect brain pH.

Nattie: But the pH change would have to be originating in the brain.

Shannon: True.

Nattie: They would be protecting the brain against itself.

Shannon: Alterations in brain pH can occur as in pumps change in certain circumstances.

Kazemi: I support Roger. I think that respiration is to support the brain. Ventilation is a very quick way to change or regulate brain pH. NMR studies show how quickly brain pH can change. Ventilation is for the benefit of the brain. This is a bold statement but it is at least in part true.

Y. Okada: Central chemoreceptors sense CO_2 and pH not only of arterial blood but of venous blood coming from the brain or brainstem. We have proposed a model based on c-fos localization of small cells located near small arteries and veins in the ventrolateral medulla in a good position to sense perivascular changes.

3. REFERENCES

1. N. R. Prabhakar. Oxygen sensing by the carotid body chemoreceptors. *J. Appl. Physiol.* 88: 2287-2295, (2000).
2. S. Lahiri and R.G. DeLaney. Stimulus interaction in the responses of carotid body chemoreceptor single afferent fibers, *Respir. Physiol.* **24,** 249-266 (1975).
3. A. Roy, C. Rozanov, A. Mokashi, and S. Lahiri. PO2-PCO2 stimulus interaction in [Ca 2+]i and CSN activity in the adult rat carotid body. *Respir. Physiol.* **122**: 15-26, (2000).
4. K.M. Spyer and T. Thomas. Sensing arterial CO_2 levels: a role for medullary P2X receptors. *J.Auton. Nerv. Syst.* **81**:228-235, (2000).

5. R.S. Fitzgerald, R.S. and D. Parks. Effect of hypoxia on carotid chemoreceptor response to carbon dioxide in cats. *Respir Physiol* **12**: 218-229, (1971).
6. Dasso, L.L, K.J. Buckler, and R.D. Vaughan-Jones. Interactions between hypoxia and hypercapnic acidosis on calcium signaling in carotid body type I cells. *Am. J. Physiol. Lung Cell Mol Physiol* **279**: L36-L42, (2000).
7. R.S. Fitzgerald. Oxygen and carotid body chemotransduction: the cholinergic hypothesis - a brief history and new evaluation. *Respir Physiol* **120**: 89-104, (2000).
8. Zhang, M., H. Zhong, C. Vollmer, and C.A. Nurse. Co-release of ATP and ACh mediates hypoxic signaling at rat carotid body chemoreceptors. *J. Physiol* (London) **525**: 143-158, (2000).
9. E. E. Nattie. Central chemoreceptors, pH, and respiratory control, in: *PH and Brain Function,* edited by K. Kaila and B. R. Ransom (John Wiley and Sons, Inc., New York, 1998), pp 535-560.
10. E. E. Nattie. CO_2 , brainstem chemoreceptors, and breathing, *Prog Neurobiol* **59**, 299-331 (1999).
11. E. Nattie. Multiple sites for central chemoreception: their roles in response sensitivity and in sleep and wakefulness, *Respir. Physiol.* **122,** 223-235, (2000).
12. A. Li, M. Randall, and E. E. Nattie. CO_2 microdialysis in the retrotrapezoid nucleus of the rat increases breathing in wakefulness but not in sleep, *J Appl Physiol* **87,** 910-919, (1999).
13. A. Li and E. E. Nattie. CO_2 microdialysis in the raphe of the unanesthetized rat increases breathing in sleep, *J Appl Physiol* (in press) (2001).
14. I. D. Clement, J.J. Pandit, D.A. Bascom, K.L. Dorrington, D.F. O'Connor, and P.A. Robbins. An assessment of central-peripheral ventilatory chemoreflex interaction using acid and bicarbonate infusions in humans, *J. Physiol.* **485,** 561-570 (1995).
15. J.H.G.M. van Beek, ,A. Berkenbosch, J. de Goede, and C.N. Olievier. Influence of peripheral O2 tension on the ventilatory response to CO2 in cats, *Respir. Physiol.* **51,** 379-390 (1983).
16. J. P. Miller, D.J.C. Cunningham, B.B. Lloyd, and J.M. Young. The transient respiratory effects in man of sudden changes in alveolar CO2 in hypoxia and in high oxygen, *Respir. Physiol.* **20,** 17-31 (1974).
17. M. E. F. Pedersen, M. Fatemian, and P.A. Robbins. Identification of fast and slow ventilatory responses to carbon dioxide under hypoxic and hyperoxic conditions in humans, *J. Physiol.* **521,** 273-287 (1999).
18. M. A. Fatemian, A. Dahan, S. Meinesz, A. van der Mey, and P.A. Robbins, Modelling the bilateral carotid body resection, in: *Frontiers in Modeling and Control of Breathing: Integration at Molecular Cellular and Systems Levels*, edited by C-S. Poon. New York: Kluwer Academic/Plenum, 2000, p. submitted.
19. J. W. Bellville, B.J. Whipp, R.D. Kaufman, G.D. Swanson, K.A. Aqleh, and D.M. Wiberg. Central and peripheral chemoreflex loop gain in normal and carotid body-resected subjects, *J. Appl. Physiol.* **46,** 843-853 (1979).
20. L. G. Pan, H.V. Forster, P. Martino, P.J. Strecker, J. Beales, A. Serra, T.F. Lowry, M.M. Forster, and A. L. Forster. Important role of carotid afferents in control of breathing. J. *Appl. Physiol.* **85,** 1299-1306 (1998).
21. T. F. Lowry, H. V. Forster, L. G. Pan, M. A. Korducki, J. Probst, R. A. Franciosi, and M. Forster. Effect of carotid body denervation on breathing in neonatal goats, *J Appl Physiol* **87**, 1026-34 (1999).
22. T. F. Lowry, H. V. Forster, L. G. Pan, A. Serra, J. Wenninger, R. Nash, D. Sheridan, and R. A. Franciosi. Effects on breathing of carotid body denervation in neonatal piglets, *J Appl Physiol* **87,** 2128-35 (1999).

Respiratory Neurons and Neural Networks

ROLE OF BRAINSTEM RESPIRATORY NEURON TYPES IN PHASE-SWITCHING PRODUCED BY AFFERENT VAGAL STIMULATION

Morton I. Cohen, Chen-Fu Shaw, Wu-Xin Huang, and Qiping Yu*

1. INTRODUCTION

A prominent feature of respiratory neural activity is the abrupt change of discharge at the phase transitions (IOS, inspiratory off-switch; EOS, expiratory off-switch). The observation of this nonlinearity has led to the search for specialized populations of brainstem neurons which change their activity immediately preceding a phase-switch. In this laboratory, we have used the method of producing perturbations of the respiratory cycle by means of phase-linked inputs and then comparing the temporal changes in the output (phrenic discharge) and in firing of different types of brainstem respiratory neurons. We here report the use of vagal (V) afferent inputs to study such responses.

Studies were done in midcollicular decerebrate, gallamine-paralyzed cats ventilated with a cycle-triggered pump, where lung inflation is delivered during neural inspiration (I) as indicated by phrenic discharge[1, 2]. The respiratory cycle was manipulated by timed changes of lung inflation: a) withdrawal of lung inflation during the I phase, which produced prolongation of I due to the removal of phasic afferent V input; b) imposition of lung inflation during the neural expiratory (E) phase, which produced prolongation of E due to increase of afferent V input. To study timing relations with greater accuracy, we delivered V afferent stimulus trains whose effects mimicked changes in output produced by manipulating lung inflation. For example, during no-inflation I phases we delivered afferent V stimulation whose strength was adjusted to produce a shortened I phase of duration comparable to that of the control (with-inflation) I phase.

* M.I. Cohen and W.-X. Huang, Albert Einstein College of Medicine, Bronx, NY 10461. C.-F. Shaw, National Sun Yat-Sen University, Kaohsiung, Taiwan. Q. Yu, Neurological Institute, New York, NY 10032.

Frontiers in Modeling and Control of Breathing, edited by Poon and Kazemi.
Kluwer Academic/Plenum Publishers, New York, 2001.

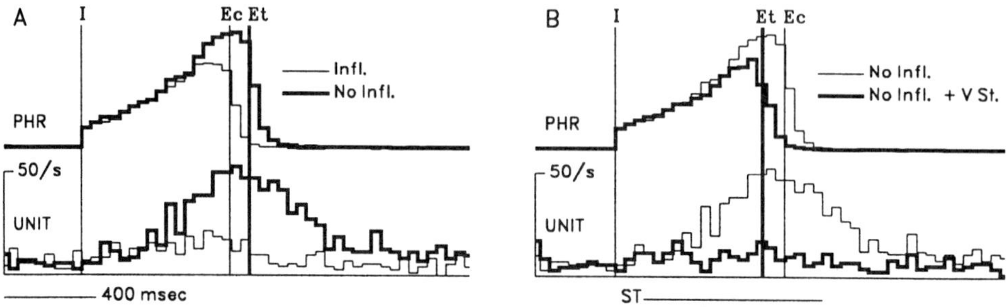

Figure 1. Pontine tonic I-modulated neuron, located in nucleus parabrachialis medialis. Effects of no-inflation (**A**) and afferent V stimulation (ST in **B**) on phrenic discharge and on unit firing. **A**: control (C), lung inflation applied during I; test (T), inflation withheld during I. **B**: C, inflation withheld during I (corresponding to test in **A**); T, V stimulation (100/s, 0.1 ms pulses), applied during an I phase when inflation was withheld. Cycle-triggered histograms (CTHs): 40-ms bins, 22 I triggers.

2. PONTINE I-MODULATED TONIC NEURONS

On the basis of lesion studies, it has been suggested that systems in the dorsolateral rostral pontine region of nucleus parabrachialis medialis and Kölliker-Fuse nucleus (formerly known as the "pneumotaxic center") are involved in control of respiratory rhythm[3]. Such lesions produce apneusis, which consists of marked weakening (retardation) of the IOS, suggesting that neurons in the region may facilitate the IOS. Therefore, we conducted studies of neurons in the region and their responses to V inputs[4-7]. These inputs affect the duration of I: afferent V electrical stimulation advances the IOS (shortens I), and withholding lung inflation during I retards the IOS (lengthens I).

The most common type of neuron in the region had a tonic discharge on which I- or IE-modulation was superimposed; and V afferent input typically produced a strong depression of activity, indicating inhibition of the I- or IE- portion of a unit's firing. It can be seen in the example of Figure 1A that withdrawal of lung inflation in I resulted in a marked increase of activity (due to disinhibition or facilitation) together with lengthening of I; whereas afferent V electrical stimulation during I (Figure 1B) produced strong depression of unit firing (due to inhibition or disfacilitation).

Analysis by means of the cumulative peristimulus histogram[7] showed that the latency from onset of the stimulus train to significant reduction of unit activity was about 170 ms, whereas the latency to the premature IOS (cessation of phrenic discharge) was about 300 ms. This timing relation is consistent with units of this type having a causative role in producing the IOS. However, the *sign* of the unit-phrenic relation was the inverse of what would be expected from such a causal relation, since a premature I termination would be expected to follow an *increase* of unit activity rather than the decrease of unit activity that was observed.

Responses such as those shown in Figure 1 were consistently found for neurons of this type (n = 23)[7]. It seems from these results that the role of such neurons in producing the IOS remains obscure. Therefore the effects of lesions in producing apneusis might have to be explained by actions on other populations (for example, tonically firing non-respiratory-modulated neurons) or by complex interactions between different populations.

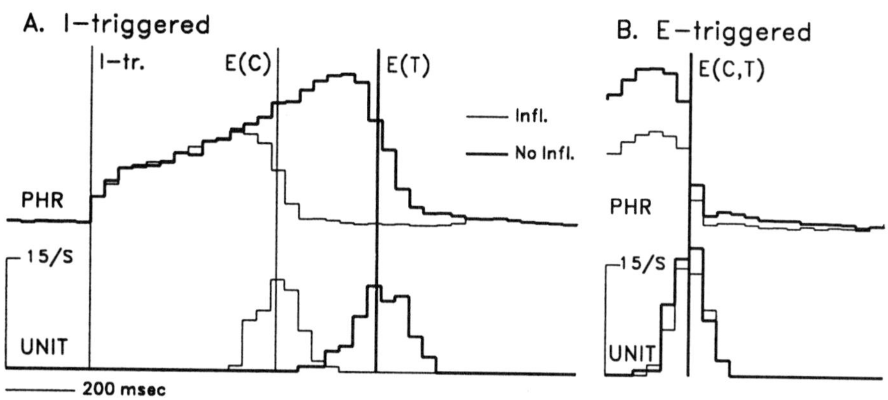

Figure 2. Late-I neuron recorded in the dorsal respiratory group. Effects of no-inflation on phrenic discharge and on unit firing. C (control), inflation applied during I; T (test), inflation withheld during I. CTHs (40 - ms bins, 36 triggers): **A**, triggered from start of I phase; **B**, triggered from start of E phase.

3. LATE-I NEURONS

A population of I neurons that start firing late in I have responses to V input consistent with a causative role in producing the IOS[8]. These neurons are excited by V input. As shown in Figure 2A, withdrawal of lung inflation results in an increased delay in onset of firing. The excitation by V input is also seen when afferent V electrical stimulation is delivered during a no-inflation I phase (not shown in a figure): the stimulation produces an earlier onset of firing.

The most crucial observation on the responses of these neurons is that their discharge in relation to the end of I is virtually unchanged when I phase duration is changed by inputs. As shown in the E-triggered histograms of Figure 2B, the discharge patterns at the end of the inflation I phase vs. the end of the no-inflation I phase are superimposable. This response differs from that of peak (end-I) phrenic amplitude and of peak (end-I) firing of other types of augmenting I neuron, which reach a larger magnitude when the I phase is lengthened (cf. phrenic trace in Figure 2B). This identity of end-I firing in late-I neuron responses during different states of I duration is also seen when I duration is changed by other types of input (e.g. aortic depressor nerve stimulation, which lengthens I).

These observations are consistent with the hypothesis that the late-I neuron population plays a causative role in producing the IOS. Their delay in firing compared to early-onset I neurons could be due to weaker excitatory inputs, higher firing thresholds, or early-I inhibitory inputs. When they reach a sufficient firing level in late I, these neurons may exert strong inhibition on other I neuron populations and thus produce the IOS.

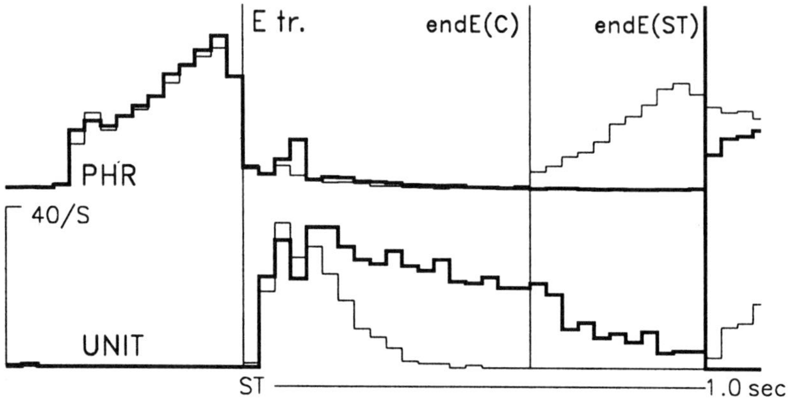

Figure 3. Early-E neuron recorded in the pre-Bötzinger region. Effects of afferent V stimulation (100/s, 0.1 ms pulses), starting 50 ms after E onset and lasting 1.0 s, on E duration and on unit firing. C, no stimulation; ST, stimulation. CTHs: 40-ms bins, 36 E triggers.

4. EARLY-E NEURONS

These neurons have their peak discharge at the start of I, followed by decrementing discharge, which in most cases persists till the end of E. Afferent V stimulation during E produces excitation of such neurons. In the example of the unit shown in Figure 3, V stimulation was started a short time after the termination of I and continued for 1.0 s. The excitation by V input was manifested as a reduction in the rate of discharge decrement, which occurred together with retardation of the EOS (increase of E duration). After stimulation was terminated, the delayed EOS occurred after a latency of about 100 ms. This response is consistent with the hypothesis that neurons of this type inhibit neurons that promote the EOS.

5. LATE-E NEURONS

These neurons start firing at various times during the E phase, have an augmenting discharge peaking near the end of E, which then gives way to an abrupt decline during the 50-100 ms before I onset (onset of phrenic discharge)[9]. V stimulation during E produces prolongation of the discharge and of E phase duration; most commonly, there is also a reduction of the slope of augmentation, but peak (end-E) frequency still increases because of the longer duration of the augmenting discharge. When V stimulation is stopped, the EOS occurs within 50-100 ms. The time course of this abrupt decline of firing before I onset is similar between control and stimulation states, even though the preceding peak frequencies may differ. The reciprocal effect of V input on early-E vs. late-E neurons suggests that interactions between the two populations are involved in control of E phase duration.

6. EXPIRATORY-INSPIRATORY (EI) NEURONS

These neurons start firing at various times during the E phase, have a slowly augmenting discharge, and sharply increase their firing just before (50-100 ms) the onset of I. Moreover, in contrast to late-E neurons, they continue firing throughout all or most of the I phase. In addition to their original designation as "EI" neurons[10], they have been recently designated as "pre-inspiratory" neurons[11]. V inputs (inflation or V nerve electrical stimulation) delivered during E produce depression of firing (even to the point of silence). As shown in the example of Figure 4A, a stimulus train applied to the afferent V nerve was started 0.5 s after onset of E and produced almost complete silence during its 1.0 s duration. When stimulation was stopped, the neuron resumed firing, but the slope of augmenting discharge was greater than in the control state, and then just before I onset (80 ms) there was a sharp increase of firing, as shown by the CTH triggered from this onset (Figure 4B). The CTHs also show that I onset occurred at the same level of unit activity in both the control and the stimulated states, i.e., although firing was delayed by the stimulus train, the initial post-stimulus activity had a larger slope than in the control state, so that firing quickly reached the same critical level at the time of I onset as in the control state. This close relation between firing level and time of I onset is consistent with neurons of this type playing a role in the EOS. The EOS action could arise from: a) excitation of a critical mass of I neurons, or b) inhibition of E neurons that allows release of the E neurons' inhibition of I neurons. The latter mechanism is suggested by the close locking at the time of I onset between late-E neurons' depression and EI neurons' excitation.

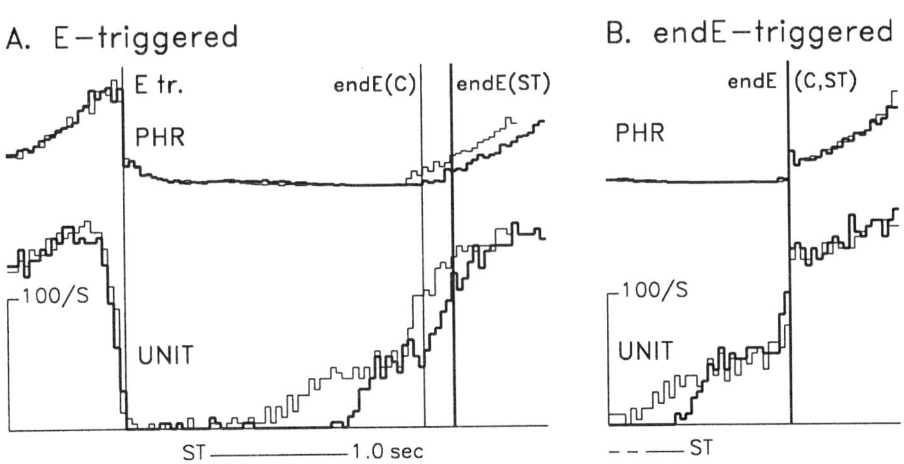

Figure 4. Expiratory-inspiratory (EI) neuron recorded in the ventral respiratory group at the obex level. Effects of afferent V stimulation (100/s, 0.1 ms pulses) applied during E on E duration and on unit firing. C, no stimulation; ST, stimulus train started 500 ms after E onset and lasting 1.0 s. Note inhibition of firing by ST and resumption of firing after end of train. CTHs: 40-ms bins, 36 E triggers (**A**), 36 end-E triggers (**B**).

7. CONCLUSIONS

In this study, we have attempted to analyze the role of different types of brainstem respiratory neurons in producing the I-E and E-I phase transitions. In order to ascertain the possible functions of different neuronal types, we have used the method of applying timed perturbations of the respiratory cycle and analyzing the relations between changes of the output (phrenic discharge) and changes of brainstem neuron firing. For this purpose, afferent V stimulation, either natural (lung inflation) or electrical, has proved useful, but we have also studied the effects of other types of input (superior laryngeal and aortic depressor). We have suggested possible functions for two types of neuron: a) the late-I neurons, which may promote the IOS; b) the EI or pre-I, which may promote the EOS.

8. ACKNOWLEDGMENT

This research was supported by N.I.H. Grant HL-27300.

9. REFERENCES

1. Feldman, J. L. and H. Gautier, Interaction of pulmonary afferents and pneumotaxic center in control of respiratory pattern in cats, *J. Neurophysiol.* **39,** 31-44 (1976).
2. Cohen, M. I. and J. L. Feldman, Discharge properties of dorsal medullary inspiratory neurons: relation to pulmonary afferent and phrenic efferent discharge, *J. Neurophysiol.* **51,** 753-776 (1984).
3. St. John, W. M., Neurogenesis of patterns of automatic ventilatory activity, *Prog. Neurobiol.* **56**, 97-117 (1998).
4. Feldman, J. L., M. I. Cohen, and P. Wolotsky, Powerful inhibition of pontine respiratory neurons by pulmonary afferent activity, *Brain Res.* **104,** 341-346 (1976).
5. Shaw, C.-F., M. I. Cohen, and R. Barnhardt, Inspiratory-modulated neurons of the rostrolateral pons: effects of pulmonary afferent input, *Brain Res.* **485,** 179-184 (1989).
6. Shaw, C.-F., *Discharge Patterns of Rostral Pontine Respiratory-Modulated Neurons in Relation to Vagal and Superior Laryngeal Afferent Input* (Ph.D. thesis, Albert Einstein College of Medicine, New York, 1990).
7. Cohen, M. I., C.-F. Shaw, and R. Barnhardt, Connectivity of rostral pontine inspiratory-modulated neurons as revealed by responses to vagal and superior laryngeal afferent stimulation, in: *Respiratory Control. Central and Peripheral Mechanisms*, edited by D. F. Speck, M. S. Dekin, W. R. Revelette, and D. T. Frazier (University Press of Kentucky, Lexington, 1993), pp. 91-94.
8. Cohen, M. I., W.-X. Huang, R. Barnhardt, and W. R. See, Timing of medullary late-inspiratory neuron discharges: vagal afferent effects indicate possible off-switch function, *J. Neurophysiol.* **69,** 1784-1787 (1993).
9. Cohen, M. I., J. L. Feldman, and D. Sommer, Caudal medullary expiratory neurone and internal intercostal nerve discharges in the cat: effects of lung inflation, *J. Physiol. (Lond.)* **368,** 147-178 (1985).
10. Cohen, M. I., Neurogenesis of respiratory rhythm in the mammal, *Physiol. Rev.* **59,** 1105-1173 (1979).
11. Schwarzacher, S. W., J. C. Smith, and D. W. Richter, Pre-Bötzinger complex in the cat, *J. Neurophysiol.* **73**, 1452-1461 (1995).

MODULATION OF THE CENTRAL RESPIRATORY EFFECTS OF 5-HT BY VAGAL AFFERENTS IN NEWBORN RAT

Jean-Charles Glérant, Frédéric Salzmann, Josette Khater-Boidin, and Bernard Duron*

1. INTRODUCTION

Recent *in vitro* studies using brain slices or an isolated preparation of brainstem and spinal cord have confirmed that serotoninergic mechanisms are involved in the central nervous control of breathing. For example, increases in respiratory frequency, via activation of 5-HT_1 receptors, have been elicited either by the use of exogenous 5-HT [1] or via the release of endogenous 5-HT after stimulation of the raphe nuclei[2]. Other workers have reported a biphasic effect of 5-HT on respiratory rate, an initial increase in inspiratory frequency (Fi) being followed by a low-frequency phase despite the continued presence of 5-HT [3, 4]. Interestingly, the effect of 5-HT on the respiratory rhythm *in vitro* appears to depend on the baseline respiratory rate[4], the inhibitory effect being more pronounced in a preparation with a higher respiratory rate.

On the contrary Khater-Boidin *et al.* [5] reported somewhat different results obtained in decerebrate newborn kitten *in vivo.* In this preparation, 5-HT applied to the floor of the IVth ventricle, acting mainly via 5-HT_2 receptors, induced a dose-related decrease in Fi and the appearence of long lasting apnoeas. Interestingly, Caroff *et al.* [6] and Cann-Moisan *et al.* [7] showed a significant increase in 5-HT metabolites in the cerebral fluid of victims of Sudden Infant Death Syndrom (SIDS). 5-HT, with other neurotransmitters, could thus be implicated in the mechanism responsible for producing apnoeas, one of the major elements of SIDS.

Moreover, in the same experimental conditions as that used in kitten the application of 5-HT and related agents to the floor of the IVth ventricle in decerebrate newborn rat induces bradypnea via the activation of 5-HT_2 receptors[8].

* Laboratoire de Neurophysiologie Clinique et Expérimentale, Faculté de Médecine, 3 rue des Louvels, 80036 Amiens Cedex France, josette.khater@.sa.u-picardie.fr

Frontiers in Modeling and Control of Breathing, edited by Poon and Kazemi.
Kluwer Academic/Plenum Publishers, New York, 2001.

Thus in spite of some interspecies differences it appears that the discrepancies between the results obtained *in vitro* and *in vivo* are probably linked to a possible role of some afferent systems. In the present study, in order to test this hypothesis, we applied 5-HT to the floor of the IVth ventricle in decerebrate bivagatomized newborn rat.

2. MATERIALS AND METHODS

Forty spontaneously breathing newborn rats (weight, 6.2 - 8.7 g; age, 0 - 3 days) were studied. This age bracket corresponds to that used to provide *in vitro* preparations, brainstem-spinal cord preparations being isolated from 0 to 4 days old rats.

2.1. Surgical Procedures

The initial surgery was performed under halothane anaesthesia with the animal breathing nasally in the supine position. By using a binocular microscope the vago-sympathetic trunks were sectioned at the cervical level.

For decerebration, the animal was placed prone with the head fixed in a small stereotaxic apparatus. The skull and dura mater were opened and the brain carefully aspirated. Under direct vision, the brainstem was sectioned at the level of the colliculi using a spatula. Haemostasis was achieved by filling the skull cavity with tiny cotton-wool balls. The anaesthesia was stopped immediately after the decerebration. The neck muscles were then dissected free to enable partial removal of the squama occipitalis along the mid-line and the dura mater was opened to expose the vermis cerebellum. This was later lifted to facilitate the placing of a catheter in the IVth ventricle.

A needle-shaped electrode (7 mm long and 0.2 mm in diameter) was inserted from a dorsal approach into the left costal portion of the diaphragm via the skin and the dorsal part of one intercostal space. At the end of the experiment, the animal received a lethal dose of anaesthetic (sodium pentobarbitone, 0.08 mg $\cdot$ (g body weight)$^{-1}$, i.p.).

2.2. Pharmacological Procedure

We used serotonin creatinine sulphate complex [0.1M 5-HT, RBI/Sigma, USA; 96.8 mg 5-HT in 2.5 ml deionized water containing 10 µl dimethylsulphoxide (DMSO, Prolabo, France)]. All solutions were prepared just prior to use (pH : 6.0 - 6.5). The choice of drug concentrations was made on the basis of our previous experiments in newborn kittens and on the basis of previous *in vitro* experiments.

5-HT (0.01 ml, 1,000 nanomoles) was administered in all newborn rats (bivagotomized, n=10; non-bivagotomized, n=10) by applying solution to the floor of the IVth ventricle via a small catheter connected to a 1 ml syringe. The duration of the injection was in all cases between 1 and 3 sec. Only one trial with one dose of 5-HT was performed in a given newborn rat to avoid any cumulative effects and any misinterpretation due to the possible saturation of serotoninergic receptors. Experiments were started 30 min (t 0) after the cessation of anaesthesia.

2.3. Data Acquisition and Analysis

Signals were filtered (30 Hz - 10 KHz), amplified (P511G, Grass), displayed on an oscilloscope (5113, Tektronic) and recorded on an electrostatic recorder (ES1000, Gould) throughout the experiment. Body temperature was checked by means of a rectal probe and maintained at 37° C by adjusting the distance of a heating lamp from the animal.

To assess the spontaneous variability in the respiratory rhythm in newborn rats, the minute inspiratory frequency (Fi) was studied in two sham groups (bivagotomized, n=10; non-bivagotomized, n=10). These animals went through all the aforementioned surgical procedures, but they did not receive 5-HT. Fi was measured every minute in all experiments starting 25 min after the end of the decerebration procedure and the cessation of anaesthesia and continuing for 35 min. The average values (± S.D.) over the first 5 minutes were calculated and used as the control (baseline) values for each member of the sham group. The experimental Fi values in a given animal are each expressed as a percentage of the corresponding baseline value in that animal to eliminate inter-individual variability.

The groups were compared in terms of their values for Fi, each expressed as a percentage of its own control and obtained every minute, using a factorial analysis of variance (ANOVA). Statistical significance ($p<0,05$) was assessed according to the number of values.

3. RESULTS

3.1. Sham Groups

Thirty minutes after decerebration and the discontinuation of anaesthesia, non-bivagatomized newborn rats (n=10) showed a rhythmic phasic inspiratory diaphragmatic discharge with a value for minute inspiratory frequency (Fi) in the range 39-123 cycles min^{-1} (mean ± sd, 80 ± 18 cycles min^{-1}). The mean Fi of bivagatomized newborn rats (n=10) was 27±7 cycles min^{-1} (range 19-37). In these two series Fi remained fairly stable although there was a slight increase in Fi at about 50 min (t 20) after decerebration and the discontinuation of anaesthesia.

3.2. 5-HT Groups

In non-bivagotomized newborn rats, placing the 5-HT solution (1,000 nmol) on the floor of the IVth ventricle caused an immediate and transient (less than 15 min) decrease in Fi (Fig. 1). In each animal, Fi showed a clear decrease within the first five minutes after 5-HT application. Four minutes after the application of 5-HT, the mean Fi for the 10 newborn rats had fallen to 41% of its control value; thereafter, the mean Fi recovered to about 99% of its control value as early as the 14th minute after 5-HT application (t 14). In fact, the mean Fi was significantly lower than that of the sham group ($p = 0.004$) from t 1 to t 9, i.e. for the first 9 min following drug application.

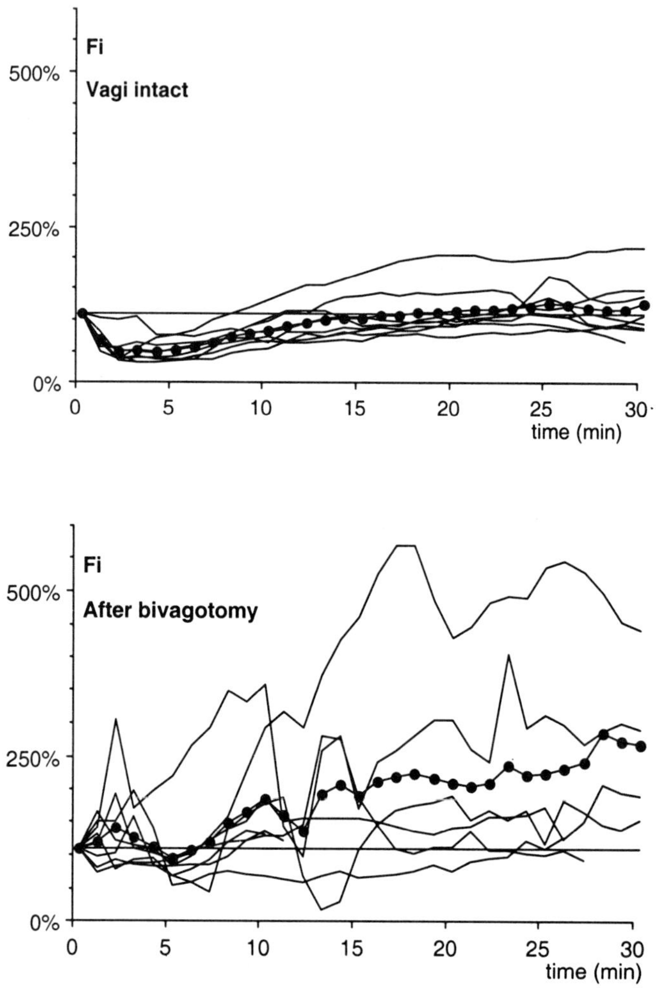

Figure 1. Respiratory effects of 5-HT. Top panel: individual responses and mean values (black dots) in non-bivagotomized new born rats. Bottom panel: individual responses and mean values (black dots) in vagotomized new born rats. Fi is expressed in percentage of control value.

In bivagotomized rats, the Fi mean value before application of 5-HT was 35±14 cycles/min (range 18-71). This mean value was similar to that obtained in the bivagotomized sham group. After application of 5-HT, we observed a progressive increase in respiratory frequency. This increase in respiratory frequency was significant (p<0,05) compared to the bivagotomized sham group. Thus, bivagotomy reverses significantly the central respiratory effect of 5-HT (Fig. 1).

4. DISCUSSION

After bivagotomy, the mean respiratory frequency of our bivagotomized decerebrate newborn rats was significantly lower than that observed in the group of decerebrate newborn rats with intact vagus. This indicates that vagal regulation exists during the early postnatal period. Moreover, vagal deafferentation reverses the central respiratory effect of 5-HT. In these *in-vivo* experimental conditions we obtained similar results as those reported *in vitro* in the isolated brainstem preparation[1, 3, 4, 9, 10, 11]. Thus, the bivagotomy modifies in the newborn rat the functional organization of the central respiratory network.

The reinforcement of the respiratory frequency observed *in vitro* is linked to an activation of 5-HT1A receptors[9]. On the contrary the decrease of respiratory frequency observed *in vivo*[8] in decerebrate newborn rat is linked to the activation of 5-HT_2 receptors. Thus we can postulate that depending on the integrity of the vagal control or not, the respiratory network responds to 5-HT via the activation of 5-HT_2 or 5-HT_{1A} receptors respectively. Different pools of neurons could be involved but at this time we do not know what are the vagal afferents (pulmonary or not) implied in the modulation of 5-HT central respiratory effect.

5. REFERENCES

1. Morin, D., Monteau, R. & Hilaire, G., 1991, 5-Hydroxytryptamine modulates central respiratory activity in the newborn rat : an in vitro study, *Eur. J. Pharmacol.* **192**, 89-95.
2. Morin, D., Hennequin, S., Monteau, R. & Hilaire, G., 1990, Depressant effect of raphe stimulation on inspiratory activity of the hypoglossal nerve: in vitro study in the newborn rat, *Neurosci. Lett.* **116**, 299-303.
3. Murakoshi, T., Suzue, T. & Tamai, S., 1985, A pharmacological study on respiratory rhythm in the isolated brainstem-spinal cord preparation of the newborn rat, *Br. J. Pharmacol.* **86**, 95-104.
4. Onimaru, H., Shamoto, A. & Homma, I., 1998, Modulation of respiratory rhythm by 5-HT in the brainstem-spinal cord preparation from newborn rat, *Pflügers Arch.* **435**, 485-494.
5. Khater-Boidin, J., Rose, D., & Duron, B., 1996, Central effects of 5-HT on activity of respiratory and hypoglossally innervated muscles in newborn kittens, *J. Physiol. (Lond)* **495.1**, 255-265.
6. Caroff, J., Girin, E., Alix, D., Cann-Moisan, C., Sizun, J. & Barthelemy, L., 1992, Neurotransmission et mort subite du nourrisson. Etude du liquide céphalo-rachidien, *C. R. Acad. Sci. Paris* **314**, 451-454.
7. Cann-Moisan, C., Girin, E., Giroux, J.D., Le Bras, P. & Caroff, J., 1999, Changes in cerebrospinal fluid monoamine metabolites, tryptophan, and gamma-aminobutyric acid during the 1st year of life in normal infants. Comparison with victims of Sudden Infant Death Syndrome, *Biol Neonate* **75** : 152-159.
8. Khater-Boidin, J., Rose, D., Glérant J.Ch., & Duron, B., 1999, Central effects of 5-HT on respiratory rythm in newborn rats *in vivo*, *Eur. J. Neurosci.* **11**, 3433-3440.
9. Morin, D., Hennequin, S., Monteau, R. & Hilaire, G., 1990, Serotonergic influences on central respiratory activity : an *in vitro* study in the newborn rat, *Brain Res.* **535**, 281-287.
10. Morin, D., Monteau, R. & Hilaire, G., 1992, Compared effects of serotonin on cervical and hypoglossal inspiratory activities : an in vitro study in the newborn rat, *J. Physiol. (Lond)* **451**, 605-629.
11. Al-Zubaidy, Z.A., Erickson, R.L. & Greer, J.J., 1996, Serotonergic and noradrenergic effects on respiratory neural discharge in the medullary slice preparation of neonatal rats, *Pflügers Arch-Eur. J. Physiol.* **431**, 942-949.

ACTIVATION OF MEDULLARY POST-INSPIRATORY RELATED NEURONS DURING CLONIDINE-INDUCED CENTRAL APNEA IN ANESTHETIZED GOATS

Ken D. O'Halloran, Jay K. Herman, and Gerald E. Bisgard*

1. INTRODUCTION

α_2-adrenoceptors (α_2-ARs) are widely distributed in the CNS including brainstem sites associated with cardiorespiratory control.[1, 2] Several lines of evidence strongly suggest a role for α_2-ARs in the regulation of respiratory rhythm. We have recently demonstrated that systemic administration of clonidine and other α_2-AR agonists cause profound disturbances in respiratory pattern in the goat that are characterized by alternating episodes of tachypnea and respiratory depression (prolonged and variable expiratory time intervals).[3-8] Dysrhythmic breathing induced by α_2-AR agonists in the goat is associated with continuous tonic activation of expiratory-related drive in laryngeal[4, 5] and pharyngeal motoneurons[6, 8] leading to active airway closure during central apnea.[6, 8]

Recent network models[9, 10] of respiratory rhythmogenesis have suggested that respiratory rhythm is highly dependent upon changes in tonic activity to cell groups exhibiting post-inspiratory (PI) activity and that changes in tonic activity to PI neurons exhibit the greatest effect on respiratory rhythm. We hypothesized that α_2-ARs are linked directly or indirectly to tonic modulation of medullary PI-related activity which contributes to the maintenance of respiratory rhythmogenesis. In this study, we wished to test the hypothesis that α_2-AR agonists such as clonidine result in the activation of medullary PI-related neurons (assessed by single-unit motor nerve recordings). Such a finding would be suggestive of a possible mechanism by which α_2-AR agonists disrupt respiratory rhythm.

* Department of Comparative Biosciences, School of Veterinary Medicine, University of Wisconsin, Madison WI 53706-1102, USA.

Frontiers in Modeling and Control of Breathing, edited by Poon and Kazemi.
Kluwer Academic/Plenum Publishers, New York, 2001.

2. METHODS

2.1. Animal Preparation

A total of 6 adult female and castrated male goats of mixed breed were used in this study. After induction of anesthesia for intubation, animals were placed in dorsal recumbancy under a thermostatically controlled heating blanket to maintain normal body temperature (38-40°C). Anesthesia was maintained with α-chloralose (10-15 mg/kg/hr) given intravenously (IV). Femoral arterial and venous catheters were implanted for monitoring systemic arterial blood pressure and for drug administration respectively. Arterial blood samples were drawn anaerobically into heparinized syringes and immediately analyzed for arterial pH, Pa_{CO_2} and Pa_{O_2}. Lactated Ringer solution was infused continuously throughout the experiment to help maintain fluid balance and arterial blood pressure. Sodium bicarbonate was administered IV as necessary. Arterial blood samples were frequently obtained during isolation and preparation of nerve fibers for recording. A C_6 phrenic nerve root was dissected free from surrounding tissue low in the neck, cut distally, desheathed and prepared for neural recording. Phrenic nerve activity served as an index of neural inspiration.

The left recurrent laryngeal nerve (RLN) was isolated through a ventral midline incision in the neck and was cut just caudal to the larynx. The RLN was positioned on a dissecting platform and submerged in mineral oil to prevent dessication. To prevent disruption of the preparation the goats were paralyzed with pancuronium (2mg IV; 1mg hourly) and mechanically ventilated. The contralateral vagus nerve was cut. Unipolar recordings were made with platinum wire electrodes from fine filaments of the RLN teased until unique unitary discharges could be readily identified. Single or pauci-fiber preparations in which spikes could clearly be distinguished and counted separately with the window discriminator were used. To verify that spikes represented the activity of a single fiber, a storage oscilloscope was used to observe the characteristic shape of the action potentials. Nerve activity was amplified and visualized on a storage oscilloscope and polygraph chart recorder. The signal was passed through a window discriminator to a rate meter and audiomonitor. The raw nerve signal, window output and arterial blood pressure and phrenic nerve signal were digitized and processed using the WINDAQ data acquisition system (Dataq Instruments, Akron, OH).

In separate studies, single-unit expiratory-related activity was recorded in the same manner from fine filaments of the pharyngeal branch of the vagus nerve supplying the middle pharyngeal constrictor muscle ($Ph\text{-}X_{MPC}$).

2.2. Protocol

Arterial blood gases were sampled frequently to ensure maintenance of blood-gas and acid-base homeostasis during mechanical ventilation. On identifying a laryngeal or pharyngeal unit, activity was recorded for several minutes during normoxic-normocapnic conditions. After recording baseline neural activities for at least 5 min, transient dose-dependent ventilatory inhibition and central apneas were induced by IV bolus injections of dopamine (10-50 μg/kg). When control conditions were re-established, the level of inspired CO_2 was lowered and if necessary, the frequency of mechanical ventilation was increased to inhibit central respiratory motor drive. Central apnea was observed when

CO_2 fell below the apneic threshold. Arterial P_{CO_2} was lowered approximately 10 mmHg below baseline conditions. When control conditions were re-established and several arterial blood samples obtained, dysrhythmic breathing episodes were induced by systemic administration of the α_2-AR agonist clonidine (0.5-3.0 μg/kg). Following cumulative doses of clonidine, the goats were treated with the selective α_2-AR antagonist SKF-86466.[11]

2.3. Drugs

All drugs were prepared on the day of each experiment. Doses of all drugs were calculated on the basis of salt weight. Dopamine HCl, clonidine and SKF-86466 were dissolved in sterile saline (0.9% NaCl) to obtain stock solutions which were further diluted in NaCl for IV administration.

3. RESULTS

Clonidine administration produced dysrhythmic breathing episodes that included prolonged and highly variable periods of central apnea characteristic of α_2-AR stimulation in anesthetized goats.[3, 4] Ventilatory disturbances induced by clonidine were clearly associated with activation of PI-related RLN units (Fig. 1). Continuous tonic activation of PI-related RLN units persisted throughout the duration of prolonged T_E intervals and central apneas induced by clonidine. Following α_2-AR blockade with SKF-86466, the discharge frequencies and patterns were restored to baseline (pre-clonidine) conditions. Similar to central apneas induced by α_2-AR stimulation with clonidine, continuous tonic activation of PI-related RLN units was consistently observed during dopamine-induced and hypocapnic-induced central apneas.

Three expiratory-related Ph-X_{MPC} units were studied. Two units displayed activity that was continuous throughout the duration of expiration whereas the other unit displayed PI-related activity similar to RLN units. Similar to RLN units, continuous tonic activation of Ph-X_{MPC} units persisted throughout the duration of prolonged T_E intervals and central apneas induced by clonidine, dopamine and hypocapnia.

4. DISCUSSION

The network oscillator model proposed by Richter and colleagues[9, 12] suggests that a single respiratory cycle is comprised of three distinct phases: inspiration, post-inspiration (PI or stage I expiration) and late stage (II) expiration (E-2). Brainstem neurons with properties correlated to one of these three phases also coincide with motoneuronal activities to respiratory muscles that regulate airflow[9, 13-15] and several studies have used motor outflow from cranial and spinal motoneurons to infer activity of the central rhythm-generating network.[14-17]

A recent computer model[10] based on the three-phase network model of respiratory rhythmogenesis provides a useful framework within which some testable predictions can be made. In this model, respiratory rhythm is highly dependent upon changes in tonic activity to cell groups exhibiting PI activity. Brainstem neurons that exhibit PI-related activity have been recorded from medullary interneurons and motoneurons mainly within the vicinity of the nucleus ambiguus.[13, 18-21] Intracellular recordings show that PI neurons exert widespread synaptic inhibition to I and E-2 brainstem neurons.[12, 18] Moreover, superior laryngeal nerve stimulation causes prolonged apnea which is associated with sustained activation of PI neurons and inhibition of I and E-2 neurons.[13, 14, 19-22] Within the context of this model, PI neurons are thought to have an overall executive role in the control of respiratory rhythm.[9, 10] Thus, this model predicts that changes in tonic activity to PI neurons exhibit the greatest effect on respiratory rhythm.

The present study clearly demonstrates that ventilatory disturbances induced by α_2-AR stimulation are associated with tonic activation of expiratory-related drive within laryngeal and pharyngeal motoneurons. Our findings suggest that α_2-AR pathways modulate tonic activity in medullary PI-related neurons and participate in the control of respiratory rhythm.

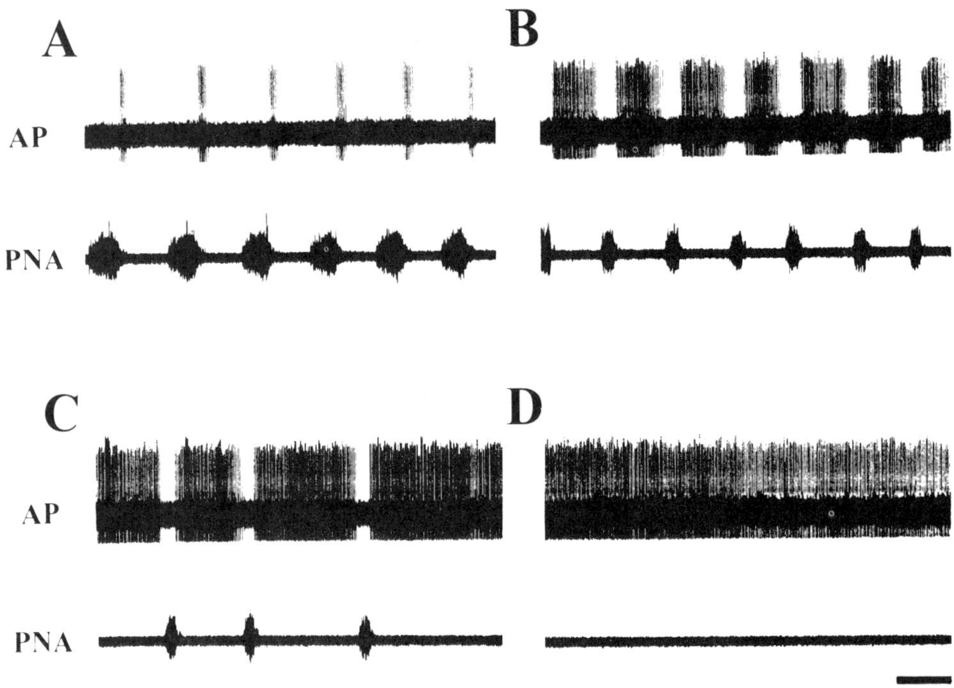

Figure 1. Single unit motoneuronal activity (AP) recorded from the RLN of an anesthetized, paralyzed, mechanically ventilated goat. Phrenic nerve activity (PNA) was used as an index of neural inspiration. Under control conditions (A) this unit is active during the post-inspiratory phase of the respiratory cycle. A single injection of clonidine (1 μg/kg IV) is associated with activation of this unit (B). Dysrhythmic breathing episodes (C) and central apnea (D) were observed following incremental doses of clonidine (3 μg/kg total cumulative dose). Note the continuous tonic activation of the RLN PI-related unit throughout the duration of prolonged TE intervals and central apneas (C and D). Time bar represents 2 seconds.

5. ACKNOWLEDGMENTS

We thank Gordon Johnson for excellent technical assistance. This work was supported by NIH Grants HL-53969 and HL-10069.

6. REFERENCES

1. A.P. Nicholas, V. Pieribone and T. Hökfelt, Distribution of alpha-2 adrenergic receptor subtypes in rat brain: an in situ hybridization study, *J. Comp. Neurol.* 328, 575-594 (1993).
2. P.G. Guyenet, R.L. Stornetta, T. Riley, F.R. Norton, D.L. Rosin and K.R. Lynch, Alpha$_{2A}$-adrenergic receptors are present in lower brainstem catecholaminergic and serotonergic neurons innervating spinal cord, *Brain Res.* 638, 285-294 (1994).
3. M.S. Hedrick, M.L. Ryan, J. Pizarro and G.E. Bisgard, Modulation of respiratory rhythm by α_2-adrenoceptors in awake and anesthetized goats, *J. Appl. Physiol.* 77, 74X-750.
4. M.S. Hedrick, M.L. Ryan and G.E. Bisgard, Recurrent laryngeal nerve activation by α_2 adrenergic agonists in goats, *Respir. Physiol.* 101, 128-137 (1995).
5. M.S. Hedrick, M.R. Dwinell, P.L. Janssen, J. Pizarro and G.E. Bisgard, Differential respiratory muscle recruitment induced by clonidine in awake goats, *J. Appl. Physiol.* 84, 1198-1207 (1998).
6. K.D. O'Halloran, J.K. Herman and G.E. Bisgard, Differential effects of clonidine on upper airway abductor and adductor muscle activity in awake goats, *J. Appl. Physiol.* 87, 590-597 (1999).
7. K.D. O'Halloran, J.K. Herman and G.E. Bisgard, Nonvagal tachypnea following α_2-adrenoceptor stimulation in awake goats, *Respir. Physiol.* 118, 15-24 (1999).
8. K.D. O'Halloran, J.K. Herman and G.E. Bisgard, Clonidine induces upper airway closure in awake goats, *Respir. Physiol.*, in press.
9. D.W. Richter, D. Ballantyne and J.E. Remmers, How is the respiratory rhythm generated? A model, *News Physiol. Sci.* 1, 109-112 (1986).
10. M.D. Ogilivie, A. Gottschalk, K. Anders, D.W. Richter and A.I. Pack, A network model of respiratory rhythmogenesis, *Am. J. Physiol.* 263, R962-R975 (1992).
11. J.P. Hieble, M. Robert, R.M. DeMarinis, P.J. Fowler and W.D. Matthews, Selective alpha-2 adrenoceptor blockade by SKF-86466: in vitro characterization of receptor selectivity, *J. Pharm. Exp. Ther.* 236, 90-96 (1986).
12. D.W. Richter, Generation and maintenance of the respiratory rhythm, *J. Exp. Biol.* 100, 93-107 (1982).
13. J.E. Remmers, D.W. Richter, D. Ballantyne, C.R. Bainton and J.P. Klein, Reflex prolongation of stage I of expiration, *Pflügers Arch.* 407, 190-198 (1986).
14. W.M. St. John and D. Zhou, Differing control of neural activities during various portions of expiration in the cat, *J. Physiol. Lond.* 424, 189-204 (1989).
15. W.M. St. John and D. Zhou, Discharge of vagal pulmonary receptors differentially alters neural activities during variuos stages of expiration in the cat, *J. Physiol. Lond.* 424, 1-12 (1990).
16. T.E. Dick, Y. Oku, J.R. Romaniuk and N.S. Cherniack, Interaction between central pattern generators for breathing and swallowing in the cat, *J. Physiol. Lond.* 465, 715-730 (1993).
17. Y. Oku, T.E. Dick and N.S. Cherniack, Phase-dependent dynamic responses of respiratory cycle motor activities following perturbation of the cycle in the cat, *J. Physiol. Lond.* 461, 321-333 (1993).
18. D.W. Richter, D. Ballantyne and J.E. Remmers, The differential organization of medullary post-inspiratory activities, *Pflügers Arch.* 410, 420-427 (1987).
19. M.F. Czyzyk-Krzeska and E.E. Lawson, Synaptic events in ventral respiratory neurones during apnoea induced by laryngeal nerve stimulation in neonatal pig, *J. Physiol. Lond.* 436, 131-147 (1991).
20. E.E. Lawson, D.W. Richter, M.F. Czyzyk-Krzeska, A. Bishoff and R.C. Rudesill, Respiratory neural activity during apnea and other breathing patterns induced by laryngeal stimulation, *J. Appl. Physiol.* 70, 2742-2749 (1991).
21. Y. Zheng, J.C. Barillot and A.L. Bianchi, Are the post-inspiratory neurons in the decerebrate rat cranial motoneurons or interneurons? *Brain Res.* 551, 256-266 (1991).
22. J.S. Jodowski and A. J. Berger, Influences from laryngeal afferents on expiratory bulbospinal neurons and motoneurons, *J. Appl. Physiol.* 64, 1337-1345 (1988).

RESPIRATORY CONTROL OF HYPOGLOSSAL MOTONEURONS

John H. Peever and James Duffin*

1. INTRODUCTION

Hypoglossal motoneurons innervate the tongue muscles,[1] which participate in a variety of motor tasks, including breathing.[2] Understanding the respiratory control of hypoglossal motoneurons is essential because of their critical involvement in maintaining airway patency during normal breathing, and their putative role in the pathogenesis of obstructive sleep apnea.[3, 4] Despite such importance in both normal and altered physiological states, virtually nothing is known about how respiratory rhythm is transmitted to them.

Neuroanatomical evidence demonstrates that the medulla, pons, and cortex contain neurons that project to hypoglossal motoneurons.[1, 5-10] In rats, hypoglossal motoneurons receive premotor projections predominantly from the lateral tegmental field.[1] In cats, it is suggested that inspiratory neurons within the lateral tegmental field and in close proximity to the ventral respiratory group (VRG) synaptically excite both phrenic and hypoglossal motoneurons.[11, 12] Based on these reports, we hypothesized that respiratory rhythm is transmitted to hypoglossal motoneurons from respiratory premotor neurons within the lateral tegmental field and VRG.

2. METHODS

The preparation used for these experiments was the vagotomized, paralyzed, ventilated, unanesthetized, decerebrate adult rat.[13] Intracellular and extracellular recordings of neuron activity in the medulla were made using microelectrodes. Hypoglossal motoneurons were identified by antidromic activation from the hypoglossal nerve. Interconnections between neurons and nerves were detected using cross-correlation of their action potential discharges. Features observed in cross-correlation and post-stimulus histograms were quantified using the *k*-ratio.[14] Peaks and troughs were

* Departments of Physiology and Anaesthesia, University of Toronto, Toronto, Ontario, M5S 1A8, Canada.

Frontiers in Modeling and Control of Breathing, edited by Poon and Kazemi.
Kluwer Academic/Plenum Publishers, New York, 2001.

tested for statistical significance at $P < 0.01$ or 0.001.[15] Explanations of these techniques have been detailed previously.[16]

3. OBERSVERATIONS AND CONCLUSIONS

3.1. Hypoglossal Motoneuron Membrane Potentials

Hypoglossal motoneurons (n = 28) antidromically activated from the ipsilateral hypoglossal nerve had resting membrane potentials that ranged from –80 to –20 mV. All showed membrane potential fluctuations with a respiratory pattern; all depolarised during inspiration and repolarized during expiration (figure 1, A). To determine whether hypoglossal motoneurons were actively inhibited, Cl^- iontophoresis was used to reverse hyperpolarization to depolarization. In 10 neurons, injection of Cl^- current had no effect on the membrane trajectories of hypoglossal motoneurons (figure 1, B). Based on this observation, we conclude that unlike phrenic motoneurons, hypoglossal motoneurons do not receive expiratory inhibition.

3.2. Inhibition of Hypoglossal Motoneurons

In rats, bulbospinal Bötzinger complex expiratory neurons actively inhibit phrenic motoneurons during expiration.[17] To further confirm our intracellular findings, we examined the hypothesis that hypoglossal and phrenic motoneurons receive expiratory inhibition from Bötzinger complex expiratory neurons. To do this, we computed post-stimulus histograms by stimulating (monopolar glass-coated tungsten microelectrodes) the axons of Bötzinger complex expiratory neurons at the border of the second and third cervical vertebrae in the spinal cord[17] while simultaneously recording the activities of hypoglossal and phrenic nerves. In 15 rats, post-stimulus histograms with hypoglossal nerves revealed no features, but showed troughs at short latencies with the phrenic nerve (figure 1, H1 and H2). These data further confirm our suggestion that unlike phrenic motoneurons, hypoglossal motoneurons are not postsynaptically inhibited during expiration.

3.3. Excitation of Hypoglossal Motoneurons

First, we examined the hypothesis that hypoglossal and phrenic motoneurons receive inspiratory drive from the same premotor neurons. In 23 rats, we computed cross-correlation histograms between the activities of ipsilateral phrenic and hypoglossal nerves, and found the cross-correlograms were featureless; no common activation was evident (data not shown). Then, we extracellularly recorded the activity of inspiratory phrenic premotor neurons in the VRG and computed cross-correlation histograms with the activities of phrenic and hypoglossal nerves. Eighteen ipsilateral and 8 contralateral inspiratory neurons were identified as premotor to phrenic motoneurons from cross-correlogram peaks (figure 1, G1 and G2). However, only 1 of the 26 displayed a cross-correlogram with a peak indicating an excitation of hypoglossal motoneurons. We conclude that very few (3.8%) inspiratory phrenic premotor neurons project to and excite hypoglossal motoneurons, and suggest that phrenic and hypoglossal motoneurons receive inspiratory drive from different premotor populations.

In 47 rats, we then tested the hypotheses that inspiratory drive is transmitted to hypoglossal motoneurons from premotor neurons in the lateral tegmental field, VRG and possibly by interneurons within the hypoglossal motor nucleus,[18, 19] by cross-correlating their activity with that of both the hypoglossal and phrenic nerves.

Of the 36 inspiratory neurons recorded within the hypoglossal motor nucleus, 9 (25%) were identified as hypoglossal motoneurons by antidromic activation of the hypoglossal nerve and from cross-correlogram features, narrow half-amplitude peaks at short latencies (figure 1, E). The cross-correlograms of those inspiratory neurons not antidromically activated (75%) featured either central peaks alone (n = 14; 39%) (figure 1, C), or central peaks followed by narrow troughs at short latencies (n = 13; 36%) (figure 1, D). We interpret these observations as evidence for interneurons within the hypoglossal motor nucleus.

Two hundred and thirty-eight inspiratory neurons were extracellularly recorded from within the lateral tegmental field and VRG. None of them were antidromically activated from either the ipsilateral hypoglossal or vagus nerves. However, 5 inspiratory neurons located within the lateral tegmental field were antidromically activated by a stimulating microelectrode placed within the ipsilateral hypoglossal motor nucleus. Although these

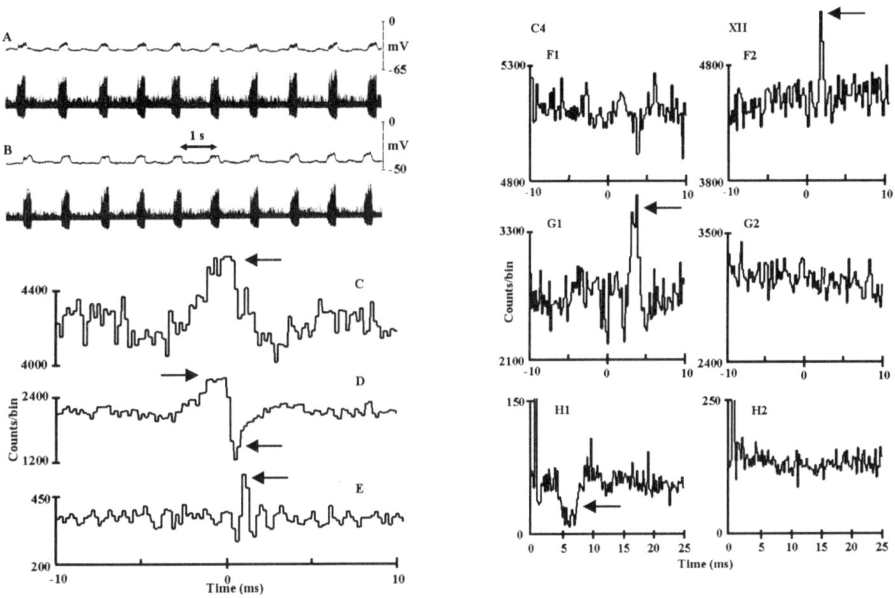

Figure 1. *Left side*: An intracellular recording from a hypoglossal motoneuron (XII mn) (upper trace), and the inspiratory activity of the XII nerve (lower trace) before (**A**) and after (**B**) 10 minutes of 10 μA Cl^- current injection into the XII mn. **C** and **D**, cross-correlograms computed between non-antidromically activated inspiratory XII neurons (interneurons) and the ipsilateral XII nerve, which display a broad central peak, and a broad central peak followed by a narrow trough at a short latency, respectively. **E**, a cross-correlogram computed between an antidromically activated XII mn and XII nerve; note the narrow peak at a short latency. *Right side*: Cross-correlograms computed between an inspiratory neuron in the lateral tegmental field and the phrenic (C4) (**F1**) and XII (**F2**) nerves. Cross-correlograms computed between an inspiratory VRG neuron and the C4 (**G1**) and XII (**G2**) nerves. Post-stimulus histograms for C4 (**H1**) and XII (**H2**) nerves; stimulus artefacts are visible at ~ 0.6 ms. All bin widths are 0.2 ms. Arrows point to either peaks or troughs.

neurons projected to the hypoglossal motor nucleus, they did not make functional connections with hypoglossal motoneurons because cross-correlation of their extracellular discharge with that of the ipsilateral hypoglossal nerve produced featureless cross-correlograms. Cross-correlation of the extracellular discharge of 19 (8 %, 15 ipsilateral, 4 contralateral) inspiratory neurons within the lateral tegmental field and hypoglossal nerve discharge produced cross-correlograms that displayed narrow peaks at short latencies, which are indicative of monosynaptic excitation of hypoglossal motoneurons (figure 1, F1 and F2). These cross-correlogram peaks were significantly wider and at longer latencies than those measured for inspiratory hypoglossal motoneurons (figure 1, E). None of the cross-correlograms computed between the discharge of these neurons and phrenic nerve activity displayed significant features (figure 1, F1). No cross-correlograms computed between the activity of neurons within the VRG and the discharge of the hypoglossal nerve showed features indicative of a monosynaptic excitation of hypoglossal motoneurons. We interpret these observations as evidence that hypoglossal premotor neurons are located diffusely throughout the lateral tegmental field.

4. DISCUSSION

The membrane potentials recorded from inspiratory hypoglossal motoneurons displayed respiratory modulation, depolarising during inspiration and rapidly repolarizing at the start of expiration. We could not detect any changes in the membrane potential trajectories of these motoneurons after Cl^- ions were injected into them, and concluded that hypoglossal motoneurons are postsynaptically excited during inspiration and are disfacilitated rather than postsynaptically inhibited during expiration. In cats, hypoglossal motoneurons exhibit similar membrane trajectories, which result from a combination of postsynaptic excitation and inhibition during inspiration, with little or no inhibition during expiration.[20, 21]

The lack of expiratory inhibition distinguishes hypoglossal motoneurons from phrenic motoneurons, which receive postsynaptic inhibition during expiration.[17] From this difference, we suggest that the principles underlying the organization of premotor pathways transmitting respiratory inputs to hypoglossal and phrenic respiratory motoneurons differ. Two lines of evidence support this contention. First, our observation that Bötzinger complex expiratory neurons, known to inhibit phrenic motoneurons[17, 22] do not inhibit hypoglossal motoneurons; and second, we found that inspiratory VRG phrenic premotor neurons do not project to and monosynaptically excite hypoglossal motoneurons. The latter observation contradicts Ono et al.[11] who report that in cats, hypoglossal and phrenic motoneurons receive inspiratory excitation from a common population of premotor neurons.

We found no evidence to support the hypothesis that premotor neurons within the VRG transmit respiratory drive to hypoglossal motoneurons. Rather, we found hypoglossal premotor neurons scattered bilaterally throughout the lateral tegmental field. This location is in agreement with electrophysiological studies in cats,[11] and neuroanatomical studies in rats,[1] which both report that hypoglossal premotor neurons are located dorsomedial to the nucleus ambiguus.

Cross-correlation of antidromically activated hypoglossal motoneurons and the hypoglossal nerve produced cross-correlograms that were markedly different from those produced by cross-correlation of non-antidromically activated hypoglossal neurons and the hypoglossal nerve. The former displayed very narrow peaks at short latencies confirming the presence of motoneuron action potentials in the hypoglossal nerve discharge. The cross-correlograms for non-activated neurons, which we contend are interneurons, did not display such peaks, but were characterised by broad central peaks. We interpreted this feature as indicating a common activation of the interneuron and hypoglossal motoneurons.

Our observations led to the following general conclusions about the source of inspiratory drive to hypoglossal motoneurons in rats. We suggest that there is a small, distinct population of inspiratory neurons scattered in the lateral tegmental field that are directly premotor to hypoglossal motoneurons. The source of inspiratory drive to hypoglossal motoneurons is separate from that for phrenic motoneurons, and the transmission of inspiratory drive may involve hypoglossal interneurons.

The latter conclusion provides an explanation reconciling the disparate observations of numerous anatomical studies that identified projections to the hypoglossal motor nucleus from specific locations,[1, 5-11] and our finding of few monosynaptic connections from scattered locations. The existence of hypoglossal interneurons is supported by our finding neurons located within the hypoglossal motor nucleus that could not be antidromically activated from the hypoglossal nerve, and whose cross-correlograms displayed features that clearly differentiate them from those of hypoglossal motoneurons.

While such observations are the first *functional* evidence for hypoglossal interneurons in rats, many investigators have speculated about their existence.[18, 19, 23-25] Neuroanatomically, hypoglossal interneurons have been distinguished from motoneurons in the cat, rat and monkey.[23-25] Electrophysiologically, interneurons have been described in intact cats based on their characteristic long latency synaptic potentials following hypoglossal nerve stimulation.[19] Although these previous studies identify putative interneurons, no specific evidence has been presented as to their function, but because the tongue participates in different motor acts, hypoglossal interneurons could serve a variety of functions. Our data demonstrate that one function of hypoglossal interneurons is to transmit respiratory drive to hypoglossal motoneurons.

In summary, we conclude that: (1) inspiratory hypoglossal motoneurons are not postsynaptically inhibited during expiration; (2) hypoglossal motoneurons are not monosynaptically excited by inspiratory phrenic premotor neurons; (3) respiratory rhythm is transmitted to hypoglossal motoneurons from a distinct population of premotor neurons scattered throughout the lateral tegmental field; (4) interneurons within the hypoglossal motor nucleus also participate in the transmission of respiratory rhythm; and (5) respiratory control of hypoglossal motoneurons is separate from that for phrenic motoneurons.

5. REFERENCES

1. E. G. Dobbins, and J. L. Feldman, Differential innervation of protruder and retractor muscles of the tongue in rat, *J. Comp. Neurol.* **357**, 376-394 (1995).
2. A. A. Lowe, The neural regulation of tongue movements, *Prog. Neurobiol.* **15**, 295-344 (1981).
3. J. E. Remmers, W. J. de Groot, E. K. Sauerland, and A. M. Anch, Pathogenesis of upper airway occlusion during sleep, *J. Appl. Physiol.* **44**, 931-938 (1978).
4. R. F. Fregosi, and D. D. Fuller, Respiratory-related control of extrinsic tongue muscle activity, *Respir. Physiol.* **110**, 295-306 (1997).

5. R. C. Borke, E. Martin, and R. L. Ringler Jr., Brain stem afferents of hypoglossal neurons in the rat, *Brain Res.* **269**, 47-55 (1983).
6. G. Ugolini, Specificity of rabies virus as a transneuronal tracer of motor networks: Transfer from hypoglossalmotoneurons to connected second-order and higher order central nervous system cells groups, *J. Comp. Neurol.* **356**, 457-480 (1995).
7. Y.-Q. Li, M. Takada, and N. Mizuno, Identification of premotor interneurons which project bilaterally to the trigeminal motor, facial or hypoglossal nuclei: A fluorescent retrograde double-labeling study in the rat, *Brain Res.* **611**, 160-164 (1993).
8. Y. Sahara, N. Hashimoto, and Y. Nakamura, Hypoglossal premotor neurons in the rostral medullary parvocellular reticular formation participate in cortically-induced rhythmical tongue movements, *Neurosci. Res.* **26**, 119-131 (1996).
9. G. Holstege, H. G. J. M. Kuypers, and J. J. Dekker, The organization of the bublar fibre connections to the hypoglossal motor nuclei, *Brain* **100**, 256-286 (1977).
10. M. Takada, K. Itoh, Y. Yasui, A. Mitani, S. Nomura, and N. Mizuno, Distribution of premotor neurons for the hypoglossal nucleus in the cat, *Neurosci. Lett.* **52**, 141-146 (1984).
11. T. Ono, Y. Ishiwata, N. Inaba, T. Kuroda, and Y. Nakamura, Hypoglossal premotor neurons with rhythmical inspiratory- related activity in the cat: Localization and projection to the phrenic nucleus, *Exp. Brain Res.* **98**, 1-12 (1994).
12. T. Ono, Y. Ishiwata, N. Inaba, T. Kuroda, and Y. Nakamura, Modulation of the inspiratory-related activity of hypoglossal premotor neurons during ingestion and rejection in the decerebrate cat, *J. Neurophysiol.* **80**, 48-58 (1998).
13. G. F. Tian, and J. Duffin, Connections from upper cervical inspiratory neurons to phrenic and intercostal motoneurons studied with cross-correlation in the decerebrate rat, *Exp. Brain Res.* **110**, 196-204 (1996).
14. T. A. Sears, and D. Stagg, Short-term synchronization of intercostal motoneurone activity, *J. Physiol.* **263**, 357-381 (1976).
15. K. Graham, and J. Duffin, Cross correlation of medullary expiratory neurons in the cat, *Exptl. Neurol.* **73**, 451-464 (1981).
16. M. A. Douse, J. Duffin, D. Brooks, and L. Fedorko, Role of upper cervical inspiratory neurons studied by cross- correlation in the cat, *Exp. Brain Res.* **90**, 153-162 (1992).
17. G. F. Tian, J. H. Peever, and J. Duffin, Bötzinger-complex expiratory neurons monosynaptically inhibit phrenic motoneurons in the decerebrate rat, *Exp. Brain Res.* **122**, 149-156 (1998).
18. J. D. Green, and K. Negishi, Membrane potentials in hypoglossal motoneurons, *J. Neurophysiol.* **26**, 835-856 (1963).
19. T. Sumi, Functional differentiation of hypoglossal neurons in cats, *Jpn. J. Physiol.* **19**, 55-67 (1969).
20. G. Woch, and L. Kubin, Non-reciprocal control of rhythmic activity in respiratory-modulated XII motoneurons, *NeuroReport* **6**, 2085-2088 (1995).
21. D. J. Withington-Wray, S. W. Mifflin, and K. M. Spyer, Intracellular analysis of respiratory-modulated hypoglossal motoneurons in the cat, *Neuroscience* **25**, 1041-1051 (1988).
22. G. F. Tian, J. H. Peever, and J. Duffin, Bötzinger-complex, bulbospinal expiratory neurons monosynaptically inhibit ventral-group respiratory neurons in the decerebrate rat, *Exp. Brain Res.* **124**, 173-180 (1999).
23. T. B. Boone, and L. D. Aldes, The ultrastructure of two distinct neuron populations in the hypoglossal nucleus of the rat, *Exp. Brain Res.* **54**, 321-326 (1984).
24. N. Takasu, and P. H. Hashimoto, Morphological identification of an interneuron in the hypoglossal nucleus of the rat: A combined golgi-electron microscopic study, *J. Comp. Neurol.* **271**, 461-471 (1988).
25. M. N. Cooper, The hypoglossal nucleus of the primate: A golgi study, *Neurosci. Lett.* **21**, 249-254 (1981).

PROJECTIONS FROM BRAINSTEM GABAERGIC NEURONS TO THE PHRENIC NUCLEUS

Gang Song and Mamoru Aoki*

1. INTRODUCTION

The phrenic motoneurons receive descending projections from premotor neurons in the brainstem.[4, 10] Projections from inspiratory neurons in the dorsal and ventral respiratory groups are excitatory and are responsible for the inspiratory activity of the diaphragm. On the other hand. projections from the nucleus raphe magnus (NRM), pontine Kölliker-Fuse nucleus (KF), and Bötzinger complex (Böt. C), are inhibitory and exert inhibitory effects on phrenic inspiratory activity.[1, 2, 7, 8, 12] Electrophysiological studies have suggested that GABA is the putative transmitter mediating inspiratory inhibition.[11] It has been reported by Aoki et al.[1, 2] that phrenic inspiratory inhibition evoked by electrical or chemical stimulation at the NRM could be blocked by microinjection into the phrenic nucleus of bicuculline (a specific $GABA_A$ receptor antagonist). Whether the GABAergic neurons (which have been reported in the NRM and Böt. C in cats[6, 9]) project to the phrenic nucleus remains unclear. In the present study, a combined method of retrograde WGA-HRP labeling and anti-GABA immunostaining was used to address this problem.

2. METHODS

Experiments were successfully performed on 8 adult cats anesthetized with pentobarbital (Nembutal, 30 mg/kg, i.p.) and weighing 2.0-2.4 kg. The C_5 spinal cord was exposed by a laminectomy and a tungsten microelectrode was inserted (at the C_5 level, 1.5-2 mm lateral to midline, and 2-4 mm from the dorsal surface) to identify the phrenic nucleus where dense inspiratory neuronal activity could be recorded. A glass micropipette (tip diameter 30 μm) filled with 5% WGA-HRP (and connected to a nanoliter pump for pressure injection) was then inserted into the phrenic nucleus.

* Gang Song, Mamoru Aoki, Department of Physiology, School of Medicine, Sapporo Medical University, S1, W17, Chuo-ku, Sapporo 060-8556, Japan

Injections of 15 or 30 nL were made at 3 sites, each separated by 2 mm, in one side within the C_{5-6} segments. The surgical incision was then sutured, and the animal allowed to recover. After a postinjection survival period of 48 hours, the animal was again anesthetized with Nembutal (50 mg/kg) and transcardially perfused with 500 mL of heparinized physiological saline and 800 mL of 1% paraformaldehyde-1.25% glutaraldehyde in 0.1M PBS (pH 7.4 and 4°C). After perfusion, the brainstem and spinal cord were removed and postfixed for 1-2 hours. The brainstem and spinal cord were then placed in 20% sucrose PBS and stored at 4°C. Serial frozen sections (40 μm thick) of the brainstem, referred to the atlas of Berman[3], were cut after 48-72 hr and stored in PBS at 4°C. The sections were first processed for a HRP reaction with the TMB method and stabilized by incubating in a PBS containing 0.05% DAB, 0.02% cobalt acetate, and 0.003% H_2O_2. Selected sections were further processed for anti-GABA immunostaining as follows. The sections were incubated in PBS-T (0.1 M PBS containing 0.3% Triton X-100) for 2-3 days at 4°C, and then incubated in rabbit polyclonal anti-GABA antibody solution (DiaSorin, 1:1000 dilution with PBS-T) for 48-72 hours at 20°C. The sections were rinsed in PBS-T (5 min x 4) and incubated in FITC (fluorescein isothiocyanate) conjugated goat anti-rabbit IgG antibody (15μg/mL) for 24 hours at 20°C, rinsed again, mounted on slides, and coverslipped. Sections were observed under a fluorescent microscope. For the identification of double-labeled neurons, the HRP labeled cells were first identified under light-field mode, and then observed under fluorescent mode for FITC double-labeling. Selected areas of the sections were photographed.

3. RESULTS

3.1. Retrogradely HRP Labeled Neurons

In TMB processed C_{5-6} sections, the HRP reaction products of 15 nL (n=4) injection sites were confined to the ipsilateral side of the C_{5-6} segments and completely covered the ventromedial part of the ventral horn (Rexed's lamina VIII-IX) where the phrenic motoneurons are concentrated. The reaction products also spread ventrally into the white matter and dorsally to lamina VI-VII. Injection of 30 nL (n=2) resulted in a larger spreading of HRP reaction products.

Injection of WGA-HRP into the phrenic nucleus retrogradely labeled neurons in several well defined respiratory nuclei in the brainstem. The structures included: the ventrolateral area of solitary nucleus; the ambiguus nucleus; medial and ventral areas of the retrofacial nucleus (Böt. C); the lateral paragigantocellular reticular nucleus; the raphe nuclei (NRM, nucleus raphe obscurus and nucleus raphe pallidus); medial and lateral parabrachial nuclei; and the KF nucleus. Labeled neurons were also observed in some non-respiratory related structures, such as the vestibular nucleus, locus ceruleus, and the medullary and pontine tegmental fields. With one exception, the labeling was bilateral with an ipsilateral dominance. The locus ceruleus was exclusively labeled ipsilaterally. Only a quantitative difference occurred in the distribution of HRP labeled neurons with smaller and larger injection volumes.

3.2. GABA Immunopositive Neurons in the Brainstem

The GABA immunopositive neurons (labeled by FITC) displayed green fluorescent light under a laser confocal microscope and were observed in structures including: the pontine KF nucleus and lateral parabrachial nucleus (15-20 cells/s); the trapezoid nucleus (8-10 cells/s); the raphe nuclei (NRM and nucleus raphe obscurus, 20-25 cells/s); the Böt. C (10-12 cells/s); the vestibular nucleus (16-22 cells/s); the lateral paragigantocellular nucleus (10-15 cells/s); the medullary lateral reticular nucleus (3-6 cells/s); and the spinal trigeminal nucleus (5-8 cells/s). Sparsely distributed positive neurons were also observed in the pontine and medullary tegmental fields (8-10 cells/s).

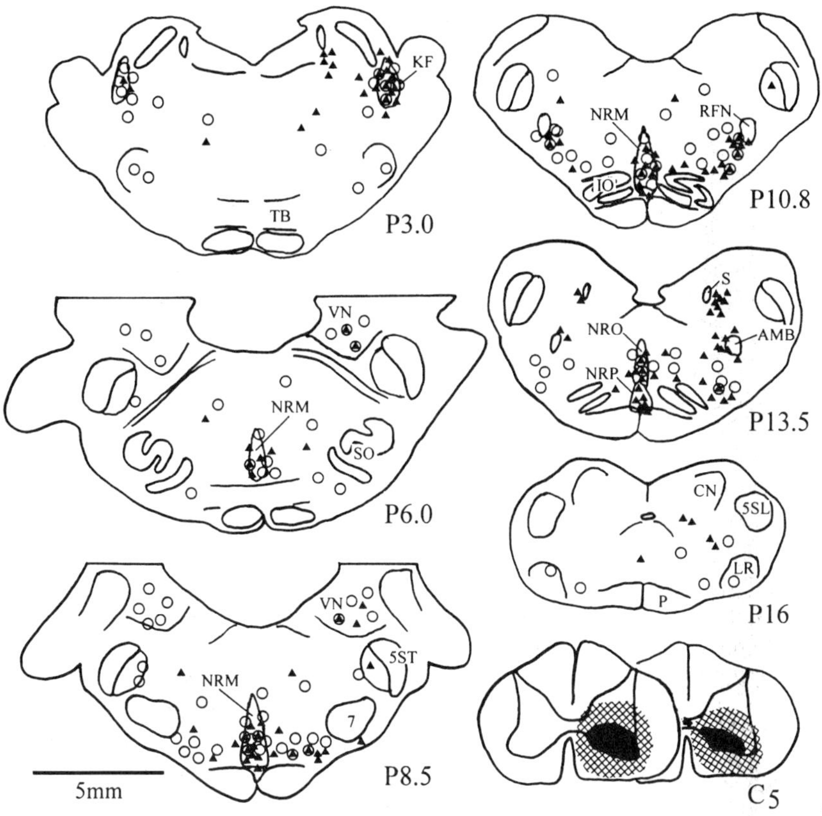

Figure 1. Schematic drawings of representative sections at six levels (P 3.0, 6.0, 8.5, 10.8, 13.5, and 16) through brainstem, showing the distribution patterns of HRP labeled neurons (triangles), GABA immunopositive neurons (open circles), and GABA-HRP double-labeled neurons (open circles with triangles inside). The sections through C_5 show the HRP injection sites in the phrenic nucleus. Left, 30 nL injection site; right, 15 nL injection site. 5SL, laminar spinal trigeminal nucleus; 5ST, spinal trigeminal tract; 7, facial nucleus; AMB, ambiguus nucleus; CN, cuneate nucleus; IO, inferior olive; KF, Kölliker-Fuse nucleus, LR, lateral reticular nucleus; NRM, nucleus raphe magnus; NRO, nucleus raphe obscurus; NRP, nucleus raphe pallidus; RFN, retrofacial nucleus; P, pyramidal tract; S, solitary tract; SO, superior olive; TB, trapezoid body; VN, vestibular nucleus.

3.3. GABA Immunopositive HRP Double-Labeled Neurons in the Brainstem

Under a laser confocal microscope neurons that contained HRP reaction products (small black granules in the cell plasma) and displayed green fluorescent light were identified as GABA-HRP double-labeled neurons. They were consistently observed (Fig. 1) in the following: NRM (4-6 cells/s) and nucleus raphe obscurus (1-3 cells/s); the KF nucleus (5-8 cells/s); the Böt.C (3-5 cells/s); the lateral reticular nucleus, specifically the ventral part (3-5 cells/s); and the vestibular nucleus (3-5 cells/s). In the NRM and nucleus raphe obscurus the distributions of GABA-HRP double-labeled neurons were roughly the same as HRP single-labeled neurons. The GABA-HRP double-labeled neurons in the NRM were triangular or multipolar neurons of medium size (25-35 μm soma diameter). Those in the nucleus raphe obscurus were small spindle-shaped neurons (15-25 μm). Although the HRP single-labeled neurons in the nucleus raphe pallidus were packed densely, only a few weakly stained GABA-HRP double-labeled neurons were observed in this nucleus. The GABA-HRP double-labeled neurons in the pontine KF nucleus were small, round, or ovoid, cells (15-20 μm). Together with HRP single-labeled neurons, they formed a densely packed, dorso-ventral cell band that extended to the ventral part of the lateral parabrachial nucleus. A few GABA-HRP double-labeled neurons were also observed in the medial parabrachial nucleus. In the medial and ventral area of retrofacial nucleus GABA-HRP double-labeled neurons, together with HRP single labeled neurons, formed a cell cluster that was identified as the Böt. C. They were deeply stained, triangular, or multipolar, neurons (30-35 μm). In the lateral reticular nucleus, GABA-HRP labeled neurons were multipolar cells of a medium to large size (30-40 μm). The vestibular nucleus contained large, round, or ovoid-shaped (40-50 μm), thickly stained, GABA-HRP double-labeled neurons. A few GABA-HRP double-labeled neurons were also distributed in the ventrolateral solitary nucleus (0-1 cells/s), the ambiguus nucleus (0-2 cells/s), and the medullary tegmental fields (1-2 cells/s).

4. DISCUSSION

In the present study we observed in the respiratory related structures a distribution pattern of HRP labeled neurons which was similar to that of previous reports[4,11]. This suggests that most of HRP labeled neurons in our study consisted of phrenic premotor neurons. In addition, the distribution pattern of GABA immunopositive neurons in the brainstem was similar to that reported by Holmes et al.[6] The distribution of HRP labeled neurons and GABA immunopositive neurons in the NRM and nucleus raphe obscurus, the KF nucleus, and the Böt. C was consistent in all of the sections which were examined. Moreover, many neurons were GABA-HRP double-labeled, suggesting that they were GABAergic neurons with axons projecting to the phrenic nucleus. The GABAergic inhibitory synaptic connections on phrenic motoneurons could be involved in the formation of the ramp-like inspiratory discharge pattern, and the maintenance of the expiratory phase. The phrenic nucleus might also receive GABAergic or other non-respiratory projections from the vestibular nucleus, the lateral reticular nucleus, and the locus ceruleus. These pathways might be related to non-respiratory functions of the diaphragm, such as the maintenance of posture and speech.

5. ACKNOWLEDGMENTS

The authors wish to thank Dr. Hajime Satomi, Department of Anatomy, who helped in processing HRP reactions; and Seiji Ohtani, Laboratory of Cell and Tissue Structure, for assistance in immunostaining. Thanks also to Drs. Kiyoji Matsuyama and Yutaka Fujito, Department of Physiology, for helpful discussion.

6. REFERENCES

1. M. Aoki, and Y. Nakazono, Raphe magnus-induced inhibition of medullary and spinal respiratory activities in the cat, in: *Control of Breathing and Its Modeling Perspective,* edited by Y. Honda, Y. Miyamoto, K. Konno and J.G. Widdicombe (Plenum Press, New York, 1992), pp. 15-23.
2. M. Aoki, Y. Sato, Y. Nakazono, and I. Kohama, Effects of GABA receptor antagonists on the raphe magnus-induced inhibition of bulbar and spinal respiratory neural activities in the cat, in: *Modeling and Control of Ventilation,* edited by S.J.G. Semple, L. Adams and B.J. Whipp (Plenum Press, New.York, 1995), pp. 33-38.
3. A.L. Berman, *The Brain Stem of the Cat, A Cytoarchitectonic Atlas with Stereotaxic Coordinates,* (Plenum Press, University of Wisconsin, Madison, 1968).
4. E.G. Dobbins, and J.L. Feldman, Brainstem networks relaying descending drive to phrenic motoneurons in rat, *J. Comp. Neurol.* **347,** 64-86 (1994).
5. L. Fedorko, C.A. Connelly, and J.E. Remmers, Neurotransmitters mediating synaptic inhibition of phrenic motoneurons, in: *Respiratory Muscles and Their Neuromotor Control,* edited by G.C. Sieck, S.C. Gandevia and W.E. Cameron (Alan R. Liss, New York, 1987), pp. 167-173.
6. C.J. Holmes, L.S. Mainville, and B.W. Jones, Distribution of cholinergic, GABAergic and serotonergic neurons in the medial medullary reticular formation and their projections studied by cytotoxic lesions in the cat, *Neurosci.* **62**, 1155-1178 (1994).
7. M. Hosogai, S. Matsuo, T. Sibahara, and Y. Kawai, Projection of respiratory neurons in rat medullary raphe nuclei to the phrenic nucleus, *Respir. Physiol.* **112**, 37-50 (1998).
8. P.M. Lalley, Responses of phrenic motoneurones of the cat to stimulation of medullary raphe nuclei, *J. Physiol.* **380**, 349-371 (1986).
9. C.A. Livingston, and A.J. Berger, Immunocytochemical localization of GABA in neurons projecting to the ventrolateral nucleus of the solitary tract, *Brain Res.* **494**, 143-150 (1989).
10. G.S. Rikard-Bell, E.K. Bystrzycka, and B.S. Nail, Brainstem projections to the phrenic nucleus: a HRP study in the cat, *Brain Res. Bull.* **12**, 469-477 (1984).
11. K. Schmid, A. Foutz, and M. Denavit-Saubie, Inhibitions mediated by glycine and $GABA_A$ receptors shape the discharge pattern of bulbar respiratory neurons, *Brain Res.* **710**, 150-160 (1996).
12. G. Song, and L. Liu, Reappraisal of the inspiratory effect of Bötzinger complex on phrenic nerve discharge, *Respir. Physiol.* **105**, 17-21 (1996).

OPTICAL RECORDING OF THE NEURONAL ACTIVITY IN THE BRAINSTEM-SPINAL CORD

Application of a voltage-sensitive dye

Yasumasa Okada, Zibin Chen, Hideaki Yoshida, Shun-ichi Kuwana, Wuhan Jiang, and Hirofumi Maruiwa*

1. INTRODUCTION

Recording of dynamic neuronal activities in the brainstem and spinal cord in the study of central respiratory control have been so far conducted mostly with conventional microelectrode techniques such as the whole-cell recording method.[1,2] However, activities of only a limited number of points in the brainstem and spinal cord can be simultaneously analyzed with conventional microelectrode techniques. In order to overcome this limitation and to analyze multiple neuronal activities simultaneously, optical recording techniques have recently been applied in the neuronal analyses of the brainstem and spinal cord. In this article we review the recent progress of this field. Although we focus on the method using voltage-sensitive dyes, several other optical methods have been also applied in the brainstem and spinal cord.

1.1. *In-Vivo* Detection of Light Reflectance Change

Changes of regional neuronal activities in response to hypercapnic and hypoxic exposure as well as to peripheral chemoreceptor stimulation with cyanide were analyzed in the exposed ventral medullary surface of *in vivo* animals.[3-5] The ventral medullary surface was illuminated, and the reflected light was collected and detected using a CCD camera. The signal with stimulation was subtracted from baseline, and neuronal response to stimulation was imaged. To avoid moving artifact in these *in vivo* experiments,

* Yasumasa Okada, Department of Medicine, Keio University Tsukigase Rehabilitation Center, Amagiyugashima-cho, Shizuoka, Japan 410-3293. Zibin Chen, Division of Biology, GlaxoSmithKline Inc., Research Triangle Park, North Carolina, 27709. Hideaki Yoshida, Department of Orthopedics, Keio University Tsukigase Rehabilitation Center, Shizuoka, Japan 410-3293. Shun-ichi Kuwana, Department of Physiology, Teikyo University, Itabashi-ku, Tokyo, Japan 173-8605. Wuhan Jiang, Lineberger Cancer Center, University of North Carolina, Chapel Hill, North Carolina, 27599. Hirofumi Maruiwa, Department of Orthopedics, Keio University School of Medicine, Tokyo, Japan 160-8582.

Frontiers in Modeling and Control of Breathing, edited by Poon and Kazemi.
Kluwer Academic/Plenum Publishers, New York, 2001.

respiratory and cardiac cycle triggering is usually necessary (see Section 4.2. Disadvantage). The neural responses detected by this method are largely influenced by changes of the regional blood flow.

1.2. *In-Vitro* Intrinsic Optical Analyses

Optical analysis of intrinsic signals does not require the use of dyes. However, only slow neuronal responses can be analyzed with this technique. Intrinsic light-scattering changes were analyzed in the embryonic chick brainstem slices, and glutamate-induced slow neuronal excitation was detected which is considered to be induced by neuronal swelling.[6] GABA-induced neuronal excitation accompanied by cell shrinkage was also detected by intrinsic light-scattering changes.[7] Each of these two experiments was conducted simultaneously with membrane potential analysis using an absorption voltage-sensitive dye. Torres *et al.* applied intrinsic optical signal imaging in the rat medullary slice preparation, and studied the functional topography of nitric oxide.[8] They reported that nitric oxide modulation of neuronal activity was primarily circumscribed to the nucleus tractus solitarius. Intrinsic optical analysis was applied also in the spinal cord slice preparation of the juvenile rat.[9] Tetanic stimulation of high-threshold primary afferent fibers in the dorsal root elicited intrinsic optical signals most prominently in the lamina II of the dorsal horn, which lasted for 30 s or longer. The origin of this intrinsic signals is considered to be the cellular swelling due to an activity-dependent rise in extracellular K^+.

1.3. Calcium Imaging

Because most neurons contain voltage-sensitive Ca^{2+} channels, and neuronal activities are accompanied by increased intracellular Ca^{2+} concentration, calcium-sensitive dyes have been used in the analyses of the brainstem and spinal cord neuronal activities. Exhibition of large changes in fluorescence is an advantage of calcium-sensitive dyes over other dyes. A novel method to use retrograde transport of the dextran conjugate of a calcium-sensitive dye (Calcium Green Dextran) to selectively label populations of the lamprey brain and spinal interneurons was reported.[10] Using this method a neuronal activity of a selected population can be analyzed as video-rate imaging. This method was applied in the en bloc spinal cord preparation of the neonatal rat by pressure-ejecting Calcium Green Dextran into ventral roots, and imaging analysis of motoneuron activity was conducted.[11] Koshiya and Smith succeeded to visualize spontaneous respiratory neuronal activities in the rat medullary slice.[12] They microinjected a calcium-sensitive dye (Calcium Green-1 AM) into the center of the slices, and inspiratory neuronal population was retrogradely labeled. Increased fluorescence intensity during the inspiratory phase of network activity was observed simultaneously with the hypoglossal nerve burst discharge. These inspiratory neurons in the pre-Botzinger complex region were considered as respiratory pacemaker cells.

2. OPTICAL RECORDING METHOD WITH VOLTAGE-SENSITIVE DYES

2.1. Brainstem and Spinal Cord Preparation

An *in vitro* optical recording technique can be applied in the en bloc or slice preparation of the brainstem and spinal cord. Preparations are made similarly as conventional electrophysiological experiments.[1, 2, 13] Briefly, animals (e.g., neonatal rats) are deeply anesthetized with diethyl ether, and the brainstem and spinal cord are isolated in oxygenated mock cerebrospinal fluid (mock CSF). For slice preparation the brainstem or spinal cord is transferred to a slicer, and medullary or spinal cord slice is prepared. The preparation is stained with a fluorescent voltage-sensitive dye (e.g., RH795 Molecular Probe, USA; concentration 0.1 mg/ml) for 15 min. After the preparation is rinsed with an oxygenated mock CSF, it is fixed in a recording chamber that is placed under an epifluorescent microscope (e.g., Optiphoto 2UD Nikon, Japan). The preparation is positioned with its interesting side up, and is continuously superfused with control mock CSF that is equilibrated with O_2-enriched gas mixture (2-5% CO_2 in O_2).

Similarly as the selective retrograde staining method of neuronal subpopulation with calcium-indicator dyes,[10] it is also possible to apply a voltage-sensitive dye to the ventral roots of the isolated embryonic chick spinal cord and to optically analyze the function of specific motoneuron subpopulation.[14]

2.2. Recording Technique

For review of a general principle of the optical recording, see the references.[15, 16] Two types of optical signals are principally used in the recording of neuronal activities; light absorption and fluorescence. Our optical recording method using voltage-sensitive dyes is described below. We analyze the change of fluorescence intensity; we illuminate the dye-stained preparation through a green filter and a shutter using a high power tungsten halogen lamp, which is driven by a stable DC current supplier. Epifluorescence through a red filter is detected using a high-sensitivity optical recording system, HR Deltaron 1700 (Fuji Film, Japan) that provides high spatial resolution (126 x 126 pixels) and high speed (maximum speed 0.6 ms/frame) incorporating a MOS-FET camera. Also, CCD cameras and photodiode arrays can be utilized as detectors in optical experiments. Neuronal activities in a recorded area are displayed as pseudocolor images. Spontaneous neuronal activity or neuronal response to electrical, pharmacological or chemical (e.g., hypercapnic) stimulation can be visualized.

3. ANALYSES OF NEURONAL ACTIVITIES

3.1. Brainstem

Optical recording technique with an absorption voltage-sensitive dye was applied in the *in vitro* embryonic chick brainstem preparations; voltage-related optical signals evoked by vagal nerve stimulation were recorded in the en bloc and slice preparations, and neuronal development and functional distribution were thoroughly investigated.[17, 18]

A technique with absorption voltage-sensitive dyes was applied also in the embryonic rat medullary slice preparation, and ionic components of action potentials in the dorsal motor nucleus were identified.[19]

Onimaru *et al.* applied an optical recording technique using a fluorescent voltage-sensitive dye (RH795) in the *in vitro* brainstem block preparation, and succeeded to visualize the excitation propagation process induced by electrical stimulation in respiration related regions.[20] We also applied an optical recording technique using RH795 in the mapping of CO_2-depolarized regions in the en bloc and slice preparations of the neonatal rat medulla.[21] By using various synaptic blockers, primary CO_2-sensitive regions were identified in the surface layer of the midline (superficial part of raphe pallidus), rostral parapyramidal, and rostral ventrolateral regions. The rostral ventrolateral chemosensitive region is more laterally extended than previously described. The distribution of these CO_2-depolarized regions corresponds well with that of surface vessels.

3.2. Spinal Cord

Recently, optical recording techniques using absorption voltage-sensitive dyes have been applied in the spinal cord. Neuronal excitation in the dorsal horn evoked by dorsal-root stimulation was visualized in the rat spinal cord slice. Stimulation to the primary-afferent A fibers in the dorsal root evoked a weak brief excitatory optical response in the laminae I and III-V. Stimulation of both A and C fibers elicited greater and longer-lasting excitatory response in the laminae I-III. These responses were contributed largely by excitatory amino acid receptors.[22] Recently, neural plasticity involved in nociception has been also investigated in the rat dorsal horn.[23]

Arai *et al.* analyzed membrane potential changes evoked by dorsal and ventral roots stimulation in the embryonic chick spinal cord slice preparations.[24] They identified a fast spike-like signal as the presynaptic action potential and a slow signal as the postsynaptic potential in the dorsal region. In the ventral region, the fast spike-like signal was identified as the antidromic action potential in motoneurons.

We have applied an optical recording technique with a fluorescent voltage-sensitive dye in the juvenile rat spinal cord, and visualized the excitation propagation processes in the transverse plane of the slice as well as on the longitudinal tract of the en bloc preparation.[25] We have been also analyzing the pathophysiological processes of excitation propagation in *in vitro* spinal cord models of acute compression and hypoxic injury.

4. CRITIQUE OF METHODS

4.1. Advantages

Optical recording methods using voltage-sensitive dyes make it possible to analyze multi-point neuronal activities without inserting multiple microelectrodes into the brain tissue. Another advantage is that the response time of fast-responding dyes (e.g., merocyanine and styryl dyes) to transmembrane changes is less than 5 μs, and this technique allows the analysis of rapid changes of transmembrane voltage. Indeed, optical signals faithfully reflect the average action potential shapes, and it is possible to visualize fast dynamic processes of excitation propagation that is difficult to analyze by calcium imaging.

4.2. Disadvantages

The following limitations must be noted in optical recording experiments. First, many of voltage-sensitive dyes bleach readily, which makes it difficult to continue the recording long time. Bleaching of dyes can be largely prevented by minimizing the light exposure using a shutter and by reducing the intensity of excitation light although the reduction of illumination intensity reduces the signal size. Second, voltage-sensitive dyes may cause phototoxic damage in the neuronal tissue. Phototoxicity is enhanced by stronger illumination, and is minimized by using a shutter. Third, generally the signal in this method is small, and the increase of signal-noise ratio is the most important issue in the actual experiments. Averaging of signals is sometimes required. Forth, although the signal in the optical recording reflects semi-quantitative neuronal membrane potential changes, it must be noted that the membrane potential change in a single pixel does not correspond to that of a single cell but an average membrane potential change of cells in a region of a recorded pixel. Fifth, optical recording techniques using voltage-sensitive dyes as well as intrinsic signals require subtraction of background signal. In *in vivo* experiments ventilatory movement and cardiac pulsation can cause artifact while subtraction processing, and respiratory and cardiac cycle triggering is usually necessary.

5. SUMMARY

Although there are several limitations, optical recording techniques are superior to multi-electrode mapping methods, as it is possible to record at large number of points in a small area without destroying the tissue and possible to know relative changes of membrane potentials. Optical recording techniques using voltage-sensitive dyes will be more importantly applied in the study of central respiratory control (e.g., mechanisms of respiratory rhythm generation) in the near future.

6. ACKNOWLEDGMENTS

The authors thank Fuji Photo Film Co. Ltd. (Tokyo) for giving an opportunity to use HR Deltaron 1700. This study was supported by the Research Grant for the Specific Diseases of the Japanese Ministry of Health and Welfare (YO) and by the Research Grant of the Japan Society for Promotion of Science (YO, HY, and HM).

7. REFERENCES

1. Y. Okada, K. Mückenhoff, and P. Scheid, Hypercania and medullary neurons in the isolated brain stem-spinal cord of the rat, *Respir. Physiol.* **93**(3), 327-336 (1993).
2. S. Kuwana, Y. Okada, and T. Natsui, Effects of extracellular calcium and magnesium on central respiratory control in the brainstem-spinal cord of neonatal rat, *Brain Res.* **786**(1-2), 194-204 (1998).
3. X. W. Dong, D. Gozal, D. M. Rector, and R. M. Harper, Ventilatory CO_2-induced optical activity changes of cat ventral medullary surface, *Am. J. Physiol.* **265**(3), R494-R503 (1993).
4. D. Gozal, X. W. Dong, D. M. Rector, and R. M. Harper, Optical imaging of the ventral medullary surface of cats: hypoxia-induced differences in neural activation, *J. Appl. Physiol.* **74**(4), 1658-1665 (1993).
5. J. L. Carroll, D. Gozal, D. M. Rector, G. Aljadeff, and R. M. Harper, Ventral medullary neuronal responses to peripheral chemoreceptor stimulation, *Neuroscience* **73**(4), 989-998 (1996).
6. K. Sato, Y. Momose-Sato, Y. Arai, A. Hirota, and K. Kamino, Optical illustration of glutamate-induced cell swelling coupled with membrane depolarization in embryonic brain stem slices, *Neuroreport* **8**(16), 3559-63 (1997).
7. Y. Momose-Sato, K. Sato, A. Hirota, and K. Kamino, GABA-induced intrinsic light-scattering changes associated with voltage-sensitive dye signals in embryonic brain stem slices: coupling of depolarization and cell shrinkage, *J. Neurophysiol.* **79**(4), 2208-2217 (1998).
8. J. E. Torres, N. R. Kreisman, and D. Gozal, Nitric oxide modulates *in vitro* intrinsic optical signal and neural activity in the nucleus tractus solitarius of the rat, *Neurosci. Lett.* **232**(3), 175-178 (1997).
9. K. Murase, T. Saka, S. Terao, H. Ikeda, and T. Asai, Slow intrinsic optical signals in the rat spinal dorsal horn in slice, *Neuroreport* **9**(16), 3663-3667 (1998).
10. D. R. McPherson, A. D. McClellan, and M. J. O'Donovan, Optical imaging of neuronal activity in tissue labeled by retrograde transport of Calcium Green Dextran, *Brain Res. Brain Res. Protocols,* **1**(2), 157-164 (1997).
11. A. Lev-Tov, and M. J. O'Donovan, Calcium imaging of motoneuron activity in the en-bloc spinal cord preparation of the neonatal rat, *J. Neurophysiol.* **74**(3), 1324-1334 (1995).
12. N. Koshiya, and J. C. Smith, Neuronal pacemaker for breathing visualized *in vitro*, *Nature* **400**(6742), 360-363 (1999).
13. Y. Okada, A. Kawai, K. Mückenhoff, and P. Scheid, Role of the pons in hypoxic respiratory depression in the neonatal rat, *Respir. Physiol.* **111**(1), 55-63 (1998).
14. P. Wenner, Y. Tsau, L. B. Cohen, M. J. O'Donovan, and Y. Dan, Voltage-sensitive dye recording using retrogradely transported dye in the chicken spinal cord: staining and signal characteristics, *J. Neurosci. Methods* **70**(2), 111-120 (1996).
15. L. B. Cohen, and B. M. Salzberg, and A. Grinvald, Optical methods for monitoring neuron activity, *Ann. Rev. Neurosci.* **1**, 171-82 (1978).
16. T. J. Ebner, and G. Chen, Use of voltage-sensitive dyes and optical recordings in the central nervous system, *Prog. Neurobiol.* **46**, 463-506 (1995).
17. K, Kamino, Y. Katoh, H. Komuro, and K. Sato, Multiple site optical monitoring of neural activity evoked by vagus nerve stimulation in the embryonic chick brain stem, *J. Physiol.* **409**, 263-283 (1989).
18. H. Komuro, T. Sakai, Y. Momose-Sato, A. Hirota, and K. Kamino, Optical detection of postsynaptic potentials evoked by vagal stimulation in the early embryonic chick brain stem slice, *J. Physiol.* **442**, 631-648 (1991).
19. Y. Momose-Sato, K. Sato, and K. Kamino, Optical identification of calcium-dependent action potentials transiently expressed in the embryonic rat brainstem, *Neuroscience* **90**(4), 1293-1310 (1999).
20. H. Onimaru H, A. Kanamaru, and I. Homma, Optical imaging of respiratory burst activity in newborn rat medullary block preparations, *Neurosci. Res.* **25**(2), 183-190 (1996).
21. Y. Okada, Z. Chen, and F. L. Eldridge, Optical recording of CO_2-sensitive cell activity with voltage sensitive dye in the neonatal rat brainstem, *FASEB J.* **11**(3), A350 (1997).
22. H. Ikeda, R. D. Ryu, J. B. Park, M. Tanifuji, T. Asai, and K. Murase, Optical responses evoked by single-pulse stimulation to the dorsal root in the rat spinal dorsal horn in slice, *Brain Res.* **812**(1-2), 81-90 (1998).
23. H. Ikeda, T. Asai, and K. Murase, Robust changes of afferent-induced excitation in the rat spinal dorsal horn after conditioning high-frequency stimulation, J. Neurophysiology, 83(4), 2412-2420 (2000).
24. Y. Arai, Y. Momose-Sato, K. Sato, and K. Kamino, Optical mapping of neural network activity in chick spinal cord at an intermediate stage of embryonic development, *J. Neurophysiol.* **81**(4), 1889-1902 (1999).
25. H. Yoshida, Y. Okada, H. Maruiwa, K. Chiba, and Y. Toyama, Excitation propagation in the rat spinal cord: optical recording using a voltage-sensitive dye, *Neurosci. Res.* (in press).

INTERFACING COMPUTER MODELS WITH REAL NEURONS: RESPIRATORY "CYBERNEURONS" CREATED WITH THE DYNAMIC CLAMP

Christopher G. Wilson, Robert J. Butera, Jr., Christopher A. Del Negro, John Rinzel, and Jeffrey C. Smith*

1. INTRODUCTION

The mechanism of respiratory rhythm generation remains an important unsolved problem. While experimental evidence has been collected about the location and distribution of respiratory neurons within the mammalian brainstem, only recently has the locus of rhythm generating neurons been discovered,[1] a functionally identified region of the brainstem known as the pre-Bötzinger complex (pre-BötC). We have just begun to investigate the membrane and network properties of pre-BötC neurons underlying the generation and transmission of inspiratory rhythm.[2,3,4,5]

We have proposed a pacemaker network model in which intrinsic membrane and synaptic conductances are responsible for inspiratory burst generation and synchronization of pacemaker neurons in the pre-BötC.[3,4] We have been exploring these mechanisms through two approaches. Our first approach relies on experimental evaluation of intrinsic and synaptic currents in pre-BötC neurons present in our in-vitro rhythmic slices.[1,2,5] Our second approach has been the use of mathematical models to evaluate hypotheses regarding the nature of the burst generating currents and synaptic coupling between neurons of the pre-BötC.[3,4] We present our recently developed approach that combines experimental and computational methods to allow direct interfacing of computer models with real neurons to explore cellular and network properties of the pre-BötC.

Within the last decade, a few investigators have made efforts to combine electrophysiology techniques with computational modeling, to develop a method of dynamically adding currents representing artificial chemical and electrical synapses into single neurons or small neural networks.[6,7] Typically, the added current is calculated

* Christopher G. Wilson, Christopher A. Del Negro, and Jeffrey C. Smith, Cellular and Systems Neurobiology Section, NINDS, NIH, Bethesda, Maryland, 20892. Robert J. Butera, Institute for Bioengineering and Biosciences, Georgia Institute of Technology, Atlanta, Georgia, 30332. John Rinzel, Center for Neural Science and Courant Institute of Mathematical Science, New York University, New York, 10013.

Frontiers in Modeling and Control of Breathing, edited by Poon and Kazemi.
Kluwer Academic/Plenum Publishers, New York, 2001.

based on a mathematical model of a synapse or ionic conductance. These "dynamic clamp" methods have relied on proprietary software and operating systems that lack the speed and flexibility necessary to investigate a wide variety of conductance mechanisms. Our laboratory has developed a very high performance, extensible dynamic clamp system that we have dubbed Model Reference Current Injection (or MRCI, pronounced mercí). We based MRCI on a free operating system (Linux) and open source tools with the intention of making it inexpensive, very extensible, and freely available.

2. BIOLOGICAL PREPARATION

Our experimental preparation consists of thin (~300 μm) slices of neonatal rat (P0-P4) brainstem containing the pre-BötC and sufficient neural circuitry to generate inspiratory network activity, recorded from the hypoglossal (XII) nerve rootlets contained in the slice.[1,5] Using infrared differential interference contrast microscopy (IR-DIC) and whole-cell patch-clamp methods, we record from inspiratory pacemaker neurons, interneurons projecting to XII motoneurons, and the XII motoneurons .

In the pilot studies presented here, we typically targeted XII motoneurons to evaluate MRCI performance. Figure 1 diagrams the experimental setup, showing the relationship between the biological elements (whole-cell patch-clamped neurons), the patch-clamp amplifier, the acquisition system, and the computer used to calculate the artificial conductance. The dynamic clamp is very straightforward: membrane voltage is read and scaled appropriately, the value of the given modeled current is calculated based on the membrane voltage and instantaneously fed back out to the neuron. We have used both Axon Instruments (Axoclamp 1/Axopatch 1D) and HEKA amplifiers (EPC-9) in our experiments; MRCI incorporates adjustable scaling to accommodate different amplifiers.

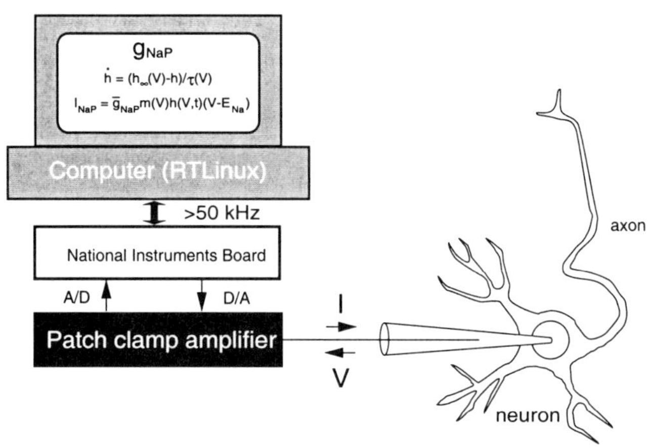

Figure 1. Simplified overview of MRCI. The real neuron, on the right, is held under whole-cell recording conditions by the patch-clamp amplifier (HEKA EPC-9 or Axon Axopatch 1D) and the membrane voltage is sampled into the computer running MRCI through a National Instruments data acquisition board (both A/D and D/A capable). The voltage is scaled appropriately and the model current calculated and injected back into the cell. Computational throughput rates of >50 KHz are possible with MRCI.

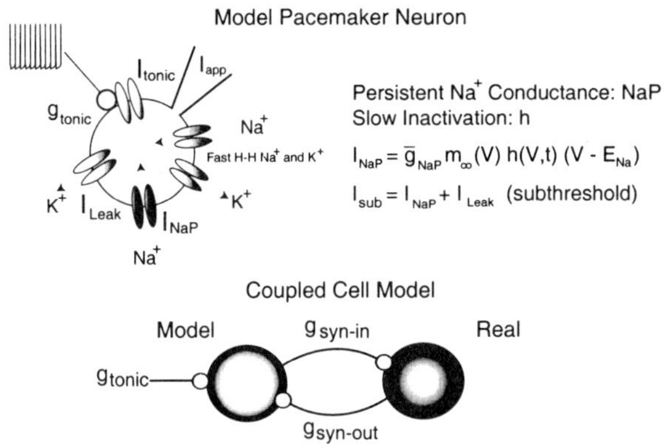

Figure 2. The model upon which MRCI is based. Our model includes a single compartment and various simulated currents/conductances including: Hodgkin-Huxley type fast Na^+ and K^+ currents, a K^+ dominated leak current (I_{Leak}), tonic inputs (I_{tonic}), an applied current as through a patch-pipette (I_{app}), and our putative burst-generating current, I_{NaP}. In addition, a simulated cell within the MRCI computer can be connected to a real neuron with artificial excitatory synapses either unidirectionally or reciprocally through synaptic conductances g_{syn} fed into the real cell ($g_{syn\text{-}in}$) and connecting the cell to the model ($g_{syn\text{-}out}$).

3. MODEL

Details of the pacemaker neuron and network model upon which MRCI is based can be found in our earlier papers.[3,4] Presented below are the relevant conductances for the single cell model and the model we use for synaptic coupling experiments (see Fig. 2). The single cell model is based on a bursting pacemaker model incorporating a slowly inactivating I_{NaP} (persistent sodium current) acting as a burst generating and burst terminating conductance. In addition to I_{NaP}, the model includes fast Hodgkin-Huxley-like action potential generating sodium (I_{Na}) and potassium (I_K) currents, a linear leak current (I_{Leak}), tonic excitatory inputs ($I_{tonic\text{-}e}$), and the ability to apply a holding or pulse current (I_{app}). MRCI allows us to "endow" an in vitro neuron with any of these intrinsic or synaptic currents, whether or not the neuron innately possesses these properties. In an alternative version, a real-time model of a pacemaker neuron is coupled to an in vitro neuron via modeled reciprocal excitatory synapses (fast glutamatergic).[4]

4. SOFTWARE

MRCI depends on real-time, meaning that the system acquires membrane voltage and injects calculated current at highly reliable timing intervals. A real-time operating system guarantees that the timing requirements of the application will be met at high rates of acquisition. Most operating systems have a variable latency (up to hundreds of milliseconds) before a task is scheduled for handling by the hardware. This varies greatly depending on the priority given to the application by the operating system and makes non-real-time operating systems inappropriate for the high speed sampling that dynamic clamping requires. To solve this problem and to allow for accurate sampling and control of the amplifier, we developed MRCI on Real-Time Linux (RTLinux, http://www.rtlinux.org), on a Pentium II based computer (266 MHz). RTLinux is a modified version of the Linux operating system that allows the MRCI application to run in real-time while the rest of the operating system runs when MRCI is idle. This design allows us to develop mission-critical real-time applications with the convenience of a free, robust development environment[8] and allows real-time adjustment of model parameters and data logging. MRCI can acquire data and inject current at up to 50 KHz with high reliability, speeds sufficient to simulate the full range of conductances in our model, including fast Hodgkin-Huxley ionic currents.

5. MRCI EXPERIMENTS

Our initial experiments included protocols in which we injected I_{NaP} into XII motoneurons that do not spike during inspiratory network activity, to show that I_{NaP} is sufficient to elicit pacemaker-like bursting without the addition of more complex currents other than those present in the cell. Figure 3 shows a neuron that did not burst until I_{NaP} was added, and then burst like a pre-BötC pacemaker. A non-pacemaker neuron will tonically spike when depolarized above action potential threshold. Conversely, pacemaker neurons generate bursts of increasingly shorter duration and higher frequency when depolarized until, at extreme depolarization, they tonically spike. We have also added I_{NaP} to cells synaptically driven by the respiratory network. Figure 4 is an example of a synaptically driven motorneuron that (top panel) tonically spikes when depolarized, only bursting when driven by the network. The lower panel, shows the same neuron after after MRCI added I_{NaP} At the same membrane potential, synaptic drive now induces a burst. This synaptic enhancement phenomenon has been previously described[9,10] and suggests that even sparsely distributed I_{NaP} can amplify synaptic input and alter the behavior of the neuron.

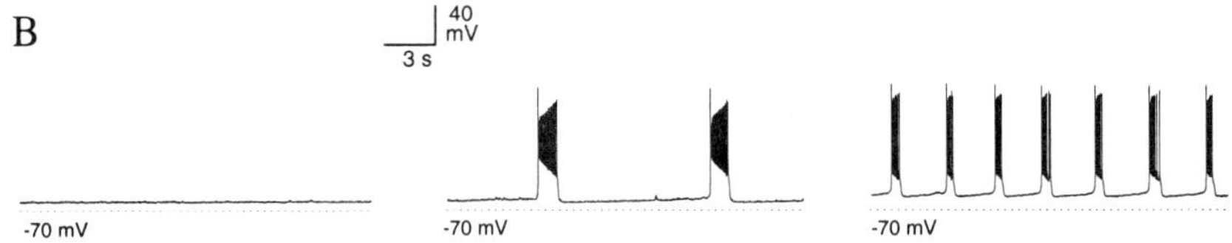

Figure 3. An example of a "mundane" motoneuron to which I_{NaP} has been added to generate pacemaker-like rhythmic bursts. In panel A (before the addition of I_{NaP} with MRCI), from left to right, as the cell is progressively depolarized with I_{app} it begins to spike tonically and does not burst. Panel B (after the addition of I_{NaP}) shows that as the cell is depolarized, the motoneuron bursts with increasing frequency (decrease in burst duration and spikes/burst) as the cell is further depolarized.

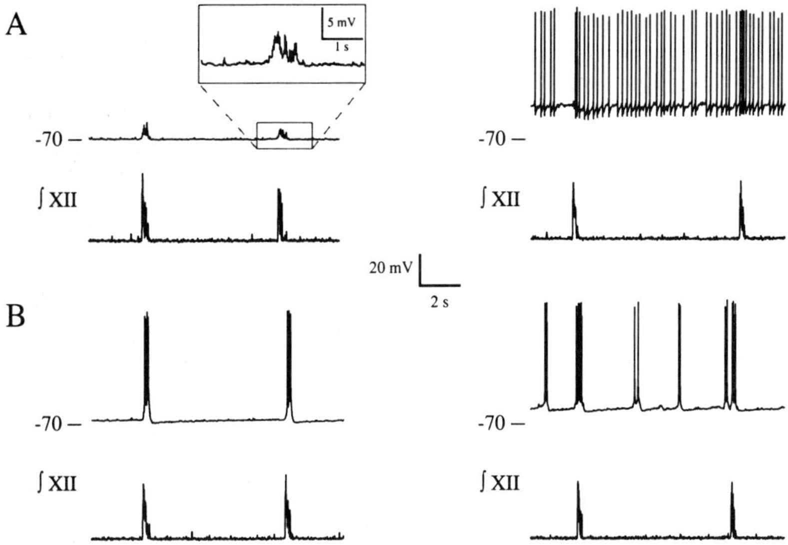

Figure 4. A synaptically driven neuron before and after MRCI added I_{NaP}. In A, the cell is depolarized and exhibits tonic spiking, with bursts occurring only during synaptic drive. In B, with the addition of I_{NaP}, the cell now discharges ectopic bursts upon depolarization. Note: at the left in both A and B, the membrane potential is the same showing an enhancement of synaptic input with the addition of I_{NaP}. Center scale bars apply for all traces except the expanded subthreshold burst in A.

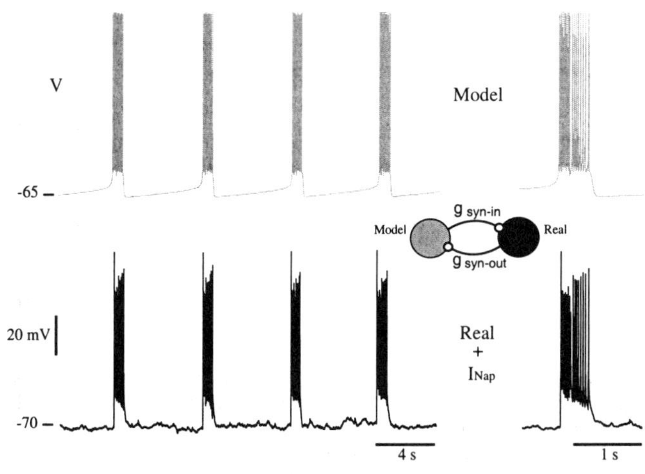

Figure 5. Coupling of a simulated bursting neuron to an MRCI generated burster. A modeled pacemaker (top panel) coupled to a real neuron (bottom panel, with added I_{NaP}) by excitatory synapses (as in the inset) provides an ideal network to test model predictions about coupling strength and it's effect on bursting. The inset indicates that these neurons are reciprocally connected with artificial excitatory synapses. The faster time scale traces on the right show the characteristic decrease in spike firing rate during the burst.

In our most recent experiments we use MRCI to simulate a respiratory pacemaker neuron and couple that modeled cell to a real neuron with MRCI-added I_{NaP}. We found that by varying the synaptic coupling strength between the model neuron and the real neuron we can synchronize the neurons and determine how coupling strength and connectivity affect pacemaker activity. We have previously shown that increased coupling strength slows pacemaker network bursting.[4] Fig. 5 shows the resulting activity when the model cell is coupled to the real neuron with it's added burst generating current.

6. CONCLUSIONS

We have developed a dramatically improved, freely available, and extensible version of the dynamic clamp using open source software and relatively common laboratory hardware. We built MRCI from the ground up as a bridge between experiment and model to test our model-driven hypotheses about burst generating currents, synaptic enhancement in bursting neurons, and coupling in pacemaker neuronal networks. Our experiments show that MRCI scales from very slow currents to very fast currents (e.g., we have reintroduced fast Na^+ and K^+ conductances in cells that have been denuded of these currents) and performs at very high sampling rates, sufficient for the majority of modeled synapses and ionic conductances. MRCI is modular and extensible using the C programming language. We use a robust and reliable open source operating system to insure that MRCI can be altered to fit many needs.

We continually improve MRCI and we hope to build a strong user community to facilitate improvement and extension of the system. We are currently adding a user-friendly interface as well as more extensive features. For more information or to download MRCI software please visit the project home page at http://users.ece.gatech.edu/~rbutera/software/mrci.html.

7. REFERENCES

1. J. C. Smith, H. H. Ellenberger, K. Ballanyi, D. W. Richter and J. L. Feldman, Pre-Bötzinger Complex: a brainstem region that may generate respiratory rhythm in mammals, *Science* **254**(5032), 726-729(1991).
2. N. Koshiya and J. C. Smith, Subthreshold-activation persistent sodium conductance (gNaP) in inspiratory neurons of the pre-Bötzinger complex in vitro, *Soc Neurosci Abstr* **25**, 696 (1999).
3. R. J. Butera, Jr., J. Rinzel, and J. C. Smith, Models of respiratory rhythm generation in the pre-Bötzinger complex. I. Bursting pacemaker neurons, *J Neurophysiol* **81**, 382-397 (1999).
4. R. J. Butera, Jr., J. Rinzel, and J. C. Smith, Models of respiratory rhythm generation in the pre- Bötzinger Complex. II. Populations of coupled pacemaker neurons, *J Neurophysiol* **81**, 382-397 (1999).
5. N. Koshiya and J. C. Smith, Neuronal pacemaker for breathing visualized in vitro, *Nature*, **400**, 360-363 (1999).
6. H. P. C. Robinson and N. Kawai, Injection of digitally synthesized synaptic conductance transients to measure the integrative properites of neurons, *J Neurosci Meth*, **49**, 157-165 (1993).
7. A. A. Sharp, M. B. O'Neil, L. F. Abbot, and E. Marder, The dynamic clamp: artificial conductances in biological neurons, *Trends in Neurosci*, **16**, 389-394 (1993).
8. V. Yodaiken and M. Barabanov, ftp://luz.cs.nmt.edu/pub/rtlinux/papers/usenix.ps.gz (1996).
9. R. R. Poznanski and J. J. Bell, A dendritic cable model for the amplification of synaptic potentials by an ensemble average of persistent sodium channels, *Math Biosci* **166**(2), 101-21 (2000).
10. J. M. Ramirez and D. W. Richter, The neuronal mechanisms of respiratory rhythm generation, *Curr Opin. Neurobiol* **6**(6), 817-825 (1996).

Respiratory Rhythm and Pattern Generation

IS THE VERTEBRATE RESPIRATORY CENTRAL PATTERN GENERATOR CONSERVED?

Insights from in-vitro and in-vivo amphibian models

M.S. Hedrick, L. Broch, M. Martinez, J.L. Powell, and R.E. Wade*

1. INTRODUCTION

There have been considerable advances in our understanding of the anatomical substrates, neural connections and modulation of the mammalian respiratory central pattern generator (CPG). Despite these advances, however, relatively little is known about the regulation of CPGs for breathing in other vertebrates. It has been proposed that CPGs supporting a variety of motor behaviors such as breathing, locomotion, scratching and chewing, have been conserved in the course of vertebrate evolution. [14, 15] At present, there is little experimental support for this hypothesis.

At first glance it would appear that breathing in 'lower' vertebrates, such as amphibians, and mammals have little in common save for the use of lungs for gas exchange. Amphibians use a positive-pressure, buccal force pump to inflate the lungs in contrast to the negative-pressure, aspiratory lung inflation mechanism present in mammals and other amniotes (birds and reptiles). In amphibians, the primary respiratory muscles responsible for generating airflow during breathing are innervated by cranial nerves, whereas in mammals the cranial nerves innervate muscles associated primarily with control of the upper airway. During the course of evolution the diaphragm, intercostals and abdominal muscles, which are innervated by spinal nerves, have assumed the role of the 'power plant' for generating airflow during breathing.

Despite these obvious differences in the mechanical aspects breathing, it is generally agreed that the peripheral structures associated with breathing are far less constrained by evolutionary change than the neural structures that support and control the peripheral elements.[15] In other words, CPGs, in general, appear to be much more conserved than the structures with which they control. Direct evidence for conserved features of the respiratory CPG is sparse in vertebrates, but there are numerous examples of invertebrate

---

* Department of Biological Sciences, California State University, Hayward, 25800 Carlos Bee Blvd., Hayward, CA 94542, U.S.A. Correspondence: Michael S. Hedrick, Phone: (510) 885-7443; FAX: (510) 885-4747; email: mhedrick@csuhayward.edu

Frontiers in Modeling and Control of Breathing, edited by Poon and Kazemi.
Kluwer Academic/Plenum Publishers, New York, 2001.

CPGs that support the hypothesis of conserved neural substrates that control specific motor behaviors.[6]

In this paper, we outline some experiments using *in vitro* and *in vivo* preparations from amphibians that are aimed at testing the hypothesis of a conserved respiratory CPG in vertebrates. In the absence of any complete identification of the neurons that comprise the respiratory CPG for any vertebrate, it is difficult to test directly any hypotheses about the evolution of the respiratory CPG. Nevertheless, a comparative approach, like the one taken here, can provide some important insights into the behavior and possible changes in the respiratory CPG that may have occurred over the course of vertebrate evolution. Although similarity alone is insufficient for adequately testing this hypothesis[14], this is a necessary first step in addressing this challenging question.

2. METHODS

2.1. In-Vitro Studies

Whole brainstem preparations from larval (tadpole) and adult North American bullfrogs (*Rana catesbeiana*) are used for *in vitro* studies. General methods for surgical preparation and neural recording for tadpoles and adults have described in detail elsewhere.[3, 5] Briefly, tadpoles and adults are anesthetized in MS-222 until breathing movements cease. The brainstem is removed and placed in a recording chamber and continuously perfused with an artificial cerebrospinal fluid (aCSF) at 22 °C. Standard extracellular nerve recording from whole cranial nerve rootlets is made with glass suction electrodes. In tadpoles, neural activity from cranial nerves (CN) V (trigeminal), VII (facial), X (vagus) and spinal nerve II (SNII = hypoglossal) are recorded and used as indices of fictive gill and lung ventilation.[3] In adults, neural activity from CN V, X and XII are used as indices of fictive lung ventilation.[5]

2.2. In-Vivo Studies

Decerebrate, paralyzed and artificially-ventilated adult cane toads (*Bufo marinus*) are used for *in vivo* studies.[16] Animals are anesthetized in MS-222 anesthesia until breathing movements cease. The animals are decerebrated, and both lungs are cannulated with PE-260 tubing for artificial ventilation. The animals are paralyzed with rocuronium (1 mg/kg) injected into the lymph space. A humidified gas mixture of 2.5% CO_2/20% O_2 is introduced unidirectionally through the lungs at a rate of 50-100 ml/min. Cranial motor nerves innervating the buccal pump muscles are placed on hook electrodes for whole nerve recording. In adult amphibians, CN V and the main branch of CN XII (XIIm) innervate the primary buccal elevator muscles whereas the sternohyoid branch of CN XII (XIIsh) innervates the primary buccal depressor muscle, and each of these branches are used as indices of fictive ventilation. Neural activity is recorded extracellularly by conventional methods.

2.3. Experimental Protocols

With the *in vitro* brainstem preparation from tadpole and adult bullfrogs, one hypothesis we wished to test was whether there is a developmental 'switch' in the mechanism used to generate respiratory rhythm. In mammals, there is evidence that respiratory rhythm is generated by a pacemaker-like mechanism in the neonate whereas in adults conventional synaptic inhibition is the primary mechanism for respiratory rhythmogenesis.[12] If the respiratory CPG is conserved, we would expect similar results during ontogenetic development in the amphibian larval-adult transition as with the mammalian neonate-adult transition. Our approach was to perfuse the tadpole and adult brainstem preparations with a Cl-free aCSF to determine if either preparation was dependent on conventional synaptic inhibition using Cl ions.

A second experimental approach using the *in vitro* preparation was to perfuse the brainstem with increasing concentrations of glycinergic and GABAergic agonists and antagonists. Our working hypothesis is that if the respiratory CPG is conserved, we should observe similar effects of GABA/glycine, and the antagonists, bicuculline and strychnine, on respiratory motor output during development that have been noted in developing mammals.[9]

Experiments with decerebrate toads involve the use of phase-resetting protocols that have been described in detail for mammals.[7, 10] The recurrent laryngeal nerve (RLN) is stimulated in brief pulses to perturb the respiratory rhythm, and the aftereffects are recorded. Amphibians characteristically breathe in an episodic manner with periods of apnea interspersed with episodes of lung ventilation that contain from 3-10 lung breaths. The RLN is stimulated during a breathing episode and at various times during the respiratory cycle as defined by the phase of the XIIsh burst. As defined in mammalian studies[17], old phase is the time of the onset of the XIIsh burst to the onset of the RLN stimulus. Cophase is the time from the offset of the stimulus to onset of the next breath. Type 0 (strong) resetting is defined as cophase having a net change of zero as old phase is varied through one complete cycle. Type 1 (weak) resetting occurs when cophase has a net change of one cycle per cycle of old phase. A phase singularity is defined as a dysrhythmic response to a discrete stimulus. Our hypothesis is that if the vertebrate respiratory CPG is conserved, we should observe similar responses to perturbation of the CPG by a brief stimulus that resets the respiratory rhythm.

3. RESULTS

Brainstem preparations from tadpoles (N=7) and adults (N=7) responded very differently to Cl-free aCSF (Fig. 1). In tadpoles, fictive lung ventilation frequency was not affected by Cl-free aCSF. By contrast, fictive lung ventilation in adult preparations was either abolished or significantly reduced by Cl-free aCSF. Addition of strychnine (2.5-25 μM) in tadpole brainstem preparations caused a significant increase in fictive breathing frequency, whereas in adult preparations fictive breathing frequency was significantly reduced (Fig. 2). Addition of GABA or glycine caused significant reductions in fictive breathing frequency, however, adult preparations were inhibited at approximately ten-fold lower concentrations of the agonists (data not shown). We interpret this as an indication of a greater sensitivity of adults to receptor agonists compared with tadpoles.

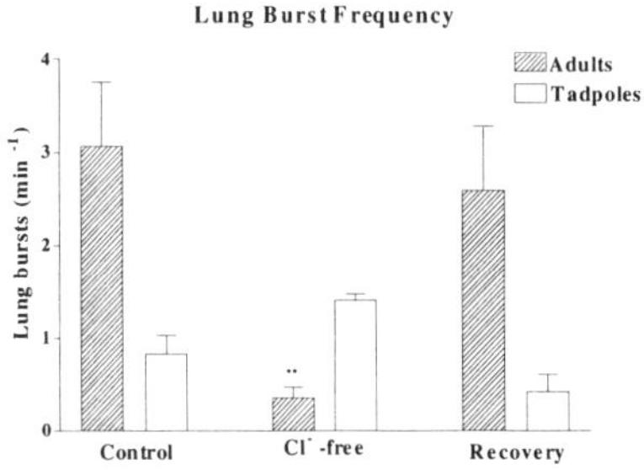

Figure 1. Effects of Cl-free aCSF on fictive breathing in tadpole and adult brainstems *in vitro* (n=7).

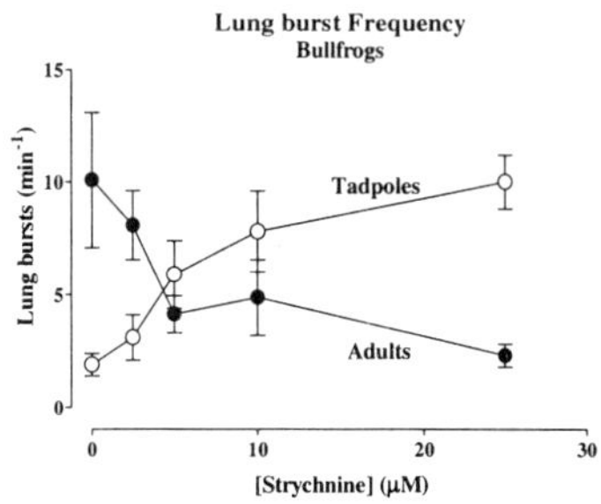

Figure 2. Effects of strychnine on fictive breathing in tadpole and adult brainstems *in vitro* (n=8).

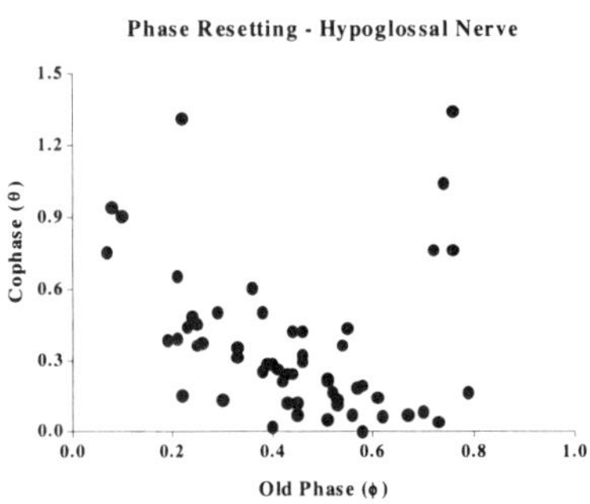

Figure 3. Effects of RLN stimulation showing Type 1 (weak) resetting with a discontinuity at 80% of the duty cycle relative to XIIsh.

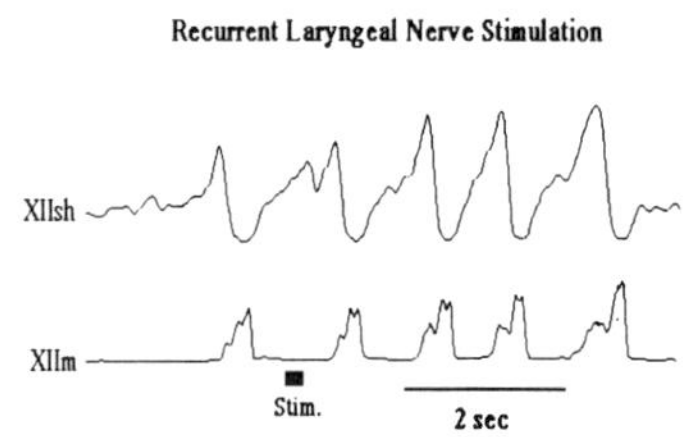

Figure 4. Effects of RLN stimulation on XIIsh and XIIm. Note phase advance in XIIsh and phase delay in XIIm.

Both Type 1 and Type 0 phase resetting responses have been observed with RLN stimulation during episodic lung bursts. The effects of RLN stimulation on phase resetting for one toad is shown in Fig. 3. Type 1 (weak) resetting occurs over most of the respiratory cycle with a discontinuity evident at a cophase of approximately 0.8. This indicates that both phase advances (Fig. 4) and phase delays occur at about 80% of the duty cycle of the XIIsh burst and occur with unpredictable results.

4. DISCUSSION

The results presented here suggest that the lung CPG of tadpoles uses a pacemaker-like mechanism, similar to that hypothesized for neonate mammals.[2] In tadpoles and neonate mammals, fictive lung ventilation persists in the absence of extracellular Cl ions.

Because conventional synaptic inhibition requires Cl ions, this is evidence for a pacemaker-driven respiratory CPG. Adult amphibians (Fig. 1) and mammals[4] appear to use conventional inhibitory synaptic inhibition because lung ventilation is abolished with Cl-free aCSF. Thus, there appears to be a developmental 'switch' in the mechanism used to generate respiratory rhythm. If the respiratory CPG has indeed been conserved, it is most parsimonious to hypothesize that common developmental mechanisms should be preserved during the course of evolution. Moreover, results from the adult lamprey, a primitive vertebrate, indicate that the respiratory CPG relies on a pacemaker-like mechanism.[13] Thus, ontogeny of the respiratory CPG may reflect the phylogeny of the respiratory CPG.

There are also similarities in the results from inhibitory amino acids in developing amphibians and mammals. If there is a developmental 'switch' from a pacemaker to a network-driven CPG, increased receptor density and sensitivity to GABA and/or glycine receptor agonists and antagonists should be a component of this process. The tadpole to adult transition in amphibians is accompanied by increasing sensitivity to GABA and glycine[1] and differential responses to the glycinergic antagonist strychnine (Fig. 2). These results are similar to those found for developing mammals[9] suggesting that common inhibitory mechanisms are found in disparate groups of vertebrates.

Phase-resetting experiments have proven useful for examining the behavior of oscillatory processes, such as breathing, to brief perturbations of the rhythmic activity.[17] These experiments are useful because the oscillator may be treated as a 'black box' without explicit knowledge of the structure of the oscillator. In mammals, phase-resetting experiments predict that precisely-timed stimuli can result in dysrhythmic breathing.[11] In toads, brief perturbations of the respiratory oscillator by RLN stimulation cause phase advances, phase delays and discontinuities in respiratory rhythm (Fig. 3,4). These preliminary results suggest that perturbations of the amphibian respiratory oscillator produce similar effects as in mammals, which argues for conservation of the respiratory CPG. Our long-term goal is to model the breathing patterns of amphibians to examine whether episodic breathing in amphibians has features in common with pathologic breathing disorders in mammals.

Direct evidence for evolutionary conservation of CPGs is generally lacking in vertebrates, but there is substantial support for this hypothesis in invertebrates where 'simple' CPGs have been described in detail. The primary reason that CPGs are highly conserved is that neural networks tend to be 'generalists' rather than 'specialists'.[6] Individual CPGs are capable of producing a variety of motor outputs that depend on neuromodulatory inputs or sensory feedback. Thus, novel CPGs need not evolve to produce new motor behaviors. A recent example of this was found for the mammalian respiratory CPG with a medullary rhythmic 'slice' containing the pre-Botzinger complex that produced a variety of motor outputs including eupnea, sighs and gasps.[8]

So, what, if anything, has changed with respect to the vertebrate respiratory CPG? It could be hypothesized that the evolution of the vertebrate respiratory CPG is associated with increased 'complexity' which may involve increased modulatory inputs, greater flexibility in the respiratory neural circuitry and more 'degrees of freedom' for respiratory motor outputs. Perhaps a greater flexibility of the respiratory CPG in mammals was associated with the evolutionary change in spinal, rather than cranial, motoneuronal control of the primary respiratory pump muscles.

5. ACKNOWLEDGMENTS

The authors' research is supported by National Institutes of Health - Minority Biomedical Research Support grant SO6 GMAI48135-04.

6. REFERENCES

1. L. Broch, A.V. Sandoval, R.D. Morales, and M.S. Hedrick, Regulation of central respiratory rhythm by GABA and glycine receptors in the isolated brainstem of larval and adult bullfrogs, *FASEB J.* **14**(4), A44 (2000).
2. J.L. Feldman and J.C. Smith, Cellular mechanisms underlying modulation of breathing pattern in mammals, *Ann. NY. Acad. Sci.* **563**, 114-130 (1989).
3. M.J. Gdovin, C.S. Torgerson, and J.E. Remmers, The fictively breathing tadpole brainstem preparation as a model for the development of respiratory pattern generation and central chemoreception, *Comp. Biochem. Physiol.* **124A**, 275-286 (1999).
4. F. Hayashi and J. Lipski, The role of inhibitory amino acids in control of respiratory motor output in an arterially-perfused rat, *Respir. Physiol.* **89**, 47-63 (1992).
5. M. S. Hedrick, R.D. Morales, J. M. Parker, and J.L.H. Pacheco, Nitric oxide modulates respiratory-related neural activity in the isolated brainstem of the bullfrog, *Neurosci. Lett.* **251**, 81-84 (1998).
6. P. S. Katz and R. Harris-Warrick, The evolution of neuronal circuits underlying species-specific behavior, *Curr. Opin. Neurobiol.* **9**, 628-633 (1999).
7. J. Lewis, M. Bachoo, C. Polosa, and L. Glass, The effects of superior laryngeal nerve stimulation on respiratory rhythm: phase-resetting and aftereffects, *Brain Res.* **517**, 44-50 (1990).
8. S.P. Lieske, M. Thoby-Brisson, P. Telgkamp, and J. M. Ramirez, Reconfiguration of the neural network controlling mutiple breathing patterns: eupnea, sighs and gasps, *Nature Neurosci.* **3**(6), 600-607 (2000).
9. J. F. R. Paton and D. W. Richter, Role of fast inhibitory synaptic mechanisms in respiratory rhythm generation in the maturing mouse, *J. Physiol.* **484**, 505-521 (1995).
10. D. Paydarfar and F. L. Eldridge, Phase resetting and dysrhythmic responses of the respiratory oscillator, *Am. J. Physiol.* **252**, R55-R62 (1987).
11. D. Paydarfar and D.M. Buerkel, Dysrhythmias of the respiratory oscillator, *Chaos* **5**, 18-29 (1995).
12. J. E. Remmers, Central Neural Control of Breathing, in: *Control of Breathing in Health and Disease*, edited by M.D. Altose and Y. Kawakami, (Marcel Dekker, New York, 1999), pp. 1-40.
13. C. M. Rovainen, Generation of respiratory activity by the lamprey brain exposed to picrotoxin and strychnine, and weak synaptic inhibition in motoneurons, *Neurosci.* **10**, 875-882 (1983).
14. K. K. Smith, Are neuromotor systems conserved in evolution? *Brain Behav. Evol.* **43**, 293-305 (1994).
15. A.J. Tierney, Evolutionary implications of neural circuit structure and function, *Behav. Proc.* **35**, 171-182 (1996).
16. T. Wang, E.W. Taylor, S. G. Reid, and W. K. Milsom, Lung deflation stimulates fictive ventilation in decerebrate, paralysed and unidirectionally ventilated toads (*Bufo marinus*), *Respir. Physiol.* 118, 181-191 (1999).
17. A. T. Winfree, *The Geometry of Biological Time*, (Springer-Verlag, New York, 1980).

UNSTABLE BREATHING RHYTHMS AND QUASIPERIODICITY IN THE PRE-BÖTZINGER COMPLEX

Christopher A. Del Negro, Christopher G. Wilson, Robert J. Butera, Naohiro Koshiya, Sheree M. Johnson, and Jeffrey C. Smith*

1. INTRODUCTION

Breathing in neonatal mammals is notoriously unstable, which can impair normal ventilatory function. Do unstable breathing patterns arise from deterministic mechanisms in the central nervous system? To address this question we examined unstable respiratory patterns using in vitro preparations from neonatal and embryonic rodents that isolate the brain stem kernel containing networks that generate and control breathing[1]. We also analyzed patterns of ventilation in neonates in vivo and in simulations of rhythm generation using a mathematical model of the kernel[2]. In all of these studies inspiratory rhythms showed evidence of quasiperiodicity, and in some cases the "quasiperiodic transition to chaos" identified in oscillatory dynamical systems by Ruelle and Takens[3]. In the quasiperiodic regime the inspiratory pattern undergoes amplitude modulation that becomes increasingly disordered and unstable as neuronal excitability is elevated, and leads to chaos-like states of irregular respiratory rhythm. We found that quasiperiodic activity is an intrinsic property of the developing respiratory network that originates in the pre-Bötzinger complex (pre-BötC), the proposed site for respiratory rhythm generation in mammals[4]. Electrophysiological and computational analyses showed that quasiperiodicity can arise in the pre-BötC from rhythm-generating neurons with intrinsic oscillatory bursting-pacemaker properties[5], thus we identify a potential mechanism for the onset of deterministic respiratory instability in immature mammals.

* Christopher A. Del Negro, Christopher G. Wilson, Naohiro Koshiya, Sheree M. Johnson, and Jeffrey C. Smith, Cellular & Systems Neurobiology Section, NINDS, NIH, Bethesda, Maryland, 20892. Robert J. Butera, Institute for Bioengineering and Biosciences, Georgia Institute of Technology, Atlanta, Georgia, 30332.

Frontiers in Modeling and Control of Breathing, edited by Poon and Kazemi.
Kluwer Academic/Plenum Publishers, New York, 2001.

2. QUASIPERIODIC BREATHING IN NEONATAL RATS IN VIVO

We recorded breathing patterns in urethane-anesthetized neonatal rats (P2-P5) using diaphragmatic electromyography (EMG). Inspiration showed cycle-to-cycle fluctuations in amplitude (Fig. 1). In deterministic systems the current state of the system is influenced by previous states. Therefore, if inspiratory amplitude fluctuations result from a deterministic process, then a Poincaré map can reveal the dynamics geometrically by successively plotting the present measure of a system variable (n+1) versus its previous value (n), for a series of such measurements taken from the system at regular intervals[6].

We obtained Poincaré maps by sampling EMG burst area (or amplitude) during the inspiratory phase (one measurement per breathing cycle)[7]. Three 10-min. epochs were recorded in each neonate. In one animal the Poincaré map consistently showed a single point cluster, indicating a stable rhythm without periodic modulation (Fig. 1 left). However, in the majority of animals tested (n=6/7) the Poincaré map formed a ring-like structure during at least one recording epoch (Fig. 1 middle), which indicates periodic modulation of inspiratory amplitude. This state is called "quasiperiodic" if the two oscillations (the breathing cycle and its periodic amplitude modulation) have independent periods. Quasiperiodic activity forms a doughnut-shaped torus in state space (the imaginary "space" of cyclic coordinate axes formed by the oscillatory variables of the system) and its Poincaré map is a ring (e.g. Fig. 3B).

To assess whether quasiperiodic behavior resulted from peripheral sensory feedback in the intact animal or could be directly attributed to respiratory rhythm and pattern-generating networks of the brain stem, we performed vagotomy to remove mechanosensory afferent feedback and repeated these measurements. After vagotomy, 3 of 6 animals continued to show ring-like Poincaré maps (Fig. 1 right), which suggests that quasiperiodic breathing could originate centrally in the respiratory rhythm generator.

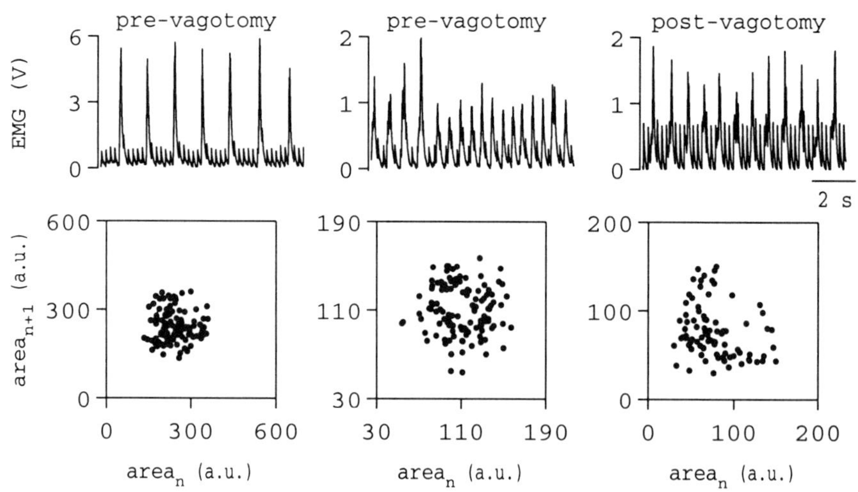

Figure 1. Breathing patterns in neonatal rats. Time series of integrated diaphragmatic EMG (top). Below are Poincaré maps of inspiratory burst area (arbitrary units) taken from the time series above.

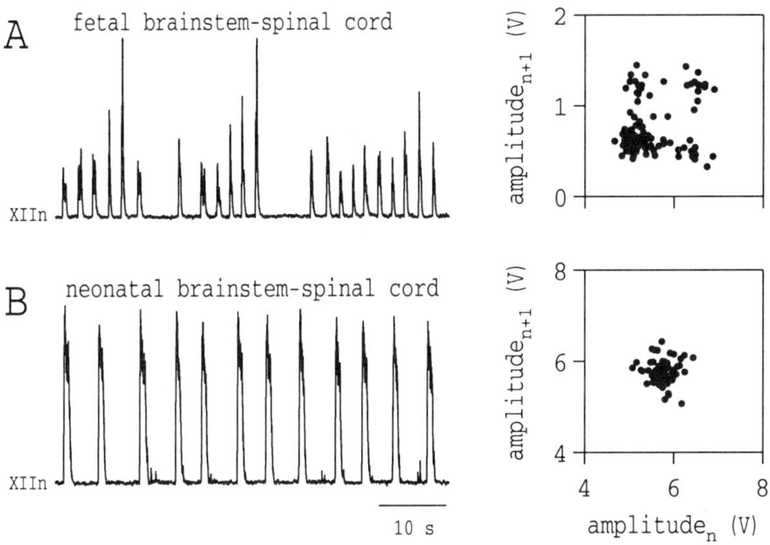

Figure 2. Inspiratory rhythm in fetal (A) and neonatal (B) brainstem-spinal cord preparations with corresponding Poincaré maps of inspiratory burst amplitude.

3. QUASIPERIODICITY IS AN INTRINSIC PROPERTY OF THE DEVELOPING RESPIRATORY NETWORK THAT ORIGINATES IN THE PRE-BÖTC

The isolated neuraxis of neonatal rodents spontaneously generates respiratory-related motor activity that can be recorded in vitro[1]. We monitored the inspiratory rhythm in brainstem-spinal cord preparations from embryonic (E17, 18) and neonatal (P0-2) mice via phrenic and hypoglossal (XIIn) nerve recordings, which conveyed equivalent signals. The embryonic respiratory pattern fluctuated quasiperiodically as shown in time series and Poincaré maps (Fig. 2A). In contrast, neonatal mice showed stable inspiratory rhythms under basal conditions in vitro: the Poincaré maps showed single point clusters (Fig. 2B), indicating that inspiratory bursts were equivalent from cycle to cycle (no periodic amplitude modulation).

Whereas the embryonic network appeared to be intrinsically quasiperiodic and unstable, the neonatal respiratory oscillator appeared to be stable in vitro under basal conditions. To test whether the neonatal network was capable of generating unstable inspiratory rhythms we examined slice preparations from neonatal rats that capture the pre-BötC, the critical locus of the ventral medulla where respiratory rhythm putatively originates[4]. Inspiratory activity can be monitored via suction electrodes on the XIIn that record motor activity, or via population-level recordings in the pre-BötC (Fig. 3A).

We manipulated neuronal excitability using extracellular K^+ concentration ($[K^+]_o$) in the bathing solution. At low excitability levels (8 mM $[K^+]_o$) the fundamental inspiratory rhythm showed no periodic amplitude modulation and the Poincaré map was a single point cluster (Fig. 3B), indicating that inspiratory rhythm was stable. However, as $[K^+]_o$ increased inspiratory burst amplitude fluctuated, creating several distinct point clusters in the Poincaré map. This behavior is called "mixed-mode oscillation" because the inspiratory magnitude returns to the same discrete values after every *N* cycles (Fig. 3B, 10 mM $[K^+]_o$). Further elevation of $[K^+]_o$ caused periodic modulation of inspiratory amplitude, which yielded ring-like Poincaré maps that reflect quasiperiodicity (Fig. 3B, 12 mM). At the highest $[K^+]_o$ tested in slices the amplitude fluctuated erratically, creating formless clusters in the Poincaré map that could reflect deterministic chaos arising from quasiperiodicity[6], or alternatively high-dimensional asynchronous tonic spiking in the network (Fig. 3D). These geometric forms in Poincaré maps were confirmed using spatial statistics[8], and were obtained in all slices tested (n=20).

The progression from stable oscillations to quasiperiodicity and chaos was first described by Ruelle and Takens[3] and occurs in a periodic system when a new oscillation arises and is coupled to the original rhythmic system. The "new" oscillation provides the periodic modulation that is visible in the time series and the ring-shaped Poincaré map. If a third oscillation is born the system becomes inherently unstable and can devolve into chaos[3, 6], although evidence of quasiperiodicity is often retained in the chaotic state[9].

4. QUASIPERIODICITY IN THE MODEL PACEMAKER NETWORK

Since manipulations of neuronal excitability destabilized inspiratory rhythms via quasiperiodicity we sought a cellular mechanism that could cause new oscillations to develop independently in the respiratory rhythm-generating network. Recently, Ca^{++} imaging studies[5] have confirmed that the pre-BötC contains a population of bursting "pacemaker" neurons. These synaptically coupled cells generate rhythm collectively as a pacemaker network[2] and can generate cellular rhythms autonomously when isolated via pharmacological blockade of synaptic receptors. Since bursting is voltage-dependent, elevations in excitability might evoke independent oscillations in pacemaker neurons.

We tested this notion using a mathematical model of the pacemaker network developed by Butera et al.[2] comprising 50 heterogeneous pacemaker neurons coupled via fast excitatory synapses. This model has been evaluated experimentally[10] and currently represents the best available mathematical model of rhythm generation in vitro. To mimic the experiment performed in the slice preparation we simulated the network at increasing levels of K^+ reversal potential (E_K) and found similar time series and Poincaré maps showing periodicity, mixed-mode oscillations, quasiperiodicity (ring-like structures in the maps), and chaos-like disordered states (Fig. 4). In the model, quasiperiodicity results from independent oscillations arising in subsets of pacemaker neurons whose activity diverges from the established network rhythm[2]. These independent oscillatory patterns develop due to the intrinsic ability of pacemaker neurons to autonomously generate bursts, which becomes more prominent when cellular excitability increases[10].

5. IMPLICATIONS FOR BREATHING IN NEONATAL MAMMALS

Evidence suggests that neurons with intrinsic bursting properties comprise the respiratory rhythm generator in the pre-BötC of neonatal mammals[4,5]. We have shown at the level of the kernel for rhythmogenesis in vitro, in mathematical models, and at the behavioral level in vivo that this system inherently possesses the machinery for quasiperiodicity and its attendant instabilities. The oscillations that modulate respiratory rhythm may be attributed to bursting-pacemaker neurons that generate rhythms autonomously within coupled pacemaker-cell networks[2,10,11]. We propose that the developing respiratory network in mammals contains pacemaker cells that tend to burst at preferred frequencies, causing independent patterns to develop in subsets of neurons within the kernel. These independent oscillations destabilize the system via quasiperiodicity.

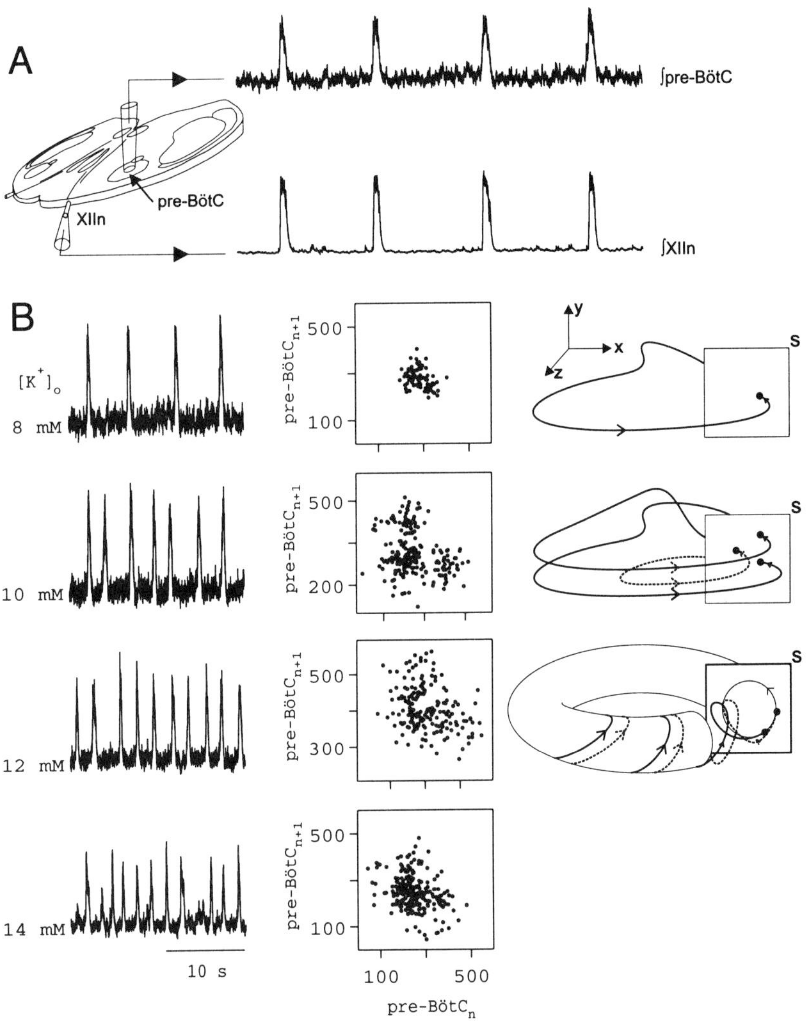

Figure 3. A, Schematic of the 350-μm-thick respiratory medullary slice preparation. Suction electrodes are used to record hypoglossal motoneuron population activity from the XIIn (lower trace) and locally from neurons in the pre-BötC (upper trace, 30-s). B, inspiratory rhythms in the pre-BötC at increasing extracellular K^+ concentrations and corresponding Poincaré maps of inspiratory burst area. State-space behaviors in periodic (top right), mixed-mode periodic (middle right) and quasiperiodic systems (lower right) illustrate how distinct geometric patterns are formed in the Poincaré maps; "s" is the planar Poincaré cross section topologically equivalent to the Poincaré map[7].

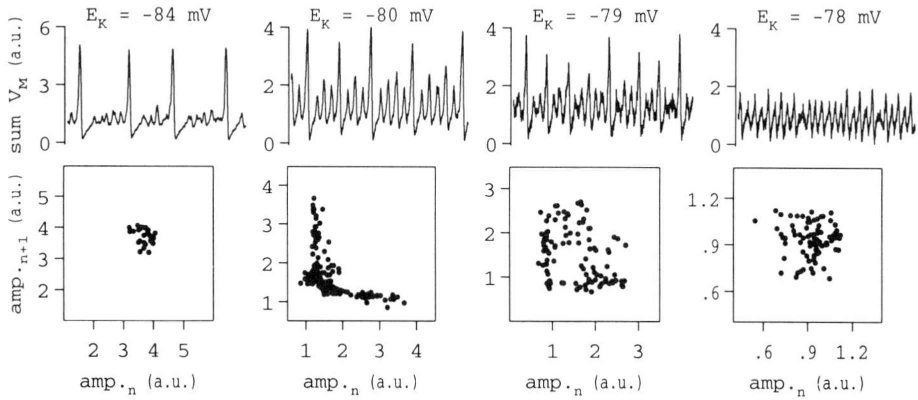

Figure 4. Samples (30 s) of the pacemaker population activity from the model network at different E_K and corresponding Poincaré maps of inspiratory population-burst amplitude.

In the fetal network this appears to be the natural state of the system. However by early postnatal development the basal state of the inspiratory rhythm is stable, as we demonstrated in neonatal brain stem-spinal cord and slice preparations. Nevertheless, perturbing excitability in vitro evoked quasiperiodicity. Breathing is regulated based on metabolic demand. If this occurs via regulation of excitable membrane properties in the kernel then quasiperiodicity and unstable breathing patterns may develop at elevated levels of cell excitability in the rhythm generator, which could impair normal ventilation.

6. REFERENCES

1. J. C. Smith, J. J. Greer, G. S. Liu and J. L. Feldman, Neural mechanisms generating respiratory pattern in mammalian brain stem-spinal cord *in vitro*. I. Spatiotemporal patterns of motor and medullary neuron activity, *J. Neurophysiol.* **64**(4), 1149-1169 (1990).
2. R. J. Butera, Jr., J. Rinzel and J. C. Smith, Models of respiratory rhythm generation in the pre-Botzinger complex. II. Populations of coupled pacemaker neurons, *J Neurophysiol* **82**(1), 398-415 (1999).
3. D. Ruelle and F. Takens, On the nature of turbulence, *Commun Math Phys* **20**, 167-192 (1971).
4. J. C. Smith, H. H. Ellenberger, K. Ballanyi, D. W. Richter and J. L. Feldman, Pre-Bötzinger complex: a brainstem region that may generate respiratory rhythm in mammals, *Science* **254**(5032), 726-729 (1991).
5. N. Koshiya and J. C. Smith, Neuronal pacemaker for breathing visualized in vitro, *Nature* **400**(6742), 360-363 (1999).
6. A. H. Nayfeh and B. Balachandran, *Applied nonlinear dynamics: analytical, computational, and experimental methods* (Wiley, New York, 1995).
7. T. Sauer, J. Yorke and M. Casdagli, Embedology, *J Stat Phys* **65**(3/4), 579-616 (1991).
8. B. D. Ripley, *Spatial Statistics* (John Wiley & Sons, New York, 1981).
9. C. Grebogi, E. Ott and J. A. Yorke, Are three-frequency quasiperiodic orbits to be expected in typical nonlinear dynamical systems, *Phys Rev Lett* **51**(5), 339-342 (1983).
10. C. A. Del Negro, S. M. Johnson, R. J. Butera and J. C. Smith, Models of respiratory rhythm generation in the pre-Bötzinger complex. III. Experimental tests of model predictions, *J Neurophysiol* (submitted for publication, 2000).
11. R. J. Butera, Jr., J. Rinzel and J. C. Smith, Models of respiratory rhythm generation in the pre-Botzinger complex. I. Bursting pacemaker neurons, *J Neurophysiol* **82**(1), 382-397 (1999).

STATIONARY ORGANOTYPIC CULTURE OF THE PRE-BÖTZINGER COMPLEX FROM THE NEWBORN RAT

H. Rigatto[*‡], C.G. Wilson[*], N. Koshiya[*], S. House[†], and J.C. Smith[*]

1. INTRODUCTION

In the early 1990s, it was suggested that the central neuronal network responsible for breathing relies on pacemaker cells to promote its rhythmic activity.[1,2,3] These pacemaker neurons were shown to be concentrated in the region of the pre-Bötzinger complex and are critical components of the network generating respiratory rhythm.[3,4] Previous studies of this neuronal population have been done primarily in fresh tissue slices in vitro which have a limited survival (less than 48 hours) and, due to thickness of the slice, offer poor accessibility to neurons of interest. Tissue culture of the isolated pre-Bötzinger complex may offer some advantages in this respect, but it is not known whether the isolated pre-Bötzinger network would maintain its functional properties. Therefore, we evaluated the viability of the stationary explant of the cultured pre-Bötzinger region ("island"). We hypothesized that the pacemaker network would maintain functional activity for extended periods in culture due to the network robust functional properties.

2. METHODS

Experiments were performed with medullary slice preparations that retain functional respiratory network properties. Details of the preparation have been previously described.[3,4] Briefly, after obtaining a slice of the brainstem from neuronal rats (0-4 days) containing the pre-Bötzinger complex, the rostral roots, and nerve tracts of the hypoglossal nerve (XII cranial nerve), a 2×2 mm piece of tissue containing the pre-Bötzinger region only was isolated.[3,5] In most experiments this region was cultured and stained with antibodies to neurofilament 160. In some other experiments the pre-Bötzinger "islands" were bathed with Ca^{2+} sensitive dye for patch-clamp and imaging studies, and later cultured and

[*]Laboratory of Neural Control, and [†]Laboratory of Neurochemistry, NINDS, National Institutes of Health, Bethesda, MD; [‡]Department of Pediatrics & Physiology, University of Manitoba, 735 Notre Dame Avenue, Winnipeg, R3E 0L8 Canada. Correspondence: H. Rigatto

Frontiers in Modeling and Control of Breathing, edited by Poon and Kazemi.
Kluwer Academic/Plenum Publishers, New York, 2001.

immunostained.

The technique of neuronal labelling was based on the use of membrane permeant Ca^{2+} sensitive dye.[4,5,6] Calcium green-1 AM (CaG molecular probes, 50 μg) was dissolved in dimethyl sulfoxide (DM50; 5 ml) containing pluronic F-127 (BASF, 25 μM) and dispersed in artifical CSF (10 μL). The CaG solution was injected with a glass pipette into the recording chamber containing the cultured pre-Bötzinger "islands". Labelling of the cell bodies was completed in 4 to 6 hours.

We used the millipore filter method for culture.[7,8] Slices were placed on a Millicell-CM filter inserts (Millipore, pore size 0.4 μm, diameter 30 mm).[8,9] Each filter insert was then placed in a Petri dish (35 mm), containing 1.1 ml of culture medium wetting the exposed surfaces of the slices, but not submerging them. Incubation of the culture was stationary at 35-36°C in 5% CO_2 enriched air. Cultures were typically maintained for 14 to 21 days in standard medium, which was replaced twice a week.

The standard culture medium was composed of Eagles basal medium (BME, Sigma, St. Louis, MO, USA) 50%, heat inactivated horse serum 25%, and Hank's balanced salt solution, 25%, 1 ml 50% glucose, 0.5 ml L-glutamine and 0.25 ml of antibiotics (penicillin, streptomycin, neomycin). In some cultures, fluorodeoxyuridine (FUDR) 0.01 mM (Sigma) was applied in the medium for the first 24 hours.

Calcium-green labelled neurons in culture were visualized with a fixed stage upright videomicroscope system (Axioskop FS, Zeiss) equipped with a 75W XENON epi-illuminator, optical filters (excitation 485 nm, emission 530 nm, and 505 nm beam splitter, OMEGA Optical) and water-immersion objectives (Zeiss Achroplan 20x, 40x, and 63x; 0.5, 0.75, and 0.9 NA, respectively).[4] Fluorescence images were captured with a CCD camera fibreoptically coupled to an image intensifier (ICCD-350F or 1000F, VideoScope Intl.).[10] Changes in fluorescence intensity (ΔF/F) were detected in real time with an image processor and quantified off-line after images were digitized via an audio-video board (Video Vision, Radius) which allows images to be formatted as a Quick Time movie file. This format enabled video-frame matching with electrophysiologic signals recorded in audio track. Voltage and current-clamp recordings were performed with an EPC-9 amplifier controlled by Pulse software (version C, HEKA, Electronic) as described previously.[12] The tissue was fixed in 4% paraformaldehyde.[11] For immunostaining, filters were excised from the inserts using a scalpel and then placed in Netwell carriers (Costar, MA, USA). Filters were rinsed in PBS, and blocked in 10% normal goat serum (NGD)/0.3% Triton X-100, for 1 hour to prevent non-specific binding. The filters were incubated in primary antiserum (neurofilament 160) overnight at 4°C. On day 2, the filters were rinsed in PBS and then incubated in biotinylated secondary antibody (Vector Laboratories, Burlingame, CA, USA) at 1:500 dilution for 1-1.5 hours, rinsed in PBS then followed by avidin-biotinylated horseradish peroxidase (1:600) for 1-1.5 hours (Vectastain Elite ABC Kit, Vector Labs). Following this incubation, the filters were rinsed in PBS and the reactive product was visualized using the chromogen 3.3' diaminobenzidine tetrahydrochloride (DAB, 5 mg/10 ml, Sigma). Tissue was eased off the filters using a No. 0 or No. 1 fine sable brush and placed on double subbed slides and allowed to dry overnight before counterstaining. Tissues were often counterstained with 0.5% methyl green (Electron Microscopy Sciences, Ft. Washington, PA, USA) and dehydrated.

3. RESULTS

The slice-explant culture method was highly effective in maintaining the neuronal network in the pre-Bötzinger complex viable for up to three weeks *in vitro*. Survival was excellent with only about 5% of the cultures being lost due to detachment or contamination. The slices typically thinned to about 80-100 μm in standard medium, so that immunocytochemical procedures could be performed directly on filters or whole mounts. We cultured 146 pre-Bötzinger "islands" successfully. The majority of these were fixed for immunostaining (130) and 16 were used for electrophysiological and fluorescence imaging.

Initially, we sliced the upper medulla rostral to caudal and stained these slices (100 μm thick) with antibodies to neurofilament 160. In the slice of interest, the pre-Bötzinger region lies adjacent to the neurons of the nucleus ambiguus. The neuronal network of the pre-Bötzinger "island" in culture shows multipolar cells (slice-explants 14 days old; Figs. 1A and B). During voltage clamp the synaptic drive is clearly present, and current clamp mode shows pattern of synaptic drive and action potential discharges during progressive depolarization (Fig. 2). Rhythmic changes in fluorescence intensity of the pre-Bötzinger cells in culture (16 days) were clearly demonstrated (Figs. 3, 4).

4. DISCUSSION

The pre-Bötzinger complex from newborn rats can be successfully cultured as an "island" and the activity of the local rhythm-generating network can be maintained for studies of cellular and network properties. This culture preparation may be useful for electrophysiologic, genetic, and immunohistochemical studies in situations where survival of the network for longer periods is required. Using electrophysiologic methods, the region immediately rostral to the obex, containing the pre-Bötzinger complex, has been identified as an important respiratory pool.[3,4] The pre-Bötzinger region was found to contain almost exclusively inspiratory pacemaker neurons, this region representing the central neuronal kernel responsible for respiratory rhythmogenesis. It has also been shown to be a significant chemosensitive region for CO_2, supporting previous work in dissociated cells.[1,2,12]

There have been no previous efforts to culture the selected neuronal pacemaker population of the pre-Bötzinger complex. However pacemaker neurons highly sensitive to carbon dioxide have been found in the dissociated neuronal cell cultures of slices taken from the region within 2 mm rostral to the obex.[2] Such cultures have been very successfully maintained. However their respiratory specificity has not been established. The identification of the pre-Bötzinger complex as the pacemaker kernel responsible for the generation of inspiratory rhythm emphasizes the need for examination of the neuronal properties of this region.[3,4]

The significant thinning of the pre-Bötzinger islands in culture allows for two immediate advantages: most neurons in this area become accessible to whole cell patch-clamp and imaging techniques, and there is more flexibility in time to do the experiments. In addition, the pre-Bötzinger island culture opens the door for dissociated neuronal cultures of this region, enabling us to create a purer pacemaker cell line, facilitating the physiologic, genetic and immunocytochemical studies of these neurons.

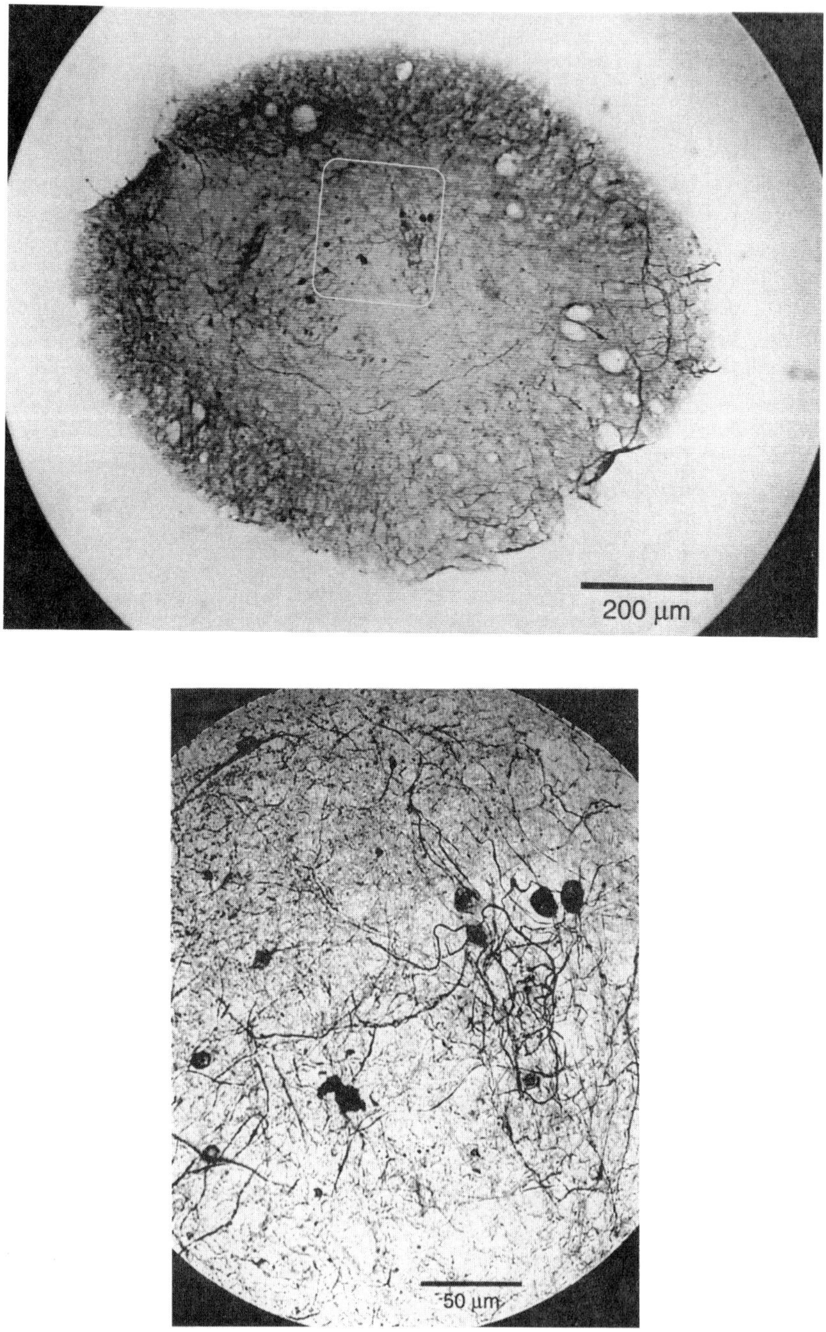

Figure 1. **A.** Cultured "island" containing the pre-Bötzinger complex (5x, 14 days, neurofilament 160 immunostaining). **B.** Higher magnification of neurofilament stained neurons, taken from highlighted region in **A,** at 14x. For a color representation of **A,** see color insert facing this page. For a color representation of **B,** see color insert facing page 143.

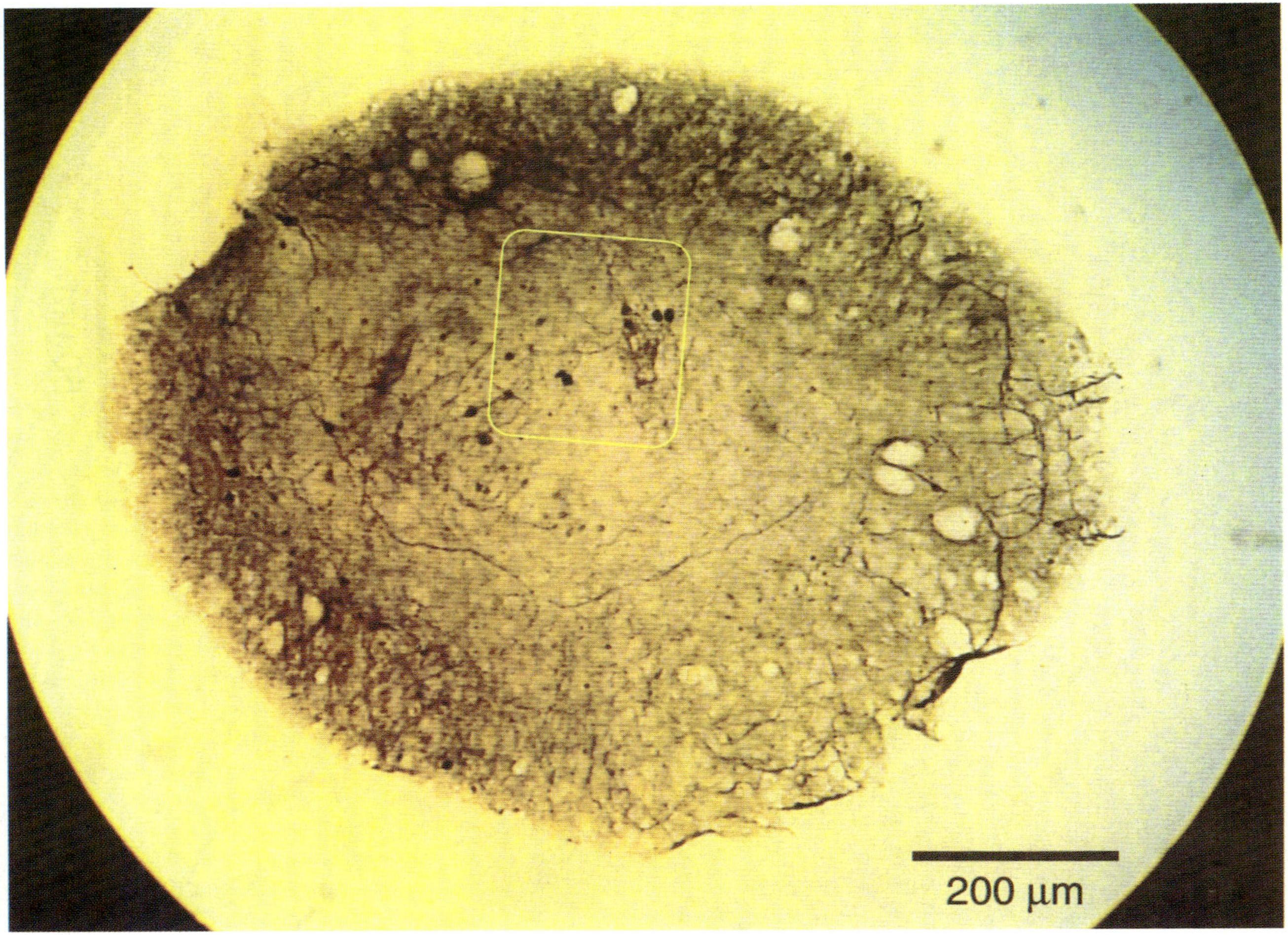

Figure 1. **A.** Cultured "island" containing the pre-Bötzinger complex (7.5x, 14 days, neurofilament 160 immunostaining).

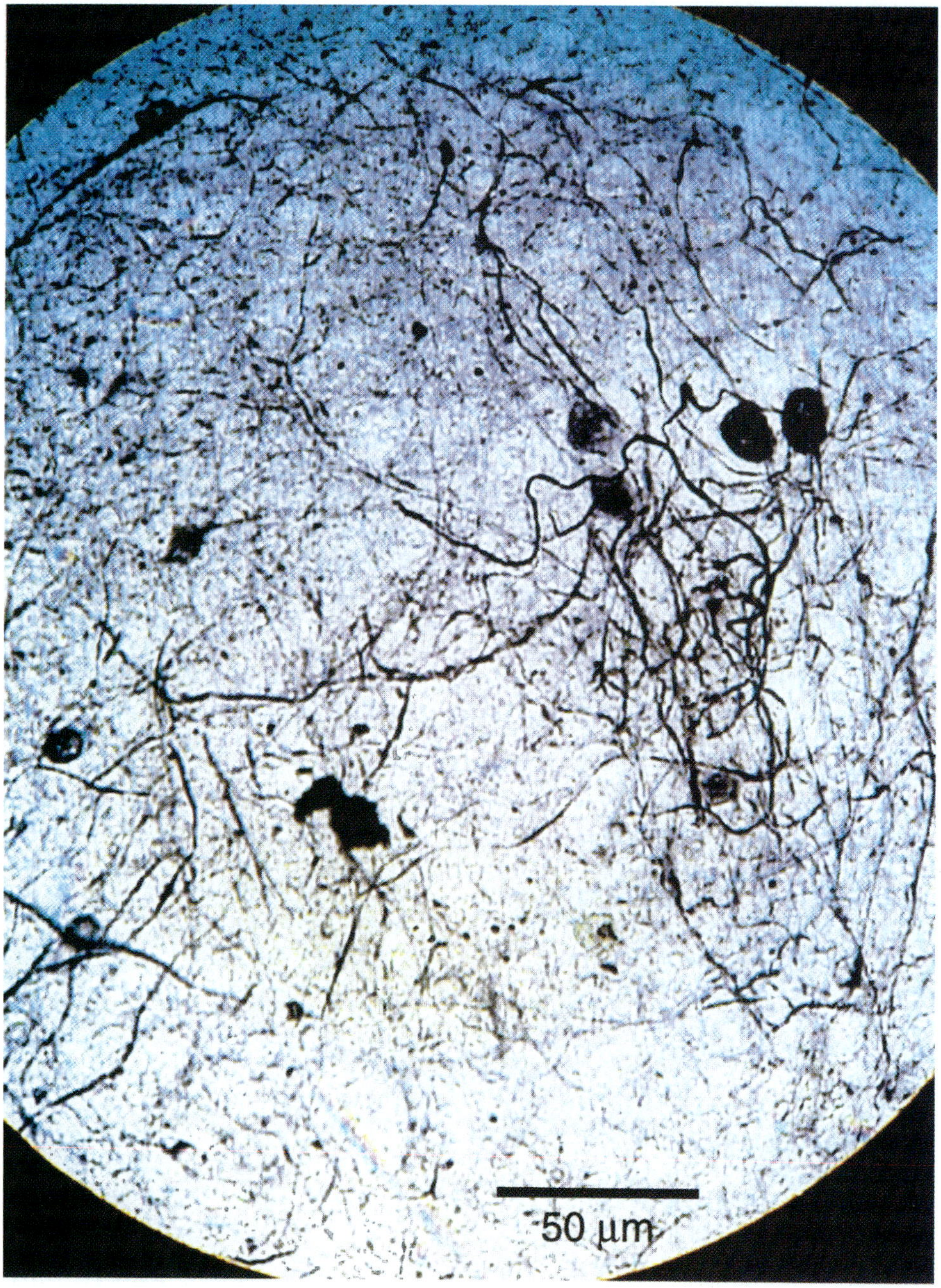

Figure 1. B. Higher magnification of neurofilament stained neurons, taken from highlighted region in **A,** at 28x.

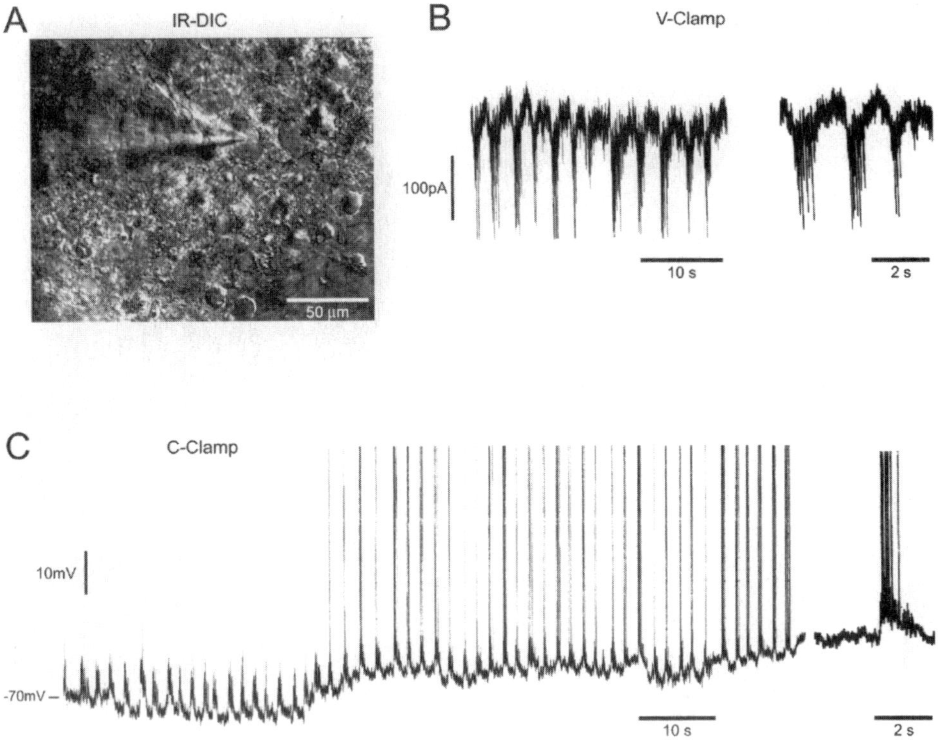

Figure 2. Whole cell patch-clamp recording from a cultured pre-Bötzinger island. **A.** High resolution IR-DIC image of patch pipette on a rhythmically active neuron (31.5x, 900nm IR filter). **B.** Voltage clamp record from cell seen in panel A. Synaptic drive is clearly preserved. **C.** Current clamp mode shows pattern of synaptic drive and action potential discharge during progressive depolarization.

The successful culture of the pre-Bötzinger complex is in keeping with preliminary observations in isolated pre-Bötzinger "islands" of fresh tissue slices, suggesting that the properties of the whole slice network is preserved in such reduced systems.[13] We have also cultured successfully the whole slice containing the respiratory network, but the numbers are small to determine whether the network properties of the whole slice are preserved in culture.

In summary, we have performed stationary organotypic cultures of the pre-Bötzinger region ("island"), obtained from newborn rats 0-4 days old. We found that this region could be successfully cultured up to three weeks, with maintenance of rhythmic neuronal activity. Immunostaining with antibodies to neurofilament 160 showed that the network is formed by a cluster of multipolar neurons. The findings suggest that this culture preparation may be useful for further characterization of the physiological properties of the neuronal network responsible for breathing.

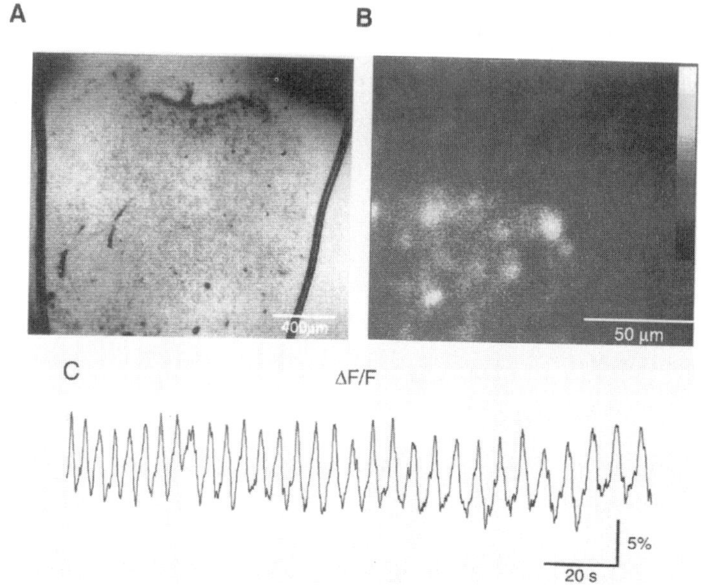

Figure 3. Ca^{2+} activity in a cultured pre-Bötzinger island. **A.** Low magnification view of the culture island (5x IR-DIC at 900 nm). **B.** High magnification (31.5) view under fluorescence showing region of rhythmic flashing (Pseudo-colored average under 0.3% incident light). **C.** Rhythmic change in calcium dependent fluorescence over several seconds. For a color representation of this figure, see color insert facing this page.

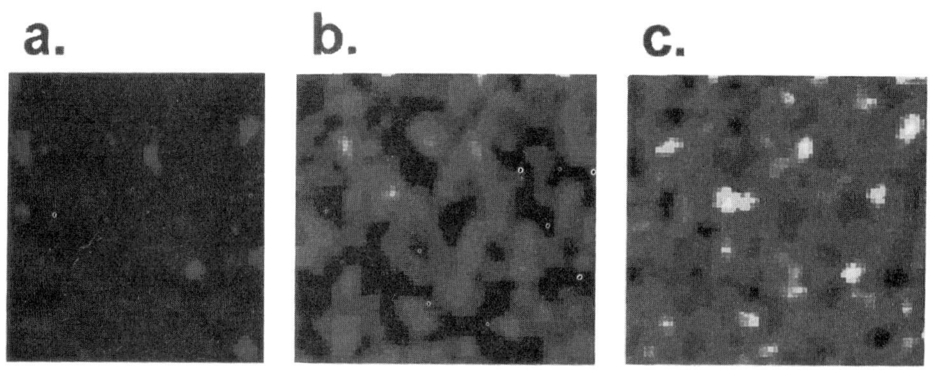

Figure 4. Real time video of the changes in fluorescence intensity with the respiratory cycle. Pre-Bötzinger island in culture for 14 days and bathed with Ca^{2+} green dye. (a) End-expiration, (b) Mid-inspiration, (c) End-inspiration. Note the pronounced increase in network activity during the inspiratory cycle. For a color representation of this figure, see color insert facing this page.

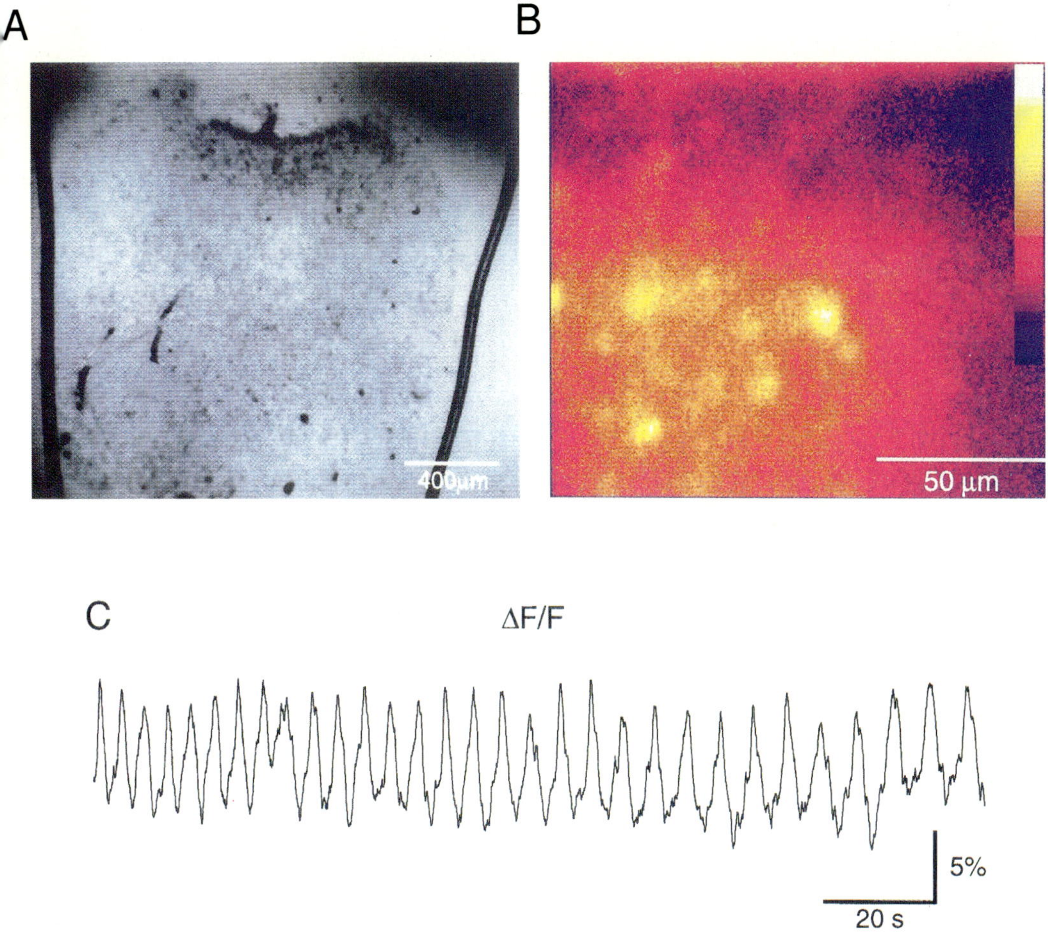

Figure 3. Ca^{2+} activity in a cultured pre-Bötzinger island. **A.** Low magnification view of the culture island (7.5x IR-DIC at 900 nm). **B.** High magnification (47.25x) view under fluorescence showing region of rhythmic flashing (Pseudo-colored average under 0.3% incident light). **C.** Rhythmic change in calcium dependent fluorescence over several seconds.

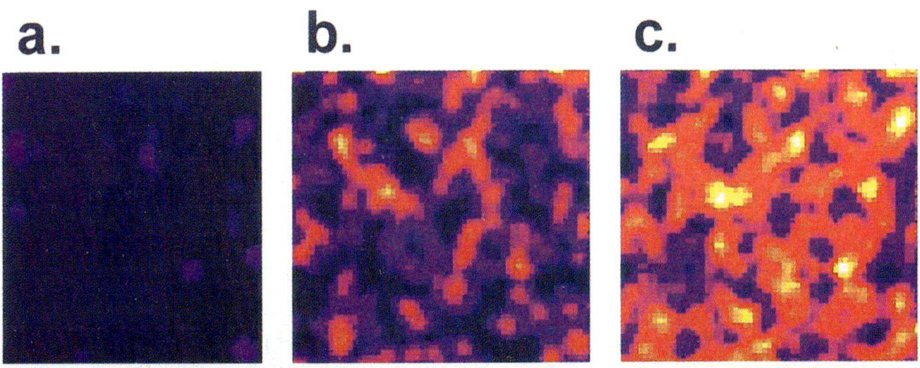

Figure 4. Real time video of the changes in fluorescence intensity with the respiratory cycle. Pre-Bötzinger island in culture for 14 days and bathed with Ca^{2+} green dye. (a) End-expiration, (b) Mid-inspiration, (c) End-inspiration. Note the pronounced increase in network activity during the inspiratory cycle.

5. ACKNOWLEDGMENT

This work was supported by The NIH and The Children's Hospital Foundation of Manitoba, Inc.

6. REFERENCES

1. S.C. Fitzgerald, M.A. Willis, C. Yu, and H. Rigatto, In search of the central respiratory neurons: I. Dissociated cell cultures of respiratory areas from the upper medulla, *J. Neurosci. Res.* **33**,579-589, 1992.
2. H. Rigatto, S.C. Fitzgerald, M.A. Willis, C. Yu, In search of the central respiratory neurons: II. Electrophysiologic studies of medullary fetal cells inherently sensitive to CO_2 and low pH, *J. Neurosci. Res.* **33**,590-597, 1992.
3. J.C. Smith, H.H. Ellenberger, K. Ballanyi, D.W. Richter, J.L. Feldman, Pre-Bötzinger complex: A brainstem region that may generate respiratory rhythm in mammals, *Science* **254**,726-729, 1991.
4. N. Koyhiya, J.C. Smith, Real time visualization of respiratory pacemaker neuron activity with calcium-sensitive dyes in the pre-Bötzinger complex in vitro, *Nature* **400**,360-363, 1999.
5. A.D. McClellan, D. McPherson, M.J. O'Donovan, Combined retrograde labelling and calcium imaging in spinal cord and brainstem neurons of the lamprey. *Brain Res.* **663**,61-68, 1994.
6. M. O'Donovan, S. Ho, W. Yee, Calcium imaging of rhythmic network activity in the developing spinal cord of the chick embryo. *J. Neurosci.* **14**,6354-6369, 1994.
7. B.H. Gahwiler, M. Capogna, D. Debanne, R.A. McKinney, S.M. Thompson, Organotypic slice cultures: a technique has come of age, *Trends Neurosci.* **20**,471-477, 1997.
8. S.B House, A. Thomas, K. Kusano, H. Gainer, Stationary organotypic cultures of oxytocin and vasopressinmagnocellular neurones from rat and mouse hypothalamus, *J. Neuroendocrinol.* **10**,849-861, 1998.
9. L. Stoppini, P.A. Buchs, D. Muller, A simple method for organotypic cultures of nervous system. *J. Neurosci. Meth.* **37**,173-182, 1991.
10. S. Inoue, K.R. Spring, Video Microscopy: The fundamentals, 2nd ed ,Plenum, New York; 1997.
11. S.M. Hsu, E. Soban, Color modification of diaminobenzidine (DAB) precipitation by metallic ions and its application for double immunohistochemistry, *J. Histochem. Cytochem.* **30**:1079-1082, 1982.
12. S.M. Johnson, C.O. Trouth, J.C. Smith, Chemosensitivity of respiratory pacemaker neurons in the pre-Bötzinger complex in vitro, *Soc. Neurosci Abstr.* **24**, 875, 1998.
13. S.M. Johnson, N. Koshiya, C.A. Del Negro, J.C. Smith, Characterization of the spontaneous respiratory rhythm generated in a novel in vitro preparation. The pre-Bötzinger complex island. *Soc. Neurosci Abstr.* **26**,927, 2000.

RESPIRATORY RHYTHM GENERATION: PREBÖTZINGER NEURON DISCHARGE PATTERNS AND PERSISTENT SODIUM CURRENT

Donald R. McCrimmon, Armelle Monnier, Krzysztof Ptak, Greer Zummo, Zhong Zhang, and George F. Alheid*

1. INTRODUCTION

Neurons within the preBötzinger complex (pBc) are postulated to have a central role in respiratory rhythm generation. This postulate is supported by studies using both *in vitro*[1] and *in vivo* preparations.[2] Consistent with a role in rhythm generation, many of the neurons within this region discharge with active periods that span the phase transitions between expiration (E) and inspiration (I).[3, 4]

A difficulty with the analysis of the preBötzinger region is distinguishing recordings made within the pBc versus adjacent respiratory regions (*i.e.*, the Bötzinger complex and rostral ventral respiratory group, rVRG). Recently, Gray et al[5] determined that neurokinin-1 (NK1) receptors provide an anatomical marker for the pBc. Several laboratories have also shown that excitatory amino acid injections into the pBc elicit stereotypic changes in respiratory motor output.[6-9] These injections typically result in a tachypnea, although the exact response has varied somewhat amongst laboratories.[7-9]

We further explored the contribution of pBc neurons to respiratory rhythm generation in two series of studies. *In vivo*, respiratory rhythm was perturbed by unilateral microinjections into the pBc of either an excitatory amino acid, DL-homocysteate (DLH), or a GABA receptor antagonist, bicuculline (Bic), while recording the activity of single respiratory neurons in the contralateral pBc. The discharge pattern of neurons with a causal role in respiratory rhythm generation were expected to maintain their phase relationship to phrenic nerve discharge despite large changes in respiratory rhythm. In contrast, this fixed relationship would be less likely in neurons with discharge patterns not causally related to rhythm generation.

A slowly inactivating persistent sodium current is postulated to underlie the pacemaker activity seen in a subset of pBc neurons.[10, 11] Koshiya and Smith[12] reported

* Department of Physiology and Institute for Neuroscience, Northwestern University Medical School Chicago, Illinois 60611-3008.

Frontiers in Modeling and Control of Breathing, edited by Poon and Kazemi. Kluwer Academic/Plenum Publishers, New York, 2001.

preliminary evidence showing the presence of persistent sodium current in pacemaker neurons within the pBc. Therefore, in a second study, we examined the potential role of persistent sodium current in generating pacemaker activity in pBc neurons. The hypothesis that a persistent sodium conductance is most strongly expressed in a subset of neurons (i.e. pacemakers) within the pBc was tested in preliminary experiments using whole-cell voltage clamp. RT-PCR was used to determine the presence or absence of mRNAs for sodium channel subunit mRNAs (Nav1.1, 1.2 and 1.6). In addition, the presence of mRNA for NK1 and μ-opioid receptors was used to confirm that the sampled neurons were derived from the pBc region.

2. RESPIRATORY NEURONAL RESPONSES TO PERTURBATION OF RESPIRATORY RHYTHM

Adult male Sprague-Dawley rats (300–500 g) were anesthetized with isoflurane and maintained with urethane (1.5g/kg). Rats were bilaterally vagotomized, mechanically ventilated with O_2 and paralyzed (0.6 mg D-tubocurarine). End-tidal CO_2 was estimated from expired CO_2 and maintained at 34-40 mm Hg. Rectal temperature was monitored and maintained at 37.5 ± 1°C. A phrenic nerve was cut and recorded with a bipolar electrode. Rats were placed in a stereotaxic frame, and an occipital craniotomy performed. Single and double barrel glass micropipettes were used for both pressure delivery of drugs and recording neuron activity. Single barrel pipettes, and 1 barrel of the double barrel pipettes, were filled with DLH (5 mM in 150 mM NaCl). The second barrel of each double barrel pipette was filled with Bic (0.5-1 mM in 150 mM NaCl). For recording and drug injection, a double barrel pipette was positioned in the pBc on one side and a single barrel pipette was inserted into the contralateral pBc. Injection volumes (2-6 nl) were monitored by measuring the movement of the fluid meniscus in the pipette using a compound microscope equipped with fine reticule.

Forty-nine respiratory neurons were recorded during injections of DLH or Bic into the contralateral pBc. Injection of ≤ 3 nl of DLH at, or immediately proximal to the recording site (≤ 100 μm) elicited tachypnea, consistent with an injection into the pBc.[9] Associated with this change in rhythm was a reduction in the peak amplitude of the phrenic nerve discharge. This finding is consistent with the prediction by Butera and coworkers[13] that increasing the excitatory drive of the rhythm generating network would decrease the peak amplitude of the motor output and increase respiratory frequency.

Similar to previous descriptions by Sun et al,[3] a large fraction of neurons (23/49, ~47%) within the pBc had discharge patterns that spanned at least one of the phase transitions between I and E. Sixteen of these (pre-I neurons) began to discharge in the second half of E with an augmenting pattern that rapidly rose to a peak frequency at the onset of I (Fig. 1). Seven neurons fired throughout the respiratory cycle but with relatively low discharge frequencies that were modulated in phase with I (n=5) or E (n=2). Of the remaining 26 neurons, 12 were I and 16 were E neurons. Nine of the E neurons had decrementing pattern of discharge (E-Dec). Of particular note, both E-Dec and augmenting E neurons discharged throughout the expiratory phase, and hence were coactive during the entire E phase. This finding is consistent with the work of Lindsey and coworkers[14] in the rVRG of cats and suggests that, at least in pBc neurons, there is not a clear distinction between phase 1 (post-I) and phase 2 (late-E).

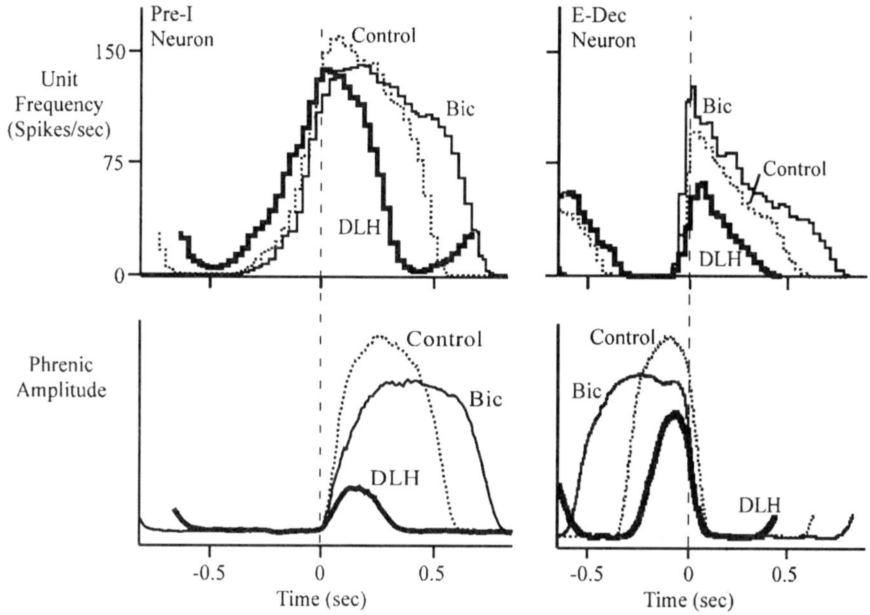

Figure 1. Maintained phase relationship of the discharge pattern of expiratory-inspiratory phase spanning (Pre-I) and expiratory-decrementing (E-Dec) neurons during marked alterations of respiratory rhythm by injection of DL-homocysteate (DLH) or bicuculline (Bic) into the pBc region.

In response to perturbation of the respiratory rhythm, most pBc neurons exhibited variable responses (*e.g.*, see Fig. 2 in McCrimmon et al[9]) with changes in both phase and magnitude of discharge. For example, neurons with an augmenting I pattern of discharge were almost completely silenced during DLH injection. In contrast, as shown in Fig. 1, substantial numbers of pre-I (10/16) and E-Dec (9/9) neurons maintained a tight phase relationship with phrenic nerve discharge. Indeed, the pre-I neurons could be divided into 2 groups, one in which the tight phase relationship was maintained. Of note in this group was that they not only maintained a tight phase relationship with phrenic nerve activity but they also exhibited little change in peak firing rate despite large changes in peak phrenic activity (Fig. 1). A subset of E-Dec neurons (4/9) not only maintained their phase relationship to phrenic nerve activity but exhibited a change in peak firing frequency that was positively correlated with expiratory duration (Fig. 1). This latter observation is consistent with their proposed role in the control of expiratory duration.[15, 16] Thus, subsets of 2 groups of pBc neurons (pre-I and E-Dec) stand out as maintaining a close phase relationship with the phrenic nerve discharge and hence are the most likely to have central roles in controlling respiratory rhythm.

3. VRG NEURONS EXPRESS PERSISTENT SODIUM CURRENT AND Nav1.1, 1.2 AND 1.6 CHANNELS

Neurons were obtained from newborn and young (0 -14 days old) Sprague-Dawley rats and acutely dissociated using a procedure described by others.[17] Animals were anaesthetized using isoflurane and decapitated. The brainstem was rapidly removed, immersed in Tyrode's solution (37° C) containing (in mM): 150 NaCl, 4 KCl, 2 $CaCl_2$, 2 $MgCl_2$, 10 HEPES, 10 glucose, pH 7.4, 300 ± 5 mOsm/l. Sagittal Vibratome slices (300μm) were placed in an oxygenated chamber containing papain (40 U/ml), 1 mM L-cysteine, 0.5 mM EDTA in HEPES-buffered (pH 7.2) Minimum Essential Medium (MEM, Life Technologies) at 37° C for 20 min. The tissue was then washed 3 times in MEM solution containing 1 mg/ml bovine serum albumin and 1 mg/ml trypsin inhibitor. The pBc region (ventral to the semicompact formation of the nucleus ambiguus) was dissected from the slice and triturated with a fire-polished Pasteur pipette.

For RT-PCR analysis of the presence of Nav 1.1, 1.2 and 1.6 subunits, tissue samples were obtained from the pBc, the facial nucleus, and the rVRG. Cellular disruption was achieved in lysis buffer with a sterile pestle and was followed by homogenization and centrifugation. The lysate was collected and RT was performed using a standard protocol (Qiagen). PCR amplification of the sodium channel α subunits Nav 1.1, 1.2 and 1.6 used the primer sets described by Maurice et al.[18]

Whole-cell recordings were made at room temperature. Borosilicate pipettes (4-6 M Ω) were filled with (in mM): 117 CsCl, 9 EGTA, 9 HEPES, 1.8 $MgCl_2$, 14 Tris-creatine PO_4, 4 MgATP, 0.3 Tris-GTP, pH 7.2 with CsOH, 260-270 mOsm/l. Sodium currents were recorded using an external solution containing (in mM): 50 NaCl, 110 TEA-Cl, 10 HEPES, 0.3 $CdCl_2$, 2 $BaCl_2$, pH 7.4 with NaOH, 300 ± 5 mOsm. To confirm that recorded currents were due to conductance through Na^+ channels, protocols were repeated with an external solution containing 300 nM tetrodotoxin (TTX).

Ventral medullary neurons (pBc = 25; rVRG = 6) were examined for the presence of both transient and persistent sodium currents. Transient Na^+ current was elicited from a holding potential of –80 mV with 40 ms steps to depolarizing voltages up to +20 mV in 5 mV increments. The peak transient current (1.7 ± 0.1 nA; n = 9) was elicited between –20 and –25 mV.

Figure 2 shows examples of the persistent Na^+ current produced with a ramp

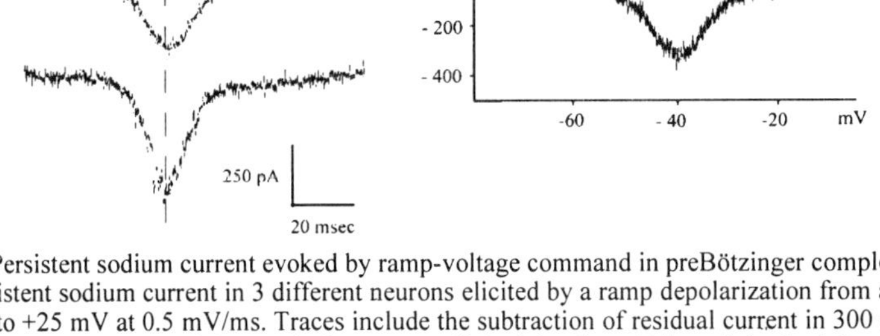

Figure 2. Persistent sodium current evoked by ramp-voltage command in preBötzinger complex neurons. Left panel: Persistent sodium current in 3 different neurons elicited by a ramp depolarization from a holding potential of –80 to +25 mV at 0.5 mV/ms. Traces include the subtraction of residual current in 300 nM TTX. Vertical line indicates –40 mV. Right panel: the current elicited for the middle cell in the left panel, before and after block of sodium currents by TTX (300 nM).

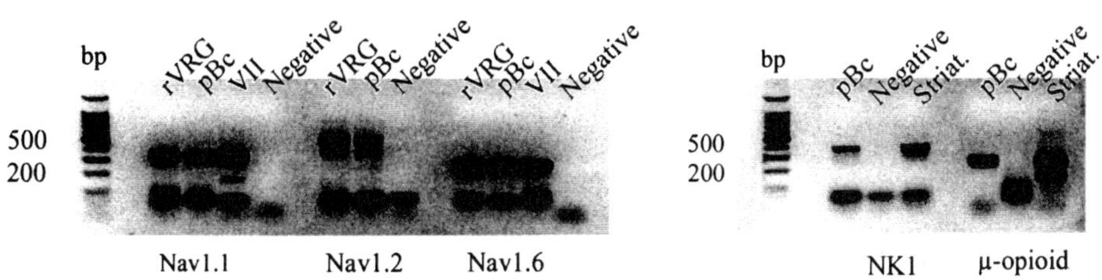

Figure 3. Expression of sodium channel mRNA in the rVRG, and preBötzinger complex (pBc) revealed by RT-PCR. Nav 1.1, 1.2 and 1.6 α Na^+ channel subunit mRNAs are coexpressed. The mRNAs for Nav 1.1 and 1.6 were also detected in the facial nucleus (VII). The mRNA for both the Neurokinin-1 receptor (NK1) and the μ-opioid receptor (μ-opioid) were also present in the pBc. Negative = the reagent control, striat = positive control from the striatum. Left lane shows 100 bp ladder.

depolarization (0.5 mV/ms) from –80 mV to +25 mV. These examples illustrate the full range (195 – 455 pA; n = 10) of current elicited in pBc neurons. An inward current was initiated at –55 mV and increased to a peak (302 ± 36 pA) at 39.0 ± 0.6 mV. This current was completely blocked by the addition of TTX (300 nM).

Tissue from the pBc, rVRG and the facial motor nucleus were examined for the presence of Nav 1.1 and 1.6 mRNAs – the α subunits that have been linked to prominent persistent currents in heterologous systems.[19] Both mRNAs were expressed in all 3 tissues (Fig. 3). The mRNA for a third α subunit known to be expressed in brain (Nav 1.2) was also present in both the pBc and rVRG. Its presence in the facial nucleus was not examined. Neurokinin-1 and μ-opioid receptors are also known to be present on pBc neurons[5] and the mRNA for both of these receptors was found in pBc tissue (Fig. 3) corresponding to the area from which neurons were sampled for the recording of persistent Na^+ current.

4. SUMMARY

Considerable evidence from several laboratories (*c.f.*, Rekling and Feldman,[20] Ramirez et al.[21]) is consistent with the concept that the pBc contains the kernel of the central rhythm generating network for breathing. The work summarized in this manuscript is also generally consistent with this notion. Of particular note is the observation that pre-I neurons and E-Dec neurons maintain a consistent phase relationship with phrenic nerve activity and maintain a similar peak discharge rate despite marked changes in the phrenic nerve rhythm and pattern. Other categories of respiratory neurons failed to maintain this relationship. Hence, the findings are consistent with pBc pre-I and E-Dec neurons having a key role in rhythm generation.

A persistent sodium current has been postulated to underlie the rhythm generating mechanism of pacemaker neurons within the pBc.[10] In the present study, a substantial persistent sodium current was documented in many neurons from the pBc and adjacent respiratory regions. This finding is not inconsistent with the postulated role in rhythm generation. However, it does suggest that other neuronal properties must act in concert with the persistent current to define a unique population of pacemaker neurons.

5. ACKNOWLEDGMENTS

This work was supported by NIH grants HL/NS-60097 and HL/NS- 60969. We express our thanks to Sasha Ulrich and Drs. D. James Surmeier and Tatiana Tkatch for help with the RT-PCR protocols and Drs. Indira Raman and N. Traverse Slater for assistance with recording from dissociated neurons.

6. REFERENCES

1. J.C. Smith, H.H. Ellenberger, K. Ballanyi, D.W. Richter, and J.L. Feldman, Pre-Bötzinger Complex: A brainstem region that may generate respiratory rhythm in mammals, *Science* **254**: 726-729 (1991).
2. J.M. Ramirez, S.W. Schwarzacher, O. Pierrefiche, B.M. Olivera, and D.W. Richter, Selective lesioning of the cat pre-Bötzinger complex *in vivo* eliminates breathing but not gasping, *J. Physiol.(Lond.)* **507**: 895-907 (1998).
3. Q.J. Sun, A.K. Goodchild, J.P. Chalmers, and P.M. Pilowsky, The pre-Bötzinger complex and phase-spanning neurons in the adult rat, *Brain Res.* **809**: 204-213 (1998).
4. S.W. Schwarzacher, J.C. Smith, and D.W. Richter, Pre-Bötzinger complex in the cat, *J. Neurophysiol.* **73**: 1452-1461 (1995).
5. P.A. Gray, J.C. Rekling, C.M. Bocchiaro and J.L. Feldman, Modulation of respiratory frequency by peptidergic input to rhythmogenic neurons in the preBötzinger complex, *Science* **286**: 1566-1568 1999.
6. F.P. Tolentino-Silva, R.R., Jr. Campos, A.K. Russo, S.L. Cravo, and O.U. Lopes, Cardiorespiratory effects of L-glutamate microinjected into the rat ventral medulla, *Respir. Physiol.* **108**: 23-33 (1997).
7. V.C. Chitravanshi and H.N. Sapru, Phrenic nerve responses to chemical stimulation of the subregions of ventral medullary respiratory neuronal group in the rat, *Brain Res.* 821: 443-460 (1999).
8. I.C. Solomon, N.H. Edelman, and J.A. Neubauer, Patterns of phrenic motor output evoked by chemical stimulation of neurons in the pre-Bötzinger complex *in vivo*, *J. Neurophysiol.* **81**:1150-1161 (1999).
9. D.R. McCrimmon, A. Monnier, F. Hayashi, and E.J. Zuperku, Pattern formation and rhythm generation in the ventral respiratory group, Clin. Exp. Pharmacol. Physiol. 27: 126-131 (2000).
10. R.J., Jr. Butera, J. Rinzel, and J.C. Smith, Models of respiratory rhythm generation in the pre-Bötzinger complex. I. Bursting pacemaker neurons, *J. Neurophysiol.* **82**: 382-397 (1999).
11. J.C. Smith, R.J. Butera, N. Koshiya, C. Del Negro, C.G. Wilson, and S.M. Johnson, Respiratory rhythm generation in neonatal and adult mammals: the hybrid pacemaker-network model, *Respir Physiol.* **122**: 131-47. 2000.
12. N. Koshiya, and J.C. Smith, Subthreshold voltage-activated persistent sodium conductance (g(NαP)) in inspiratory pacemaker neurons of pre-Bötzinger complex, *FASEB J.* **14**: A642. (2000).
13. R.J., Jr. Butera, J. Rinzel, and J.C. Smith, Models of respiratory rhythm generation in the pre-Bötzinger complex. II. Populations of coupled pacemaker neurons, *J. Neurophysiol.* **82**: 398-415 (1999).
14. B.G. Lindsey, L.S. Segers, and R. Shannon, Discharge patterns of rostrolateral medullary expiratory neurons in the cat: Regulation by concurrent network processes, *J. Neurophysiol.* **61**: 1185-1196 (1989).
15. D.W. Richter, D. Ballantyne, and J.E. Remmers, How is the respiratory pattern generated? A model, *News Physiol. Sci.* **1**: 109-112 (1986).
16. F. Hayashi, S.K. Coles, and D.R. McCrimmon, Respiratory neurons mediating the Breuer-Hering reflex prolongation of expiration in rat, *J. Neurosci.* **16**: 6526-6536 (1996).
17. I.M. Raman, A.E. Gustafson, and D. Padgett, Ionic currents and spontaneous firing in neurons isolated from the cerebellar nuclei. *J. Neurosci.* **20:** 9004-9016 (2000).
18. N. Maurice, T. Tkatch, M. Meisler, L. Sprunger, and D.J. Surmeier, D1/D5 Dopamine receptor activation differentially modulates rapidly inactivating and persistent sodium currents in prefrontal cortex pyramidal neurons. *J. Neurosci.* (Submitted).
19. A.L. Goldin, Diversity of mammalian voltage-gated sodium channels, *Ann. NY Acad. Sci.* **868**: 38-50 (1999).
20. J.C. Rekling and J.L. Feldman, PreBötzinger complex and pacemaker neurons: Hypothesized site and kernel for respiratory rhythm generation, *Annu. Rev. Physiol.* **60**: 385-405 (1998).
21. J.M. Ramirez, P. Telgkamp, F.P. Elsen, U.J.A. Quellmalz, and D.W. Richter, Respiratory rhythm generation in mammals: synaptic and membrane properties, *Respir. Physiol.* **110**: 1-85 (1997).

ROLES OF THE BÖTZINGER COMPLEX IN THE FORMATION OF RESPIRATORY RHYTHM

Gang Song, Qin Li, and Ming Lü*

1. INTRODUCTION

The Bötzinger complex (Böt.C) is a group of expiratory neurons in the vicinity of retrofacial nucleus.[1-5] Electrophysiological and histological studies have shown that this group of neurons is inhibitory in nature and has widespread axonal projections to medullary and spinal respiratory neurons.[6-8] Electrical and chemical stimulation of the Böt.C resulted in strong inhibition of inspiration.[9, 10] Unilateral focal cold block of this structure in cat produced shortening of expiratory duration and increase of respiratory rate.[11] These findings suggest that the Böt.C may be of critical importance in the formation of respiratory rhythm.[12, 13] In order to test this hypothesis, we examined the roles of the Böt.C in the respiratory phase transition[14] and expiratory phase formation.[15]

2. INSPIRATORY INHIBITION AND "REBOUND" DISCHARGES OF PHRENIC NERVE CAUSED BY ELECTRICAL STIMULATION AT THE BÖT.C

Experiments were done on 8 pentobarbital (Nembutal™, initial dosage of 30-40 mg/kg, i.p., supplemented by 3-5 mg/kg per h, i.v.) anesthetized adult rabbits of either sex, weighing 2.0-3.0 kg. After tracheal cannulation, the head of the animal was fixed on a stereotaxic frame. Then the dorsal surface of the brain stem was exposed by cutting the posterior rim of the occipital bone and removing the central part of cerebellum by suction. Tungsten microelectrodes (shaft diameter 70 µm, tip electrically etched to 1-3 µm) were used for extracellular recording and stimulation. The C_5 branch of phrenic nerve was exposed, desheathed, and mounted on a bipolar silver electrode for recording discharges. Blood pressure was monitored and maintained at 100-120 mmHg (mean). Rectal temperature was kept at 36-38°C.

At the end of each experiment, direct current was passed through the tungsten microelectrode to mark the stimulating points. Then the animals were sacrificed by an overdose of pentobarbital (60 mg/kg, i.v.) and transcardially perfused with 10% formalin.

* Department of Physiology, Shandong University, Wenhua Xilu-44, Jinan, Shandong, P.R. China.

Frontiers in Modeling and Control of Breathing, edited by Poon and Kazemi.
Kluwer Academic/Plenum Publishers, New York, 2001.

The brains were removed, post-fixed, and cut into 100-μm frozen sections for histological verification of stimulating points.

The Böt.C was first delineated with extracellular recording. Electrical stimulation was then delivered at sites where multi-units expiratory discharges were recorded. The results were as follows. 1) Short train (5 pulses, 100 Hz, 0.010-0.10 mA) stimulation delivered during inspiratory phase could cause transient or complete termination ("inspiratory off-switch") of the phrenic inspiratory discharges. There was a negative linear correlation between the threshold intensity and the delivery time of the stimulating train (r=-0.88, P<0.001). The latency (from the beginning of the stimulation to the complete termination of phrenic discharges) of this "inspiratory off-switch" was 4.7±0.3 msec. After the stimulation, the phrenic nerve usually showed a short lasting (25-35 msec) "rebound" discharges, which could also be inhibited by another properly timed short train stimulation. The latency (from the beginning of the stimulation to the beginning of the "rebound" discharge) of this "rebound" discharges was 86±3.7 msec. 2) Short train stimulation delivered in the expiratory phase caused a short-lasting "rebound" discharges similar to that mentioned above. The amplitude and duration of the "rebound" discharges that caused by stimulating train delivered in late expiration were larger and longer than that caused by stimulating train delivered in early expiration. (Figure 1)

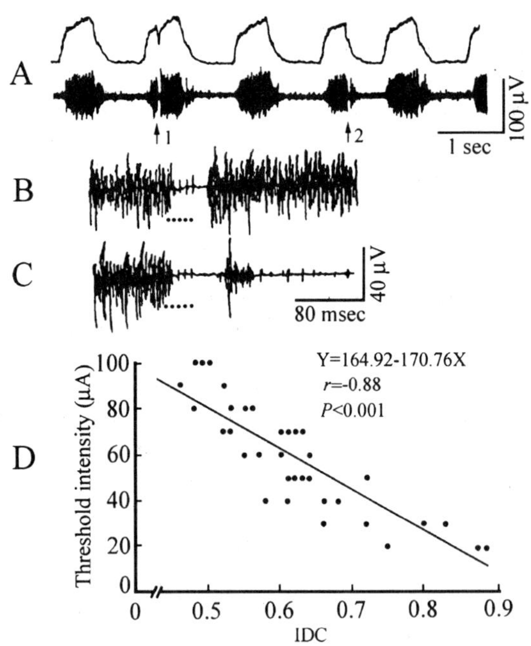

Figure 1. Transient inhibition of phrenic discharge and "inspiratory off-switch" by short train stimulation at the Böt.C in inspiratory phase. A, short train stimulation (0.030 mA, 5 pulses) at 0.2 sec (arrow 1) and 0.4 sec (arrow 2) after the onset of inspiration. Upper trace: integrated phrenic discharge; lower trace: original phrenic discharge. B, a higher sweep speed showing the transient inhibition of A (arrow 1). Dot indicates stimulating pulse. C, a higher sweep speed showing "inspiratory off-switch" of A (arrow 2). D, relation between the delivery time of stimulus and threshold for "inspiratory off-switch". IDC (inspiratory delay constant) = time from the onset of inspiration / inspiratory time.

3. TONIC PHRENIC DISCHARGES FOLLOWING BILATERAL MICRO-INJECTION OF GLYCINE OR LIDOCAINE INTO THE BÖT.C

Experiments were done on 16 pentobarbital (Nembutal™, initial dosage of 30-40 mg/kg, i.p., supplemented by 3-5 mg/kg per h, i.v.) anesthetized adult rabbits. Animals were paralyzed with gallamine triethiodide and artificially ventilated. Other preparation was the same as described above. In some animals, the hypoglossal nerve was also exposed and recordings made. For microinjection, a double-barreled glass micropipette was made, in which one barrel of the pipette was filled with glycine or lidocaine solution and connected to a 1 μL Hamilton microsyringe. A tungsten filament was sealed into the other barrel with the tip exposed and was used for extracellular recording of neuronal activity. The Böt.C was first delineated with extracellular recording. Microinjections were made at sites where multi-units expiratory discharges were recorded. Responses of phrenic discharges to unilateral and bilateral microinjection of glycine at the Böt.C were observed. The results were as follows.

Unilateral microinjection of glycine (Sigma, 50-100 nL, 0.1 mol/L, in ACSF) into the Böt.C shortened the expiratory phase and increased the phrenic expiratory discharges, and had no significant effect on the phrenic inspiratory discharges. Another injection into the contralateral Böt.C caused further increases in phrenic expiratory discharges at first, and then made the expiratory phase undistinguishable. Now the phrenic nerve showed tonic discharges with no observable rhythm. The amplitude of the tonic phrenic discharges was about 40-70% of the control level. The hypoglossal nerve, which showed respiratory rhythmic discharges before injection, also lost its respiratory rhythm and became continuously discharging. Phrenic nerve discharges showed similar responses to microinjection of lidocaine into unilateral or bilateral Böt.C. (Figure 2)

Pre-microinjection of strychnine (Sigma, 100-300 nL, 2.5 mmol/L) into the Böt.C blocked the effect of glycine, but not the lidocaine. Microinjection of strychnine itself decreased the inspiratory amplitude and increased the expiratory duration. (Table 1).

Control injections into sites 2 mm lateral to the Böt.C had no significant effects on phrenic discharges. Control injections of ACSF (with equal osmolarity and pH to the glycine solution) into unilateral or bilateral Böt.C caused transient fluctuation in respiration, but no statistically significant change was observed.

Table 1. Effects of microinjection of strychnine into bilateral Böt.Cs on different respiratory parameters

	Percentage of changes (n=12)				
	30 s	5 min	10 min	15 min	40 min
T_I	-0.46±1.63	1.83±5.60	4.86±7.19	1.33±2.74	0.98±2.00
T_E	3.08±3.11*	10.91±2.84**	17.38±3.91**	14.77±3.41**	0.73±3.06
iPhr	4.89±4.14*	-12.79±4.83**	-14.40±5.26**	-13.10±5.43**	-0.15±4.48
RR	-1.47±1.50	-6.42±2.19**	-10.65±2.37**	-8.12±1.17**	-0.75±2.29

All values are means±SD. *P<0.05, **P<0.01. T_I, inspiratory time (s); T_E, expiratory time (s); iPhr, integrated phrenic nerve discharges (arbitrary units); RR, respiratory rate (breaths/min).

Long train electrical stimulation (lasting 3-6 sec, 0.05 mA) at the Böt.C could inhibit the tonic phrenic discharges caused by bilateral microinjection of glycine. After the cessation of the stimulation, several cycles of rhythmic discharges appeared. Then the phrenic nerve resumed tonic discharges. Electrical stimulation at the central end of the cervical vague nerve also inhibited the tonic phrenic discharges, but rhythmic discharges were not observed after the cessation of the stimulation.

In the present study we found that bilateral inhibition[16-18] or blocking of the Böt.Cs could eliminate expiratory phase and produce inspiratory apnea. Artificial activation of the Böt.C could transiently resume the respiratory rhythm. These results suggested that the Böt.C was of critical importance in the formation of expiratory phase as well as respiratory rhythm. The Böt.C may also be of importance in the phase transition from expiration to inspiration. The abrupt cessation of activity of the Böt.C that occurs naturally at the phase transition from expiration to inspiration when Bötzinger augmenting expiratory neurons suddenly stop discharging, or in artificial condition such as the cessation of stimulation, liberates the inspiratory neurons from the postsynaptic inhibition. This will result in "rebound" excitation of the inspiratory premotor or motor neurons,[13, 14] hence the starting of inspiration.

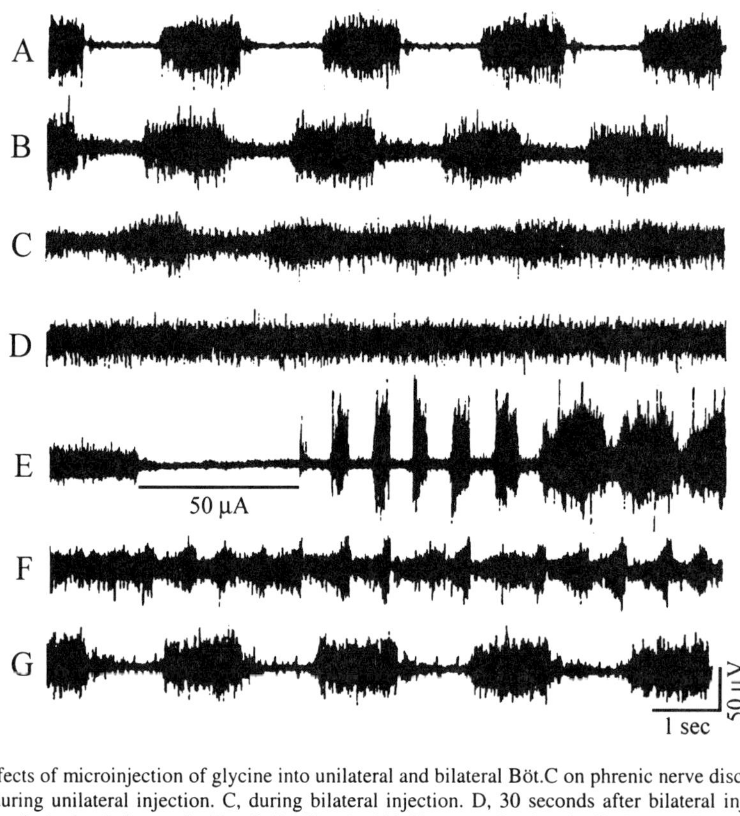

Figure 2. Effects of microinjection of glycine into unilateral and bilateral Böt.C on phrenic nerve discharges. A, control. B, during unilateral injection. C, during bilateral injection. D, 30 seconds after bilateral injection. E, effects of electrical stimulation at the Böt.C (0.050 mA, 100 Hz) on tonic phrenic discharges. F and G, 20 and 60 minutes after bilateral injections, respectively.

4. ACKNOWLEDGMENT

The authors thank Dr. Y.T. Wang (University of Toronto) for critically reading this article. Project was supported by a grant (No. 39970278) from China Natural Science Foundation.

5. REFERENCES

1. M. Kalia, J. L. Feldman, and M. I. Cohen, Afferent projection to the inspiratory region of the ventrolateral nucleus of the tractus solitarius, *Brain Res.* **171**, 135-141 (1979).
2. J. Lipski, and E. G. Merrill, Electrophysiological demonstration of the projection from expiratory neurones in rostral medulla to contralateral dorsal respiratory group, *Brain Res.* **197**, 521-524 (1980).
3. E. G. Merrill, J. Lipski, J. Kubin, and L. Fedorko, Origin of the expiratory inhibition of nucleus tractus solitarius inspiratory neurons, *Brain Res.* **263**, 43-50 (1983).
4. K. Otake, H. Sasaki, H. Mannen, and K. Ezure, Morphology of expiratory neurons of the Bötzinger complex: an HRP study in the cat, *J. Comp. Neurol.* **258**, 565-579 (1987).
5. C. Jiang, and J. Lipski, Extensive monosynaptic inhibition of ventral respiratory group neurons by augmenting neurons in the Bötzinger complex in the cat, *Exp. Brain Res.* **81**, 639-648 (1990).
6. G. Song, Y. Sato, I. Kohama, and M. Aoki, Afferent projections to the Bötzinger complex from the upper cervical cord and other respiratory related structures in the brainstem in cats: retrograde WGA-HRP tracing, *J. Auton. Nerv. Sys.* **56**, 1-7 (1995).
7. G. Song, A. Watanabe, and M. Aoki, in: *Advances in Modeling and Control of Ventilation*, edited by Hughson et al (Plenum Press, New York, 1998), pp. 67-72.
8. G. F. Tian, and J. Duffin, Spinal connections of ventral-group bulbospinal inspiratory neurons studied with cross-correlation in the decerebrated rat, *Exp. Brain Res.* **111**, 178-186 (1996).
9. G. Song, L. Liu, Reappraisal of the inspiratory effect of Bötzinger complex on phrenic nerve discharge, *Respir. Physiol.* **105**, 17-21 (1996).
10. F. Bongianni, G. Fontana, and T. Pantaleo, Effects of electrical and chemical stimulation of the Bötzinger complex on respiratory activity in the cat, *Brain Res.* **445**, 254-261 (1988).
11. K. Budzinska, C. von Euler, F.F. Kao, T. Pantaleo, and Y. Yamato, Effects of graded focal cold block in rostral area of the medulla, *Acta Physiol. Scand.* **124**, 239-340 (1985).
12. G. Song, Bötzinger complex and the formation of respiratory rhythm, *Progress Physiol. Sci.* **30**, 237-240 (1999).
13. A. L. Bianchi, M. Denavit-Saubie, and J. Champagnat, Central control of breathing in mammals: neuronal circuitry, membrane properties, and neurotransmitters, *Physiol. Rev.* **75**, 1-45 (1995).
14. Q. Li, M. Lü, G. Song, H. Zhang, L. Liu, Studies on respiratory phase switching effects of Bötzinger complex, *Acta Physiol. Sin.* **51**, 96-100 (1999).
15. M. Lü, Q. Li, G. Song, H. Zhang, L. Liu, Effects of glycine and strychnine microinjected into unilateral and bilateral Bötzinger complex on phrenic nerve discharges in rabbits, *Acta Physiol. Sin..* **50**, 693-697 (1998).
16. A. Haji, R. Takeda, and J.E. Remmers, Evidence that glycine and GABA mediated postsynaptic inhibition of bulbar respiratory neurons in the cat, *J. Appl. Physiol.* **73**, 2333-2342 (1992).
17. K. Schmid, G. Bohmer, and K. Gebauer, Glycine receptor-mediated fast synaptic inhibition in the brainstem respiratory system, *Respir. Physiol.* **84**, 351-361 (1991).
18. A. C. Bonham, Neurotransmitters in the CNS control of breathing. *Respir. Physiol.* **101**, 219-230 (1995).

MODELS OF NEURONAL BURSTING BEHAVIOR: IMPLICATIONS FOR *IN-VIVO* VERSUS *IN-VITRO* RESPIRATORY RHYTHMOGENESIS

Ilya A. Rybak, Walter M. St. John, and Julian F. R. Paton*

1. INTRODUCTION

Neonatal *in-vitro* preparations can generate respiratory-like oscillations based on the intrinsic bursting or pacemaker properties in some neurons within the medullary pre-Bötzinger complex (PBC)[1-3]. This discovery provided a basis for a *hybrid pacemaker-network* theory that suggests that pacemaker PBC neurons form a kernel of the central pattern generator and hence provide a *necessary* contribution to the genesis of the respiratory rhythm[4]. However, it is controversial whether a pacemaker-based mechanism necessarily operates *in vivo.* Specifically, the pacemaker-based models face principal problems in providing explanations to systems-level phenomena such as an independent regulation of respiratory phases, respiratory reflexes (e.g. the Hering-Breuer reflex), various phase resetting phenomena. On the other hand, the *network* theories and models based on *in-vivo* data can explain these phenomena[5, 6], but demand the respiratory phase transitions to be based on the reciprocal inhibitory interactions between respiratory neurons. Therefore, the network-based models have so far been unable to explain a pacemaker-driven rhythm recorded *in vitro* which is resistant to blockade of synaptic inhibition[7]. The current state of knowledge in the field requires a comprehensive computational study of the relationships between *in-vivo* and *in-vitro* data. The objective of this study was to explore the concept that the respiratory network can generate a breathing pattern by *either* a network *or* a hybrid pacemaker-network mechanism, which is state-dependent.

* Ilya A. Rybak, School of Biomedical Engineering, Science and Health Systems, Drexel University, Philadelphia, PA 19104. Walter M. St. John, Department of Physiology, Dartmouth Medical School, Lebanon, NH 03756. Julian F. R. Paton, Department of Physiology, School of Medical Sciences, University of Bristol, Bristol BS8 1TD, UK

Frontiers in Modeling and Control of Breathing, edited by Poon and Kazemi.
Kluwer Academic/Plenum Publishers, New York, 2001.

2. INSIGHTS FROM *IN-VITRO* STUDIES

In vitro preparations typically generate a rhythmic output if the external potassium concentration $[K^+]$out is increased to at least 8-9 *mM.*[2] The question is: *why is the increase of $[K^+]$out necessary for bursting behavior?* The explanation that it compensates a reduction of neuronal excitability *in vitro* is unlikely, since both experimental[2] and modeling[8] studies showed that an increase of excitability (e.g. by applying additional depolarization) changed neuronal activity from bursting to tonic firing, i.e. produced an opposite effect. On the other hand, according to the Nernst equation an increase in $[K^+]$out to 8-9 *mM* shifts the reversal potential for potassium (Ek) 10-20 *mV* more positive and hence reduces all potassium currents by about 2 times. We suggest that *in vitro* this suppression of potassium currents (e.g. the transient K_A^+) provides the necessary conditions for bursting behavior in PBC neurons and therefore allows the generation of a pacemaker-driven rhythm. In contrast, under normal *in-vivo* conditions the K_A^+ current and phasic inhibition prevents the generation of pacemaker-driven oscillations.

3. INSIGHTS FROM MODELING STUDIES

Butera et al.[8] developed the first models of pacemaker neurons in the PBC, which closely captured the endogenous bursting activity recorded *in vitro*. These models however contained only a minimal set of ionic currents necessary for bursting behavior (Na_{fast}^+, Na_P^+, K_{DR}^+) but neither K_A^+ nor calcium currents. The intrinsic bursting was based on the activation of the persistent sodium current (Na_P^+). While developing our models, we used the models of Butera et al.[8] as a basis, but shifted the value of half-activated voltage for Na_P^+ current ($V_{1/2}$) from –40 *mV* [8] to the physiologically more realistic value of -49 *mV* [9]. In addition, we incorporated the K_A^+ current and in one case (Model 3) calcium (Ca_L^{2+}, Ca_T^{2+}) and calcium-dependent potassium ($K^+(Ca^{2+})$) currents.

Three distinct models of pacemaker neurons were developed (Figure 1). All models contained the Na_P^+ current whose activation could potentially initiate bursting. The major difference between models was the mechanism for burst termination. In Model 1 (Figure 1A1), burst termination was provided by slow inactivation of the Na_P^+ conductance (similar to Model 1 by Butera et al.[8]). In Model 2 (Figure 1B1), this termination was provided by the activation of the slowly activating potassium current K_S^+ (similar to Model 2 by Butera et al.[8]). In Model 3 (Figure 1C1), the burst termination mechanism was based on the high-threshold CaL^{2+} current (providing calcium influx and accumulation) and subsequent activation of the calcium-dependent potassium $K^+(Ca^{2+})$ current.

Shifting the $V_{1/2}$ for Na_P^+ to a more negative value (in comparison with Butera et al.[8] models) did not affect the neuron's ability to generate bursts. However, since Na_P^+ current was activated at a lower voltage level, it was found to interact with the K_A^+ current. As a result, at lower levels of $[K^+]$out (6 *mM*, Ek = -84 *mV*) the K_A^+ current suppressed the Na_P^+-dependent bursting behavior. The increase of $[K^+]$out to 9 *mM* (Ek = -72 *mV*) reduced the K_A^+ current and hence released bursting activity in all three models (see Figures 1A2, 1B2, and 1C2).

Our simulations also demonstrated that this bursting behavior might be disturbed by a preceding synaptic inhibition or cellular hyperpolarization in all models.

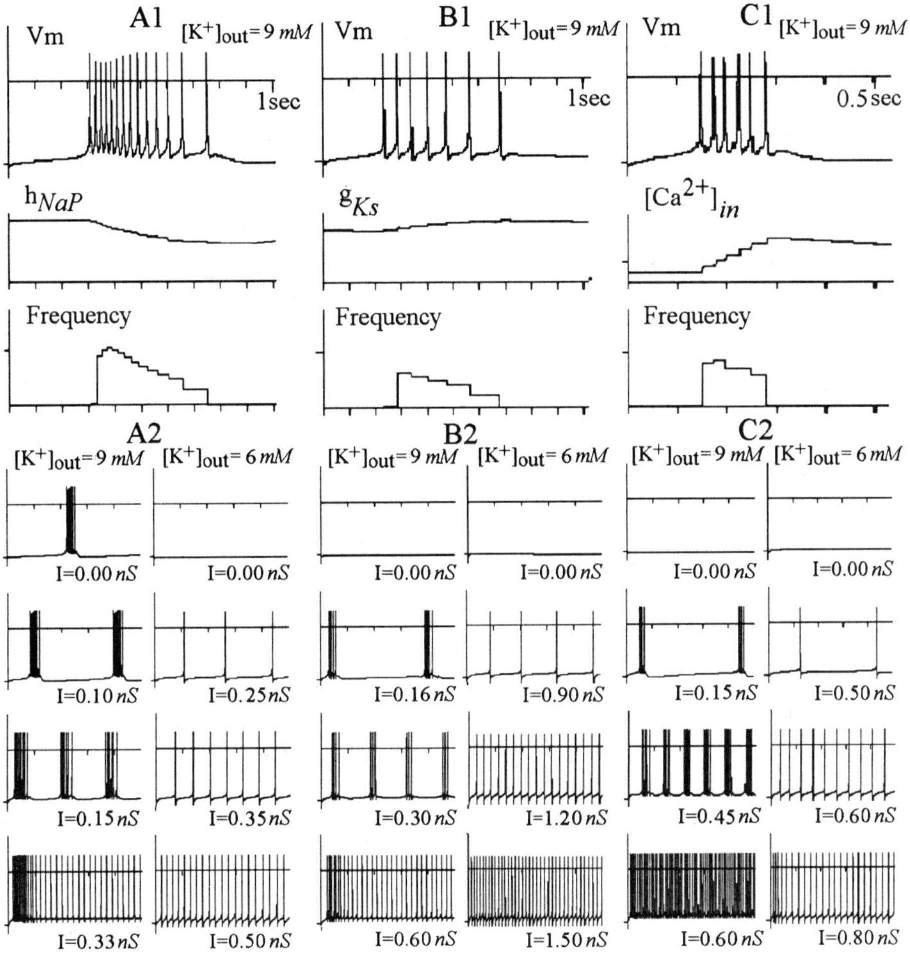

Figure 1. Firing patterns produced by different models of the pacemaker neuron: Model 1 (**A1**, **A2**), Model 2 (**B1**, **B2**), and Model 3 (**C1**, **C2**). Burst termination in Model 1 was provided by the slow inactivation (h_{NaP}) of the Na_P^+ conductance, which decreased during bursts and increased between bursts (**A1**). Burst termination in Model 2 was provided by the activation of the slowly activating potassium channels whose conductance - g_{Ks} increased during the bursts and decreased between bursts (**B1**). Burst termination in Model 3 was produced by the high-threshold calcium and calcium-dependent potassium currents. The intracellular calcium was accumulated during spikes and decayed between the spikes (**C1**). At 6 *mM* $[K^+]$out (Ek = -84 *mV*), all three models demonstrated tonic firing whose frequency increased with increased excitatory drive I (**A2-C2**, right columns). At 9mM $[K^+]$out (Ek = -72 *mV*), all models generate bursting (**A2-C2**, left columns). The burst frequency increased with increased excitatory drive. In all models, the bursting activity was switched to tonic firing at some critical value of input drive (**A2-C2**, left columns).

4. SWITCHING RESPIRATORY RHYTHMOGENESIS FROM A NETWORK TO A HYBRID PACEMAKER-NETWORK MECHANISM

A *switching* concept, explored in this study, suggests that under normal *in-vivo* conditions the respiratory network generates oscillations by a network mechanism. However, rhythmogenesis may be switched to a hybrid pacemaker-network mechanism under certain conditions. We believe that the value of $[K^+]$out is a critical parameter defining this switch at the cellular level. The normal value of $[K^+]$out, the transient K_A^+ and phasic inhibition all suppress the Na_P^+-dependent bursting activity in PBC pacemaker neurons. The increase of $[K^+]$out in *in-vitro* preparations reduces the K_A^+ and other potassium currents and hence releases endogenous bursting behavior.

At the network level the effect of this switching depends on the interconnections between the conditional pacemakers and other respiratory neurons. It is generally accepted that the respiratory network receives excitatory tonic drive from different sources including central and peripheral chemoreceptors (see Figure 2). The central chemoreceptors provide a major portion of this drive[10]. The tonic drive provides a necessary condition for the network-based oscillations. We suggest that the central chemoreceptors *are* the conditional pacemakers. *In vivo* they provide tonic excitatory drive to the respiratory network, which therefore generate a respiratory pattern by a *network* mechanism. Lifting the $[K^+]$out *in vitro* switches the central chemoreceptors to the bursting mode. The synchronized activity of these central chemoreceptors provides bursting excitatory drive to the network, which now operates in a "forced mode" producing oscillations defined by the bursting drive (a *hybrid* mechanism). This transition is illustrated in Figure 2.

Our computer simulations at the network level supported the possibility of the above transition (see Figure 3). We used a modified version of our previous network model[5] and incorporated conditional pacemakers performing the function of central chemoreceptors. Importantly, our model reproduced the physiologically realistic changes in the shape of phrenic burst from an augmenting pattern during the network-based oscillations (Figure 3A) to a higher amplitude decrementing (gasping-like) pattern during the pacemaker-driven oscillations (Figures 3B and 3C).

Our experimental studies have provided additional support to our theoretical concept. Using an *in situ* perfused preparation of rat, we simulated *in-vitro* conditions. By increasing $[K^+]$out to 10 *mM* within the perfusate, and in combination with 4-AP (an K_A^+ channel blocker) and a low dose of strychnine (to reduce synaptic inhibition), the regular augmenting pattern of phrenic nerve discharge was changed to a decrementing pattern similar to that recorded *in vitro* (Figure 4).

We believe that the transition from network-based to pacemaker-driven oscillations may also occur *in vivo* under conditions similar to those in *in-vitro* preparations (e.g. when $[K^+]$out is increased and inhibition is reduced). Specifically, the transition to a pacemaker-driven rhythm may occur *in vivo* during hypoxia[11]. This is supported by the fact that hypoxia causes both a significant increase in $[K^+]$out[12] and activation of the Na_P^+ current[13]. This is consistent with the finding that the PBC has intrinsic hypoxic chemosensitivity and may play a role in hypoxia-induced gasping[14].

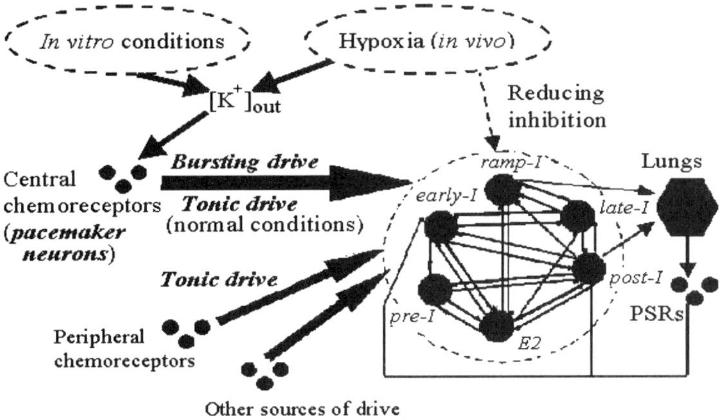

Figure 2. Schematic of the transition of the respiratory pattern generation from a network to a hybrid pacemaker-network mechanism. The respiratory network receives excitatory drive from central chemoreceptors. Under normal *in-vitro* conditions this drive is tonic, and respiratory oscillations are generated by the respiratory network similarly to our earlier network model[5]. An increase in $[K^+]$out *in vitro* (or *in vivo* during hypoxia) switches the central chemoreceptor drive to bursting. This forces the respiratory network to generate oscillations defined by the bursting drive from chemoreceptors.

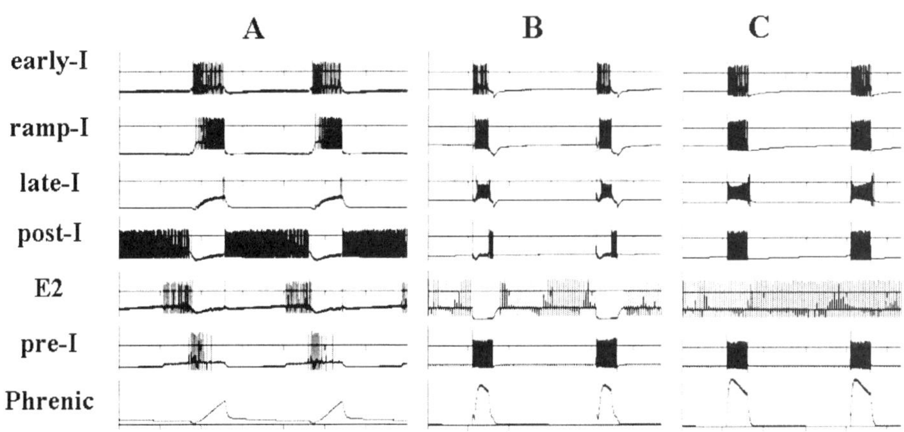

Figure 3. The results of our computer simulation of the transition in respiratory pattern generation from the network mechanism (the normal three-phase respiratory pattern in **A**) to a hybrid pacemaker-network mechanism (in **B** and **C**). All neurons in the model except E2 received excitatory drive from central chemoreceptors (conditional pacemakers). An increase in $[K^+]$out (to simulate *in-vitro* conditions) changed drive from tonic to bursting and switched rhythmogenesis to the hybrid pacemaker-network mechanism (**B**). The post-I and E2 neurons were still inhibited during inspiration by inspiratory neurons. An additional blockade of inhibition was applied to stimulate hypoxia-induced gasping (**C**). It produced a shift the discharge of post-I neurons to the inspiratory phase and eliminated the respiratory modulation of E2 firing. Note the changes in the shape of phrenic burst from an augmenting pattern during the network-based oscillations (**A**) to higher amplitude decrementing (gasping-like) pattern during the pacemaker-driven oscillations (**B** and **C**).

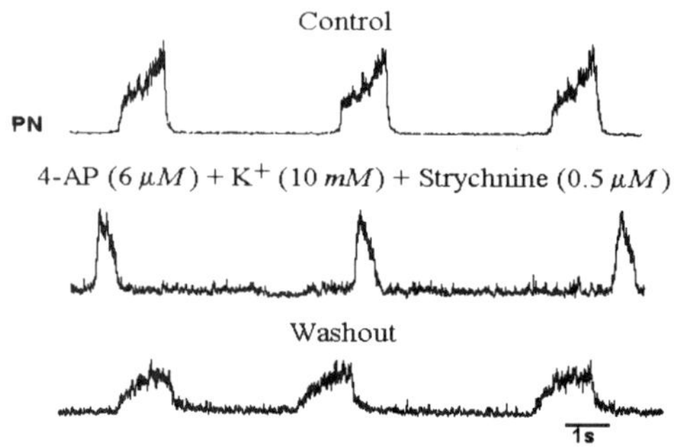

Figure 4. Experimental simulation of *in-vitro* conditions in an *in situ* preparation of rat. We used an *in situ* arterially perfused, decerebrate rat preparation. Phrenic nerve activity was recorded via a suction electrode. In order to *simulate in-vitro* conditions we added to the perfusate the following: 10 *mM* potassium, 4-AP to block K_A^+ channels, and a low dose of strychnine to antagonize glycine receptors. The effect was dramatic and changed the augmenting patterns of phrenic nerve activity (see "Control") to a decrementing gasping-like discharges similar to those recorded *in vitro* (middle trace). This effect was fully reversible following washout of the drugs and returning the $[K^+]$out to the control levels (see "Washout").

5. REFERENCES

1. S. M. Johnson, J. C. Smith, G. D. Funk, and J. L. Feldman, Pacemaker behavior of respiratory neurons in medullary slices from neonatal rat. *J. Neurophysiol.* **72**, 2598-2608 (1994).
2. N. Koshiya and J. C. Smith, Neuronal pacemaker for breathing visualized *in vitro. Nature* **400** (6742), 360-363 (1999).
3. J. C. Smith, H. Ellenberger, K. Ballanyi, D. W. Richter, and J. L. Feldman, Pre-Bötzinger complex: a brain stem region that may generate respiratory rhythm in mammals. *Science* **254**, 726-729 (1991).
4. J. C. Smith, in: *Neurons, Networks, and Motor Behavior,* edited by P. Stain, S. Grillner, A. I. Selverston, and D. G. Stuart (MIT Press, Cambridge, MA, 1997), pp. 97-104.
5. I. A. Rybak, J. F. R. Paton, and J. S. Schwaber, Modeling neural mechanisms for genesis of respiratory rhythm and pattern: II. Network models of the central respiratory pattern generator, *J. Neurophysiol.* **77**, 2007-2026 (1997).
6. I. A. Rybak, J. F. R. Paton, and J. S. Schwaber, Modeling neural mechanisms for genesis of respiratory rhythm and pattern: III. Comparison of model performances during afferent nerve stimulation, *J. Neurophysiol.* **77**, 2027-2039 (1997).
7. X. M. Shao and J. L. Feldman, Respiratory rhythm generation and synaptic inhibition of expiratory neurons in pre-Bötzinger complex: differential roles of glycinergic and gabaergic neural transmission, *J. Neurophysiol.* **77**, 1853-1860 (1997).
8. R. J. Butera, J. R. Rinzel, and J. C. Smith, Models of respiratory rhythm generation in the pre-Bötzinger complex: I. Bursting pacemaker neurons, *J. Neurophysiol.* **82**, 382-397 (1999).
9. C. R. French, P. Sah, K. J. Buckett, and P. W. Gage, A voltage-dependent persistent sodium current in mammalian hippocampal neurons, *J. Gen. Physiol.* **95**, 1139-1157 (1990).
10. E. Nattie, CO_2, brainstem chemoreceptors and breathing, *Progr. in Neurobiol.* **59**, 299-331 (1999).
11. W. M. St.-John, Rostral medullary respiratory neuronal activity of decelebrated cats in eupnea, apneusis and gasping, *Resp. Physiol.* **116**, 47-65 (1999).
12. J. E. Melton, S. C. Kadia, Q. P. Yu, J. A. Neubauer, and N. H. Edelman, Respiratory and sympathetic activity during recovery from hypoxic depression and gasping in cats, *J. Appl. Physiol.* 80, 1940-1948 (1996).
13. A. K. Hammarström and P. W. Gage, Inhibition of oxidative metabolism increases persistent sodium current in rat CA1 hippocampal neurons, *J. Physiol.* **510**, 735-741 (1998).
14. I. C. Solomon, N. H. Edelman, and J. A. Neubauer, Pre-Bötzinger complex functions as a central hypoxia chemosensor for respiration *in vivo*, *J. Neurophysiol.* **83**, 2854-2868 (2000).

NEUROGENESIS OF THE RESPIRATORY PATTERN: INSIGHTS FROM COMPUTATIONAL MODELING

Ilya A. Rybak, Julian F. R. Paton, and James S. Schwaber*

1. INTRODUCTION

The primary respiratory rhythm generator is located in a relatively small area of the lower brainstem and can be defined by the neuronal properties and synaptic interactions within this limited area. The genesis and control of the respiratory motor pattern involve a complex cross-level integration of cellular, network and systems mechanisms. Computational modeling is a powerful method that allows linking experimental data related to different levels of system organization. Therefore, a comprehensive computational model can provide useful insights for understanding the multilevel neural mechanisms involved in generation and control of the respiratory pattern. Our ultimate goal is to develop such a model.

2. MODELING RESPIRATORY NEURONS (CELLULAR LEVEL)

We developed models of respiratory neurons in the Hodgkin-Huxley style and matched their behavior to *in vivo* intracellular recordings of respiratory cells. We have shown that a slow adaptation of firing observed in early-inspiratory (early-I) and post-inspiratory (post-I) neurons can be produced by interactions of the high threshold calcium (Ca_L^{2+}) and calcium-dependent potassium ($K^+(Ca^{2+})$) currents (see Figure 1A). Further, interactions between the low threshold (Ca_T^{2+}) and $K^+(Ca^{2+})$ currents can produce the ramp-firing pattern seen in inspiratory neurons such as ramp-I (see Figure 1B). These slow monotonic changes in spiking frequency of neuronal firing patterns are essential elements of the phase transition mechanisms. Specifically, they may contribute to control of both the duration of respiratory phases and the timing of phase transitions.

* Ilya A. Rybak, School of Biomedical Engineering, Science and Health Systems, Drexel University, Philadelphia, PA 19104. Julian F. R. Paton, Department of Physiology, School of Medical Sciences, University of Bristol, Bristol BS8 1TD, UK. James S. Schwaber, Department of Pathology, Anatomy, and Cell Biology, Thomas Jefferson Medical School, Philadelphia, PA 19107.

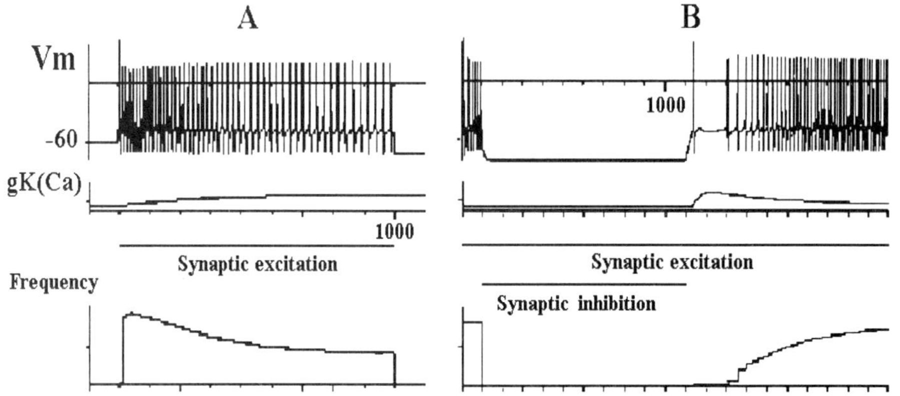

Figure 1. **A.** The adaptive response of the model of decrementing respiratory neurons (early-I and post-I) to a stepwise excitatory synaptic input. The slow decrease of spiking frequency is produced by the increase of gK(Ca) conductance, which is activated by the increasing concentration of intracellular calcium entering the cell though Ca_L^{2+} channels. **B.** The augmenting firing pattern generated by the model of ramp-I neurons. In this model, the ramp increase of frequency is provided by the slow decay of the gK(Ca) conductance. The conductance decays after its activation by an increase in intracellular concentration of calcium entering though Ca_T^{2+} channels when the neuron has been released from synaptic inhibition.

3. RESPIRATORY PHASE TRANSITIONS MECHANISMS AT THE NETWORK LEVEL

The network-based respiratory rhythm is generated as a result of the sequential transitions between phases of the respiratory cycle, which comprises inspiration, postinspiration and expiration[1, 2]. To build a network model of the central respiratory pattern generator (CRPG), we started with the analysis and simulation of possible phase transition mechanisms. In our model, the transition from inspiration to posinspiration operated via late-I neurons, which were considered inspiratory off-switching neurons[2-5]. The transition from expiration to inspiration (expiratory off-switching) is poorly understood. We considered several mechanisms for the expiratory off-switching mechanism including that proposed by Klages et al.[6] and Richter[1] where this transition was provided by pre-I neurons.

We found that the structure and performance of the CRPG model is essentially dependent on (1) the number of respiratory phases which can be controlled independently (two vs. three), (2) the type of expiratory off-switching mechanism, and (3) the mechanism for frequency augmentation in the firing patterns of the late-expiratory neurons (E2). Further, in light of the current insufficiency in specific data, a number of plausible network-based CRPG models were proposed based on different combinations of possible phase transition mechanisms. This resulted in the development of a series of two- and three-phase CRPG models, each of which could generate a robust and stable respiratory rhythm exhibiting realistic membrane potential trajectories of respiratory neurons as recorded *in vivo*.

4. FINDING A REALISTIC NETWORK ARCHITECTURE FOR THE CRPG: MODEL TESTING AT THE SYSTEMS LEVEL

An approach to assess the realism of our CPRG models was to test their ability to reproduce a large body of data in the literature concerning various changes, reorganization or cessation of the respiratory pattern produced by various stimulations/ perturbations applied via major peripheral nerves. This included stimulation of vagus, carotid sinus (CS) and superior laryngeal (SLN) nerves as well as activation of the Hering-Breuer reflex. The ability of a CRPG model to reproduce these evoked systems-level responses depended on the model's network architecture. Comparative testing of different CRPG models in computational experiments allowed us to evaluate the plausibility of each model *per se*, as well as the plausibility of used neural architectures and embedded neural mechanisms. The results of this testing have allowed us to determine a CRPG model with the most realistic network architecture.

The schematic of the model is shown in Figure 2. Under normal conditions the model generated a three-phase respiratory rhythm and demonstrated realistic firing patterns of respiratory neurons (Figure 3A). Disconnecting the pulmonary vagal feedback (Figure 3B) caused both an increase in the duration and amplitude of ramp-I neuron and phrenic discharges (which reflected the loss of the Hering-Breuer reflex[3, 5, 7]) and a decrease in the duration of post-I neuron bursts and postinspiratory phase (which was consistent with Richter et al.[8]).

A large body of experimental data concerning reorganizations of the respiratory rhythm and pattern during various types of stimulation applied to the SLN, vagus, and CS nerves were considered in simulation to test the model at the systems level. Examples of these model experiments are shown in Figure 4A-C.

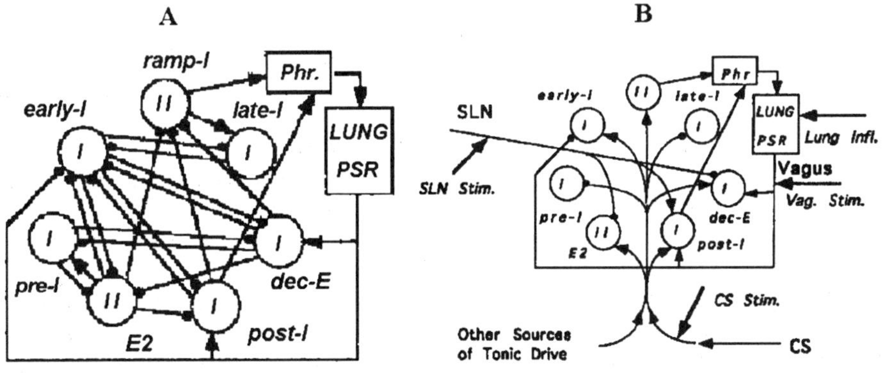

Figure 2. A. Schematics of the CRPG model. The model includes a network of respiratory neuron types (early-I, ramp-I, late-I, dec-E, post-I, E2, and pre-I) and simplified models of lung and pulmonary stretch receptors (PSR) that provide feedback to the respiratory network. The large circles represent neurons. The numbers inside the circles indicate the neuron types (I - decrementing, II - augmenting). The small filled circles are inhibitory synapses. The arrows are excitatory synapses. Tonic input to all neurons is not shown. **B.** Afferent inputs from peripheral nerves: SLN, vagus and CS. The large arrows indicate points to which stimuli were applied in different model experiments at the systems level.

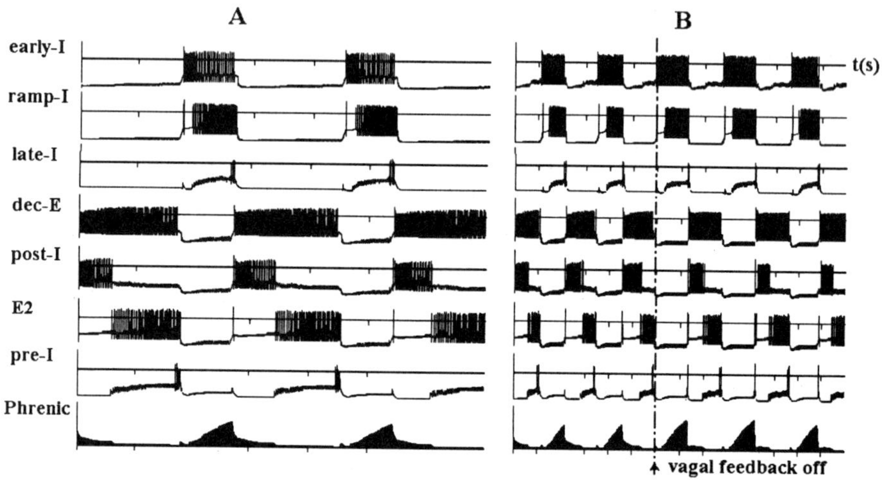

Figure 3. A. Model performance (membrane potential trajectories of all neurons in the respiratory network and the dynamics of phrenic nerve discharges). **B.** Effect of removing vagal feedback ("vagotomy"). See text for details.

A short stimulus applied to the vagus nerve during inspiration terminated this phase if the stimulus amplitude exceeded some threshold. In Figure 4A, three stimuli were given. The first, large amplitude stimulus was applied at the beginning of the inspiratory phase and terminated inspiration. The second stimulus, which had smaller amplitude, but was applied at the same phase, did not terminate inspiration. The third stimulus had the same amplitude as the second stimulus, but being delivered during a later stage of the inspiratory phase, terminated inspiration. Thus, the threshold for inspiratory termination decreased during the course of the inspiratory phase. These simulation results are consistent with the experimental measurements showing that the threshold for inspiratory termination decreases during the course of the inspiratory phase with an approximately hyperbolic dependence[3, 7, 9].

Short stimuli applied to SLN reset the respiratory cycle to inspiration (Figure 4B). This is consistent with *in vivo* data by Paydarfar et al.[10] obtained in studies of phase-resetting effects during electrical stimulation of the SLN in cats.

Model performance during phasic stimulation of the CS nerve is shown in Figure 4C. Stimuli delivered at the beginning of the inspiratory phase produced a shortening of this phase (first stimulus) whereas stimuli applied in the middle or at the end of inspiration prolonged the inspiratory phase and caused an increase in the amplitude of the phrenic burst (second stimulus). Stimuli delivered in the post-inspiratory phase produced a prolongation of expiration (third stimulus). These modeling results are consistent with the experimental findings of Eldridge[11, 12] and Remmers et al.[13] However, the current model could not demonstrate a shortening of expiration when CS stimuli were delivered in the late expiratory phase[11-13].

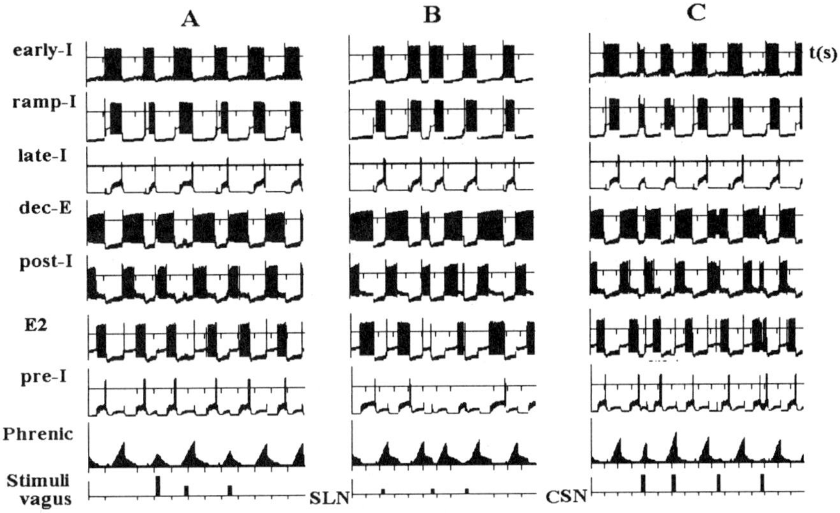

Figure 4. Effects of short duration stimuli applied to the vagus nerve (**A**), SLN (**B**), and CSN (**C**). See text for details.

5. CROSS-LEVEL FUNCTION OF THE INTRINSIC NEURONAL PROPERTIES

Analysis of our model shows that the dynamics of calcium and calcium-dependent potassium currents in early-I and ramp-I neurons may directly define the duration of inspiration and timing of the inspiratory off-switch (see Figure 5A). Similarly, the dynamics of calcium and calcium-dependent potassium currents in dec-E neurons can significantly contribute to the control of expiratory duration and timing of the expiratory off-switch.

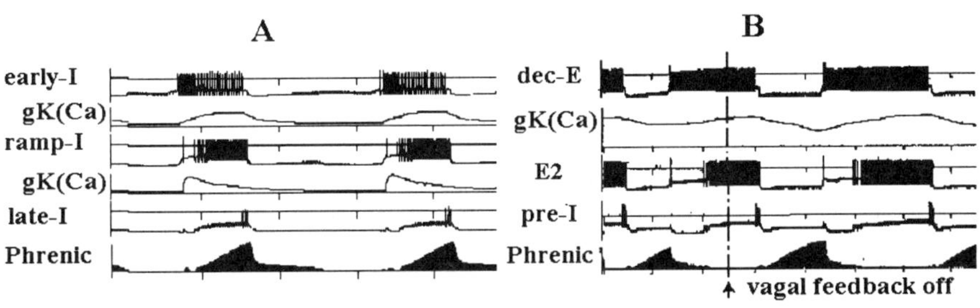

Figure 5. A. Membrane potential trajectories of inspiratory neurons and dynamics of gK(Ca) in early-I and ramp-I neurons; this conductance provides the slow frequency changes in firing patterns of these neurons. The timing of inspiratory off-switching and the duration of inspiration depend on both the rate of frequency decrement in firing of the early-I neuron (which inhibits the switching late-I neuron) and the ramp of frequency increase in firing of the ramp-I neuron (which excites the late-I neuron). **B.** The dynamics of gK(Ca) in the dec-E neuron maintains the relationship between the durations of inspiration and expiration.

Our model predicts that the intrinsic properties of respiratory neurons can contribute to the control of breathing at the systems level. An example is shown in Figure 5B. It is known that expiratory duration (*Te*) is dependent on the duration of preceding inspiration (*Ti*) with an approximately linear relationship, so that when *Ti* has been shortened or lengthened the succeeding *Te* is reduced or increased respectively[3, 9]. The exact mechanism providing this relationship is unknown. Interestingly, our model demonstrated the above functional relationship. The mechanism for changing *Te* following a change in *Ti* in our model is based on the dynamics of the $K^+(Ca^{2+})$ channel conductance in dec-E neurons (gK(Ca)). Figure 5B shows that this conductance decreases during inspiration and increases during expiration. The prolongation of inspiration was produced by disconnecting vagal feedback. When *Ti* increased the value of gK(Ca) in dec-E neurons decreased for a longer period of time and reached a lower level. Therefore, the inhibitory effect of $K^+(Ca^{2+})$ current on the activity of dec-E neuron was diminished. This produced an increase in the spike frequency of dec-E, thereby increasing the inhibition of E2 by dec-E during the following expiratory phase. Consequently, the spike frequency of E2 neurons increased for longer period, which delayed both the firing of pre-I neuron and termination of expiration so prolonging the duration of expiration (*Te*). We predict that this intrinsic mechanism operates *in vivo* to maintain the relationship between the inspiratory and expiratory durations.

In conclusion, our computational studies have provided novel and important insights into the integration of cellular, network and systems mechanisms involved in the generation and control of respiratory motor pattern.

6. REFERENCES

1. D. W. Richter, in: *Comprehensive Human Physiology*, edited by R. Gregor and U. Windhorst (Springer-Verlag, Berlin, 1996, vol. II), pp. 2079-2095.
2. D. Richter, D. Ballantyne, and J. E. Remmers, How is the respiratory rhythm generated? A model, *News Physiol. Sci.* **1**, 109-112 (1986).
3. M. I. Cohen, Neurogenesis of respiratory rhythm in the mammal, *Physiol. Rev.* **59**, 1105-1173 (1979).
4. M. I. Cohen and J. L. Feldman, Models of respiratory phase-switching, *Federation Proc.* **36**, 2367-2374 (1977).
5. J. L. Feldman, in: *Handbook of Physiology,* edited by F. E. Bloom (Am. Physiol. Soc., Bethesda, MD, 1986, sec. 1, vol. 4), pp. 463-524.
6. S. Klages, M. C. Bellingham, and D. W. Richter, Late expiratory inhibition of stage 2 expiratory neurons in the cat - A correlate of expiratory termination, *J. Physiol. Lond.* **70**, 1307-1315 (1993).
7. C. von Euler, in: *Handbook of Physiology. The Respiratory System II*, edited by N. S. Cherniack and J. G. Widdicombe (Am. Physiol. Soc., Washington, DC, 1986), pp. 1-67.
8. D. Richter, D. Ballantyne, and J. E. Remmers, The differential organization of medullary post-inspiratory activities, *Pflügers Arch.* **410**, 420-427 (1987).
9. F. J. Clark and C. von Euler, On the regulation of depth and rate of breathing, *J. Physiol. Lond.* **222**, 267-295 (1972).
10. D. Paydarfar, F. L. Eldridge, and J. P. Kiley, Resetting of mammalian respiratory rhythm: existence of a phase singularity, *Am. J. Physiol.* **250**, R721-R727 (1986).
11. F. L. Eldridge, The importance of timing on the respiratory effects of intermittent carotid sinus nerve stimulation, *J. Physiol. Lond.* **222,** 297-318 (1972).
12. F. L. Eldridge, Expiratory effects of brief carotid sinus nerve and carotid body stimulations, *Resp. Physiol.* **26**, 395-410 (1976).
13. J. E. Remmers, D. W. Richter,.D. Ballantyne

RECONFIGURATION OF THE CENTRAL RESPIRATORY NETWORK UNDER NORMOXIC AND HYPOXIC CONDITIONS

Steven P. Lieske, Muriel Thoby-Brisson, and Jan-Marino Ramirez*

1. INTRODUCTION

Apart from the obvious medical interest in understanding breathing, the respiratory rhythm is ideally suited for the study of state dependence and modifiability in central pattern generators (CPGs) in general. The list of variables with dramatic effects on breathing is impressive, including temperature, sleep/wake state, O_2 and CO_2 concentrations, airway irritation, conscious control, and a variety of pathological conditions. The ongoing breathing rhythm can be interrupted by coughing, sneezing, hiccoughing, swallowing, or vomiting; it can also be suspended in the post-inspiratory phase by vocalization. The pattern of breathing itself can take on several qualitatively distinct patterns, including eupneic ("normal") breathing, panting, sighing, periodic breathing (during sleep), and gasping (in hypoxia).

Here we focus on two of these respiratory patterns, eupnea and gasping, and summarize recent evidence from *in vivo* and *in vitro* studies suggesting that both respiratory patterns are generated by the same neural circuit through a process of network reconfiguration. We will also discuss possible mechanisms by which this reconfiguration is accomplished.

2. DEFINITION OF EUPNEA AND GASPING (*IN VIVO*)

The breathing pattern that is observed in control conditions *in vivo*—normoxia, normocapnea, and normothermia—is termed eupnea. Hypoxic conditions lead to an initial augmentation in eupneic frequency, followed by a depression and apnea, and the re-emergence of breathing in a qualitatively distinct pattern known as gasping[1]. Although details of the gasping pattern can vary, there is a core set of features that reliably differentiate gasping from eupnea: a larger peak inspiratory amplitude, an inspiration that

* Committee on Neurobiology and the Department of Organismal Biology and Anatomy, University of Chicago, Chicago, Illinois, 60637.

Frontiers in Modeling and Control of Breathing, edited by Poon and Kazemi.
Kluwer Academic/Plenum Publishers, New York, 2001.

is square or decrementing in shape rather than augmenting (i.e. in which peak amplitude occurs relatively earlier in the inspiratory phase), usually a decreased inspiratory time, and a slower respiratory frequency[2-4].

Although each of these parameters significantly differs between eupnea and gasping, the range of values observed for both patterns is highly dependent upon the specifics of the preparation: the species and age of animal used, anesthetic and/or decerebration, maintenance of normal blood gases, temperature and recording site. *As a consequence, it is only when both patterns can be directly compared in the same preparation that a differential identification of respiratory pattern is straightforward.* This point becomes critical when discussing results obtained *in vitro*.

The application of the term "gasping" to a respiratory pattern generated by a reduced preparation in the absence of hypoxia originated with Lumsden[5] in 1923. Following pontomedullary transection in the cat, the breathing pattern was indeed slower, larger in amplitude, and more rapid in onset than when the animal was merely anesthetized. For many years, this result was misinterpreted as a demonstration that the respiratory nuclei of the medulla play a role only in gasping and not in eupnea. More recently, even as it became clear from work done both *in vivo* and *in vitro* that medullary areas are essential for eupnea as well as gasping, a number of lesion studies raised the possibility that *different* medullary areas are active in each of these patterns[4,6-9]. As a consequence, the relationship between the normal eupneic rhythm *in vivo* and the respiratory rhythm obtained from the various *in vitro* preparations has continued to come under scrutiny.

3. EUPNEA AND GASPING *IN VITRO*

Not surprisingly, much of this discussion has focused on differences in burst shape and frequency between the most commonly used *in vitro* preparations and the corresponding *in vivo* animals of the same age. Indeed, several authors have gone so far as to suggest that the respiratory rhythm observed *in vitro* is, without exception, gasping[10-11]. Because these variables could well be affected by various attributes of the *in vitro* preparations, such as a reduced temperature and the lack of afferent feedback, however, a direct comparison of *in vitro* and *in vivo* data is not necessarily meaningful[12]. As discussed above, the classification of respiratory pattern on the basis of these criteria only succeeds when both eupnea and gasping can be elicited, and contrasted against each other.

Using transverse medullary slices of mice, we rigorously compared the population activity of the ventral respiratory group (VRG) during normoxia and hypoxia. When the oxygen in the superfused saline is replaced by nitrogen (anoxia), the respiratory activity undergoes a transformation very similar to the *in vivo* hypoxic response, consisting of an initial augmentation and a subsequent depression. The depression phase was associated with significant reductions in burst duration, rise time and frequency, consistent with the definition of the gasping pattern obtained *in vivo*[13] (fig. 1). The amplitude of VRG activity tended to decrease, consistent with the observation that many respiratory neurons ceased to discharge during hypoxia *in vivo*[14-17]. Burst amplitude in the hypoglossal motor output, however, increased dramatically during anoxia *in vitro*[18], again consistent with *in*

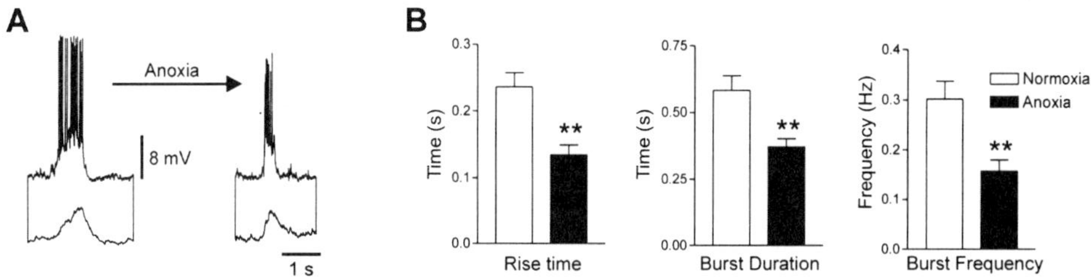

Figure 1. Comparison of fictive eupnea and fictive gasping. A. Intracellular recording of an inspiratory neuron (top trace) and simultaneous integrated VRG activity (bottom trace) during fictive eupnea (left) and fictive gasping (right). B. Mean values (+SEM) of population (integrated VRG) burst rise time (left) duration (middle), and frequency (right). Asterisks denote values significantly different from control (** $p < 0.01$).

vivo observations[15]. Thus, the respiratory rhythm generated by the transverse medullary slice *in vitro* exhibits a transformation during the late phase of the anoxic response very similar to the transformation that defines gasping as distinct from eupnea *in vivo*.

The possibility to distinguish in the same preparation gasping in hypoxia from eupnea under control conditions indicates that the defining elements for both patterns are present *in vitro*. The preservation of eupnea under control conditions is further supported by recent demonstrations of an inspiratory-shortening (Breuer-Hering) reflex[19] and expiratory motor output activity[20] in other *in vitro* preparations. We therefore refer to the respiratory patterns generated *in vitro* in normoxia and anoxia as fictive eupnea and fictive gasping, respectively.

Two important caveats bear mention. First, it is well established that most, if not all, known rhythmic motor behaviors in vertebrates and invertebrates depend not only on the activity generated by the central rhythm generating network, but also on sensory feedback[21-23], interactions with muscle activity[24,25] and various modulatory influences from other areas of the CNS[26,27]. Slices lack the connections to the lungs, peripheral chemoreceptors and other respiratory nuclei of the pons and medulla. Inevitably, therefore, the fictive activity generated *in vitro* will differ from the behavior of the whole animal. Second, the presence of distinct fictive eupnea and gasping activities in a reduced preparation does not preclude the possibility that, in the intact network, other areas may also contribute to the genesis of these patterns. It is possible that the extent of the respiratory rhythm-generating network differs during eupneic and gasping activity. Indeed, there is evidence for such an arrangement in other motor systems, in which the number of active elements in a rhythm generating neural network is clearly variable and state-dependent[28,29].

The presence of distinct eupneic and gasping patterns in a transverse medullary slice raises exciting issues that have to be resolved in the future: What are the rhythm generating mechanisms that differentiate eupneic from gasping activity? What are the cellular mechanisms that transform the network from a eupneic into a gasping state during hypoxia? What is the role of other respiratory-related areas in modulating these two motor patterns?

4. NEUROGENESIS OF THE EUPNEIC RHYTHM

The presumed central pattern generator for breathing is located in a small area of the medulla known as the pre-Bötzinger Complex[30] (PBC), within the Ventral Respiratory Group (VRG). Neurons within this complex exhibit inspiratory, post-inspiratory, and expiratory activities, which are characteristic for eupneic activity[31]. Synaptic inhibition is essential for establishing these phases *in vivo* and *in vitro*[32]. Following the blockade of synaptic inhibition, expiratory neurons become tonically active[32-34], post-inspiratory neurons rhythmically active in phase with inspiration[32]. Similarly, electrical activity recorded from an internal intercostal muscle was shifted from the early portion of the expiratory phase to the inspiratory phase in the presence of strychnine in a brainstem—spinal cord—rib preparation from the neonatal rat[20]. The inspiratory phase is also affected by the blockade of synaptic inhibition, as indicated by the observed reductions in inspiratory burst duration and rise time. This is consistent with recordings from individual inspiratory cells *in vivo* and *in vitro*[13,32]. Thus, under both *in vivo* and *in vitro* conditions defining aspects of the eupneic activity pattern depend on inhibitory interactions.

5. RECONFIGURATION FROM A "EUPNEIC" TO A "GASPING" NETWORK

During the initial phase of the anoxic response, expiratory neurons become tonically active and post-inspiratory neurons shift to an inspiratory discharge pattern; both cell types become quiescent during prolonged anoxia. At the same time, inspiratory activities are qualitatively changed, from an augmenting depolarization pattern under control conditions, to a rapidly peaking pattern during anoxia[13].

The reconfiguration of all three cell types is consistent with gasping and can be explained in large part by a progressive loss of phasic inhibition, which takes place early in anoxia (fig. 2 and 3). The phasic inhibition of expiratory cells during inspiration is lost during gasping (fig. 2). The shift from tonic activity early in anoxia to quiescence later could perhaps relate to the different types of gasping reported by some authors at different time points in anoxia[35]. The phasic excitation exhibited by post-inspiratory cells during the initial portion of the anoxic response is present during eupnea as well, but overwhelmed by a stronger synaptic inhibition[32]. Indeed, this excitation might help create a post-inspiratory rebound when the inspiratory burst ends and the inhibition is relieved. When phasic inhibition is suppressed in anoxia, only the excitation remains, and post-inspiratory cells shift to an inspiratory pattern of discharge. Inspiratory cells also receive mixed excitatory and inhibitory inputs during the inspiratory phase of eupnea[32]. This combination of positive and negative feedback could form part of the explanation for the augmenting shape of the depolarization of these cells (fig. 3); the loss of inhibitory inputs during gasping would leave the excitatory feedback unopposed, resulting in the rapidly peaking burst shape observed during gasping (fig. 1A). Thus we see that the suppression of inhibition apparent in recordings from expiratory cells is coincident with the transition in burst shape from eupnea-like to gasp-like *in vitro* (fig. 2B). Furthermore, the pharmacological blockade of glycinergic inhibition with strychnine induces changes in burst shape very similar to what is observed during gasping[13].

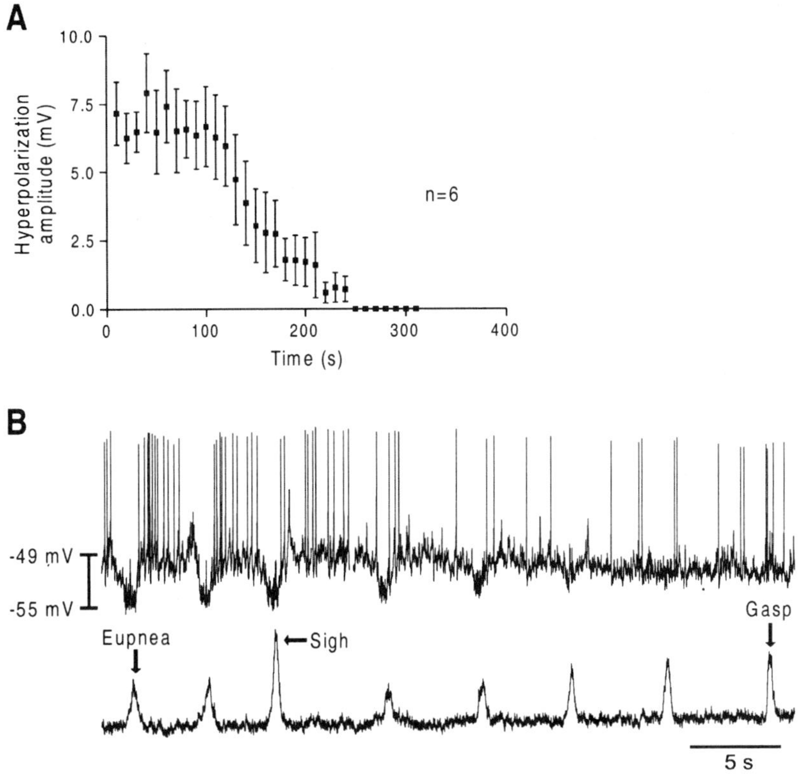

Figure 2. Suppression of synaptic inhibition in anoxia. A. Mean amplitude (±SEM) of the phasic inspiratory hyperpolarization in expiratory cells of the VRG, versus time in anoxia (n = 6). B. Intracellular recording of a VRG expiratory neuron (top trace) and simultaneous integrated VRG activity (bottom trace) during the transition from fictive eupnea to fictive gasping in anoxia. Note that the suppression of synaptic inhibition of this cell is coincident with the change in population burst shape from eupnea-like to gasp-like. For a discussion of sighs (large amplitude burst denoted with arrow) in this preparation, see Lieske *et al.* (2000)[13].

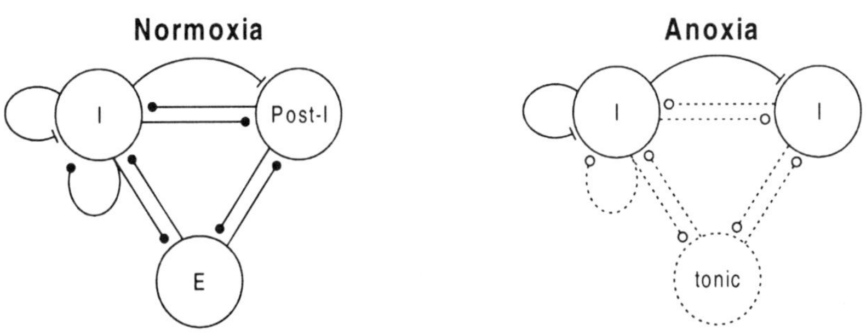

Figure 3. Hypothetical model of respiratory network during fictive eupnea and fictive gasping. Filled circles depict inhibitory and bars depict excitatory connections. The population of inspiratory cells (top left) includes both pacemaker neurons and followers.

Unlike gasping, however, the burst frequency in strychnine is not significantly altered from control conditions[13], so while anoxic suppression of synaptic inhibition could explain the changes in respiratory pattern, additional mechanisms are required to explain the concurrent changes in frequency. One approach to identifying these mechanisms is to investigate the anoxic response of individual cells[36]. Because expiratory neurons were tonically active early in anoxia and quiescent later, these cells cannot be the source of the changes in respiratory frequency. Similarly, non-pacemaker inspiratory neurons also fell silent during prolonged anoxia and cannot be responsible for the frequency changes[36]. In contrast, pacemaker neurons showed an increase in bursting frequency early in anoxia, even when isolated from the network[36]. This frequency increase of isolated pacemaker neurons was very similar to the increase in respiratory frequency during the augmentation phase of the intact network. The onset of the depression phase in the intact network coincided temporally with a cessation of rhythmic activity late in anoxia in one group of pacemaker neurons. This is an intriguing finding as pacemaker neurons are thought to be important for driving the respiratory rhythm[37]. Thoby-Brisson and Ramirez (2000) identified a second group of pacemaker neurons that maintained rhythmic activity throughout anoxia[36]. Although these neurons could be candidates for driving a gasping rhythm in anoxia, the frequency of the isolated neurons was much faster than that of the intact network during fictive gasping. Thus, while an attractive hypothesis, the possibility that eupneic and gasping activity are driven by separate populations of pacemaker neurons is too simple.

6. CONCLUSIONS

The demands of respiratory control are different during normoxic and hypoxic conditions, the former presumably emphasizing efficiency with respect to energy use, and the latter placing a premium on resuscitation and reoxygenation at any cost. Nonetheless, the muscles available for generating respiratory movements do not change, implying that the same motoneurons must ultimately be activated. Network reconfiguration provides a powerful mechanism for generating a family of such related motor patterns, well-known in the study of invertebrate central pattern generators[26]. Our data suggest that there is but a single network responsible for respiratory rhythmogenesis, the output of which is state-dependent, being modifiable through a process of network reconfiguration.

This raises the possibility that other forms of respiration might arise in a similar manner, via the effects of other modulatory influences. Indeed, we have recently shown that a third respiratory pattern, sighing, is also generated by this same network[13]. Although less is known about the origin of the other respiratory-related motor patterns listed in the introduction—coughing, sneezing, hiccoughing, etc.—each of these interrupts the pattern of respiration[38], and each arises in an area overlapping with the ventral respiratory group. Thus, it is possible that the cells of the respiratory CPG are also involved in generating these patterns, raising the interesting issue of how these other reconfigurations might be accomplished.

The respiratory CPG is also subject to a high degree of control from other areas of the CNS: speaking suspends the respiratory rhythm in the post-inspiratory phase, sleeping has marked effects on respiratory pattern and airway muscle tone, both the initiation and the termination of inspiration can be forced deliberately, and it is possible to "hold one's breath" at any level of lung inflation. What is particularly intriguing about these examples is that they can also be overridden when inspiratory drive becomes sufficiently strong; breath cannot be "held" indefinitely. All of this suggests that the respiratory network is particularly rich with possibilities for modulation, which is essential to adjust breathing in a state-dependent manner to the continuously changing requirements in a behaving organism.

7. REFERENCES

1. J. A. Neubauer, J. E. Melton, and N. H. Edelman, Modulation of respiration during brain hypoxia, *J. Appl. Physiol.* **68**(2), 441-451 (1990).
2. Wang, W., Fung, M. L., Darnall, R. A. & St. John, W. M. Characterizations and comparisons of eupnoea and gasping in neonatal rats. *J. Physiol. (London)* **490**, 277-292 (1996).
3. Wang, W., Fung, M. L. & St. John, W. M. Pontile regulation of ventilatory activity in the adult rat. *J. Appl. Physiol.* **74**, 2801-2811 (1993).
4. Fung, M. L., Wang, W. & St. John W. M. Medullary loci critical for expression of gasping in adult rats. *J. Physiol. (London)* **480**, 597-611 (1994).
5. T. Lumsden, Observations on the respiratory centres in the cat, *J. Physiol. (London)* **57**, 153-160 (1923).
6. D. R. McCrimmon, A. Monnier, F. Hayashi, and E. J. Zuperku, Pattern formation and rhythm generation in the ventral respiratory group, *Clin. Exp. Pharmacol. Physiol.* **27**(1-2), 126-131 (2000).
7. N. Koshiya and P. G. Guyenet, Tonic sympathetic reflex after blockade of respiratory rhythmogenesis in the cat, *J. Physiol. (London)* **491**, 859-869 (1996).
8. Q. Huang,, D. Zhou, and W. M. St. John, Lesions of regions for in vitro ventilatory genesis eliminate gasping but not eupnea, *Respir. Physiol.* **107**, 111-123 (1997).
9. J. M. Ramirez, S. W. Schwarzacher, O. Pierrefiche, B. M. Olivera, and D. W. Richter, Selective lesioning of the cat pre-Bötzinger complex in vivo eliminates breathing but not gasping. *J. Physiol. (London)* **507**, 895-907 (1998).
10. W. M. St. John, Neurogenesis of patterns of automatic ventilatory activity, *Prog. Neurobiol.* **56**(1), 97-117 (1998).
11. W. M. St. John, and J. F. Paton, Characterizations of eupnea, apneusis and gasping in a perfused rat preparation, *Respir. Physiol.* **123**(3), 201-213 (2000)
12. J. C. Smith, J. J. Greer, G. S. Liu, and J. L. Feldman, Neural mechanisms generating respiratory pattern in mammalian brain stem-spinal cord in vitro. I. Spatiotemporal patterns of motor and medullary neuron activity, *J. Neurophysiol.* **64**(4), 1149-1169 (1990).
13. S. P. Lieske, M. Thoby-Brisson, P. Telgkamp, J. M. Ramirez, Reconfiguration of the neural network controlling multiple breathing patterns: eupnea, sighs and gasps, *Nat. Neurosci.* **3**(6), 600-607 (2000).
14. W. M. St John, Rostral medullary respiratory neuronal activities of decerebrate cats in eupnea, apneusis and gasping, *Respir. Physiol.*, **116**(1), 47-65 (1999).
15. W. M. St. John and A. L. Bianchi, Responses of bulbospinal and laryngeal respiratory neurons to hypercapnia and hypoxia, *J. Appl. Physiol.* **59**, 1201-1207 (1985).
16. D. W. Richter, A. Bischoff, K. Anders, M. Bellingham, and U. Windhorst, Response of the medullary respiratory network of the cat to hypoxia, *J. Physiol. (London)* **443**, 231-256 (1991).
17. S. J. England, J. E. Melton, M. A. Douse, and J. Duffin, Activity of respiratory neurons during hypoxia in the chemodenervated cat, *J. Appl. Physiol.* **78**, 856-861 (1995).
18. P. Telgkamp and J. M. Ramirez, Differential response of respiratory nuclei to anoxia in rhythmic brainstem slices of mice, *J. Neurophysiol.* **82**(5), 2163-2170 (1999).
19. N. M. Mellen and J. L. Feldman, Phasic lung inflation shortens inspiration and respiratory period in the lung-attached neonate rat brain stem spinal cord, *J. Neurophysiol.* **83**(5), 3165-3168 (2000).
20. M. Iizuka, Intercostal expiratory activity in an in vitro brainstem-spinal cord-rib preparation from the neonatal rat, *J. Physiol. (London)* **520**(Pt 1), 293-302 (1999).
21. K. G. Pearson and J. M. Ramirez, in: *Neurons, Networks and Motor Behaviour*, edited by P. S. G. Stein, S. Grillner, A. Selverston, and D. Stuart (MIT Press, Cambridge, Massachusetts, London, England, 1997), pp. 225-237.
22. F. Hayashi, S. K. Coles, and D. R. McCrimmon, Respiratory neurons mediating the Breuer-Hering reflex prolongation of expiration in rat, *J. Neurosci.* **16**(20), 6526-6536 (1996).

23. A. C. Bonham, S. K. Coles, and D. R. McCrimmon, Pulmonary stretch receptor afferents activate excitatory amino acid receptors in the nucleus tractus solitarii in rats, *J. Physiol. (London)* **464**, 725-745 (1993).
24. L. G. Morris, J. B. Thuma, and S. L. Hooper SL, Muscles express motor patterns of non-innervating neural networks by filtering broad-band input, *Nat. Neurosci.* **3**(3), 245-250 (2000).
25. S. L. Hooper, Central pattern generators, *Curr. Biol.* **10**(5), R176 (2000).
26. E. Marder and R. L. Calabrese, Principles of rhythmic motor pattern generation, *Physiol. Rev.* **76**(3), 687-717 (1996).
27. P. S. Katz, Intrinsic and extrinsic neuromodulation of motor circuits, *Curr. Opin. Neurobiol.* **5**(6), 799-808 (1995).
28. K. G. Pearson and J. M. Ramirez, in: *Dynamic Biological Networks. The Stomatogastric Ganglion*, edited by R. M. Harris-Warrick, E. Marder, A. I. Selverston, M. Moulins (MIT Press, Cambridge, Massachusetts, London, England, 1992), pp. 263-281.
29. J. M. Ramirez, Reconfiguration of the respiratory network at the onset of locust flight, *J. Neurophysiol.* **80**(6), 3137-3147 (1998).
30. J. C. Smith, H. H. Ellenberger, K. Ballanyi, D. W. Richter, and J. L. Feldman, Pre-Bötzinger complex: a brainstem region that may generate respiratory rhythm in mammals, *Science* **254**(5032), 726-729 (1991).
31. S. W. Schwarzacher, J. C. Smith, and D. W. Richter, Pre-Botzinger complex in the cat, *J. Neurophysiol.* **73**(4), 1452-1461 (1995).
32. J. M. Ramirez, P. Telgkamp, F. Elsen, and D. W. Richter, Synaptic and membrane properties: Insights into the mechanisms of respiratory rhythm in mammals, *Respiration Physiology* **110**, 71-85. (1997).
33. X. M. Shao and J. L. Feldman, Respiratory rhythm generation and synaptic inhibition of expiratory neurons in pre-Botzinger complex: differential roles of glycinergic and GABAergic neural transmission. *J. Neurophysiol.* **77**(4), 1853-1860 (1997).
34. J. Brockhaus and K. Ballanyi, Synaptic inhibition in the isolated respiratory network of neonatal rats. *Eur. J. Neurosci.* **10**(12), 3823-3839 (1998).
35. D. Gozal, J. E. Torres, Y. M. Gozal, and T. J. Nuckton, Characterization and developmental aspects of anoxia-induced gasping in the rat, *Biol. Neonate.*, **70**(5), 280-288 (1996).
36. M. Thoby-Brisson and J. M. Ramirez, Role of inspiratory pacemaker neurons in mediating the hypoxic response of the respiratory network in vitro, *J. Neurosci.* **20**(15), 5858-5866 (2000).
37. J. C. Smith, R. J. Butera, N. Koshiya, C. Del Negro, C. G. Wilson, and S. M. Johnson, Respiratory rhythm generation in neonatal and adult mammals: the hybrid pacemaker-network model. *Respir Physiol.* **122**(2-3), 131-147 (2000).
38. R. Shannon, D. M. Baekey, K. F. Morris, Z. Li, and B. G. Lindsey, Functional connectivity among ventrolateral medullary respiratory neurones and responses during fictive cough in the cat. *J. Physiol. (London)*, **525**, 207-24 (1997).

HOW IS THE RESPIRATORY CENTRAL PATTERN GENERATOR CONFIGURED AND RECONFIGURED?

A workshop summary

Workshop Chair: Bruce G. Lindsey*

1. INTRODUCTION

The workshop briefly considered some current views on the location of the respiratory central pattern generator, its network and cellular properties, the reconfiguration of central pattern generators, and some future directions for research.

2. THE ROLE OF THE PRE-BÖTZINGER COMPLEX IN RESPIRATORY RHYTHM GENERATION (D. R. McCrimmon*)

Much of the support for the concept that neurons within the pre-Bötzinger complex (pBc) have an essential role in respiratory rhythm generation is derived from work using *in vitro* preparations. Nevertheless, several excellent studies support a similar conclusion *in vivo*. Bilateral chemical interruption of neural activity in the pBc by very localized injections of Ω-conotoxin GVIA and TTX reversibly interrupts respiratory rhythm generation[1]. Similarly, bilateral interruption of synaptic inhibition within the pBc abolishes rhythmic respiratory activity in cats[2]. In contrast, early studies employing extensive electrolytic lesions of the VRG in cats failed to disrupt respiratory rhythm generation[3].

In preliminary studies addressing this issue in anesthetized adult Sprague-Dawley rats, we have some surprising observations. Consistent with a necessary role for the pBc in rhythm generation we find that small (2-6 nl), unilateral injections of the potent $GABA_A$ receptor agonist muscimol (2 mM; total injection amount = 4-12 pmol) reversibly abolishes respiratory rhythm. Interestingly, similar (or only slightly larger, < 10 nl) injections placed 500-1000 μm further caudally within the rostral ventral

* Bruce G. Lindsey, University of South Florida, College of Medicine, Tampa, FL 33612 (workshop chair). Discussants: M.I Cohen, Albert Einstein College of Medicine, Bronx, NY 10461; D.R. McCrimmon, Northwestern University Medical School, Chicago, IL 60611; J.M. Ramirez, University of Chicago, Chicago, IL 60637; I.A. Rybak, Drexel University, Philadelphia, PA 19104.

Frontiers in Modeling and Control of Breathing, edited by Poon and Kazemi.
Kluwer Academic/Plenum Publishers, New York, 2001.

respiratory group (rVRG) elicit a similar, bilateral abolition of phrenic nerve activity. These injections are enough caudal to the pBc that they should have little or no direct influence on pBc activity.

In marked contrast, we have made large bilateral electrolytic lesions of the pre-Bötzinger complex that do not abolish respiratory rhythm *in vivo.* The lesions were histologically verified to encompass about a 1 mm rostrocaudal extent and were centered within the pBc region. These findings are consistent with earlier observations[3], but seem to contradict the pharmacological data. Given that the ineffective electrolytic lesions were bilateral and very large, while effective muscimol injections could be unilateral and small, we do not believe that the explanation for the difference is that the drug injections were influencing a larger volume of tissue. However, one clear difference between the approaches was that electrolytic lesions destroyed axons of passage while chemical injections did not.

In speculating about a possible explanation for the discrepancies, we considered the possibility that chemical inactivation of the pBc (or even the rVRG) may have disinhibited an inhibitory input to neurons in this region. Neurons within the Bötzinger complex could provide this type of inhibition. As a preliminary test of this possibility, we made an initial, unilateral injection of 3 nl of muscimol into the pre-Bötzinger complex of a rat. This abolished spontaneous respiratory rhythm on the phrenic nerve. However, we then made a second, larger (30 nl) injection of muscimol. Coincident with the large injection was the restoration of weak rhythmic activity on the phrenic nerve.

These findings are very preliminary and the interpretation highly speculative. Nevertheless, the role of pre-Bötzinger neurons in respiratory rhythm generation and pattern formation is likely to be quite complex. While much data indicate that there is something special about the pBc in respiratory rhythm generation, other results indicate that, at least *in vivo*, respiratory rhythm can continue in the absence of an intact pBc.

3. THE CONTROL OF RESPIRATORY PHASE TIMING AND SHAPE OF NEURONAL AND PHRENIC DISCHARGE PATTERNS UNDER VARYING CONDITIONS (I. A. Rybak[*])

A while ago, we developed a network model of the respiratory pattern generation.[4] Respiratory phase timing in the model was completely defined and could be separately controlled by the intrinsic properties of respiratory neurons, network connections, and drives from major afferent nerves. The model provided theoretical support for the previous experiment-driven conclusions that control of the respiratory phase durations is provided by a *cross-level integration of cellular, network and systems mechanisms*.

In addition, this *network* model reproduced a series of systems-level experimental data, such as independent regulation of the duration of different respiratory phases, respiratory reflexes (e.g. the Hering-Breuer reflex), and various phase resetting phenomena produced by stimulation of afferent nerves[4]. It is interesting to learn *whether and how such systems-level data can be reproduced and explained in frames of a hybrid pacemaker-network paradigm*? This paradigm suggests that a *pattern-formatting network* is driven by a *pacemaker-based kernel*. The neurons in the pattern-formatting network cannot (by definition) control the timing of phase transition. Therefore, in order to produce phase resetting or to control phase duration, the afferent nerves (e.g. vagus and SLN) should project to the kernel and somehow control the pacemaker-driven basic rhythm. However, such a suggestion does not have any experimental support yet, and so far no attempt has been made to find an explanation for the above systems-level phenomena in frames of the hybrid theory.

I also believe that two questions are extremely important for understanding the difference between the respiratory oscillations *in vivo* vs. *in vitro*, as well as for understanding the neural mechanisms for respiratory motor pattern generation *per se*: (1) *what mechanism provides a ramp increase of spiking frequency in the bursts of ramp-I (AUG-I) neurons and in phrenic bursts as seen in vivo*, and (2) *why does the shape of patterns of these ramp-I (AUG-I) neurons and phrenic bursts change from an augmenting type to a decrementing type in vitro, as well as during hypoxic gasping*[5] *and after blockade of glycinergic inhibition in vivo*[6]*?*

We found that a combination of the low-threshold calcium channels with calcium-dependent potassium channels could produce frequency augmenting (ramp) firing patterns when the neuron released from hyperpolarization[4]. The ramp firing pattern in the model was similar to that recorded from ramp-I neurons *in vivo*. Therefore, we suggested that *the augmenting firing patterns in vivo are produced by the interaction of the intrinsic neuronal properties (calcium and calcium-dependent membrane currents) and phasic inhibition.*

Now, why does the shape of inspiratory bursts change from augmenting type to decrementing type in some *in vitro* preparations and during gasping? In order to explain these phenomena (and also to explain why the respiratory rhythm *in vitro* is resistant to blockade of inhibition) we proposed a *switching concept*. This concept suggests that under normal *in vivo* conditions the respiratory network generates oscillations by a network mechanism (and therefore, the augmenting patterns are produced by the combination of intrinsic properties of inspiratory neurons and phasic inhibition). However, the respiratory network may operate normally only when receiving excitatory drive. We believe that a significant part of this drive is provided by a population of conditional pacemaker neurons in the pre-Bötzinger complex (which may play the role of central chemoreceptors). We suggest that *in vivo* these neurons provide *tonic* excitatory drive to the respiratory network, which therefore generates a respiratory pattern by a *network mechanism*. However under certain conditions (e. g. because of an increase in the extracellular potassium concentration *in vitro*[7], or during hypoxia *in vivo*[8]), the conditional pacemaker neurons *switch* to a bursting mode. The synchronized endogenous bursting activity of these neurons now provides bursting excitatory drive to the respiratory network, which therefore operates in a "forced mode" producing oscillations defined by that bursting input drive (a *hybrid pacemaker-network mechanism*). After switching from the network mode (eupnea) to the hybrid mode (*in vitro* or during hypoxia *in vivo*) *the same* premotor inspiratory neurons (and the phrenic nerve) demonstrate decrementing bursts that simply reproduce the shape of pacemaker bursts that drive the respiratory network activity. In addition, such "forced" pacemaker-driven rhythm is resistant to blockade of inhibition.

The switching concept allows maintenance of the advantages of earlier network theories and models providing explanations for various respiratory reflexes and other systems-level phenomena, and, at the same time, does not contradict *in vitro* data. This concept may provide a bridge between the network and hybrid pacemaker-network theories.

4. POSSIBLE ROLES OF DIFFERENT BRAINSTEM NEURON POPULATIONS IN RESPIRATORY PHASE-SWITCHING (M. I. Cohen*)

We have used excitation of volume related vagal afferents (by inflation or by electrical nerve stimulation delivered at precise times in the respiratory cycle) to produce changes of cycle duration: vagal inputs during neural inspiration produced shortening of inspiratory (I) duration, whereas such inputs during neural expiration produced lengthening of expiratory (E) duration. The patterns of a unit's activity immediately before a delayed or retarded phase-switch, compared to such activity in a non-stimulated phase, were interpreted as indicators of the role of a unit in producing the phase switch.

Late-I neurons. Neurons of this type, located in both dorsal and ventral respiratory groups, start firing late in the I phase, have a rapidly augmenting discharge at this time, and are excited by vagal afferent inputs[9]. Electrical vagus stimulation advances their discharge onset, whereas withholding lung inflation retards discharge onset. A crucial feature is that the discharge is closely linked to the end of I (start of E): cycle-triggered histograms (CTHs) triggered at this time point in control vs. test cycles are superimposable around the time of I termination and expiration onset. This property, which is not seen in phrenic discharge or in discharge of other types of augmenting I neuron, is consistent with the late-I neurons having an inspiratory off-switch function.

Expiratory-inspiratory (EI) neurons. These neurons (also designated as pre-I neurons), which are found at various levels of the ventral respiratory group, start firing during the expiratory phase, have a gradually augmenting discharge, and have an abrupt increase of discharge just before the end of E. Vagal inputs during E (lung inflation or electrical nerve stimulation) produce depression of discharge, often to the point of silence. When stimulation ceases, the augmenting discharge resumes and increases abruptly just before the end of expiration (start of I). The CTHs triggered at this time point are superimposable in control vs. test cycles around the time of expiration termination, regardless of the level of preceding activity. This property is consistent with such neurons having an expiratory off-switch function.

5. THE RECONFIGURATION OF RESPIRATORY CPGS AND THE DIFFICULTIES OF IDENTIFYING NEURONS THAT ARE ELEMENTS OF MAMMALIAN CPGS (J. M. Ramirez*)

Despite intense research aimed at unraveling the mechanisms of respiratory rhythm generation, it appears surprising that we still do not know which neural elements are responsible for respiratory rhythm generation. Models based on pacemaker properties, network properties and pacemaker-network hybrids can explain respiratory rhythm generation, but these models remain hypothetical. Why is it so difficult to prove or falsify these models? In order to understand this difficulty, it is important to discuss criteria that define an element of a rhythm-generating network. I would like to address criteria that have been established in invertebrate neuronal networks, and discuss how these criteria apply to the generation of the mammalian respiratory rhythm.

A neuron can be considered to be an element of a rhythm generator, if it fulfils two criteria[10]: (1) its cell membrane should be rhythmically polarized in phase with the rhythmic activity, and (2) current injection into this neuron should alter the generation of the rhythm by shifting the phase of the rhythm. The second criterion is usually associated with the ability to speed or slow rhythmic activity. In contrast, neurons that are located downstream of a rhythm generator (follower neurons), will neither influence phase nor frequency of the rhythm. However, these follower neurons may alter amplitude or shape of a rhythmic activity (motor pattern). While these criteria can easily be applied to small invertebrate networks, difficulties arise in larger mammalian networks. If rhythm generation depends on the activity of larger neuronal populations, then stimulation of a single neuron will not be sufficient to reset the rhythm or alter the speed of the rhythm. In this case, the absence of a resetting property is meaningless.

For mammalian respiration there is compelling evidence for the existence of pacemaker neurons. However, since stimulation of single pacemaker neurons does not alter the overall network activity, it remains uncertain whether these neurons are elements of the rhythm-generating network. Until it is technically feasible to stimulate populations of pacemaker neurons, it will, therefore, be necessary to rely on indirect evidence. It has been demonstrated[11], for example, that hypoxia depolarized only isolated pacemaker, but not non-pacemaker neurons. The depolarization in the pharmacologically isolated pacemaker neurons resulted in a frequency increase strikingly similar to the frequency change of the intact network, suggesting that pacemaker neurons play an important role in regulating the frequency of the respiratory rhythm.

The identification of rhythm generating neurons is also essential for understanding the organization of the mammalian respiratory network. The demonstration that the pre-Bötzinger complex is both essential and sufficient to generate respiratory activity[14] does not exclude the possibility that also neurons in other areas of the mammalian brain contribute to respiratory rhythm generation. In fact, as suggested by McCrimmon (see comment above) it is very likely that other areas also contribute to rhythm generation. However, to demonstrate this possibility, it is essential to identify rhythm-generating neurons in areas outside the pre-Bötzinger complex, which hinges on the same problem: the difficulty to identify rhythm-generating neurons in a mammalian network. Rhythm generating neurons can much easier be identified in invertebrates. In analogy to the pre-Bötzinger complex, the metathoracic ganglion in the locust respiratory network seems to be essential and sufficient for respiratory rhythm generation ("kernel"). Rhythm generating neurons were identified not only in the metathoracic ganglion, but also outside of the kernel. These neurons were (1) active during respiration and (2) could reset the respiratory rhythm and alter the respiratory frequency[11]. In fact, the neurons outside of the metathoracic ganglion were more important for reconfiguring the respiratory network during locomotion than were the neurons located within the kernel[12]. These experiments have important implications as they indicate that the number of active elements in a rhythm-generating network can be variable and state-dependent: in the locust, the respiratory network was reconfigured in a state-dependent manner during locomotion.

Recent *in vivo* and *in vitro* studies suggest that the mammalian respiratory network is also reconfigured, with the same neurons participating in generation of eupneic breathing and other motor patterns such as coughing[13] and gasping[14]. All recorded neurons (pacemaker and non-pacemaker neurons) in the isolated respiratory network participated in the reconfiguration from fictive eupneic activity into gasping. Inspiratory neurons changed in vitro from an augmenting inspiratory to a decrementing activity, which was accompanied by a suppression of synaptic inhibition[16]. However, it is uncertain, which of these neurons are essential for this network reconfiguration. This uncertainty remains as long as we can't prove which neuron type is an element of the rhythm generator and which is a follower neuron.

6. SOME FUTURE DIRECTIONS

Participants indicated that better characterization of ion channels and receptors in respiratory neurons is needed. In this regard, other promising approaches described elsewhere at the meeting were noted. J. Champagnat reviewed genetic and developmental models for the neural control of breathing and embryonic rhythm generation in the mouse. C. G. Tankersley described genetic linkage and DNA sequence analysis of genes that code neurotransmitter receptor sub-units and implications for the genetic control of the respiratory motor pattern. D. McCrimmon presented preliminary RT-PCR studies of VRG neurons with the goal of identifying ion channels and receptors expressed in different "types" of neurons with proposed roles in respiratory rhythm generation.

At the behavioral level, J. Orem reminded conference participants that the control of the respiratory central pattern generator is poorly understood. He described neural correlates of changes in the respiratory pattern during different sleep states. Efforts should be directed toward the development of a comprehensive model of respiratory rhythmogenesis and the neural control of breathing that brings experimental data and current hypotheses into a unified framework.

7. REFERENCES

1. Ramirez, J.M., Schwarzacher, S.W., Pierrefiche, O., Olivera, B.M., and Richter, D.W., 1998, Selective lesioning of the cat pre-Botzinger complex *in vivo* eliminates breathing but not gasping. *J. Physiol.(Lond.).* **507**: 895-907.
2. Pierrefiche, O., Schwarzacher, S.W., Bischoff, A.M., and Richter, D.W., 1998, Blockade of synaptic inhibition within the pre-Botzinger complex in the cat suppresses respiratory rhythm generation *in vivo. J. Physiol.(Lond.)* **509**: 245-254.
3. Speck, D.F., and Feldman, J.L., 1982, The effects of microstimulation and microlesions in the ventral and dorsal respiratory groups in medulla of cat. *J .Neurosci.* **2**: 744-757.
4. Rybak, I.A., Paton, J.F., and Schwaber, J.S., 1997, Modeling neural mechanisms for genesis of respiratory rhythm and pattern. I-III. *J. Neurophysiol.* 77:1994-2039.
5. St.-John, W.M., 1999, Rostral medullary respiratory neuronal activity of decerebrate cats in eupnea, apneusis and gasping, *Respirat. Physiol.* **116**:47-65.
6. Büsselberg, D., Bischoff, A.M., and Richter, D.W., 1998, Blocking of inhibitory synaptic transmission in the respiratory network changes the activity pattern of respiratory neurons. *Soc. Neurosci. Abstr.* **346.10**.
7. Koshiya, N., and Smith, J.C., 1999, Neuronal pacemaker for breathing visualized in vitro. *Nature* **400**:360-363.
8. Melton, J.E., Kadia, S.C., Yu, Q.P., Neubauer, J.A., and Edelman, N.H., 1996, Respiratory and sympathetic activity during recovery from hypoxic depression and gasping in cats. *J. Appl. Physiol.* **80**:1940-1948.
9. Cohen, M.I., Huang, W.X., Barnhardt, R., and See, W.R., 1993, Timing of medullary late-inspiratory neuron discharges: vagal afferent effects indicate possible off-switch function. *J. Neurophysiol.* **69**:1784-1787.
10 Ramirez, J. M. and Pearson, K. G., 1989, Distribution of intersegmental interneurones that can reset the respiratory rhythm of the locust. *J. Exp. Biol.* **141**: 151-176.
11 Thoby-Brisson, M. and Ramirez, J. M., 2000, Role of inspiratory pacemaker neurons in mediating the hypoxic response of the respiratory network in vitro. *J. Neurosci.* **20**:5858-5866.
12 Ramirez, J.M., 1998, Reconfiguration of the respiratory network at the onset of locust flight. *J. Neurophysiol.* **80(6)**:3137-3147.
13 Shannon, R., Baekey, D. M., Morris, K. F., Li, Z., and Lindsey, B. G., 2000, Functional connectivity among ventrolateral medullary respiratory neurones and responses during fictive cough in the cat. *J. Physiol.* (London) **525**:207-224.
14 Lieske, S.P., Thoby-Brisson, M., Telgkamp, P., Ramirez, J.M., 2000, Reconfiguration of the neural network controlling multiple breathing patterns: eupnea, sighs and gasps. *Nat. Neurosci.* **3**:600-607.

Development and Plasticity of Respiratory Control

RESPIRATORY SUPPRESSION INDUCED BY NICOTINE WITHDRAWAL IN THE NEONATAL RAT BRAINSTEM

Implications in the SIDS risk factor

Yasumasa Okada, Shun-ichi Kuwana, and Masaki Iwanami*

1. INTRODUCTION

Maternal smoking is known to be one of the SIDS risk factors[1]. With maternal smoking, nicotine is transported through the placenta to a fetus, and nicotine concentration fluctuates in fetal blood[2, 3]. Also smoking of a mother makes nicotine concentration fluctuates in infantile blood through breast milk or cigarette smoke[4, 5]. Because nicotine has been known to stimulate respiratory neuronal network in the brainstem[6, 7], fluctuation in the fetal or infantile blood nicotine concentration may induce instability of fetal respiration-like activity or infantile respiratory activity, and may be related to the pathophysiology of SIDS.

The purpose of the present study is to test our hypothesis that intermittent perinatal nicotine exposure is related to infantile respiratory instability, and we have studied the effect of nicotine withdrawal on central respiratory control in the perinatal mammals. In order to exclude the influence from circulatory and peripheral effects, we used the isolated brainstem spinal cord preparation of the neonatal rat. The effects of nicotine exposure on respiratory output and medullary neuronal activities have been currently in preparation for publication independent of the present paper[7], and in the present study we focused on the effect of nicotine withdrawal.

* Yasumasa Okada, Department of Medicine, Keio University Tsukigase Rehabilitation Center, Amagiyugashima-cho, Shizuoka, Japan 410-3293. Shun-ichi Kuwana, Department of Physiology, Teikyo University, Itabashi-ku, Tokyo, Japan 173-8605. Masaki Iwanami, Department of Microbiology, Jikei Medical University, Minato-ku, Tokyo, Japan 105-8461.

Frontiers in Modeling and Control of Breathing, edited by Poon and Kazemi.
Kluwer Academic/Plenum Publishers, New York, 2001.

2. METHODS

2.1. Brainstem-Spinal Cord Preparation

The isolated brainstem-spinal cord preparation has been described elsewhere[8, 9, 10]. In brief, 58 neonatal Wister rats (1-4 day old) were used. The rat was deeply anaesthetized with diethyl ether, and the brainstem with cervical spinal cord was isolated in a dissecting chamber filled with oxygenated mock cerebrospinal fluid (mock CSF). The pons and cerebellum were ablated. The preparation was transferred to a recording chamber and fixed with miniature pins on a silicon rubber base with the ventral side up. The preparation was superfused at 26 C with oxygenated control mock CSF, which was equilibrated with a control gas mixture (2 % CO_2 in O_2; mock CSF pH = 7.8). With such "alkaline" superfusate, tissue pH of the superficial (< 400 μm) medullary layer could be maintained in the physiological range[8]. The composition of the control mock CSF was (in mM): 125 NaCl, 4 KCl, 2 $CaCl_2$, 1$MgSO_4$, 0.5 NaH_2PO_4, 26 $NaHCO_3$ and 30 glucose. Central respiratory activity was recorded from the C4 ventral roots with glass suction electrode and integrated. The signals were recorded on a chart recorder. The respiratory frequency count was based on the C4 burst frequency. A respiratory activity was assumed to be augmented or suppressed when respiratory frequency was increased or decreased more than 10 % of a control value respectively.

2.2. Neuronal Recording

Activities of neurons in the superficial (< 400 μm) rostral ventrolateral medulla were recorded intracellularly using a perforated patch configuration as described elsewhere[10]. Briefly, a glass pipette (GC100-TF-10; Clark; Reading, UK) was pulled with a horizontal puller (PA-91; Narishige; Tokyo, Japan) to a tip size of approximately 2 μm. Electrode resistance was about 12 MΩ when filled with a solution containing (mM) 130 K-gluconate, 10 EGTA, 10 HEPES, 1 $CaCl_2$, 1 $MgCl_2$ and nystatin (100mg / ml); pH was adjusted to 7.2-7.3 using KOH. The micropipettes were inserted into the rostral ventrolateral medulla with a manual hydraulic micromanipulator. Membrane potentials were recorded with a whole cell patch amplifier (CEZ 3100; Nihon Koden; Tokyo, Japan). Neurons were searched by applying positive pressure (10-20 cm H_2O) inside the pipette. After a giga-ohm seal was obtained, the recorded membrane potential became gradually negative, and was stabilized in about 10 minutes. A membrane potential was presented without correcting the liquid junction potential.

2.3. Protocol

The effect of nicotine withdrawal on respiratory frequency was observed for 40 minutes, after the preparation was exposed to nicotine containing mock CSF (nicotine concentration 0.01 - 1 μM) for 15 minutes. To analyzed the neuronal response to nicotine withdrawal, neuronal activities were recorded with a whole cell patch technique after preparations were exposed to 10 μM nicotine for 5 minutes. Statistical difference of data was analyzed with paired *t*-test. $P < 0.05$ was considered to be significant. Data were presented as mean ± S.D.

Table 1. Effect of nicotine withdrawal on respiratory frequency

nicotine concentration (before withdrawal)	0.01	0.1	1.0	10 (μM)
fR (control)	5.6 ± 2.3	4.5 ± 2.7	6.3 ± 2.7	5.9 ± 1.9
fR (withdrawal)	5.6 ± 3.3	4.2 ± 3.3	6.8 ± 3.1	7.3 ± 2.3
n	13	14	14	13
Statistics	NS	NS	NS	P<0.03

fR: respiratory frequency (bursts/min) NS: not significant

3. RESULTS

3.1. Effect of Nicotine Withdrawal on Respiratory Output

Statistical difference was not found in the comparison between respiratory frequencies of control and withdrawal states at nicotine concentrations 0.01, 0.1 and 1 μM. However, at a nicotine concentration 10 μM respiratory frequency of a withdrawal state was significantly higher than that of control (Table 1).

Although respiratory frequency was not statistically different from that of control at nicotine concentration 0.01, 0.1 and 1 μM, marked suppression was found in some preparations even at the lowest nicotine concentration (0.01 μM). Especially at nicotine concentration 0.1 μM, among 14 preparations 4 showed decreases of respiratory frequency, including 3 preparations with respiratory cessation (Figure 1).

An example of a preparation which exhibited a prolonged respiratory suppression by nicotine withdrawal (1 μM) is shown in Figure 2.

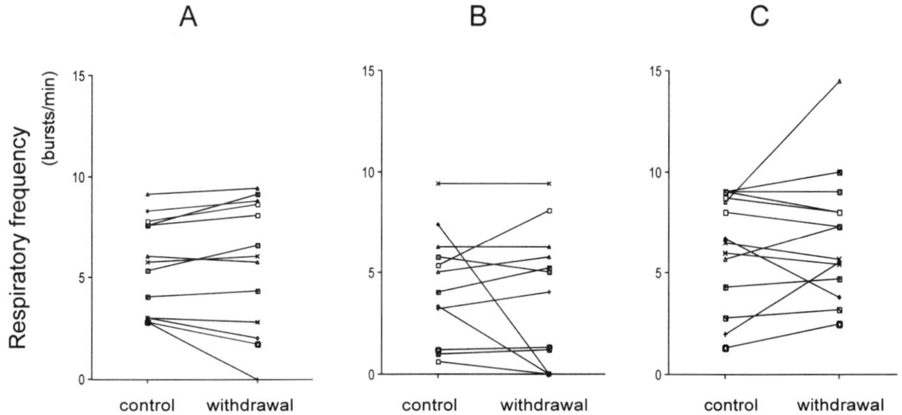

Figure 1. Effect of nicotine withdrawal on respiratory frequency. **A.** Nicotine concentration 0.01 μM. Among 13 preparations 3 showed a decrease of respiratory frequency. **B.** nicotine concentration 0.1 μM. Among 14 preparations 4 showed a decrease of respiratory frequency including 3 preparations with respiratory cessation. **C.** nicotine concentration 1μM. Among 14 preparations 4 showed a decrease of respiratory frequency.

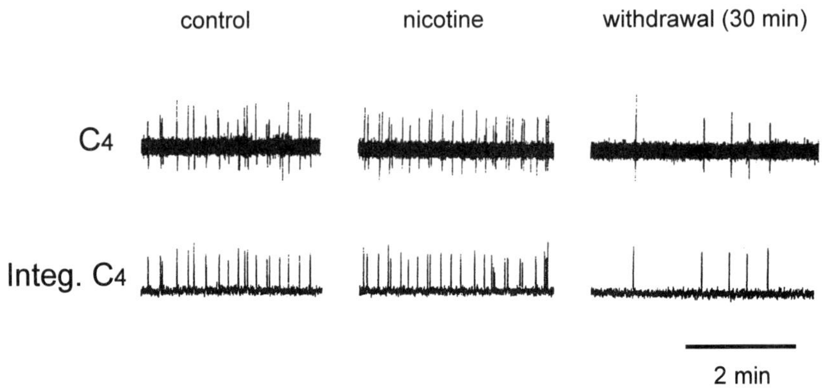

Figure 2. Example of a preparation which exhibited a marked decrease of respiratory frequency by nicotine withdrawal (1 μM). Integ.C4: integrated C4 activity.

3.2. Effect of Nicotine Withdrawal on Neuronal Activities

We successively whole cell recorded 17 neurons in the rostral ventrolateral medulla. Among these 17 neurons 7 were tonic non-respiratory, 6 were inspiratory, and 4 were expiratory. Their membrane potentials were -56.7 ± 6.3 (mV). At 10 minutes after the onset of nicotine withdrawal (10 μM), 86% of tonic non-respiratory and 67 % of inspiratory neurons persisted to show depolarization (> 5 mV depolarized than control). However, none of expiratory neurons showed membrane potential response to nicotine withdrawal. An example of a response to nicotine withdrawal is shown in Figure 3. As shown in this figure, nicotine withdrawal suppressed the activity of inspiratory neurons; a firing frequency during an inspiratory burst decreased concomitant with a decrease of respiratory frequency, although a membrane potential remained depolarized.

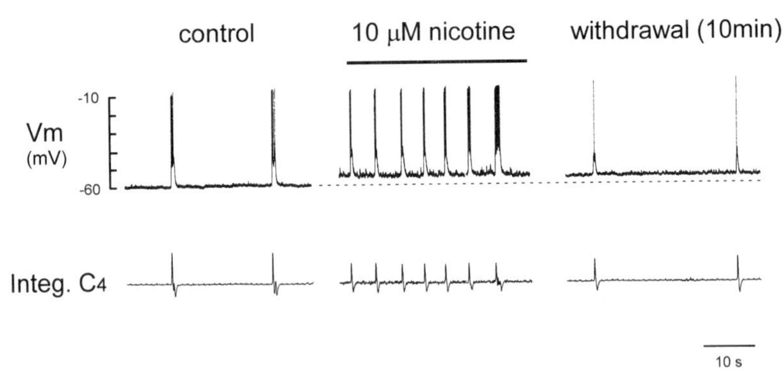

Figure 3. Example of membrane potential responses to nicotine and nicotine withdrawal. Nicotine (10 μM) depolarized this inspiratory neuron and increased excitatory postsynaptic potentials (EPSPs) with an increase of a respiratory frequency. Withdrawal of nicotine suppressed the neuronal activity; a firing frequency during an inspiratory burst and EPSPs were decreased with a decrease of respiratory frequency. However, the membrane potential remained depolarized. Vm: membrane potential. Integ.C4: integrated C4 activity.

In 4 tonic non-respiratory neurons responses of membrane potentials to various current injection were successively analyzed in control and nicotine withdrawal states. Although they showed a marked reduction in membrane resistances with nicotine (10 μM) (data not shown), nicotine withdrawal restored membrane resistance in 20 minutes in 3 preparations. However, one preparation persistently showed an increase of firing threshold with positive current injection and a decrease of membrane resistance in 40 minutes after the start of nicotine withdrawal (Figure 4).

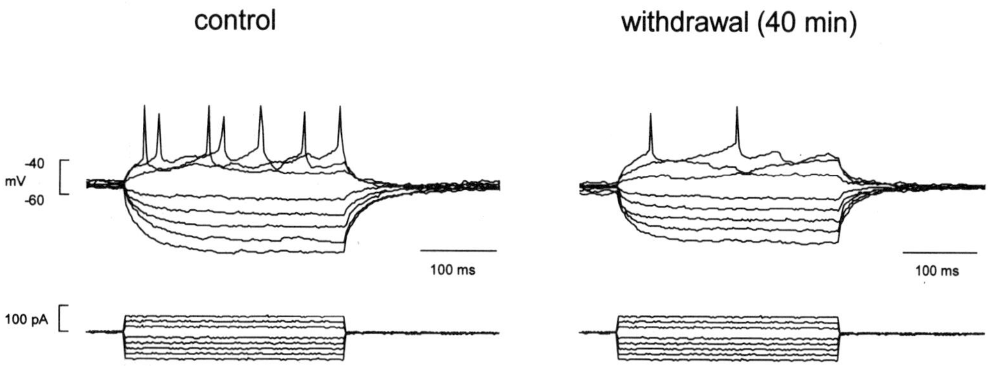

Figure 4. Example of membrane potential responses in a tonic non-respiratory neuron to various current injection with current clamp mode. Withdrawal from nicotine (10 μM) decreased firing frequency with positive current injection. Voltage response to various current injections were diminished by nicotine withdrawal.

4. DISCUSSION

We have demonstrated that nicotine withdrawal occasionally suppresses a respiratory activity in the neonatal rat brainstem. Nicotine withdrawal induced respiratory suppression was observed even with a concentration of nicotine (0.01 and 0.1 μM) which is lower than a smoker's blood nicotine concentration.

4.1. Critique of Methods

In the present study we applied the isolated brainstem-spinal cord preparation of the neonatal rat[8, 9, 10]. In order to study the central effects of nicotine withdrawal, influence on peripheral chemoreceptors and cardiovascular system must be excluded[11]. By applying the isolated brainstem-spinal cord preparation, it is possible to exclusively study the central effects of nicotine withdrawal. One possible critique may be that respiratory suppression observed during nicotine withdrawal is not by nicotine withdrawal itself but by spontaneous inactivation of the respiratory control mechanism. This is not the case because in our preliminary study a respiratory activity of the isolated brainstem-spinal cord preparation is always well stable at least 3 hours. Also when respiratory activity was suppressed by nicotine withdrawal and the preparation was re-exposed to nicotine, respiratory activity could be restored (Figure 5).

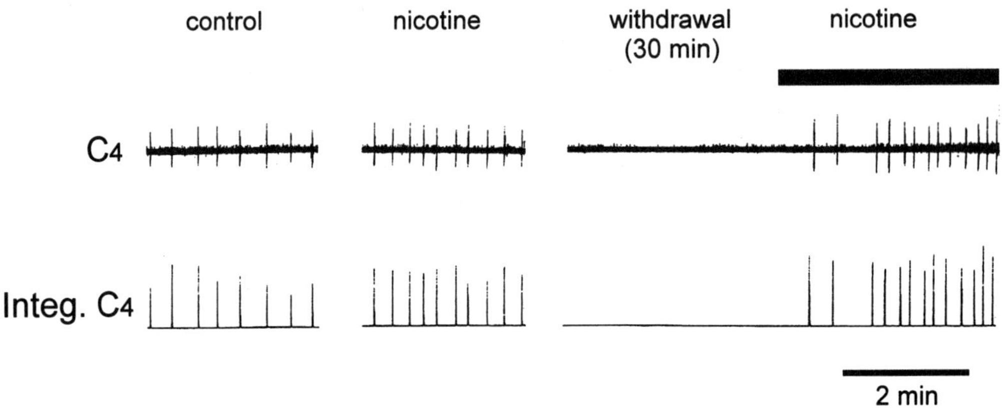

Figure 5. Example of a preparation which exhibited an increase of respiratory frequency by nicotine (0.01 μM) followed by disappearance of a respiratory activity by nicotine withdrawal. Re-exposure of the preparation to nicotine restored a respiratory activity.

Our analyses of nicotine withdrawal responses to neuronal activities may be still preliminary, because the number of whole cell recorded neurons are not enough. However, in neurons that were excited by nicotine (especially inspiratory neurons and tonic non-respiratory neurons), nicotine withdrawal apparently suppressed firings although even these neurons were still depolarized (Figures 3, 4). One possible explanation of this observed neuronal suppression is due to sodium channel inactivation.

4.2. Implications in the SIDS Risk Factor

Maternal smoking is known to be one of the SIDS risk factors[1]. It has been shown in experiments with neonatal rats that maternal nicotine depresses eupneic ventilation[12].This report suggests that infantile respiratory suppression related to maternal smoking can partly explain the causative relationship between maternal nicotine exposure and respiratory failure in SIDS victims. The actual mechanism to increase the risk of SIDS by perinatal nicotine exposure may be complex. Several mechanisms may be involved, including abnormality of nicotinic acetylcholine receptor development,[13] direct respiratory suppressive effect of nicotine[14], and attenuation of hypoxic ventilatory response[15, 16]. There have been reports on the effect of maternal nicotine exposure in ewes on fetal respiratory movement[17] and on the effect of maternal cigarette smoking on human fetal respiratory movement[18, 19]. In these reports cigarette smoking and nicotine exposure showed direct suppressive effects on respiratory movement in fetuses. It must be noted that nicotine has an apparent respiratory stimulatory effect. However, nicotine also has a direct respiratory suppressive effect especially when animals are exposed to a high dose of nicotine[19, 14, 20]. Interestingly, it was reported with a real-time ultrasound system that after maternal smoking fetal breathing movements as well as a number of apneic epochs were increased in humans[21]. Their report indicates that maternal smoking transiently stimulates the fetal respiratory control system but subsequently depresses the system. This abnormal stimulation is repeated in the fetal respiratory control system during the entire period of intra-uterine development, and may be partly responsible for respiratory depression in SIDS victims. It is suggested that nicotine induced respiratory suppression is due to excessive depolarization of respiratory neurons[7]. We have also shown in the present study that some respiratory neurons remain inactivated even after nicotine is withdrawn (Figure 3). This seems to be one of the mechanisms for the respiratory suppression by nicotine withdrawal.

It is concluded that nicotine withdrawal from acute nicotine exposure in the isolated brainstem-spinal cord preparation of the neonatal rat induced occasional respiratory suppression including even apnea. It is suggested that nicotine withdrawal induced respiratory suppression is related to instability of respiratory control mechanism in human infants whose parents smoke. Effects of cigarette smoke exposure on human infantile respiratory control need to be studied.

5. ACKNOWLEDGMENTS

This study was supported by the Research Grant for the Specific Diseases of the Japanese Ministry of Health and Welfare (YO) and by the Research Grant of the Japan Society for Promotion of Science (YO).

6. REFERENCES

1. P. S. Blair, P. J. Fleming, D. Bensley, I. Smith, C. Bacon, E. Taylor, J. Berry, J. Golding, and J. Tripp, Smoking and the sudden infant death syndrome: results from the 1993-5 case-control study for confidential inquiry into stillbirths and deaths in infancy, *Br. Med. J.* **313**(7051), 195-198 (1996).
2. W. Luck, H. Nau, R. Hansen, and R. Steldinger, Extent of nicotine and cotinine transfer to the human fetus, placenta and amniotic fluid of smoking mothers, *Dev. Pharmacol. Ther.* **8**(6), 384-95 (1985).
3. B. V. Sastry, M. B. Chance, M. E. Hemontolor, and T. A. Goddijn-Wessel, Formation and retention of cotinine during placental transfer of nicotine in human placental cotyledon, *Pharmacology* **57**(2), 104-16 (1998).
4. W. Luck, and H. Nau, Nicotine and cotinine concentrations in the milk of smoking mothers: influence of cigarette consumption and diurnal variation, *Eur. J. Pediatr.* **146**(1), 21-6 (1987).
5. R. A. Greenberg, N. J. Haley, R. A. Etzel, and F. A. Loda, Measuring the exposure of infants to tobacco smoke. Nicotine and cotinine in urine and saliva, *N. Engl. J. Med.* **310**(17), 1075-8 (1984).
6. N. B. Dev, and H. H. Loeschcke, Topography of the respiratory and circulatory responses to acetylcholine and nicotine on the ventral surface of the medulla oblongata, *Pflügers Arch.* **379**(1), 19-27 (1979).
7. S. Kuwana, Y. Okada, M. Iwanami, and T. Natsui, Effects of nicotine and nicotinic receptor antagonist on central respiratory control in the isolated brain stem-spinal cord of neonatal rat, *Jpn. J. Physiol.* (Abstract) (in press)
8. Y. Okada, K. Mückenhoff, G. Holtermann, H. Acker, and P. Scheid, Depth profiles of pH and PO_2 in the isolated brain stem-spinal cord of the neonatal rat, *Respir. Physiol.* 93(3), 315-326 (1993).
9. Y. Okada, A. Kawai, K. Mückenhoff, and P. Scheid, Role of the pons in hypoxic respiratory depression in the neonatal rat, *Respir. Physiol.* 111(1), 55-63 (1998).
10. S. Kuwana, Y. Okada, and T. Natsui, Effects of extracellular calcium and magnesium on central respiratory control in the brainstem-spinal cord of neonatal rat, *Brain Res.* **786**(1-2), 194-204 (1998).
11. P. Zapata, A. Zuazo, and F. Llados, Respiratory and circulatory reflexes induced by nicotine injections: role of carotid body chemoreceptors, *Arch. Internat. Pharmacodyn. Ther.* **219**(1), 128-139 (1976).
12. W. M. St.-John, and L. C. Leiter, Maternal nicotine depresses eupneic ventilation of neonatal rats, *Neurosci. Lett.* **267**(3), 206-208 (1999).

13. H. A. Navarro, F. J. Seidler, J. P. Eylers, F. E. Baker, S. S. Dobbins, S. E. Lappi, and T. A. Slotkin, Effects of prenatal nicotine exposure on development of central and peripheral cholinergic neurotransmitter systems. Evidence for cholinergic trophic influences in developing brain, *J. Pharmacol. Exp. Ther.* **251**(3), 894-900 (1989).
14. J. W. Sloan, W. R. Martin, and M. Bostwick, Mechanisms involved in the respiratory depressant actions of nicotine in anesthetized rats, *Pharmacol. Biochem. Behav.* **34**(3), 559-564 (1989).
15. T. A. Slotkin, S. E. Lappi, E. C. McCook, B. A. Lorber, and F. J. Seidler, Loss of neonatal hypoxia tolerance after prenatal nicotine exposure: implications for sudden infant death syndrome, *Brain Res. Bull.* **38**(1), 69-75 (1995).
16. J. Milerad, H. Larsson, J. Lin, and H. W. Sundell, Nicotine attenuates the ventilatory response to hypoxia in the developing lamb, *Pediatr. Res.* **37**(5), 652-660 (1995).
17. F. Manning, D. Walker, and C. Feyerabend, The effect of nicotine on fetal breathing movements in conscious pregnant ewes, *Obstet. Gynecol.* **52**(5), 563-568 (1978).
18. G. Gennser, K. Marsal, and B. Brantmark, Maternal smoking and fetal breathing movements, *Am. J. Obstet. Gynecol.* **123**(8), 861-867 (1975).
19. F. A. Manning, and C. Feyerabend, Cigarette smoking and fetal breathing movements, *Brit. J. Obstet. Gynaecol.* **83**(4), 262-270 (1976).
20. M. Sasaki, M. Yamaya, W. Hida, M. Nakamura, T. Sasaki, H. Sasaki, and T. Takishima, Effect of hypercapnia on ventilatory response to intravenous nicotine administration in anesthetized dogs, *Respir. Physiol.* **78**(2), 177-186 (1989).
21. P. S. Eriksen, G. Gennser, O. Lofgren, and K. Nilsson, Acute effects of maternal smoking on fetal breathing and movements, *Obstet. Gynecol.* **61**(3), 367-372 (1983).

EFFECT OF DNA HYPOMETHYLATION ON NEURAL CONTROL OF RESPIRATION: A GENETIC MODEL

Guoping Fan[*], Marina Siniaia[†], Chi-Sang Poon[†], and Rudolf Jaenisch[*‡]

1. INTRODUCTION

DNA methylation represents one form of epigenetic factors that can influence gene activities without a change in DNA sequence[1]. In mammals, methylation of DNA occurs at the C5 position of the cytosine residues, primarily at the CpG dinucleotides. A family of DNA methyltransferases carries out the covalent reaction of cytosine methylation. The first methylase gene, *Dnmt1,* encodes a maintenance DNA methyltransferase (Dnmt1, EC 2.1.1.37) which preferentially methylates hemi-methylated DNA produced after DNA replication[2]. Two newly discovered methylase genes, *Dnmt3a* and *3b*, are the *de novo* methyltransferases that can methylate native DNA substrates[3, 4]. The essential role for these enzymes in establishing and maintaining DNA methylation has been demonstrated by targeted mutation of DNA methyltransferases in mice. Mutant mice lacking either Dnmt1 or Dnmt3a and 3b exhibit significant demethylation in the genome, and die at embryonic day (E) 8-10 just after gastrulation[4-6]. These results indicate that DNA methylation is essential for mammalian embryonic development. At the cellular level, DNA hypomethylation also perturbs the events of genomic imprinting, X-chromosome inactivation, and suppression of endogenous retroviruses[7-10].

The mammalian brain contains high levels of Dnmt1 both during development and in adulthood[11-14] and the level of DNA methylation seen in the adult brain is also higher than that in other tissues[15-17]. At the perinatal stages, the level of DNA methylation is actively regulated in the brain[16], suggesting a role for DNA methylation in the neuronal differentiation and maturation. A recent study also connects DNA methylation with a human neurodevelopmental disease, Rett syndrome, which is caused by mutations in the *MECP2* gene[18]. MECP2 encodes a methylcytosine binding protein that is also abundant in the central nervous system (CNS). Since a major function of the MECP2 protein is to mediate methylation-induced gene suppression[19], the effect of *MECP2* mutations on brain development suggests that changes of DNA methylation may also affect the postnatal CNS function. It is of interests to mention that patients with Rett syndrome

* Whitehead Institute for Biomedical Research and ‡ Department of Biology, Massachusetts Institute of Technology, 9 Cambridge Center, Cambridge, MA 02142 USA; † Harvard-MIT Division of Health Sciences and Technology, MIT, Cambridge, MA 02139 USA

Frontiers in Modeling and Control of Breathing, edited by Poon and Kazemi.
Kluwer Academic/Plenum Publishers, New York, 2001.

often exhibit abnormality in respiration[20, 21]. This implies that methylation-mediated gene regulation may be important for the respiratory control.

We set out to investigate the role of DNA methylation in the CNS function with a focus on neural control of respiration. Since the previously generated Dnmt1 mutant mice die at the midgestation stage before neuronal differentiation[5, 6], they are not useful for our study. We therefore generated a line of conditional mutant mice in which DNA methylation is reduced primarily in the CNS. For this, we used the cre/loxP binary system to delete the *Dnmt1* gene tissue-specifically in brain precursor cells at the midgestational stage. *Dnmt1* gene deletion in brain precursor cells resulted in significant DNA hypomethylation in progeny cells, including descendant postmitotic neurons. We found that DNA hypomethylation in the brain apparently affects the vital functions of CNS control of respiration, because all conditional mutant mice died at birth due to respiratory failure.

2. MATERIALS AND METHODS

2.1. Brain-Specific Dnmt1 Conditional Mutant Mice

We used the cre/loxP binary system to generate Dnmt1 conditional mutants. Briefly, a conditional Dnmt1^{2lox}allele was generated by inserting 34bp loxP sites flanking the exons 4 and 5 of the *Dnmt1* gene[22]. In the presence of the cre recombinase, the two Dnmt1 exons would be deleted, resulting null Dnmt1^{1lox} alleles. To obtain Dnmt1 gene deletion in CNS precursor cells *in vivo*, we crossed the female mice carrying the *Dnmt1*$^{2lox/2lox}$ conditional alleles with male *Dnmt1*$^{+/2lox}$ mice carrying the Nestin-cre transgene[23, 24]. It has been demonstrated that Nestin-cre mediated gene deletion is primarily in the CNS. Southern blot analysis and PCR reactions were used for analyzing the efficiency of gene deletion, global DNA methylation, and genotyping mice.

2.2. Histology and Electrophysiology

Embryos or newborn mice were fixed in 10% formalin for over 24 hours, and processed for paraffin sections. Tissue sections were visualized with the staining of hematoxylin/eosin.

Newborn mice were anesthetized with a light dose of sodium pentobarbital (Nembutal) (25-30 mg/kg i. p.)[25] and were placed supine on a heating pad. Central respiratory activity was monitored by recording hypoglossal nerve discharge. A hypoglossal (XII) nerve (which innervates both the genioglossus muscle and tongue retractors) was isolated by blunt dissection with a ventral approach at the cervical level, cut distally and mounted on custom-made bipolar silver-wire electrodes and kept in a warm mineral oil pool. Inspiratory-related efferent neuronal discharges of the hypoglossal nerve were amplified (Axon Instruments, CyberAmp 380), time-averaged with a leaky integrator (Paynter filter, time constant, 15 ms) and recorded (Thermal Array Recorder, Nihon Kohden). In some experiments small custom-made bipolar fish hook electrodes were implanted into the diaphragm to record an electromyogram (EMG). Immediately after birth, small stainless steel electrodes were implanted subdermally on both sides of the chest to record an electrocardiograph (EKG). In some experiments the EKG was recorded in normal and mutant pups *in utero* with the mother under urethane anesthesia (1.2 g/kg i.p.).

3. RESULTS

3.1. Dnmt1 Gene Deletion in CNS Precursor Cells Causes Global DNA Hypomethylation in the Brain and Neonatal Lethality of Mutant Animals

To achieve the *Dnmt1* gene deletion in CNS precursor cells *in vivo*, we generated transgenic mice carrying the *Dnmt1*$^{2lox/2lox}$ conditional alleles and a nestin promoter-driven cre transgene (*Nestin-cre*)[23]. *Nestin-cre* is actively in brain precursor cells at E9-10, and leads to virtually complete gene deletion in the CNS at the midgestational stage[23, 24]. *Nestin-cre; Dnmt1* conditional mutant mice were obtained at the expected frequency at all embryonic stages examined (E10.5-19.5). Southern blot analysis of E12.5-19.5 mutant embryos showed that the *Dnmt1* gene deletion occurred in approximately 95% of the cells in the brain. Moderate levels of gene deletion (less than 30%) were observed in muscles and kidney, and a very low frequency of gene deletion was seen in lung, heart, and liver. Western blot analysis showed that levels of Dnmt1 proteins in the CNS were already very low at E12.5 and were undetectable at E15.5, suggesting that the Dnmt1 protein turned over rapidly following gene deletion.

Although conditional mutant mice were readily obtained during embryonic stages, we did not recover any mutant animals alive throughout the postnatal ages. In fact, when we carefully followed up the newborn mutant animals, we found that all mutant animals died immediately after birth (see below 3.2.).

To assess the effect of Dnmt1 deficiency on global DNA methylation, we examined the methylation level in the centromeric region and intra-cisternal A particle sequences that are highly methylated in wild type cells. Brain DNA was digested with methylation sensitive restriction enzyme and subjected to Southern blot analysis as previously described[22]. Samples from E12.5-19.5 mutant brains exhibited extensive DNA hypomethylation, indicating that Dnmt1 is still required for maintaining DNA methylation at the midgestational stages.

3.2. DNA Hypomethylation Disrupts Neural Control of Respiration

To examine when and how the conditional mutant mice die at the neonatal age, we carefully observed the behaviors of naturally- or Cesarean-delivered pups. In contrast to the control mice that survived shortly after delivery, all mutant mice died within one hour after birth. Mutant animals only occasionally gasped, but never initiated any coordinated rhythmic breathing, indicating that they were suffering from respiratory distress. Postmortem examination showed that the lungs in mutant animals failed to inflate, confirming respiratory failure (Fig. 1). Multiple causes could lead to respiratory failure, including a defect in neural rhythmogenesis or motor output of the respiratory pattern generator, occlusion of the respiratory tract, delay in lung development, or abnormalities secondary to the failure of the cardiovascular or other vital systems. When we examined the behavior of mutant fetuses within the uterus, we found that they had normal body

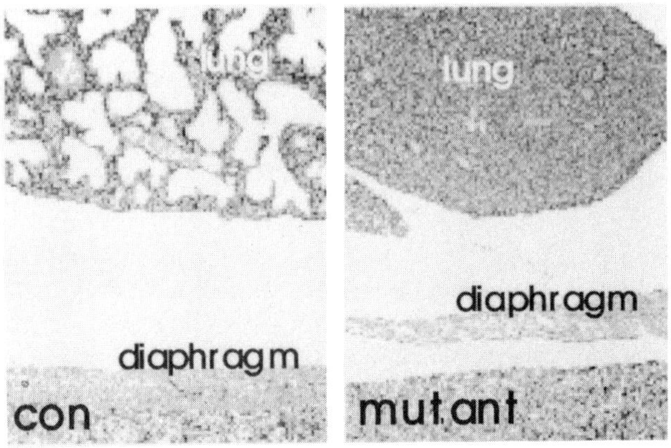

Fig. 1. Parasagittal sections of chest area of C-section delivered control (con) and mutant mice. Note that the lung in mutant mice was still compressed.

movement, either spontaneous or in response to external mechanical stimuli, suggesting that they had a normal sensorimotor reflex. We also observed EKGs of the heart *in utero,* and found a normal heart rate in mutant mice (control: 356/min; mutant: 346-364/min). This suggests that respiratory failure in mutant animals is not caused by malfunction of the cardiovascular system. Since virtually no *Dnmt1* deletion was detected in the lung of mutant mice, lung cell development was likely normal. Histological examination did not show any obvious abnormalities in respiratory muscle groups, including the intercostal and diaphragm muscles. In addition, diaphragm muscles were innervated normally by the respiratory phrenic nerve originating from the cervical spinal cord. These observations suggest that the respiratory failure in the mutant is not triggered by abnormalities of the respiratory motor apparatus in the viscera.

Since the conditional Dnmtl mutant mice showed extensive DNA hypomethylation in the CNS, we examined whether DNA hypomethylation could affect the survival and function of CNS neurons, resulting in abnormal neural control of respiration. No obvious morphological defect was observed in the CNS of E18.5-19.5 mutant mice. Neuronal survival appeared to be normal in mutant mice as assessed by Nissl and TUNEL staining in brain sections at E15.5 and E18.5-19.5. To investigate whether the respiratory distress observed in Dnmtl mutants could reflect a functional defect in the CNS control of respiration, we decided to monitor the neuronal respiratory activity in mutant mice. We recorded the respiratory activity in the 12th cranial hypoglossal nerve and performed diaphragmatic electromyogram in E18.5 and E19.5 mice delivered by Cesarean section. The majority of hypoglossal neuronal discharges are inspiratory, with a bursting period similar to that of phrenic nerve activity. Indeed, many hypoglossal motor neurons receive innervations from pre-motor inspiratory neurons located within the respiratory pattern generator in the ventrolateral medulla[26-28]. We found that the hypoglossal nerve activity in normal mice displayed a rhythmic but gasping-like pattern approximately 5 minutes after birth (frequency, 12/min), which was converted to an eupneic pattern with increased respiratory frequency (50/min) and heart rate (260/min) within 10-20 min. In contrast, mutant mice after delivery only produced sporadic gasping activity, which never evolved into a normal breathing pattern. The ability of the mutant mice to produce gasping activity in the hypoglossal nerve and diaphragm suggests that their respiratory motor efferent pathway was functional. The hypoglossal motor nucleus in the brainstem of mutant mice was morphologically normal. We propose that the diminished respiratory discharge in hypoglossal nerve may result from a defect in the respiratory rhythomogenesis or neurotransmissoin at the brain stem and medulla respiration center.

4. DISCUSSION

We generated a genetic model in which DNA methylation in the developing brain is significantly reduced. DNA hypomethylation did not affect the proceeding of CNS development, but apparently perturbed the neural control of respiration. The neural mechanism underlying this defect is still not clear. The absence of rhythmic respiratory discharge in the mutant hypoglossal nerve suggests that DNA hypomethylation may disrupt respiratory rhythmogenesis or neurotransmission in the mutant brain. This possibility will be examined with experiments to record the rhythmic discharge using *in vitro en bloc* brainstem and spinal cord preparation as well as slice preparations. It will be also of interests to determine whether DNA hypomethylation impairs neuronal transmission or activity at single neuronal cell level. In this case, a combination of electrophysiological and pharmacological approach can be used to monitor neuronal cultures or slice preparations from mutant and control mice. The effect of various neurotransmitters on modulating the respiratory discharges of mutant and control cells can also be tested.

At the molecular level, we intend to understand the effect of DNA hypomethylation on neuronal gene expression. We are currently using cDNA array and gene chip technology to look for potential candidate genes that are deregulated in hypomethylated CNS neurons. DNA methylation is required for controlling proper expression of imprinting genes, the X-linked *Xist* gene, and so on. Interestingly, some of the imprinting genes such as *Necdin* are known to be important for respiratory control[29]. Understanding the change of neuronal gene expression in hypomethylated neurons may also shed light on the neural mechanism underlying the abnormal respiratory control of the hypomethylated brain.

5. REFERENCES

1. R. Jaenisch. DNA methylation and imprinting: why bother? *Trends Genet.* **13**, 323-329 (1997).
2. T. Bestor , A. Laudano, R. Mattaliano, and V. Ingram. Cloning and sequencing of a cDNA encoding DNA methyltransferase of mouse cells. The carboxyl-terminal domain of the mammalian enzymes is related to bacterial restriction methyltransferases. *J Mol Biol* **203**, 971-983 (1988).
3. M. Okano, S. Xie, and E. Li. Cloning and characterization of a family of novel mammalian DNA (cytosine-5) methyltransferases. *Nature Genetics* **19**, 219-220 (1998).
4. M. Okano, D. W. Bell, D. A. Haber, and E. Li. DNA methyltransferases Dnmt3a and Dnmt3b are essential for de novo methylation and mammalian development. *Cell* **99**, 247-257 (1999).
5. E. Li, T. Bestor, R. Jaenisch. Targeted mutation of the DNA methyltransferase gene results in embryonic lethality. *Cell* **69**, 915-926 (1992).
6. H. Lei, S. Oh , M. Okano, R. Juttermann, K. Goss, R. Jaenisch, and E. Li,. De novo DNA cytosine methyltransferase activities in mouse embryonic stem cells. *Development* **122**, 3195-3205 (1996).
7. E. Li, C. Beard, and R. Jaenisch. Role for DNA methylation in genomic imprinting. *Nature* **366**, 362-365 (1993).

8. C. Beard, E. Li, and R. Jaenisch. Loss of methylation activates Xist in somatic but not in embryonic cells. *Genes Dev* **9**, 2325-2334 (1995).
9. B. Panning, and R. Jaenisch. DNA hypomethylation can activate Xist expression and silence X-linked genes. *Genes Dev* **10**, 1991-2002 (1996).
10. C. P. Walsh, J. R. Chaillet, and T. H. Bestor. Transcription of IAP endogenous retroviruses is constrained by cytosine methylation. *Nat Genet* **20**, 116-117 (1998).
11. K. Goto, M. Numata, J. I. Komura, T. Ono, T. H. Bestor, and H. Kondo. Expression of DNA methyltransferase gene in mature and immature neurons as well as proliferating cells in mice. *Differentiation* **56**, 39-44 1994).
12. P. J. Brooks, C. Marietta, and D. Goldman. DNA mismatch repair and DNA methylation in adult brain neurons. *J Neurosci* **16**, 939-945 (1996).
13. J. M. Trasler, D. G. Trasler, T. H. Bestor, E. Li, and F. Ghibu. DNA methyltransferase in normal and Dnmtn/Dnmtn mouse embryos. *Dev Dyn* **206**, 239-247 (1996).
14. K. Inano, I. Suetake, T. Ueda, Y. Miyake, M. Nakamura, M. Okada, and S. Tajima. Maintenance-type DNA methyltransferase is highly expressed in post-mitotic neurons and localized in the cytoplasmic compartment. *J Biochem.* **128**, 315-321 (2000)
15. V. L. Wilson, R. A. Smith, S. Ma, and R. G. Cutler. Genomic 5-methyldeoxycytidine decreases with age. *J Biol Chem* **262**, 9948-9951 (1987)
16. R. Tawa, T. Ono, A. Kurishita, S. Okada, and S. Hirose. Changes of DNA methylation level during pre- and postnatal periods in mice. *Differentiation* **45**, 44-48 (1990).
17. T. Ono T, Y. Uehara, A. Kurishita, R. Tawa, and H. Sakurai. Biological significance of DNA methylation in the ageing process. *Age Ageing* **22**, S34-43 (1993).
18. R. E. Amir, I. B. Van Den Veyver, M. Wan M, C. Q. Tran, U. Francke, and H. Y. Zoghbi. Rett syndrome is caused by mutations in X-linked MECP2, encoding methyl- CpG-binding protein 2. *Nat Genet* **23**, 185-188 (1999).
19. H. H. Ng, and A. Bird. DNA methylation and chromatin modification. *Curr Opin Genet Dev* **9**, 158-163 (1999).
20. A.M. Kerr. A review of the respiratory disorder in the Rett syndrome. *Brain Dev* **14**, 43-45 (Suppl) (1992).
21. D. D. Armstrong, J. K. Dunn, R. J. Schultz, D. A. Herbert, D. G. Glaze, and K. J. Motil. Organ growth in Rett syndrome, a postmortem examination analysis. *Pediatr Neurol* **20**, 125-129 (1999).
22. L. Jackson-Grusby, C. Beard, R. Possemato, D. Fambrough, G. Csankovszki, J. Dausman, P. Lee, C. B. Wilson, E. Lander, and R. Jaenisch. Loss of Genomic Methylation cause p53-dependent apoptosis and epigenetic deregulation. *Nature Genet.* **27**, 31-9 (2001).
23. B. Bates, M. Rios, A. Trumpp, C. Chen, G. Fan, J. M. Bishop, and R. Jaenisch. Neurotrophin-3 is required for proper cerebellar development. *Nat Neurosci* **2**, 115-117 (1999).
24. A. Trumpp, M. J. Depew, J. L. R. Rubenstein, J. M. Bishop, and G. R. Martin. Cre-mediated gene inactivation demonstrates that FGF8 is required for cell survival and patterning of the first banchial arch. *Genes Dev* **13**, 3136-48 (1999).
25. J. F. Paton, J. M. Ramirez, and D. W. Richter. Functionally intact in vitro preparation generating respiratory activity in neonatal and mature mammals. *Pflugers Arch* **428**, 250-260 (1994).
26. D. J. Withington-Wray, S. W. Mifflin, K. M. Spyer. Intracellular analysis of respiratory-modulated hypoglossal motoneurons in the cat. *Neuroscience* **25**, 1041-1051 (1988).
27. T. Ono, Y. Ishiwata, N. Inaba, T. Kuroda, and YNakamura. Hypoglossal premotor neurons with rhythmical inspiratory-related activity in the cat, localization and projection to the phrenic nucleus. *Exp Brain Res* **98**, 1-12 (1994).
28. W. St. John. Neurogenesis of patterns of automatic ventilatory activity. *Progress in Neurobiol.* **56**, 97-117 (1998).
29. M. Gerard, L. Hernandez, R. Wevrick, and C. L. Stewart. Disruption of the mouse necdin gene results in early post-natal lethality. *Nat Genet.* **23**, 199-202 (1999).

RESPIRATORY PATTERN AND HYPOXIC VENTILATORY RESPONSE IN MICE FUNCTIONALLY LACKING α_{2A}-ADRENERGIC RECEPTORS

John M. Bissonnette, Sharon J. Knopp, Danielle M. Wright, and Leigh B. MacMillan*

1. INTRODUCTION

In newborn mammals including man the respiratory pattern undergoes changes during the neonatal period[10, 11, 26, 27, 36]. These developmental changes are characterized by an increase in the duration of inspiratory time (TI), which in species born after relatively short gestations such as opossum[10] and mouse[27] can be quite marked. In addition to developmental differences in metabolism, pulmonary mechanics and stimulatory input from the periphery the intrinsic membrane properties of inspiratory neurons may contribute to this lengthening of TI. Spike frequency adaptation and afterhyperpolarization is present as early as postnatal day 1 (P1) in rat hypoglossal motoneurons[35] and PO in mouse inspiratory neurons located in the pre-Bötzinger complex (PBC)[28]. Entry of calcium through high-voltage-activated (HVA) calcium channels has been shown to be essential for the generation of the AHP in PBC neurons[9], ventral nucleus of the solitary tract[29] and hypoglossal motoneurons[32, 33] by activating slow-conductance calcium-activated potassium channels[29, 34]. Therefore, the inhibiting modulation of HVA calcium channels during development is a potential mechanism by which TI could be prolonged. α_{2a}-adrenergic receptors which are coupled by pertussis toxin sensitive G proteins to various effectors, including inhibition of voltage activated Ca^{2+} channels[20], and have been demonstrated in cardiorespiratory areas of the medulla[1, 31] are candidates for this role. In order to evaluate their importance respiratory timing in the postnatal period was examined in mice who functionally lack α_{2a}-adrenergic receptors.

The animals used have mutated α_{2a}-adrenergic receptors in which asparagine is substituted for aspartic acid at position 79. Aspartic acid at this site within the punative second transmembrane region is highly conserved among G protein coupled receptors[18].

* John M. Bissonnette, Sharon J. Knopp, and Daniel M. Wright. Departments of Obstetrics and Gynecology and of Cell and Developmental Biology, Oregon Health Sciences University, Portland, OR, 97201. Leigh B. MacMillan Department of Pharmacology, Vanderbilt University, Nashville, TN 37232

Frontiers in Modeling and Control of Breathing, edited by Poon and Kazemi.
Kluwer Academic/Plenum Publishers, New York, 2001.

When the mutated receptor (D79N) was transfected into AtT20 mouse pituitary tumor cells, electrophysiological studies showed that coupling to inwardly rectifying potassium currents was disrupted. The inhibition of voltage-activated Ca^{2+}channels and of adenyl cyclase was retained[30]. In gene targeted mice, which carry the D79N mutated α_{2a}-adrenergic receptor, radioligand binding studies, however, revealed that membrane preparations from the brain had only 20 percent of the receptor level of WT animals. The loss in receptor density was not due to reduced transcription as mRNA for the α_{2a}-adrenergic receptor was the same in D79N and WT mice[19, 20]. In addition electrophysiological studies using locus coeruleus neurons demonstrated that α_{2a}-adrenergic agonists did not inhibit Ca^{++} currents nor activate K^+ channels in D79N mice[17]. Taken together these results indicate that when expressed in mice this point mutation results in a functional null mutation.

Newborn and juvenile mammals show a biphasic response to acute reductions in oxygen tension (see (23) for a review), in which an initial increase in minute ventilation is followed by a depression towards or below the prehypoxic level. In adult animals α_2-adrenergic agonists depress respiration primarily by lengthening the duration of TE [13, 14, 21]. Experiments in fetal sheep have shown that activation of α_2-adrenergic receptors inhibits respiration early in development as well[3, 4]. In addition, blocking α_2-adrenergic receptors prevented the respiratory depression associated with hypoxia in the fetus[2]. These results suggest that activation of α_2-adrenergic receptors may play a prominent role in the depressive phase of the ventilatory response to hypoxia. Accordingly, studies were carried out to compare the response to acute hypoxia in D79N mice compared to WT.

2. METHODS

Pregnant C57BL/6 mice were obtained from Harlan Sprague Dawley, Inc. (Indianapolis, IN) at 12-14 days gestation. D79N breeders on a pure C57BL/6 background for the initial studies were obtained from Vanderbilt University where they had been generated by LBM[17, 19]. For subsequent studies D79N breeders were obtained from Jackson Laboratories (Bar Harbor, ME) where they had been made available by LBM and associates (designated *Adra2a*^{TmLel} , #2-777).

The body plethysmograph method used to measure respiratory activity has been described by others[8, 22, 25]. Briefly, mice were studied in a 45 ml chamber with Parafilm fitted around their neck. Flow out of and into the chamber was measured with a pneumotachograph connected to a differential pressure transducer (Model PT5A, Grass Instrument Co., West Warwick, RI, USA). The pressure signal was integrated to give tidal volume (VT). Volume changes were calibrated by injecting known amounts of air into the chamber. This system allowed for determination of inspiratory time (TI), expiratory time (TE), total respiratory cycle duration (TTOT) and VT. Breath frequency (f) was calculated from 1/TTOT x 60 and minute ventilation from f x VT. The chamber was kept at 32.5-33.5° C by means of a lamp for animals at P1-P17 and 30.5-31.5° C for older animals. The choice of this ambient temperature in the younger animals is based on studies in 2 day old rats which showed that CO_2 production (and oxygen consumption) during 10% hypoxia more closely approached that in normoxia at 33-34° C.[24]

For the guanfacine experiments baseline measurements were made over a 5 to 9 minute period. The animal was then removed from the chamber and an intraperitoneal injection (70 to 100 µl at 174 mM, in mice with weights between 6.2 and 9.7g, P14-P17) was given. The animal was returned to the chamber and respiratory variables studied for an additional 20 to 35 minutes.

For the hypoxia experiments, which were performed in animals between P14 and P17, respiratory variables were measured for a five minute baseline period, while warmed humidified air was passed through a cylinder, which fit loosely over the head of the chamber. The fractional concentration of oxygen was then rapidly changed to 8 percent (from a premixed cylinder) for five minutes and this was followed by five minutes in air. Thirty to fifty consecutive breaths were analyzed for rate and tidal volume in the final minute of the baseline control and in each of the subsequent 10 minutes.

Results are given as mean ± SD. Significant differences were determined using two-way analysis of variance (ANOVA) with strain and postnatal age used as the two factors. The hypoxia experiments were analyzed using two-way ANOVA for repeated measures. Subsequent comparisons were made with the Tukey test. A P value of <0.05 was used to denote a significant difference.

3. RESULTS

Litter size of D79N mice used in these studies (5.2 ± 0.7) did not differ from WT (7.4 ± 1.0; P=0.10). Animal weights between P0.5 and P31 for mice with a mutated α_{2a}-adrenergic receptor were similar to WT (figure 1).

Respiratory variables from birth to adult for D79N and C57BL/6 (WT) mice are given in Table 1. TI in D79N mice did not show a lengthening with advancing age as seen in WT in which inspiratory time at P14-17 was the longest when compared to that at P1-P3. As a result TI in D79N animals (68 ± 2 ms) was shorter than WT (83 ± 2 ms; P<0.001) and this difference did not depend on what level of age was present. TE and TTOT were longer at P1-P3 than at any subsequent age and these differences were independent of strain (P values between 0.016 and < 0.001). TE for mice functionally lacking α_{2a}-adrenergic receptors was shorter than WT (208 ± 6 ms versus 225 ± 5 ms; P=0.047) as was TTOT (276 ± 7 vs. 307 ± 7 ms; P=0.02). In addition TI/TTOT in D79N mice (0.252 ± 0.004) was shorter than that in WT (0.274 ± 0.004; P<0.001).

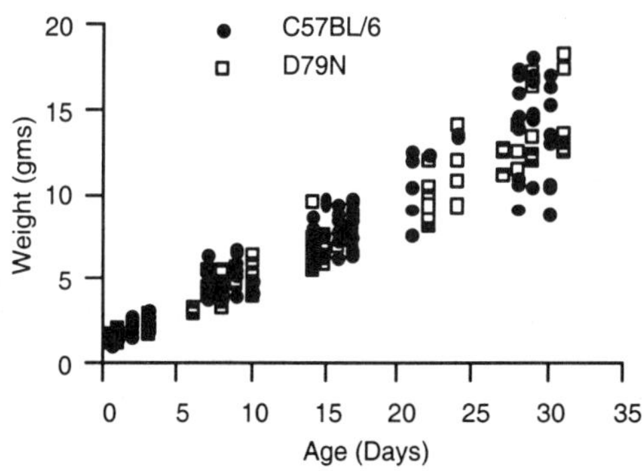

Figure 1. Postnatal weight gain in C57BL/6 (solid circles) and D7NN (open squares) mice.

TABLE 1. Respiratory variables in D79N and WT mice between P1 and adult

Strain	Postnatal Age (days)	N	TI (ms)	TE (ms)	TTOT (ms)	TI/TTOT
WT	P1-P3	10	79 ± 24	273 ± 85	352 ± 104	0.228 ± 0.001
D79N		4	71 ± 12	287 ± 67	358 ± 77	0.202 ± 0.022
WT	P7-P10	17	74 ± 12	229 ± 42	303 ± 52	0.250 ± 0.024
D79N		19	67 ± 11	220 ± 36	287 ± 45	0.233 ± 0.016
WT	P14-P17	24	91 ± 15	235 ± 44	326 ± 55	0.282 ± 0.025
D79N		22	77 ± 8	215 ± 30	293 ± 34	0.266 ± 0.026
WT	P21-P24	9	88 ± 18	216 ± 83	304 ± 98	0.298 ± 0.044
D79N		12	72 ± 13	175 ± 22	247 ± 31	0.247 ± 0.031
WT	P28-P31	8	79 ± 20	192 ± 26	271 ± 39	0.291 ± 0.050
D79N		16	59 ± 9	179 ± 43	237 ± 50	0.251 ± 0.030
WT	Adult	11	83 ± 12	203 ± 33	286 ± 41	0.294 ± 0.028
D79N		6	63 ± 8	174 ± 18	237 ± 15	0.268 ± 0.040

In order to demonstrate that D79N mice do not have a respiratory response to α_2-adrenergic agonists the effects of guanfacine (579 ± 89 μg/kg ip) were examined in animals between P14 and P17. In WT animals the agonist resulted in a fall in respiratory frequency from 209 ± 19 min^{-1} to 148 ± 16 min^{-1} ($P<0.001$). In D79N mice quanfacine did not affect respiratory rate (figure 2). The slowing of frequency in WT animals was entirely due to a lengthening of TE (figure 2) as there was no significant change in TI.

The response to a five min exposure to 8 percent O_2 was examined in WT and D79N mice at P14 to P17. Minute ventilation (VE) in WT animals increased from 9.8 ± 1.8 to 13.9 ± 1.7 ml min^{-1} in the first min of hypoxia (P=0.001) and then declined in the 5th min to 12.8 ± 1.6 ml.min^{-1} (P=0.014 compared to control). In contrast hypoxia failed to increase ventilation in the first min of hypoxia in D79N mice (10.7 ± 1.6 vs. 10.9 ± 1.7 ml.min^{-1}) and by the 5th min VE had dropped to 7.1 ± 1.6 ml.min^{-1} (P=0.007 vs. control and P= 0.011 vs. 1 min of hypoxia). VE at 5 min of hypoxia in D79N was significantly different from that in WT (P=0.003). In WT mice the roll off in ventilation was due to a decline in tidal volume (VT) which fell from 56.3 ± 5.1 μl at 1 min to 48.6 ± 4.9 μl in the 5th min of hypoxia (P=0.027). D79N animals showed a progressive fall in VT from 50.6 ± 5.0 to 42.2 ± 5.1 μl at one min (P=0.044) and 33.4 ± 4.9 at 5 min ($P<0.001$ compared to control and P=0.034 compared to 1 min).

Respiratory frequency in response to hypoxia also differed in D79N mice compared to WT (figure 3). Animals with intact α_{2a}-adrenergic receptors showed a sustained increase in frequency for the 5 min of hypoxia. In contrast in those who functionally lack the receptor the initial increase in f (213 ± 30 to 256 ± 36, P=0.033) was followed by a decline. Respiratory frequency in D79N mice at 5 min of hypoxia was not different from their control value and was significantly less than WT (217 ± 30 vs. 271 ± 35; P=0.001). The decrease in TI during hypoxia, (approximately 18%) was similar in both strains and persisted for the 5 min exposure. In contrast while WT mice showed a sustained shortening of TE which averaged 171 ± 53 ms in the fifth min that in D79N returned to a value (231 ± 81 ms; $P<0.001$ compared to WT) longer than their control (212 ± 30 ms).

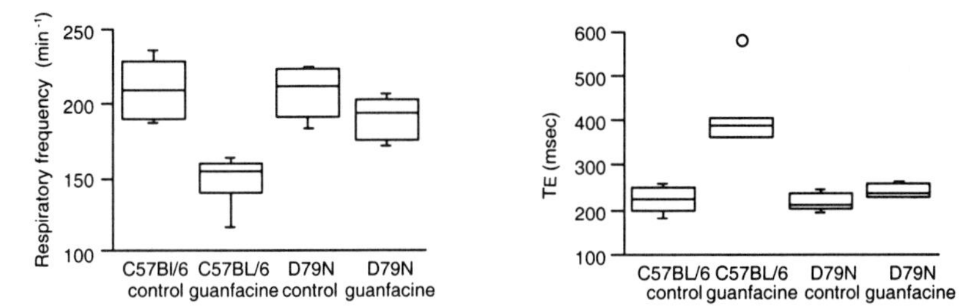

Figure 2. Respiratory response to guanfacine. Left Panel: box plots of respiratory rate before and after quanfacine for C57BL/6 and D79N mice. Right panel: box plots of expiratory time (TE). Bar within box is median value; box describes 25th to 75th percentile and wiskers describe 10th to 90th percentile.

4. DISCUSSION

Two main conclusions can be drawn from this series of experiments. First, mice functionally lacking α_{2a}-adrenergic receptors have an altered postnatal development of their respiratory pattern. This is characterized by a failure to lengthen TI and reduce TI/TTOT with advancing age as was seen in WT mice (Table 1). Second, animals lacking α_{2a}-adrenoreceptors have an altered response to acute hypoxia in which minute ventilation remained at control levels in the first min and declined to below control by the fifth min. In contrast WT mice showed an increase in minute ventilation which remained above control at five min.

The lengthening of TI in WT C57BL/6 mice reached its peak (P14-P17) at the same time as previously reported for MRI-1 and Baharbor strains[27]. This developmental change in timing has also been observed in infants[11], rabbits[36], cat[26], and opossum[10]. Recordings from phrenic nerve and diaphragmatic electromyogram in opossum[10] and mouse[27] showed that early in development the inspiratory pattern is characterized by a rapid onset short plateau and rapid decline. In more mature animals and adults the burst is prolonged and shows a ramp component.

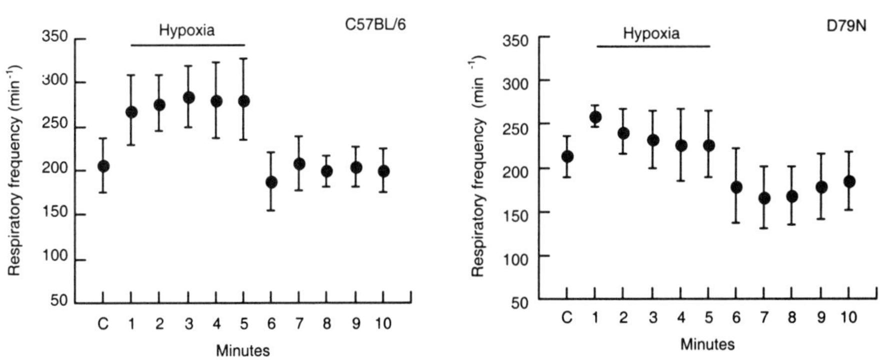

Figure 3. Respiratory frequency response to acute hypoxia. Left panel frequency for C57BL/6 mice during control (C), 5 min of 8% O_2 breathing followed by 5 min of return to air. Right panel, same data for D79N mice. Error bars are ± SD.

Inspiratory neurons in the early neonates of these two species show few spikes per burst which then increase with advancing age[10, 27]. The mechanisms underlying these developmental changes are not completely understood but the failure of TI to lengthen in mice functionally lacking α_{2a}-adrenergic receptors suggests some possible explanations. In early (<P4) rat hypoglossal neurons which show a decrementing or adapting firing pattern to injected current the amplitude of the calcium dependent afterhyperpolarization increases with successive action potentials which would lead to termination of the burst[35]. Recordings from dispersed locus coeruleus neurons of D79N mice have demonstrated that α_{2a}-adrenergic agonists fail to inhibit calcium currents[17]. Thus in the mutant mice intracellular calcium concentrations may increase more rapidly than WT leading to summation of the afterhyperpolarizations and termination of the burst. The onset of TI lengthening in WT mice (~P15) coincides with the time that thyrotropin-releasing hormone (TRH) mRNA expression in raphe neurons, TRH immunoreactivity in raphe fibers projecting to hypoglossal motoneurons, and radiolabeling of TRH receptors on hypoglossal neurons all increase[6]. Application of TRH to hypoglossal neurons lengthens their burst duration[5]. If the maturation of the components of the TRH system in mouse inspiratory neurons follows a similar time course it may well contribute to the lengthening of TI. The present data suggest that this must be coupled with α_{2a}-adrenergic receptor mediated inhibition of calcium conductances to retard afterhyperpolarization summation and arrest of inspiratory bursts.

The observations that α_2-adrenergic receptor agonists are respiratory depressants in the fetus[3, 4] and adult,[13, 14, 21] coupled with those that α_2-adrenergic antagonists prevent hypoxic respiratory depression in the fetus[2], suggest that α_2-adrenergic receptor stimulation may play a role in the respiratory depression seen with hypoxia. The imidozoline receptor agonist moxomidine did not affect hypoglossal or phrenic nerve activity in adult cat experiments[12]. This same study[12] and others[13, 14, 21] have found that clonidine which has agonist activity at both imidozoline and α_2-adrenergic receptors depresses respiration primarily by a lengthening of TE. The observation that animals functionally lacking α_{2a}-adrenergic receptors do not show a respiratory response to an α_2-adrenergic agonist (figure 2) confirms previous conclusions that the respiratory effects are not mediated by imidazoline receptors. In addition, these studies demonstrate that it is the α_{2a} as opposed to the α_{2b} or α_{2c}-adrenergic receptor subtypes which effects the respiratory depression.

The present results examining the hypoxic ventilatory response (HVR) in mice lacking α_{2a}-adrenergic receptors differ from those in which α_2-adrenergic receptor antagonists were used. In fetal sheep L-657, 743 (Merk) prevented the apnea produced by hypoxia. In addition, the antagonist SKF-86466 has been shown to potentiate the hypoxic augmentation of carotid sinus nerve discharge frequency in adult cats[16]. While it is possible that these antagonists are acting at α_{2b} or α_{2c}-adrenergic receptors, the observation that α_2-adrenergic agonists effect their respiratory depression at α_{2a} receptors (figure 2) makes this explanation less attractive. At a minimum the present studies do demonstrate that the role off in HVR is not due to activation of α_{2a}-adrenergic receptors.

Mice lacking functional α_{2a}-adrenergic receptors showed a greater depressive phase in their respiratory response to hypoxia than WT animals. α_2-adrenergic receptors are located presynaptically where they inhibit neurotransmitter release and postsynaptically where there are coupled to potassium channels (enhanced activity); voltage-regulated calcium channels (decreased activity) and cyclic AMP (decreased activity).

Immunohistochemistry methods have demonstrated that the α_{2a}-adrenergic receptor is located both pre and post-synaptically in areas of the pons and medulla involved in respiratory control[1, 31]. Thus the exaggerated depression in D79N mice could be attributed to either or both pre and post-synaptic mechanisms. Experiments using mice with targeted disruption of either the α_{2a}, α_{2b} or α_{2c} subtype of the α_2-adrenergic receptor plus a combined α_{2a},α_{2c} knockout strain have shown that both the α_{2a} and α_{2c} receptors are required for presynaptic autoinhibition[15]; therefore it is most likely that the observed altered HVR in D7NN animals is due to post-synaptic events. The failure of adrenergic agonists to inhibit voltage-activated calcium channels in neurons of these mice[17] could lead to a more rapid and sustained rise in intracellular calcium compared to WT. This in turn may cause a prolonged activation of calcium-activated slow-conductance potassium channels (SK) which would prevent the shortening of TE during hypoxia. This proposed mechanism is strengthened by the recent observation that mice which overexpress SK channels have an altered HVR characterized by failure to sustain an increase in respiratory frequency[7].

In summary, these studies demonstrate that α_{2a}-adrenergic receptors play a prominent role in the postnatal development of the respiratory pattern and that the α_{2a} receptor is the subtype involved in the respiratory depressive effects of α_2-adrenergic agonists. In addition, the α_{2a}-adrenoreceptor is required to maintain an increase in respiratory frequency during an acute hypoxic challenge.

5. ACKNOWLEDGMENTS

The authors thank Ms. Sharon Woods and Ms. Keri Stepper for preparation of the manuscript. This study was supported by grants from the National Institutes of Health (JMB, HD 38316 and LBM, HL43671), the Medical Research Foundation of Oregon and the American Lung Association, Oregon Affiliate.

6. REFERENCES

1. Aoki,C., C.-G.Go, C.Venkatesan, and H.Kurose. Perikaryal and synaptic localization of α_{2A}-adrenergic receptor-like immunoreactivity. *Brain Res.* 650: 181-204, 1994.
2. Bamford,O. and R.L.Hawkins. Central effects of an α_2-adrenergic antagonist on fetal lambs: a possible mechanism for hypoxic apnea. *J.Devel.Physiol.* 13: 353-358, 1990.
3. Bamford,O.S., G.S.Dawes, R.Denny, and R.A.Ward. Effects of the α_2-adrenergic agonist clonidine and its antagonist idazoxan on the fetal lamb. *J.Physiol.* 381: 29-37, 1986.
4. Bamford,O.S., R.L.Hawkins, and C.E.Blanco. Effects of clonidine on breathing movements and electrocortical activity in the fetal lamb. *Am.J.Obstet.Gynecol.* 163: 661-668, 1990.
5. Bayliss,D.A., F.Viana, and A.J.Berger. Thyrotropin-releasing hormone causes excitation of rat hypoglossal motoneurons in vitro. *Sleep* 16: 49-52, 1993.
6. Bayliss,D.A., F.Viana, R.K.Kanter, C.L.Szymeczek-Seay, A.J.Berger, and D.E.Millhorn. Early postnatal development of thyrotropin-releasing hormone (TRH) expression, TRH receptor binding, and TRH responses in neurons of rat brainstem. *J.Neurosci.* 14: 821-833, 1994.
7. Bond,C.T., R.Sprengel, J.M.Bissonnette, W.A.Kaufmann, D.Pribnow, T.Neelands, T.Storck, M.Baetscher, J.Jerecic, J.Maylie, H.-G.Knaus, P.H.Seeburg, and J.P.Adelman. Respiration and parturition affected by conditional overexpression of the Ca^{2+}-activated K^+ channel subunit, SK3. *Science* 298: 1942-1946, 2000.
8. Eden,G.J. and M.A.Hanson. Maturation of the respiratory response to acute hypoxia in the newborn rat. *J.Physiol.* 392: 1-9, 1987.

9. Elsen,F.P. and J.M.Ramirez. Calcium currents of rhythmic neurons recorded in the isolated respiratory network of neonatal mice. *J.Neurosci.* 18: 10652-10662, 1998.
10. Farber,J.P. Medullary inspiratory activity during opossum development. *Am.J.Physiol.* 254:R578-584, 1988.
11. Haddad,G.G., R.A.Epstein, M.A.F.Epstein, H.L.Leistner, P.A.Marino, and R.B.Mellins. Maturation of ventilation and ventilatory pattern in normal sleeping infants. *J.Appl.Physiol.* 46: 998-1002, 1979.
12. Haxhiu,M.A., I.A.Dreshaj, B.Erokwu, L.A.Collins, and P.Ernsberger. Effect of I_1-Imidazoline receptor activation of responses of hypoglossal and phrenic nerve to chemical stimulation. *Ann.NY.Acad.Sci.* 763: 445-462, 1995.
13. Hedrick,M.S., M.L.Ryan, and G.E.Bisgard. Recurrent laryngeal nerve activation by α_2 adrenergic agonists in goats. *Respir.Physiol.* 101: 129-137, 1995.
14. Hedrick,M.S., M.L.Ryan, J.Pizarro, and G.E.Bisgard. Modulation of respiratory rhythm by α_2-adrenoceptors in awake and anesthetized goats. *J.Appl.Physiol.* 77: 742-750, 1994.
15. Hein,L., J.D.Altman, and B.K.Kobilka. Two functionally distinct α_2-adrenergic receptors regulate sympathetic neurotransmission. *Nature* 402: 181-184, 1999.
16. Kou,Y.R., P.Ernsberger, P.A.Cragg, N.S.Cherniack, and N.R.Prabhakar. Role of α_2-adrenergic receptors in the carotid body response to isocapnic hypoxia. *Respir.Physiol.* 83: 353-164, 1991.
17. Lakhlani,P.P., L.B.MacMillan, T.Z.Guo, B.A.McCool, D.M.Lovinger, M.Maze, and L.E.Limbird. Substitution of a mutant α_{2a}-adrenergic receptor via "hit and run" gene targeting reveals the role of this subtype in sedative, analgesic, and anesthetic-sparing responses *in vivo*. *PNAS USA* 94: 9950-9955, 1997.
18. MacDonald,E., B.K.Kobilka, and M.Scheinin. Gene targeting - horming in on a_2-adrenoceptor-subtype function. *TIPS* 18: 211-219, 1997.
19. MacMillan,L.B., L.Hein, M.S.Smith, M.T.Piascik, and L.E.Limbird. Central hypotensive effects of the α_2a-adrenergic receptor subtype. *Science* 273: 801-803, 1996.
20. MacMillan,L.B., P.P.Lakhlani, D.M.Lovinger, and L.E.Limbird. Alpha$_2$-Adrenergic receptor subtypes: Subtle mutation of the α_{2A}-Adrenergic receptor *in vivo* in gene targeting strategies reveals the role of this sybtype in multiple physiological settings. *Recent Progress in Hormone Research* 53: 25-42, 1998.
21. McCrimmon,D.R. and P.M.Lalley. Inhibition of respiratory neural discharges by clonidine and 5-hydroxytryptophan. *J.Pharmacol.Exp.Ther.* 222: 771-777, 1982.
22. Mortola,J.P. Breathing pattern in newborns. *J.Appl.Physiol.* 56: 1533-1540, 1984.
23. Mortola,J.P. Ventilatory responses to hypoxia in mammals. In: *Tissue Oxygen Deprivation*, edited by G.G.Haddad and G.Lister. New York: Marcel Dekker, Inc., 1996, p. 433-477.
24. Mortola,J.P. and A.Dotta. Effects of hypoxia and ambient temperature on gaseous metabolism of newborn rats. *Am.J.Physiol.* 263: R267-R272, 1992.
25. Mortola,J.P. and A.Noworaj. Two-sidearm tracheal cannula for respiratory airflow measurments in small animals. *J.Appl.Physiol.* 55: 250-253, 1983.
26. Parot,S., M.Bonora, H.Gautier, and D.Marlot. Developmental changes in ventilation and breathing pattern in unanesthetized kittens. *Respir.Physiol.* 58: 253-262, 1984.
27. Paton,J.F.R. and D.W.Richter. Maturational changes in the respiratory rhythm generator of the mouse. *Pflugers Arch.* 430: 115-124, 1995.
28. Ramirez,J.M., U.J.A.Quellmalz, and D.W.Richter. Postnatal changes in the mammalian respiratory network as revealed by the transverse brainstem slice of mice. *J.Physiol.* 491: 799-812, 1996.
29. Richter,D.W., J.Champagnat, T.Jacquin, and R.Benacka. Calcium currents and calcium-dependent potassium currents in mammalian medullary respiratory neurones. *J.Physiol.* 470: 23-33, 1993.
30. Surprenant,A., D.A.Horstman, H.Akbarali, and L.E.Limbird. A point mutation of the α_2-adrenoceptor that blocks coupling to potassium but not calcium currents. *Science* 257: 977-980, 1992.
31. Talley,E.M., D.L.Rosin, A.Lee, P.G.Guyenet, and K.R.Lynch. Distribution of α_{2A}-adrenergic receptor-like immunoreactivity in the rat central nervous system. *J.Comp.Neurol.* 372: 111-134, 1996.
32. Umemiya,M. and A.J.Berger. Properties and function of low- and high- voltage-activated Ca^{2+} channels in hypoglossal motoneurons. *J.Neurosci.* 14: 5652-5660, 1994.
33. Viana,F., D.A.Bayliss, and A.J.Berger. Calcium conductances and their role in the firing behavior of neonatal rat hypoglossal motoneurons. *J.Neurphysiol.* 69: 2137-2149, 1993.
34. Viana,F., D.A.Bayliss, and A.J.Berger. Multiple potassium conductances and their role in action potential repolarization and repetitive firing behavior of neonatal rat hypoglossal motoneurons. *J.Neurophysiol.* 69: 2150-2163, 1993.
35. Viana,F., D.A.Bayliss, and A.J.Berger. Repetitive firing properties of developing rat brainstem motoneurones. *J.Physiol.* 486: 745-761, 1995.
36. Wyszogrodski,I., B.T.Thach, and J.Milic-Emili. Maturation of respiratory control in unanesthetized newborn rabbits. *J.Appl.Physiol.* 44: 304-310, 1978.

DO GENETIC FACTORS INFLUENCE THE DEJOURS PHENOMENON?

Shyam Subramanian, Fang Han, Bernadette O. Erokwu, Thomas E. Dick, and Kingman P. Strohl*

1. INTRODUCTION

Ventilation in response to breathing 100% oxygen is characterized in mammalian species by a decrease in minute ventilation[1]. This effect, which we will term the "Dejours phenomenon", is especially evident in the period immediately following exposure to hypoxia. Though the physiologic basis for this phenomenon is not completely understood, it is believed to reflect an acute reduction in peripheral chemoreceptor input, and the magnitude of respiratory depression is considered an index of chemoreceptor drive from the carotid body.

Another component of this response is a phenomenon referred to as post-hypoxic frequency decline, originally described in anesthetized rats as a decrease in respiratory frequency immediately following a brief exposure to hypoxia relative to the baseline frequency before exposure to hypoxia. The ventrolateral pontine neurons are thought to play a central role in orchestrating this response[2].

Many traits in respiration are likely to be under genetic as well as environmental control[3]. For instance, the ventilation and pattern of breathing at rest and in hypoxia or hypercapnia may differ among mouse[4] and rat[5] strains. Such observations indicate that there are significant genetic effects on ventilatory behavior. Furthermore, it is proposed by Tankerseley et al.[7] that in the mouse there are a limited number of genes that determine the level of ventilation in hypoxia.

The Dejours phenomenon is very clearly demonstrated under non-steady state conditions involving rapid transitions from room air to hypoxia and then to hyperoxia. It is possible that such extreme and unusual environmental manipulation, rather than genetic influence, is the predominant factor influencing this response. To test this hypothesis, we exposed two rat strains known to differ in ventilatory behavior[5] to oxygen after exposure to hypoxic gas, recording values describing the Dejours phenomenon.

* Division of Pulmonary and Critical Care Medicine, Department of Medicine, Louis Stokes Cleveland VA Medical Center, Case Western Reserve University, Cleveland 44107. *Corresponding author:* Shyam Subramanian, MD, Email: shyam_md@ yahoo.com. Tel/ Fax: +1 216-731-0871.

Frontiers in Modeling and Control of Breathing, edited by Poon and Kazemi.
Kluwer Academic/Plenum Publishers, New York, 2001.

2. METHODS

Adult Sprague Dawley (SD/harlan and SD/harlan/cwru) rats (n = 38, 20 males and 18 females) and Brown Norway (BN/harlan, BN/harlan/mcw, and BN/harlan/mcw/cwru) rats (n = 55, 20 males and 35 females) between the ages of 12 and 22 weeks were tested. Animals were housed in our animal facility and fed rat chow and water ad libitum.

2.1. Measurement of Ventilation

Ventilation and metabolism were assessed by whole body plethysmography via the open circuit method, as previously described[5]. Briefly, the chamber consisted of a 14 cm diameter Plexiglas cylinder of 8.4 liter volume, with air intake and output ports to allow for different gas mixtures to be flushed through the chamber at a rate of 30 l/min, and a low continuous flow of the gas to be drawn through the chamber during the testing period at a rate of 600 ml/min to prevent CO_2 build-up and to maintain chamber temperature at ambient. A small opening in the top of the chamber permitted sampling of chamber air for assessment of oxygen and carbon dioxide concentration within the chamber. Ventilatory parameters included minute ventilation (V_E), tidal volume (Vt), frequency (*f*), and metabolic parameters: oxygen consumption (VO_2) and carbon dioxide production (VCO_2). Two set-ups allowed for 2 animals to be tested at a time.

2.2. Protocol

Testing was done between 10:00 a.m. and 1:00 p.m. The animals were brought to the laboratory at 9 a.m. and allowed 45 min to acclimatize to the testing chamber. Baseline resting ventilation and carbon dioxide and oxygen concentrations were continuously recorded during this time. At the end of the acclimatization period, five measurements were taken over a 15-min period to provide resting values for ventilation and metabolism. The animal was then exposed to a 5 min presentation of the test gases in the following order: 10% O_2/balance N_2 followed rapidly by 100% O_2. During the challenges, ventilatory parameters were continuously recorded. Representative values of frequency (f), tidal volume (Vt) and minute ventilation (Ve) were obtained at baseline on room air (a1) and at the following time points: end of each minute on 10% O_2/90% N_2 at 1, 2, 3, 4,and 5 minutes (h1, h2, h3, h4, h5) and, after the switch to 100%O_2, at 1, 2, 3, 4, and 5 minutes (o1, o2, o3, o4, and o5).

Breaths were scored by computer. Sniffs or sighs were not included in the calculations of tidal volume and frequency. A mean value under each condition was entered for each animal. At the end of testing, the animal was removed from the chamber, and body weight and body length obtained

2.3. Statistical Analysis

ANOVA and Levine tests with Students t-test models were used for comparisons of ventilation and its components, metabolism and derivative values (e.g., minute ventilation/VCO_2). Models for statistical analysis were developed using the SPSS[R] software.

3. RESULTS

At baseline, there were significance differences between strains in O_2 consumption, CO_2 production, weight, and tidal volume, but no significant differences in age, respiratory frequency, and minute ventilation corrected for CO_2 production (Table 1).

In the response to 100% O_2 after a brief hypoxic exposure, there were significant differences between strains in absolute values for frequency and minute ventilation at the end of 1 minute of hyperoxia (Table 2). In the BN, there occurred a large drop in both frequency ($p<0.0001$) and minute ventilation ($p<0.0001$) in the first minute of exposure to hyperoxia as compared to values for these traits prior to hypoxic exposure. In contrast, in the SD, the minute ventilation in the first minute of hyperoxia was similar to baseline values, and the frequency response in the first minute of exposure to hyperoxia was significantly higher ($p<0.0001$) as compared to baseline values before hypoxic exposure.

When viewed in terms of percent change from baseline to 1st minute of hyperoxia (Table 3), the values for frequency and minute ventilation in the BN were qualitatively

Table 1. Baseline characteristics of the Sprague Dawley and Brown Norway rats

Measure	BN	SD	P
Age	16.51 ± 4.57	17.47 ±3.7	0.177
VCO_2	3.56 ± .63	4.24 ±.51	< 0.01
VO_2	5.47 ± .63	6.2 ±1	< 0.01
Weight	220.69± 75.02	347.97±21.03	< 0.0001
Freq.a1	94.41 ± 20.37	102.08±25.08	0.106
Vt.a1.	.789 ± .153	.956 ± .166	< 0.0001
$Ve.VCO_2$	21.07 ± 5.54	22.73 ±5.02	0.140

Table 2. Change in frequency and minute ventilation from baseline to hyperoxia

Measure	BN	SD
Ve.a1	73.20 ±15.21	95.64 ±20.20
Ve.01	51.89 ±19.05	100.79 ±26.37
P	< 0.0001	0.208
Freq.a1	94.41 ±20.37	102.08 ±25.08
Freq.o1	73.14 ±19.85	119.10 ±33.58
P	< 0.0001	0.008

Table 3. Ventilation in response to 100%oxygen following brief hypoxic exposure in rats

Measure	BN	SD	P
% change Vt a1-o1	9.24 ±20.87	8.94 ±13.45	0.591
% change f a1-o1	-21.98 ±15.19	+21.47 ±41.38	< 0.0001
% change Ve a1-o1	-28.56 ±22.26	7.11 ±26.73	< 0.0001

and quantitatively different than the SD ($p < 0.0001$). However, the proportional change in tidal volume was similar. The strain differences were equally apparent when results from male and female animals were examined separately (data not shown).

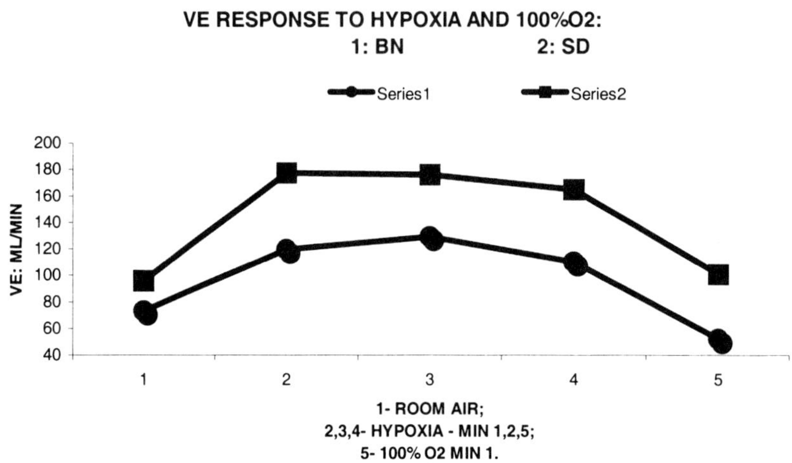

Figure 1. Shown are the mean values for each strain for minute ventilation (ml/minute). Points along the horizontal axis represent mean values at baseline (Point 1); during hypoxia at minutes 1 (Point 2), 2 (Point 3), and 5 (Point 4); and with re-oxygenation at the end of the first minute of rapid transition from hypoxia to 100% oxygen (Point 5). BN = closed circles. SD = closed squares.

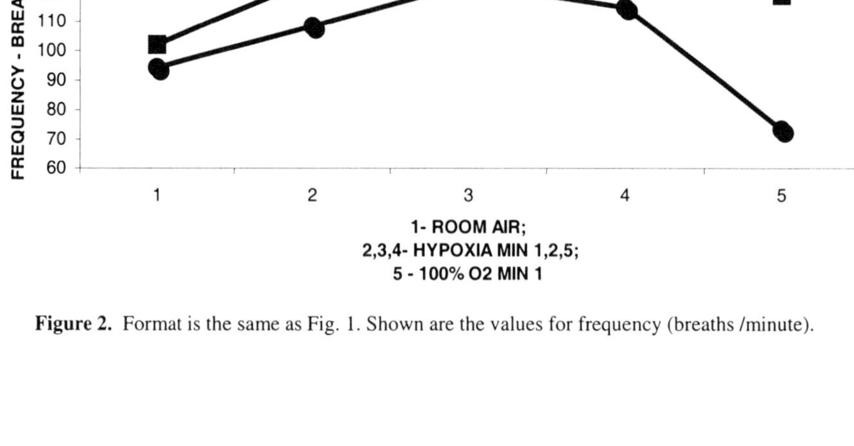

Figure 2. Format is the same as Fig. 1. Shown are the values for frequency (breaths /minute).

4. DISCUSSION

The results indicate significant differences in the pattern and magnitude of the ventilatory response to hyperoxic re-oxygenation between two strains of rats and clearly disprove the null hypothesis. Thus, genetic factors influence the expression of the Dejours phenomenon. Additionally, in the unanesthetised SD we find mean values for frequency in the Dejours phenomenon that are significantly higher from values at baseline. In comparison to previous work[3], this suggests that the phenomenon of hyperoxic depression and posthypoxic frequency decline is more consistently observed in the anesthetized as compared to the unanesthetised SD rat.

The physiologic basis for the difference in the Dejours phenomenon between strains is not clear at this point. The baseline frequency and ventilation/ carbon dioxide production are not significantly different in these strains, suggesting that dead space or ventilation/perfusion play a minor role in explaining the strain differences. The significant decrease in ventilation seen in the BN rat in response to reoxygenation would be interpreted as indicating a significant contribution to ventilatory drive from arterial chemoreceptors during normoxic (baseline) breathing. However, a previous description of strain differences indicated that the level of ventilation in the BN is lower than the SD animal[5]. In contrast, the SD rat would be considered to have little peripheral chemoreceptor component to the respiratory drive under baseline conditions. Carotid ablation studies may be helpful in resolving this issue, but this remains to be demonstrated.

The ventilatory response to acute hypoxia in mammalian species is biphasic; an initial hyperventilatory response is followed by a reduction in ventilation within 2-3 min below the peak level ("roll-off"). Central pathways, including the nucleus tractus solitarius and neurotransmitters, including GABA, are believed to play an important role in this phenomenon[7]. There were significant differences between strains in both the hypoxic ventilatory response and in the degree and the fall in ventilation during hypoxia (roll-off). However, there was no correlation between magnitude of the hypoxic ventilatory response or the degree of roll-off, in either strain, with the magnitude of the Dejours phenomenon suggesting that strain differences involve physiologic mechanisms that are not directly coupled to events that influence ventilation during the preceding hypoxic exposure.

It is possible that gene polymorphisms influence one or more key physiologic and/or biochemical pathways in the carotid bodies of the two rat strains and are responsible for the observed variability in the Dejours phenomenon. Both surgical and chemical manipulations of the carotid body have been shown to produce changes in the ventilatory response to breathing 100% O_2 [8,9,10]. Response to such maneuvers can differ among species, but whether they differ among strains is not known. Such a finding would further implicate genetic influences in peripheral chemoreception[3]. Another factor to consider would be the characteristics of the "controlled system" — lungs, chest wall, and upper airway. The static and dynamic mechanical properties of these elements may be under genetic control. The current study does not directly address the possible distinctions between chemosensory networks and the controlled elements among strains.

Variability in hyperoxic depression has been described in human neonates, premature infants, and adults[11,12,13]. Both prior experience and genetic polymorphisms might operate to produce such variability[3]. Observations from rodent models will permit more rapid application of genetic principles to understanding variations in ventilatory behavior in health as well as in disorders of respiratory control, including SIDS, sleep disordered breathing and patients with chronic lung disease, and hypoventilation syndromes.

5. ACKNOWLEDGMENTS

This work was supported by the VA Medical Research Service and by the National Institutes of Health grants HL 58844 (Kingman P. Strohl) and HL 25830 (Thomas E. Dick), and a Sleep Academic Award (HL 97015, Kingman P. Strohl).

6. REFERENCES

1. P. Dejours, Chemoreflexes in breathing. Physiol Rev **42**(3) 335-358 (1962).
2. S. Coles, T. E. Dick, Neurones in the ventrolateral pons are required for posthypoxic frequency decline in rats. J Physiol. **497**: 79-93 (1996).
3. F. Han, K. P. Strohl, Inheritance of ventilatory behavior in rodent models. Respir Physiol. **121**:247-256 (2000).
4. C. G. Tankersely, R. S. Fitzgerald, S. R. Kleeberger, Differential control of breathing among inbred strains of mice. Am J. Physiol. **267**: R1371-R1377 (1994).
5. K. P. Strohl, A. J. Thomas, P. St.Jean, E. H. Schlenker, R. J. Koletsky, N. J. Schork, Ventilation and metabolism among rat strains J Appl. Physiol. **82**: 317-323 (1997).
6. C. G. Tankerseley, R. C. Elston, A. H. Schnell, Genetic determinants of acute hypoxic ventilation: patterns of inheritance in mice. J Appl. Physiol. **88**: 2310-2318 (2000).
7. B. Hoop, J. L. Beagle, T. J. Maher, H. Kazemi, Brainstem amino acid neurotransmitters and hypoxic ventilatory response. Respir Physiol. **118**(2-3):117-29 (1999).
8. E. B. Olson, E. H. Vidruk, J. A. Dempsey, Carotid body excision significantly changes ventilatory control in awake rats. J. Appl. Physiol. **64**: 666-671 (1988).
9. M.J.Miller, S.M.Tenney, Hyperoxic hyperventilation in carotid deafferented cats. Respir. Physiol. **23**: 23-30 (1975).
10. H. Holgert, T. Hokfelt, T. Hertzberg, H. Lagercrantz, Functional and developmental studies of the peripheral arterial chemoreceptors in rat: phenomenons of nicotine and possible relation to sudden infant death syndrome. Proc Natl Acad Sci. **92**: 7575-7579. (1995).
11. B. Bouferrache, S. Filtchev, A. Leke, Q. Marbaix-Li, M. Freville, C. Gaultier, The hyperoxic test in infants reinvestigated. Am J Respir Crit Care. **161**: 160-5. (2000).
12. G. Cross, G. Oppe, The phenomenon of inhalation of high and low oxygen concentrations on the respiration of the premature infant. J. Physiol. **117**: 38-55 (1952).
13. D. Reinstorff, A. Fenner, Ventilatory response to hyperoxia in premature and newborn infants during the first three days of life. Respir Physiol; **15**: 159 – 165 (1972).

ROLE OF NITRIC OXIDE IN SHORT-TERM POTENTIATION AND LONG-TERM FACILITATION

Involvement of NO in breathing stability

David D. Kline and Nanduri R. Prabhakar*

1. INTRODUCTION

It has been well documented that increases in breathing in response to a variety of stimuli may often outlast the actual period of stimulation in various species (see references[1, 2]). For instance, after a brief hypoxic challenge, ventilation slowly returns to baseline values over a period of minutes, a phenomenon often termed as "short-term potentiation (STP)." On the other hand, after episodes of recurrent hypoxic challenges, depending on species, breathing may remain elevated as long as an hour. This long lasting enhancement of respiration that follows recurrent episodes of hypoxia is referred to as "long-term facilitation (LTF)." It has been proposed that STP and LTF are critical for maintaining stability of breathing, especially in situations involving acute changes in arterial blood gases[2], and may involve brainstem neurons[1,2]. Recent studies on hippocampal neurons suggest that nitric oxide (NO) is critical for the development of long term potentiation (LTP) associated with learning and memory[3, 4]. Given that NOS-1, the enzyme that generates NO, is co-localized with 5-HT in raphe neurons[5, 6], which has been shown to be important for the generation of LTF [2], as well as our recent observation that NOS-1 mutant mice exhibit breathing instability during hypoxia[7], prompted us to examine whether NO generated by NOS-1 plays a role in the development STP and/or LTF.

2. METHODS

Experiments were performed on age-matched wild-type (Wt, n=9) and NOS-1 mutant (n=8) mice of either sex. Respiration was monitored in awake animals by whole body plethysmography as described previously by us[7]. To examine STP, after recording

* Department of Physiology and Biophysics, Case Western Reserve University, Cleveland, OH 44106.
Tel: 216-368-8636, Fax: 216-368-1693, email: nrp@po.cwru.edu

base line breathing under normoxia, mice were challenged with 7% O_2 (balanced nitrogen) for five minutes. Following termination of the hypoxic challenge, breathing was monitored for an additional five minutes, and was compared with pre-hypoxic controls. To study the phenomenon of LTF, baseline breathing was recorded for five minutes while the animals breathed room air (21% O_2). Subsequently, they were challenged with three episodes of hypoxia (7% O_2), each lasting five minutes and interspersed with room air breathing. Following the last episode of hypoxia, animals were allowed to breath room air, and breathing was continuously monitored for 60 min. In addition to breathing, oxygen consumption (VO_2), and CO_2 production (VCO_2) was also monitored in the same animals as above. The protocols were repeated in an additional eight Wt mice, before and after administration of 7-nitroindazole (7-NI, 50mg/kg; i.p.), a putative inhibitor of NOS-1. The following variables were analyzed: respiratory rate, tidal volume, minute ventilation/gm of body weight, VO_2 and VCO_2 . Results are expressed as percent changes from baseline values (mean ± SEM). Statistical analysis was performed using paired *t*-test and p values less than 0.05 were considered significant.

3. RESULTS

Respiration following a brief hypoxic challenge in Wt and NOS-1 mutant mice is depicted in Figure 1. Following termination of the hypoxic challenge, breathing remained elevated by 58 ±17% of pre-hypoxic controls in Wt mice ($p<0.05$, paired t-test). The increase in breathing was primarily due to increases in respiratory rate. By contrast, in NOS-1 mutant mice, respiration promptly returned to controls after terminating hypoxic challenge ($p > 0.05$, paired t-test). STP was ablated or markedly attenuated in Wt mice treated with 7-NI ($p<0.05$, paired t-test). Furthermore, VO_2 and VCO_2 were comparable between Wt and NOS-1 mutant mice ($p > 0.05$, t-test).

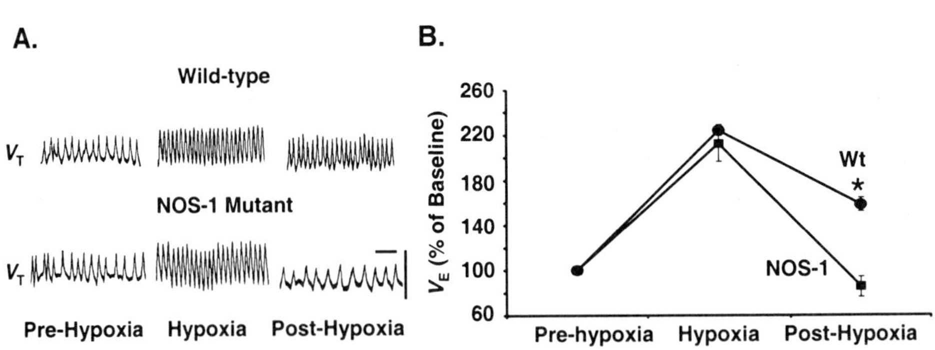

Figure 1. Lack of STP in NOS-1 mutant mice. ***A***. Post-hypoxia denotes breathing during the first minute of 21% O_2. V_T denotes tidal volume. Horizontal and vertical bars represent 1 sec and 100 μL, respectively. ***B.*** Results are presented as percent change of baseline and are presented as mean ± SEM. * denotes p values less than 0.05 (paired *t*-test). In Wt mice, respiration increased during hypoxia, however, upon returning to normoxia, breathing remained significantly elevated. On the other hand, respiration in NOS-1 mutant mice promptly returned back to control values.

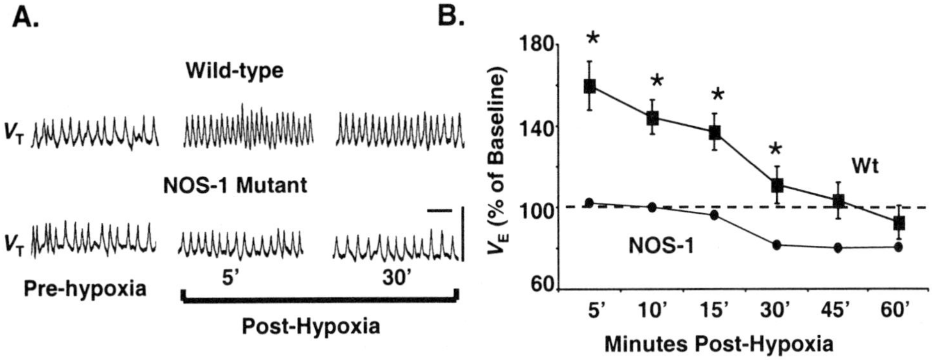

Figure 2. Lack of LTF in NOS-1 mutant mice. ***A.*** Pre-hypoxia represents respiration during 21% O_2 prior to 3 episodes of hypoxia, 5' and 30' denotes minutes during 21% O_2 following hypoxia. V_T denotes tidal volume. Horizontal and vertical bars represent 1 sec and 100 μL, respectively. Respiration was significantly elevated following the last hypoxic period for up to 30 minutes in Wt mice. On the other hand, respiration in NOS-1 mutant mice promptly returned back to control values. ***B.*** Results are presented as percent change of pre-hypoxia and are presented as mean ± SEM. 5', 10' and etc represent minutes post -hypoxia. * denotes p values less than 0.05 (paired *t*-test) Note that respiration was significantly elevated for up to 30 minutes in Wt mice, whereas it promptly returned back to control levels in NOS-1 mutant mice.

An example of respiration before and after 3 episodes of hypoxia (7% O_2) in a Wt and NOS-1 mutant mouse is depicted in Figure 2. Following the hypoxic challenges, breathing remained elevated over control respiration for up to 30 minutes in Wt mice, suggesting LTF ($p < 0.05$, paired t-test). The elevation of breathing was due to increases in respiratory rate. By contrast, in NOS-1 mutant mice, respiration returned to control within 5 minutes, suggesting the absence of LTF. In Wt mice treated with the putative NOS-1 inhibitor 7-nitroindazole, LTF was either absent or markedly attenuated ($p < 0.05$, paired t-test). Oxygen consumption and CO_2 production were comparable between Wt and NOS-1 mutant mice ($p < 0.05$, t-test).

4. DISCUSSION

The results of the present study demonstrate that Wt mice exhibit the phenomenon of STP and LTF following single as well as recurrent episodes of hypoxia, similar to that reported in other species[1,2]. More importantly, our data demonstrates that STP and LTF are either absent or markedly attenuated in mice deficient in NOS-1, which suggests NO generated from NOS-1 has an essential role in the expression of STP and LTF.

The mechanism(s) associated with STP of breathing following a single hypoxic challenge are not known. It has been proposed, however, that changes in intracellular calcium contribute to STP of breathing[1, 8]. The present results demonstrate that STP is absent in NOS-1 mutant mice, suggesting the importance of NO in the expression of STP. How might NO contribute to expression of STP? First of all, NOS-1 is expressed in brainstem neurons, such as the nucleus tractuus sollitari[9, 10], that have been implicated in the generation of STP[11]. McCrimmon *et al.*[12] demonstrated that inspiratory bulbospinal neurons exhibit STP which is mediated by glutamatergic NMDA receptors. In addition,

Poon *et al.*,[13] have recently demonstrated that STP of phrenic nerve activity following carotid sinus nerve stimulation can be attenuated by systemic blockade of NMDA receptors. Since NOS-1 and the NMDA receptor have been co-localized, and activation of the NMDA receptor increases intracellular calcium levels to promote the generation of NO [14, 15], NO may contribute to the generation of STP. It has been suggested that STP plays a critical role in the stability of breathing[1, 2] and in the absence of STP, breathing may become unstable resulting in waxing and waning of breathing such as that seen in Chyne-Stokes pattern of breathing. We have previously observed that NOS-1 mutant mice exhibit periodic increases in breathing during normoxia and hypoxia[7]. These observations, taken together with the present finding that STP is absent in NOS-1 mutant mice, suggest that NO contributes to breathing stability during hypoxia via STP.

Another major observation of the present study is that the phenomenon of LTF following recurrent episodes of hypoxia is completely absent in mice deficient in NOS-1. These findings suggest that NO is necessary not only for STP, but also for LTF. This notion is further supported by the finding that 7-nitroindazole, a putative inhibitor of NOS-1, markedly attenuated or abolished LTF in Wt mice. Because changes in VO_2 or VCO_2 and core body temperature are comparable between both groups of mice (present study and ref. 7), we believe that NO mediates LTF by acting on neuronal substrates such as the midline raphe nuclei that mediate LTF. How might NO from NOS-1 contribute to the development of LTF following episodic hypoxia? It has been proposed that 5-HT released from raphe neurons is critical for the development of LTF[1, 2]. It has also been shown that NOS-1 is co-localized with 5-HT in raphe neurons in mice as well as in rats[5, 6]. NO might either favor the release of 5-HT and/or inhibit its uptake, thus amplifying neuronal activity responsible for LTF expression. Such a notion seem to be supported by the observation that NO potentiates or prolongs the action of 5-HT in midbrain periaqueductal (PAG) neurons[16-18]. As regards to its functional significance, it has been suggested that LTF reduces the periods of obstructive sleep apnea[1]. Thus, it is likely that NO generated by NOS-1 may contribute to stability of breathing during sleep apneas by way of promoting LTF.

5. CONCLUSION

In *summary*, in the present study we examined whether NO from NOS-1 modulates STP and LTF of respiration using mutant mice deficient in NOS-1. Our results showed that wild-type mice challenged with either single or multiple hypoxic episodes exhibited elevation of breathing following removal of the hypoxic stimulus. STP and LTF were absent in mutant mice deficient in NOS-1. Taken together, these observations suggest that NO is an important chemical messenger for the expression of STP and LTF and contributes to the stability of breathing.

6. ACKNOWLEDGMENTS

We would like to thank Dr. Paul L. Huang of the Harvard Medical School for providing the NOS-1 mutant mice. This work is supported by National Heart, Lung, and Blood Institute Grant HL-25830 (N. R. Prabhakar); D. D. Kline is supported by training grant T32-HL-07887.

7. REFERENCES

1. F.L.Powell, W.K.Milsom, and G.S.Mitchell, Time domains of the hypoxic ventilatory response., *Respir. Physiol.* **112**:123 (1998).
2. F.L.Eldridge and D.E.Millhorn, Oscillation, gating, and memory in the respiratory control system, *in:* "The Respiratory System; Control of Breathing", N.S.Cherniack and J.G.Widdicombe, eds., American Physiological Society, Bethesda (1986).
3. J.C.McEachern and C.A.Shaw, The plasticity-pathology continuum: defining a role for the LTP phenomenon, *J. Neurosci. Res.* **58**:42 (1999).
4. G.Y.Ko and P.T.Kelly, Nitric oxide acts as a postsynaptic signaling molecule in calcium/calmodulin-induced synaptic potentiation in hippocampal CA1 pyramidal neurons, *J. Neurosci.* **19**:6784 (1999).
5. L.Leger, Y.Charnay, S.Burlet, N.Gay, N.Schaad, C.Bouras, and R.Cespuglio, Comparative distribution of nitric oxide synthase-and serotonin-containing neurons in the raphe nuclei of four mammalian species, *Histochem. Cell Biol.* **110**:517 (1998).
6. L.Leger, N.Gay, S.Burlet, Y.Charnay, and R.Cespuglio, Localization of nitric oxide-synthesizing neurons sending projections to the dorsal raphe nucleus of the rat, *Neurosci. Lett.* **257**:147 (1998).
7. D.D.Kline, T.Yang, P.L.Huang, and N.R.Prabhakar, Altered respiratory responses to hypoxia in mutant mice deficient in neuronal nitric oxide synthase., *J. Physiol.(Lond.)* **511**:273 (1998).
8. P.G.Wagner and F.L.Eldridge, Development of short-term potentiation of respiration., *Respir. Physiol.* **83**:129 (1991).
9. M.A.Haxhiu, C.H.Chang, I.A.Dreshaj, B.Erokwu, N.R.Prabhakar, and N.S.Cherniack, Nitric oxide and ventilatory response to hypoxia, *Respir. Physiol.* **101**:257 (1995).
10. S.R.Vincent and H.Kimura, Histochemical mapping of nitric oxide synthase in the rat brain, *Neuroscience* **46**:755 (1992).
11. S.W.Mifflin, Short-term potentiation of carotid sinus nerve inputs to neurons in the nucleus of the solitary tract., *Respir. Physiol.* **110**:229 (1997).
12. D.R.McCrimmon, E.J.Zuperku, F.Hayashi, Z.Dogas, C.F.Hinrichsen, E.A.Stuth, M.Tonkovic-Capin, M.Krolo, and F.A.Hopp, Modulation of the synaptic drive to respiratory premotor and motor neurons., *Respir. Physiol.* **110**:161 (1997).
13. C.S.Poon, M.S.Siniaia, D.L.Young, and F.L.Eldridge, Short-term potentiation of carotid chemoreflex: an NMDAR-dependent neural integrator, *Neuroreport* **10**:2261 (1999).
14. K.S.Christopherson and D.S.Bredt, Nitric oxide in excitable tissues: physiological roles and disease., *J. Clin. Invest.* **100**:2424 (1997).
15. S.H.Snyder, S.R.Jaffrey, and R.Zakhary, Nitric oxide and carbon monoxide: parallel roles as neural messengers, *Brain Res. Brain Res. Rev.* **26**:167 (1998).
16. M.M.Hamalainen and T.A.Lovick, Role of nitric oxide and serotonin in modulation of the cardiovascular defense response evoked by stimulation in the periaqueductal grey matter in rats, *Neurosci. Lett.* **229**:105 (1997).
17. M.M.Hamalainen and T.A.Lovick, Involvement of nitric oxide and serotonin in modulation of antinociception and pressor responses evoked by stimulation in the dorsolateral region of the periaqueductal gray matter in the rat, *Neuroscience* **80**:821 (1997).
18. T.A.Lovick, Role of nitric oxide in medullary raphe-evoked inhibition of neuronal activity in the periaqueductal gray matter, *Neuroscience* **75**:1203 (1996).

INDUCTION OF LONG-TERM MODULATION OF THE EXERCISE VENTILATORY RESPONSE IN MAN

James W. Reed and James C. Coates*

1. INTRODUCTION

Current views on the control of ventilation include not only the classical reflexogenic mechanisms of afferent feedback, both chemical and neurogenic, but also "feedforward" input from higher centres in the brain, commonly termed central command[1]. Even so, it is apparent that there is still a failure to account for exercise ventilation, when there would appear to be no error signal to regulate breathing[2].

It has recently become apparent that there is a degree of modulation or plasticity in the respiratory control system, with several reports of both short and long term respiratory "memory". For example, following the cessation of carotid sinus nerve stimulation in the cat, phrenic nerve activity does not immediately return to pre-stimulus levels but declines exponentially over several minutes. This observation, originally referred to as the "afterdischarge" has been repeatedly demonstrated[3] and is now an accepted phenomenon. It is currently referred to as short term potentiation (STP; i.e. within the time of the experiment). A similar effect has been found in humans[4, 5]. It has been suggested that STP contributes to the ventilatory response to exercise[6].

Evidence for long term modulation (LTM; i.e. evident for hours or days post-stimulation) is less well documented. In goats, an enhanced ventilatory response to steady-state exercise following repeated short exposures to hypercapnic exercise has been interpreted as LTM[7, 8, 9]. Although there are few relevant studies in man, it has been suggested that LTM of ventilation in humans can be brought about both at rest[10] and during exercise[11]. This LTM of exercise ventilation in man has been postulated to be uniquely dependant upon an associative conditioning paradigm of repetitive exposure to the paired ventilatory stimuli of exercise and hypercapnia[12].

The purpose of the present study was to test this latter hypothesis by examining the ventilatory response to exercise following isocapnic hyperpnœic trials carried out at rest, thus removing both any potentially deterministic exercise element, and any hypercapnic component.

* Department of Physiological Sciences, University of Newcastle, The Medical School, Framlington Place, Newcastle upon Tyne, NE2 4HH, UK.

Frontiers in Modeling and Control of Breathing, edited by Poon and Kazemi.
Kluwer Academic/Plenum Publishers, New York, 2001.

2. EXPERIMENTAL DESIGN AND METHODS

Subjects were healthy volunteers (n = 20, 10 male, mean age 21 ± 1 yr) drawn from staff and students of the University. To comply with the requirements of the Institutional Ethical Committee the procedures were explained in detail, but the subjects were naive as to the main purpose of the study and the possible outcome.

Lung function was assessed as the forced vital capacity (FVC) and the forced expiratory volume in 1 s (FEV_1); maximal inspiratory and expiratory mouth pressures (PImax, PEmax) were measured as surrogates for respiratory muscle strength. All subjects performed a progressive exercise test (20 watt increases in workload each minute after an initial rest period of two minutes) on a cycle ergometer, during which ventilation (Ve) and oxygen uptake (VO_2) were obtained breath-by-breath.. The test was terminated when the heart rate (fC) reached approximately 80% of predicted maximum (fCmax = 210 – [0.65.age]).

Ten of the subjects (6 male) then undertook a period of isocapnic hyperventilation, breathing at 70% of predicted maximal voluntary ventilation (MVV; calculated as FEV_1 x 35), with a tidal volume of 60% FVC and an appropriate frequency. The CO_2 levels were kept constant at normocapnic levels (variation less than 3 torr of target value) by having the subjects breathe through a dynamically variable deadspace. The training regimen was 20 mins a day, five days a week for 4 weeks. The remaining 10 subjects served as a control group and did not take part in the training sessions.

Resting lung function and exercise performance for all subjects was re-assessed after completion of the 4 week training regimen. For both groups, assessment was carried out on the day following the last training session.

3. RESULTS

The isocapnic hyperventilation regimen proved to be effective in increasing both PImax and PEmax, indicative of a respiratory muscle training effect (Table 1). However, the submaximal exercise ventilation post-training was significantly increased compared to the initial assessment. This relative exercise hyperventilation was as a result of an increased respiratory frequency, and despite a significant reduction in the exercise tidal volume. The equivalent indices for the control group were neither materially nor statistically significantly changed. There were no changes in the measured indices of ventilatory capacity in either group.

4. DISCUSSION

The existence of LTM, and any method of induction, has profound implications for our understanding of the development of respiratory control in man. Such a mechanism would allow for the exercise ventilatory response to be modified by experience. In control theory, this has been termed adaptive feed-forward control[13], and is consistent with at least part of the observed exercise hyperpnoea being a learned response[2, 14]. The phenomenon has been described in conscious exercising goats[9], and is generally accepted

Table 1. Effect of isocapnic hyperventilation: intervention group (n = 10)

	Ve[a] (l.min^{-1})		Vt[b] (l)		fR[c] (b.min^{-1})		PImax (cms H_20)		PEmax (cms H_20)	
	mean	sd	mean	sd	mean	sd	mean	sd	mean	sd
Pre	27.5	2.6	1553	515	20.5	4.3	112	37.8	158	53
Post	31.9	3.4	1374	449	23.2	4.8	123	31.4	183	67
p =	0.01		0.007		0.006		0.058		0.023	

Pre = pre-isocapnic hyperventilation. Post = post - isocapnic hyperventilation.
[a] =ventilation at an oxygen uptake of 1 l.min^{-1}. [b] =at a ventilation of 30 l.min^{-1}. [c] =at a ventilation of 30 l.min^{-1}.
PImax = maximal inspiratory pressure measured from residual volume.
PEmax = maximal inspiratory pressure measured from total lung capacity.

as such. The role of LTM of the exercise ventilatory response in humans is less clear. There are few studies extant, and the conclusions from those studies, which employed the same associative conditioning technique, are conflicting. Turner et al.[11, 12] have reported an exercise hyperventilation associated with a relative hypocapnia. An earlier study[15] was initially interpreted as being consistent with LTM, but subsequent analysis has shown that the reported exercise hypocapnia was in fact a time-order effect and not dependant upon the conditioning experiments (Moosavi, personal communication). More recently, Cathcart et al.[16] have found no evidence of changes in either exercise ventilation nor CO_2 response following "conditioning" and have therefore been unable to confirm the existence of long term modulation or plasticity.

The protocol used by all the quoted studies involved exposure to the conditioning stimuli during short periods (5 min) of steady-state exercise. Although the methodology differed, the ventilatory response we have observed following our hyperventilation procedures is qualitatively similar to those observed in the conditioning experiments of Martin and Mitchell in their LTM experiments on goats.

The methodology we adopted, isocapnic hyperventilation, did result in a measurable respiratory muscle training effect over the period of study. Respiratory muscle training might be expected to increase breathing efficiency and therefore result in an increased tidal volume and reduced respiratory frequency at submaximal levels of exercise, with the equivalent exercise ventilation either being unaffected or slightly lower.

In the event, the changes in the exercise ventilatory response were the opposite: the submaximal exercise ventilation was increased as a consequence of increased respirator frequency, and exercise tidal volume was actually reduced. An increase in exercise ventilation following respiratory training manœuvres has been reported previously[17] but was attributed to respiratory muscle fatigue. This could not be the explanation for our observations, as the effect was seen 24 hours after the last training procedure, and respiratory muscle strength was increased at the time of testing. Whilst there is the possibility that our rather extended procedures did in some way influence the results, the most likely explanation is that our observations are of a manifestation of LTM of exercise ventilatory control.

It is possible that the underlying mechanism may be that of self-tuning adaptive control[18]. Poon's model suggests that coupled oscillations in the input signals to, and the output signals from, the respiratory controller may give rise to an exercise hyperpnoea through adaptive enhancement of neural transmission within the respiratory controller.

This is consistent with the hypothesis that LTM of ventilation may be a learned or conditioned response. However, it is evident that such a response is not dependant on an associative conditioning paradigm; our data suggest that long term modulation of the ventilatory response to exercise in man may be induced by the relatively simple technique of repetitive reinforcement learning without the need for a specific exercise stimulus or hypercapnia, either individually or in association.

5. REFERENCES

1. B. J.Whipp, S. A.Ward, Determinants and control of breathing during muscular exercise. *Br J Sports Med* **32**: 199-211 (1998).
2. G. G. Somjen, The missing error signal - regulation beyond negative feedback. *NIPS* **7**: 184-185 (1992).
3. F. L. Eldridge and P. Gill-Kumar, Central neural respiratory drive and afterdischarge. *Respir. Physiol.* **40**: 49-63 (1980).
4. G. D. Swanson, D. S. Ward and J. W. Bellville, Posthyperventilation isocapnic hyperpnea. *J. Appl. Physiol* **40**: 592-596 (1976).
5. R. F. Fregosi, Short-term potentiation of breathing in humans. *J.Appl. Physiol* **73**: 892-899 (1991).
6. F. L. Eldridge and T. G. Waldrop, Neural control of breathing during exercise. In: *Exercise, Pulmonary Physiology and Pathophysiology*. New York, Dekker, pp 309-370 (1991).
7. G. S. Mitchell, K. Foley, S. McGuirk, B. D. Selby, S. L. Schaefer and K. J. Lange, Effects of chronic thoracic dorsal rhizotomy (TDR) on ventilatory control during mild exercise in goats. *FASEB J* **2**: A1508 (1988).
8. G. S. Mitchell, M. A. Douse and K. T. Foley, Receptor interactions in modulating ventilatory activity. *American Journal of Physiology* **259** : 911-920 (1990).
9. P. A. Martin and G. S. Mitchell, Long-term Modulation of the exercise Ventilatory Response in Goats. *J. Physiol.* **470**: 601-617 (1993).
10. J. Gallego and P. Perruchet, Classical conditioning of ventilatory responses in humans. *J Appl Physiol*; **70**: 676-682 (1991).
11. D. L. Turner, J. R. Greenwell, H. Lawrence, P. Lyons, M. R. Taylor and Z. M. Iqbal, Long term modulation of ventilatory control in exercising humans. *Soc. Neurosci. Abstracts* **22:** 1602 (1996).
12. D. L. Turner, K. B. Bach, P. A. Martin, E. B. Olsen, M. Brownfield, K. T. Foley and G. S. Mitchell, Modulation of ventilatory control during exercise. *Respiration Physiology* **110:** 277-285 (1997).
13. J. C. Houk, Control strategies in physiological systems. *FASEB J.* **2:** 97-107 (1988).
14. J. A. Dempsey, H.V. Forster and D. M. Ainsworth, Regulation of hyperpnœa, hyperventilation, and respiratory muscle recruitment during exercise. In: Dempsey JA, Pack AI, eds. *Regulation of Breathing* 2nd edition, Lung Biology in Health and Disease, New York, Marcel Dekker, pp1065-1133 (1995).
15. L. Adams, S. Moosavi, A. Guz, Ventilatory response to exercise in man increases by prior conditioning of breathing with added deadspace. *Amer. Rev. Resp. Disease* **145**: A882 (1992).
16. A. J. Cathcart, N. Herrold, A. P. Turner, J. Wilson and S. A. Ward, Absence of long-term modulation in response to external dead space loading during moderate exercise in humans. *J Physiol* **528P**:44P (2000).
17. J. Kohl, E. A. Koller, M. Brandenberger, M. Cardenas and U. Boutellier, Effect of exercise-induced hyperventilation on airway resistance and cycling endurance. *Eur J Appl Physiol* **75**: 305-311 (1997).
18. C. S. Poon, Self-tuning optimal regulation of respiratory motor output by Hebbian covariance learning *Neural Networks* **9**: 1367-1383 (1996).

METABOLIC CONSEQUENCES OF HYPOXIC CONDITIONING IN *LYMNAEA STAGNALIS*

Barbara E. Taylor, Kim Smyth, John E. Remmers and Ken Lukowiak*

1. INTRODUCTION

The ventilatory behavior of *Lymnaea stagnalis* (surfacing and opening the respiratory orifice, the pneumostome) can be operantly conditioned[1]. Snails placed in severely hypoxic water (< 1 ml $O_2 \cdot L^{-1}$; created by gassing with nitrogen) are normally driven to increase their rate of ventilatory behavior[1]. However, if a noxious tactile stimulus is applied to the pneumostome at the time of ventilation, the animals can be conditioned not to breathe. They learn to self-impose hypoxia, and they remember to do so whenever they are placed in the nitrogen-gassed water. How is *Lymnaea* able to do so; most animals typically exhibit a hypoxic response designed to rectify oxygen limitation?

In most animals, exposure to hypoxia typically evokes a biphasic response of first compensatory changes in ventilation followed by adaptive changes in both ventilation and metabolism[2]. Compensatory changes involve increases in ventilatory behavior representing the attempt to continue to meet cellular demands for oxygen. Persistent or increased hypoxia, however, evokes adaptive changes; ventilatory behavior and metabolism are reduced in an attempt to minimize oxygen requirements, and thereby lessen or avoid the detrimental effects of hypoxia[3,4]. In our conditioning paradigm *Lymnaea* elects to tolerate the hypoxic conditions of the training in order to avoid the noxious stimulus. Clearly, the conditioning process altered the hypoxic response. Conditioning eliminated the option of compensatory changes, begging the question whether this accelerated the animals into the adaptive phase of their hypoxic response.

The nature of *Lymnaea's* hypoxic response under the conditioning regime is investigated here by comparing the metabolism of animals subjected to the training regime with that of naïve animals. Any difference identified in such a comparison is put into context by comparison with changes in metabolism evoked by short- and long-term changes in environmental temperature. We hypothesized that conditioned animals would exhibit a lower metabolism in the conditioning environment than naïve animals, just as

* B.E. Taylor, Department of Physiology, Dartmouth Medical School, Borwell Bldg., 1 Medical Center Drive, Lebanon, NH 03756-0001, USA. K. Smyth, J. Remmers and K. Lukowiak, Neuroscience and Respiratory Research Groups, Faculty of Medicine, University of Calgary, 3330 Hospital Drive NW, Calgary Alberta, Canada, T2N 4N1.

Frontiers in Modeling and Control of Breathing, edited by Poon and Kazemi.
Kluwer Academic/Plenum Publishers, New York, 2001.

animals from a 'cold' environment would exhibit a lower metabolism than those from a 'hot' environment.

2. MATERIALS AND METHODS

2.1. Animals

Sixty-three laboratory-reared *Lymnaea stagnalis* (L.) were used in this study. All animals were 1.0 - 2.5 g in mass (20 - 35 mm in shell length). Animals were maintained at three temperatures. The aquaria that housed the animals were either held at room temperature (20 - 22 °C), heated with submersible heaters to be 'hot' (28 – 30 °C), or kept in a refrigerator to be 'cold' (3 – 5 °C). Regardless of holding temperature, animals had continual access to food (lettuce supplied twice per week).

2.2. Operant-Conditioning Paradigm

The conditioning regime employed here has been used previously in studies of learning in *Lymnaea*[1]. *Lymnaea* respiration is bimodal; gas exchange occurs across the snail's skin and across its rudimentary lung, which has access to the environment through the pneumostome. Under normoxic water conditions, animals surface and open their pneumostomes once every 15 – 20 minutes, and when animals are held in water made hypoxic by gassing with nitrogen, the frequency of their ventilatory behavior significantly increases[1]. During training sessions, pneumostome closure was evoked by an investigator's application of a tactile stimulus to the pneumostome area with a sharpened wooden stick. The intensity of the stimulus was not quantified, however, it was always sufficient to elicit pneumostome closure without evoking a whole-animal withdrawal response.

Animals for this investigation were randomly assigned to one of the following treatments: (1) baseline control; (2) conditioned; (3) conditioning control; (4) short-term hot; (5) long-term hot; (6) short-term cold; (7) long-term cold. The baseline-control group (N = 15) was an untreated group used to establish a typical metabolism for *Lymnaea*. The conditioned group (N = 7) received two 30-minute training sessions, with sessions separated by 1 hour. This conditioning regime has been shown to produce learning and memory that lasts for 24 hours[5]. This was the interval that elapsed before metabolism was measured. The conditioning-control group (N = 10) received two 30-minute exposures to nitrogen-gassed water with sessions separated by 1 hour, but no stimuli were delivered to their pneumostomes. These three groups were maintained at room temperature (20 – 22 °C). The short-term hot group (N = 8) had metabolism measured immediately (within < 5 minutes) upon being exposed to water at 28 – 30 °C, whereas the long-term hot group (N = 7) was held at this temperature for 7 days prior to measurement of metabolism. Similarly, the short-term cold group (N = 8) had metabolism measured immediately upon being exposed to water at 3 – 5 °C, whereas the long-term cold group (N = 8) was held at this temperature for 7 days prior to measurement of metabolism.

2.3. Metabolism

Metabolism was measured indirectly as oxygen consumption (VO_2) of animals held individually in an aquatic respirometer without access to air. The pond water within the respirometer was either normoxic (for the baseline-control, and the short- and long-term hot and cold groups) or hypoxic (for the conditioning-control and conditioned groups). Confinement to pond water forced the snails to respire cutaneously, which they are able to do without altering their metabolic rate (unpublished data). O_2 concentration of the respirometer medium was measured with a Clark-type O_2 electrode linked to a WinDAQ data acquisition system. VO_2 was measured as $\mu l\ O_2 \cdot g$ live mass$^{-1} \cdot h^{-1}$ over a 20-minute period with the animals in water maintained at the temperatures previously defined as room, hot or cold and stirred continuously with a magnetic stir bar.

2.4. Statistical Analyses

VO_2 data were analyzed using Analysis of Variance coupled with Tukey Multiple Comparison Test (Tukey MCT). Values presented in the text are means ± standard error.

3. RESULTS

Figure 1 illustrates metabolism, measured indirectly as oxygen consumption (VO_2), of *Lymnaea stagnalis* from the various treatment groups employed in this study. The typical or baseline metabolic rate for these snails was reflected by a VO_2 of $92 \pm 6\ \mu l\ O_2 \cdot$ g live mass$^{-1} \cdot h^{-1}$. Neither the conditioning-control group, which was exposed to hypoxic water but not stimulated upon opening their pneumostome, nor the conditioned group, which was stimulated upon opening their pneumostome, had a significantly different VO_2 compared to baseline ($F_{62,6} = 19.9$; $p < 0.001$; Tukey MCT). The conditioned animals exhibited a significant reduction in ventilatory behavior (failure to surface and open their pneumostome; data not presented) reminiscent of previous studies[1]. Furthermore, this level of learning has previously been shown sufficient to produce memory for 24 hours[5]. Animals exposed to either hot or cold had a significantly different VO_2 ($F_{62,6} = 19.9$; $p < 0.001$; Tukey MCT) regardless of the duration of their exposure to the temperature change. Animals exposed to hot water immediately underwent an increase in metabolism. VO_2 was $157 \pm 22\ \mu l\ O_2 \cdot$ g live mass$^{-1} \cdot h^{-1}$, and this increase was maintained without significant change (Tukey MCT) for at least 7 days when VO_2 was $159 \pm 10\ \mu l\ O_2 \cdot$ g live mass$^{-1} \cdot h^{-1}$. Animals exposed to cold water immediately underwent a decrease in metabolism. VO_2 was $18 \pm 4\ \mu l\ O_2 \cdot$ g live mass$^{-1} \cdot h^{-1}$, and this increase was maintained without significant change (Tukey MCT) for at least 7 days when VO_2 was $37 \pm 5\ \mu l\ O_2 \cdot$ g live mass$^{-1} \cdot h^{-1}$.

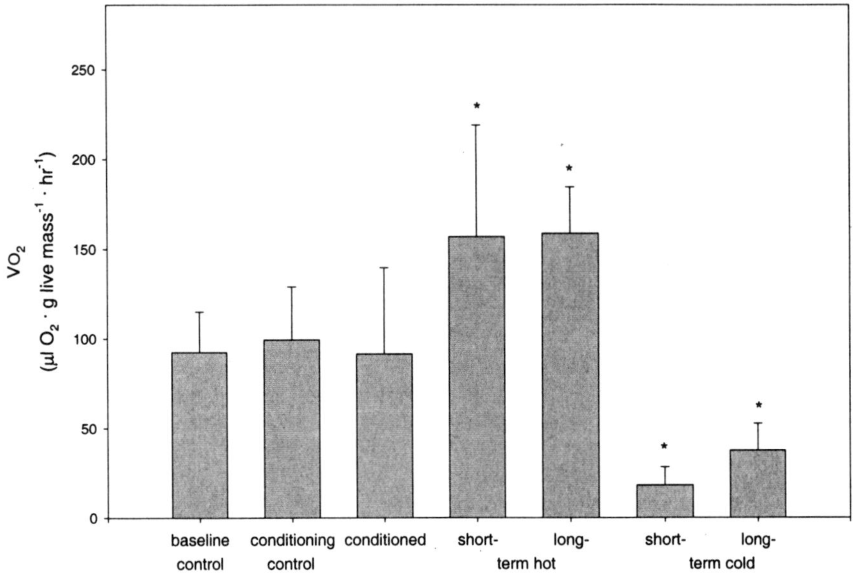

Figure 1. Metabolic rate of snails (*Lymnaea stagnalis*), measured as VO_2. The snails were either treated or untreated (baseline control). Those that were treated were either subjected to or conditioned in hypoxia, or they were held in hot (28 – 30 °C) or cold (3 – 5 °C) pond water for short (< 5 minutes) or long-term (7 days). N = 7 – 15 for each treatment group. * indicates a significant difference from baseline control.

4. DISCUSSION

The results of the present study indicate that operantly conditioned *Lymnaea stagnalis* that have learned to self-impose hypoxia do not cope with this hypoxia by reducing their metabolic rate. Their metabolic rate is maintained at a typical baseline level. This suggests that an oxygen debt is incurred during the self-imposed hypoxia experienced during the training period; however, this remains to be demonstrated. Thus, the reduced ventilatory behavior exhibited by trained snails when they are returned to the nitrogen-gassed training chamber is certainly evidence of learning and memory and not a result of hypoxic hypometabolism.

In this study conditioned *Lymnaea* did not exhibit a typical response to hypoxia. The snails learned not to increase ventilation, and suppression of this compensatory response did not facilitate the adaptive response of decreased metabolism. This is evidence for the plasticity of hypoxic response. Developmental plasticity has been demonstrated in the hypoxic ventilatory response of rats; neonatal exposure to hypoxia influenced adult response to hypoxia[6,7]. Furthermore, Thomas and colleagues[8,9] reported that associations of hypoxia and auditory stimulation during the first month of life influenced ventilatory patterns during sleep in adult rats. The present study further characterizes the plasticity of hypoxic response; it demonstrates plasticity associated with behavioral modification, but not in association with development.

The results of metabolic measurements on animals subjected to short- and long-term temperature changes have a dual role in the present discussion. The temperature experiments were initially included in the study to create a frame of reference for the metabolic changes that were expected to be associated with hypoxic conditioning. To this end the results show an expected sensitivity of metabolism to temperature. In addition they indicate an impressive responsiveness of *Lymnaea*'s metabolism to environmental change. Snails exposed to either hot or cold temperatures made a rapid adjustment in metabolic rate. The rapidness of this response did not compromise the appropriateness of its magnitude, because it was unnecessary for the animals to further adjust their metabolism when the temperature change was maintained for seven days.

5. ACKNOWLEDGMENTS

We thank Michael Harris for his useful comments on the study design and manuscript. Medical Research Council of Canada Grants to KL and JER, and an Alberta Heritage Foundation for Medical Research Summer Student Award to KS funded this study.

6. REFERENCES

1. Lukowiak, K., Ringseis, E., Spencer, G., Wildering, W., Syed, N., 1996, Operant conditioning of aerial respiratory behaviour in *Lymnaea stagnalis, J. Exp. Biol.* **199**:683.
2. Powell, F.L., Milsom, W.K. and Mitchell, G.S., 1998, Time domains in the hypoxic ventilatory response, *Respir. Physiol.* **112**:123.
3. Hochachka, P.W., Buck, L.T., Doll, C., and Land, S.C., 1996, Unifying theory of hypoxia tolerance: molecular metabolic defense and rescue mechanisms for surviving oxygen lack, *Proc. Natl. Acad. Sci. USA* **51**:9493.
4. Bickler, P.E. and Buck, L.T., 1998, Adaptations of vertebrate neurons to hypoxia and anoxia: maintaining critical Ca 2+ concentrations, *J. Exp. Biol.* **201**:1141.
5. Lukowiak, K., Adatia, N., Krygier, D., Syed, N., 2000, Operant conditioning in *Lymnaea*: evidence for intermediate- and long-term memory, *Learning & Memory* **7**:140.
6. Okubo, S. and Mortola, J.P.Jr., 1990, Control of ventilation in adult rats hypoxic in the neonatal period, *Am. J. Physiol.* **259**:R836.
7. Ling, L., Olson, E.B.Jr., Vidruk, E.H. and Mitchell, G.S., 1997, Developmental plasticity of the hypoxic ventilatory response, *Respir. Physiol.* **110**:261.
8. Thomas, A.J., Austin, W., Friedman, L. and Strohl, K.P., 1992, A model of ventilatory instability in the unrestrained rat, *J. Appl. Physiol.* **73**:1530.
9. Thomas, A.J., Friedman, L., MacKenzie, C.N. and Strohl, K.P., 1995, Modification of conditioned apneas in rats: Evidence for cortical involvement, *J. Appl. Physiol.* **78**:1215.

THE *LYMNAEA* RESPIRATORY SYSTEM:

Where are we going with learning?

Ken Lukowiak*

1. INTRODUCTION

Lymnaea has proven to be a fantastic model system in which to elucidate the neuronal mechanisms that underlie the control of respiratory behaviour[1]. In this paper I will attempt to convince you that it is also a fantastic model in which to study the neuronal mechanisms that constitute learning and memory.

2. THE RESPIRATORY CENTRAL PATTERN GENERATOR

The central pattern generator (CPG) that controls aerial respiration in the pond snail, *Lymnaea* (Fig. 1) is made up of 3 neurons and it produces rhythmic activity as a result of emergent network properties. The interested reader is directed to a number of recent reviews outlining how this CPG functions[2, 3]. Suffice to say here that this is one of the few circuits where both the sufficiency and necessity of the neurons for the mediation of a behaviour has actually been demonstrated. A second neural circuit is also represented here. This is the whole body withdrawal circuit (WBWC). While only a single neuron is illustrated here, it is a circuit of electrically coupled neurons that acts to withdraw the body of the snail into the shell. This behaviour, withdrawal into the shell, is hierarchically the most important behaviour possessed by the snail, as it takes precedence over all other behaviour, including aerial respiration. Activity in the WBWC directly (via monosynaptic inputs) shuts off air breathing. It accomplishes this by inhibiting the ability of the circuit to generate rhythmicity and by activating closure of the pneumostome, the respiratory orifice[4].

* Ken Lukowiak, Department of Physiology & Biophysics, University of Calgary, Calgary, Alberta, Canada, T2N 4N1

Frontiers in Modeling and Control of Breathing, edited by Poon and Kazemi.
Kluwer Academic/Plenum Publishers, New York, 2001.

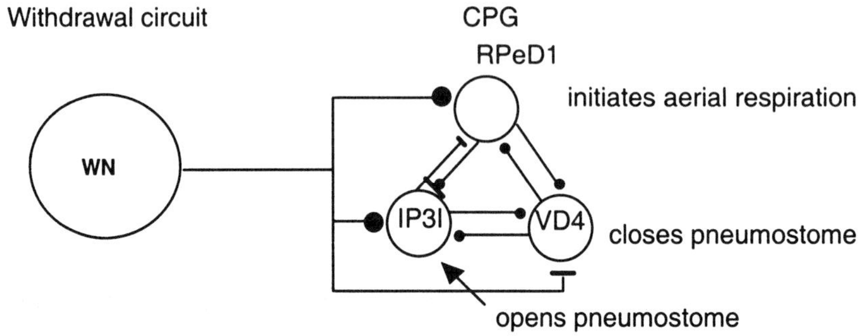

Figure 1. The whole body withdrawal circuit (WBWC) and the respiratory central pattern generator (CPG). Shown in this diagram are the neurons and synaptic connections that comprise the CPG underlying aerial respiration and the circuit representing the WBWC. The respiratory CPG is made up of 3 interneurons. RPeD1 initiates the activity of the circuit whilst neurons VD4 and IP3I act to respectively close and open the pneumostome (the respiratory orifice). Rhythmicity is the result of emergent network properties. Only one neuron is shown representing the WBWC since the circuit is composed of neurons connected via electrical synapses. The WBWC via WN makes monosynaptic inhibitory synaptic connections (filled circles) with RPeD1 and IP3I and a monosynaptic excitatory synapse (bar) with VD4. Thus activity in WN directly inhibits rhythmic activity and closes the pneumostome.

3. OPERANT CONDITIONING

We can operantly condition a form of associative learning, aerial respiratory behaviour. In our conditioning procedure the snails learn not to air-breathe in a hypoxic environment! Hypoxia normally increases the drive to air-breathe. This might seem to be a bad thing. But remember, these snails can also breathe across their skin, thus learning not to air-breathe does not result in death. We can demonstrate that the snail is capable of remembering this learning for long periods of time (up to 4 weeks). That is, it has the ability to form long-term memory (LTM)[5].

Such a model system puts us into a position where we can begin to perform experiments that could directly determine causal mechanisms of learning and memory. However, to be able to do this we have to successfully complete a number of important experiments. I will go over these below. This constitutes the "Where are we going".

4. NEURAL CORRELATES OF LEARNING AND MEMORY

We first have to show that there are changes in the activity of the CPG neurons that are consistent with the snail not air-breathing in the hypoxic environment where conventional wisdom predicts that it should. We have shown that there are correlated changes in the activity of one of the CPG neurons, RPeD1[6]. However, we performed this in isolated ganglia preparations following operant conditioning of air breathing. We now have to show that the observed changes have anything to do with the actual behaviour of the animal. To accomplish this we have developed a semi-intact preparation where we can simultaneously record neuronal activity and pneumostome activity. The schematic diagram in figure 2 summarizes our recent findings in these semi-intact preparations.

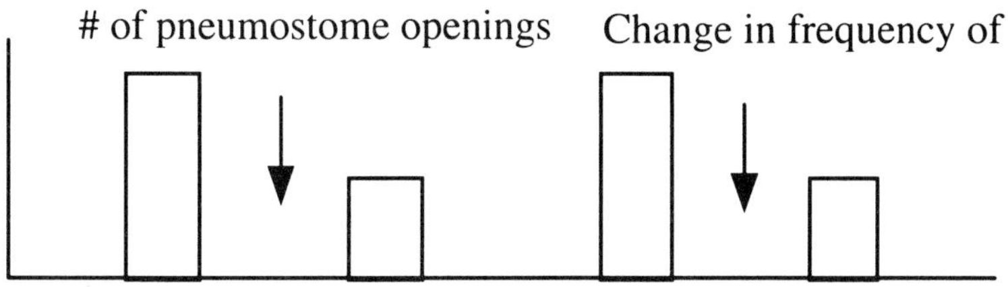

Figure 2. Neural correlates of learning and memory in a semi-intact preparation. Diagrammatic representation of data that have been obtained in our on-going studies. Plotted are the number of pneumostome openings and concomitant activity of RPeD1 on two training sessions (10-min duration) following training in the intact animal. A reinforcing stimulus (at arrow) was applied to the pneumostome as it attempted to open following the first 10-minute session. As can be seen both the number of pneumostome openings and the spontaneous activity of the neuron were both significantly less following the presentation of the reinforcing stimulus. In yoked control animals there are no significant changes in either the number of pneumostome openings or in RPeD1 activity following the presentation of the reinforcing stimulus.

As can be seen, following the presentation of the reinforcing stimulus to the semi-intact preparation there was a concomitant significant decrease (you have to trust me on this) in the number of pneumostome openings and the frequency of RPeD1 activity. This of course is still just a correlation. We now have to show that the 'activity' of RPeD1 is causative to the associative learning and its memory. In order to accomplish this we will make use of the different properties of memory and the special anatomy of the molluscan nervous system.

5. TYPES OF MEMORY

Memory can be categorized according to how long it persists. Long-term memory (LTM) lasts for at least 24h and may last for years. LTM is also dependent on altered gene activity and changes in protein synthesis. Another form of memory, known as intermediate term memory (ITM) only persists for about 3h and it is *not* dependent on altered gene activity but is dependent on a change in protein synthesis[7]. That is, the mRNA necessary for the memory is already present.

We have been able to demonstrate that these two forms of memory exist in our *Lymnaea* model system. ITM is evoked if we decrease the interval between the training sessions from 1h to 0.5h. LTM is seen if we have the 1h interval between the training sessions. Thus we can differentially produce either ITM or LTM.

6. SPECIAL ATTRIBUTES OF MOLLUSCAN NEURONS

Figure 3, another schematic diagram, shows that the synaptic inputs to and from molluscan neurons occur on specializations on the primary neurite. Thus, it is possible to remove the soma of the cell and not interrupt the synaptic organization of the neuron. Amazingly, the isolated primary neurite survives this procedure and functions normally for up to 1 month. We have ablated the soma of RPeD1 in the otherwise intact, freely moving snail and asked whether the snail could continue to perform aerial respiration. If

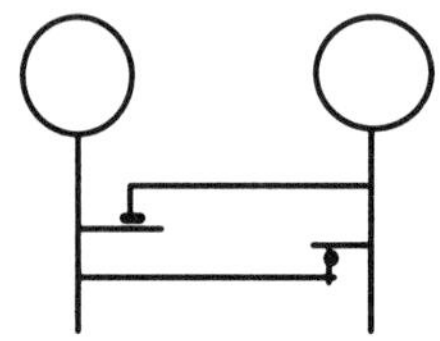

Figure 3. Representation of the anatomy of *Lymnaea* central neurons. The central neurons of *Lymnaea* neurons are monopolar. Both the pre- and post-synaptic specializations occur on the primary neurite. Thus it is possible to remove the soma and leave behind a functioning primary neurite. In our experiments the surviving primary neurite remains viable for up to 4 weeks. The filled circle represent an inhibitory synapse whilst the bar represents an excitatory synapse.

we kill RPeD1 by the injection of pronase into it, the snail can no longer air-breathe. However, just removing the soma does not prevent aerial respiratory behaviour. We therefore can now begin experiments to directly determine if RPeD1 (or any other neuron for that matter) is necessary for the storage of memory. The experiments are conceptually simple. We remove RPeD1's soma and then operantly condition the snail. If our hypothesis that RPeD1 is a site for LTM storage were correct we would expect to see the following results.

7. EFFECTS ON MEMORY OF ABLATING THE SOMA OF RPeD1

We would expect to see learning, since learning occurs even if protein synthesis is blocked. This point also serves to remind us that learning and memory, although related, are two different processes. We would also expect to see ITM, since ITM is only dependent on protein synthesis. You might hesitate at this point because conventional wisdom states that protein synthesis occurs within the soma. However, we have recently shown that isolated (i.e. w/o the soma) primary neurites are capable of *de novo* protein synthesis[8]. The machinery necessary to both translate mRNA and integrate it into a functional protein within the membrane is present in the primary neurite. We do not however expect to see LTM. LTM is dependent on both altered gene activity (i.e. transcription) and protein synthesis. Since the nucleus has been removed, there are no genes, thus no ability to produce new message. Our preliminary data (Figure 4) show that these results can be obtained. As a control we remove the soma of a neuron that is of similar size but that is not involved in the mediation of aerial respiratory behaviour. Thus, we may be able (if the new data are consistent with the preliminary data) that RPeD1 is a site of LTM storage.

We will also have to demonstrate that a translational protein blocker will inhibit both ITM and LTM whilst a transcriptional blocker will only inhibit the formation of LTM. These experiments are now being performed.

Another series of experiments that follow from the RPeD1 soma ablation studies will test whether forgetting or extinction involves altered gene activity and protein synthesis. It may well turn out that forgetting is not a passive process but rather than active process. Support for this notion comes from a recent paper that shows that when a LTM memory is being recalled it can be disrupted by protein synthesis blockers[9].

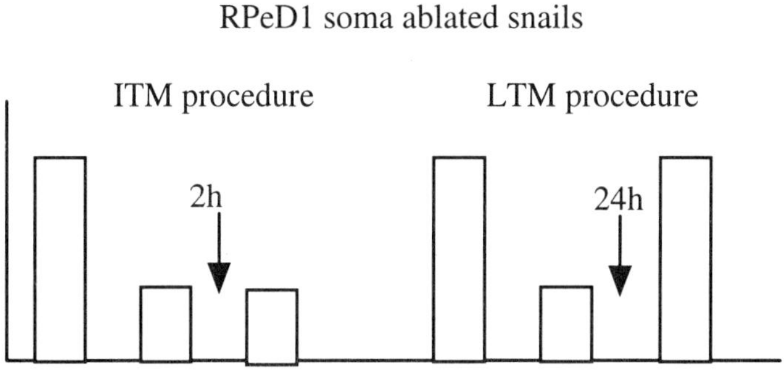

Figure 4. Intermediate term (ITM) memory but not long-term (LTM) memory occurs in RPeD1 soma-ablated *Lymnaea*. RPeD1's soma was ablated 2 days previously. Since the snails can still perform aerial respiration the primary neurite of RPeD1 must still be functional. The number of attempted pneumostome openings in 45 min training sessions has been plotted. The test for memory took place either 2h after the last training session for ITM or 24h after the last training session for LTM. In the training procedure used to produce only ITM the training sessions are separated by 30 min whilst for LTM a 1h interval was used.

8. ULTIMATE CHALLENGES STILL TO COME

Our ultimate challenge will be to show that learning and memory reside in the CPG by transplanting neurons from an "educated' animal to a naïve animal and show that the naïve animal now exhibits learning and memory.

Our first task will have to show that LTM can survive the regeneration process. That is, if we crush the axon of RPeD1 in a trained animal, following regeneration of RPeD1's axon and the re-establishment of its proper synaptic connections will LTM still be present? Our preliminary data show that LTM is resistant to the cell biological processes that underlie regeneration.

We will initially just attempt to transplant RPeD1. Thus the experiment will be to train an animal, then remove its RPeD1 and transplant it into a naïve animal whose own RPeD1 had been removed. Following regeneration of RPeD1 and the establishment of the proper synaptic connectivity we would expect the animal to behave as though it had been trained and had memory. If these experiments are successful, then we could confidently begin studies designed to elucidate the molecular mechanisms by which learning and memory are formed and maintained.

We believe that our *Lymnaea* model system is not only a great system in which to study how respiration is controlled; but it is also one that may allow us to begin to unlock the mysteries of how learning and memory are mediated.

9. REFERENCES

1. Syed, N., Bulloch, A., and Lukowiak, K., 1990, In vitro reconstruction of the respiratory central pattern generator of the mollusk *Lymnaea, Science* **250:** 282
2. Lukowiak, K. and Syed, N., 1999 Learning, memory and a respiratory central pattern generator. *Comp. Biochem. Physiol. A.* **124**: 265.
3. Taylor, B., and Lukowiak, K., 2000, The respiratory central pattern generator of *Lymnaea:* a model, measured and malleable. *Respiration Physiology,* **122**: 197

4. Inoue, T., Tasasaki, M., Lukowiak, K., and Syed, N., 1996, Inhibition of the respiratory pattern-generating neurons by an identified whole-body withdrawal interneurone of *Lymnaea stagnalis, J. Exp. Biol.* **199**:1887
5. Lukowiak, K., Ringseis, E., Spencer, G., Wildering, W., and Syed, N., 1996, Operant conditioning of aerial respiratory behaviour in *Lymnaea stagnalis, J. Exp. Biol.* **199**:683
6. Spencer, G., Syed, N., and Lukowiak, K., 1999, Neural changes after operant conditioning of the aerial respiratory behaviour in *Lymnaea stagnalis. J. Neuroscience.***19:** 1836
7. Lukowiak, K., Adatia, N., Krygier, D., and Syed, N.., 2000, Operant conditioning in Lymnaea: Evidence for intermediate and long-term memory. *Learning and Memory* **7**: 140
8. Spencer, G., Syed, N., van Kesteren E, Lukowiak, K., Geraerts, W.P.M and van Minnen (2000). Synthesis and Functional Integration of a Neurotransmitter Receptor in Isolated Axons. J. Neurobiol. 44, 72 – 81.
9. Nader, K., Schafe, G., and Le Doux, J., 2000, Fear memories require protein synthesis in the amygdala for reconsolidation after retrieval. *Nature* **406**: 722

POSSIBLE MECHANISMS THAT MAY DETERMINE THE SET POINT AND SENSITIVITIES OF THE CHEMOREFLEXES

Peter A. Robbins*

1. ROLE OF CO_2 IN THE CONTROL OF BREATHING

A century ago, Haldane and Priestley demonstrated the exquisite sensitivity of ventilation ($\dot{V}_E$) to small variations in alveolar P_{CO_2} [6]. This belief in the primacy of CO_2 over O_2 in the regulation of $\dot{V}_E$ has persisted through into modern physiology, for example, in his textbook on Respiratory Physiology[15], West writes, "The most important factor in the control of ventilation under normal conditions is the P_{CO_2} of the arterial blood." Indeed, a reasonable estimate is that ventilation is ~30 fold more sensitive to acute variations in P_{CO_2} compared with P_{O_2} for blood gas tensions around sea level.

There are, however, problems with a simple "CO_2-stat" model of pulmonary regulation. In chronic lung disease, there is widespread variation in the respiratory controller's response to essentially similar pathological processes occurring in the lung[3]. At one extreme are the pink puffers, who are making considerable respiratory efforts with values for arterial P_{CO_2} that are normal or low. From where does this respiratory drive originate? At the other extreme are the blue bloaters, who make little respiratory effort with values for arterial P_{CO_2} that are often alarmingly high. Why is there a failure of stimulation by carbon dioxide under these conditions?

These clinical observations suggest that, over the long term, the ventilatory response to CO_2 may adapt. The ventilatory response to CO_2 could thus provide a high gain response to acute challenges to the respiratory system, but not provide a mechanism for setting the basal level of ventilation long term. Such function would be analogous to the role of baroreceptors in regulating blood pressure.

* University Laboratory of Physiology, University of Oxford, Parks Road, Oxford OX1 3PT, UK

Frontiers in Modeling and Control of Breathing, edited by Poon and Kazemi.
Kluwer Academic/Plenum Publishers, New York, 2001.

2. PUTATIVE ROLE OF O_2 IN THE CONTROL OF BREATHING

If the P_{CO_2} of the arterial blood is not the signal that regulates basal ventilation over the long term, what does? In a sense, the ventilation of the lungs only has two metabolic feedback outputs, alveolar P_{CO_2} and alveolar P_{O_2}. Can alveolar P_{O_2} act as the major signal for determining basal ventilation over the long term? Clearly, there would have to be considerable variability in the response because pink puffers are considerably less hypoxic than blue bloaters. However, chronic hypoxic exposure, as determined by studies at altitude, exerts a much more powerful influence over basal ventilation than does acute hypoxic exposure[11]. These acclimatization effects are due to, at least in part, the hypoxic exposure per se, as opposed to the associated alkalosis[7, 14]. More recently, we have shown that very mild sustained hypoxia can still induce marked acclimatization[5], and that a sustained exposure to hyperoxia induces the reverse of most of these changes[13]. These latter observations suggest that the mechanism underlying "acclimatization to altitude" may be active through the range of values of P_{O_2} normally associated with sea-level residence. This raises the possibility that, rather than being seen as an evolutionary adaptation to life at various altitudes, the phenomenon of ventilatory acclimatization to altitude may arise through the effects of hypoxia on a normal process regulating ventilation at sea level. An analogy to this would be the erythropoietic response to high altitude, which now seems to arise as a consequence of the effects of hypoxic exposure on the system responsible for regulating haematocrit at sea level.

3. ADAPTIVE CONTROL IN THE REGULATION OF BREATHING

If O_2 were the major factor in setting the long-term stability of pulmonary ventilation, then the question arises as to how such a control mechanism interacts with short-term chemoreflex control loops. It has long been recognised that exposure to high altitude results in an increase in sensitivity of the chemoreflex response to CO_2 [8, 12]. On the other hand, carotid body resection appears to reduce chemoreflex sensitivity to CO_2 [2, 4], an effect which seems to be most striking ~5 days after loss of carotid body afferent activity[9]. These results suggest that, over longer time courses, O_2 acts to modify the acute chemoreflex (predominantly CO_2) responses. This type of scheme may be modelled as an adaptive controller[1, 10]. Such a control scheme has an advantage over a static feedback system in which reflex sensitivities are pre-ordained, in the sense that it can cope with more major variations in the plant, such as occur physiologically during growth and pathologically during lung and musculo-skeletal disease.

4. CONCLUSIONS

We have a considerable understanding of the properties of the acute chemoreflex responses, and the primary role of CO_2 in these. Our understanding of the long-term determinants of pulmonary ventilation is much less advanced, and as such it tends to be an extrapolation of what we know from experiments of shorter duration. The purpose of this article is to present some of the evidence in favour of a significant role for oxygen sensing in the long-term regulation of breathing at sea level in humans.

5. REFERENCES

1. Astrom, K. J. & Wittenmark, B. ,1995. *Adaptive Control.* Reading: Addison-Wesley.
2. Bellville, J. W., Whipp, B. J., Kaufman, R. D., Swanson, G. D., Aqleh, K. A. & Wiberg, D. M., 1979. Central and peripheral chemoreflex loop gain in normal and carotid body-resected subjects. *Journal of Applied Physiology* **46,** 843-853.
3. Cherniack, N. S., 1999. Control of breathing in chronic obstructive pulmonary disease. In M. D. Altose & Y. Kawakami (Eds.), *Control of Breathing in Health and Disease* (pp. 423-437). New York: Marcel Dekker.
4. Fatemian, M., Dahan, A., Meinesz, S., Mey, A. v. d. & Robbins, P. A., 2000a. Modelling the ventilatory response to variations in end-tidal PCO_2 in patients who have undergone bilateral carotid body resection. in: *Frontiers in Modeling and Control of Breathing: Integration at Molecular, Cellular and Systems Levels,* edited by C.-S. Poon and H. Kazemi (Kluwer Academic/Plenum Press, New York, 2001), this volume.
5. Fatemian, M., Kim, D. Y., Poulin, M. J. & Robbins, P. A., 2000b. Long-haul flights may induce respiratory changes similar to ventilatory acclimatisation to altitude. in: *Frontiers in Modeling and Control of Breathing: Integration at Molecular, Cellular and Systems Levels,* edited by C.-S. Poon and H. Kazemi (Kluwer Academic/Plenum Press, New York, 2001), this volume.
6. Haldane, J. S. & Priestley, J. G., 1905. The regulation of the lung-ventilation. *Journal of Physiology* **32,** 225-266.
7. Howard, L. S. G. E. & Robbins, P. A., 1995. Ventilatory response to 8 h of isocapnic and poikilocapnic hypoxia in humans. *Journal of Applied Physiology* **78,** 1092-1097.
8. Michel, C. C. & Milledge, J. S., 1963. Respiratory regulation in man during acclimatisation to high altitude. *Journal of Physiology* **168,** 631-643.
9. Pan, L. G., Forster, H. V., Martino, P., Strecker, P. J., Beales, J., Serra, A., Lowry, T. F., Forster, M. M. & Forster, A. L., 1998. Important role of carotid afferents in control of breathing. *Journal of Applied Physiology* **85,** 1299-1306.
10. Poon, C.-S. & Siniaia, M. S., 2000. Plasticity of cardiorespiratory neural processing: classification and computational functions. *Respiration Physiology* **122,** 83-109.
11. Rahn, H. & Otis, A. B., 1949. Man's respiratory response during and after acclimatization to high altitude. *American Journal of Physiology* **157,** 445-462.
12. Rahn, H., Stroud, R. C., Tenney, S. M. & Mithoefer, J. C., 1953. Adaptation to high altitude: respiratory response to CO_2 and O_2. *Journal of Applied Physiology* **6,** 158-162.
13. Ren, X., Fatemian, M. & Robbins, P. A., 2000. Changes in respiratory control in humans induced by 8 h of hyperoxia. *Journal of Applied Physiology* **89,** 655-662.
14. Tansley, J. G., Fatemian, M., Howard, L. S. G. E., Poulin, M. J. & Robbins, P. A., 1998. Changes in respiratory control during and after 48 h of isocapnic and poikilocapnic hypoxia in humans. *Journal of Applied Physiology* **85,** 2125-2134.
15. West, J. B., 2000. *Respiratory Physiology.* Philadelphia: Lippincott Williams & Wilkins.

MODELING RESPIRATORY ADAPTATIONS IN HUMANS

Safraaz Mahamed* and James Duffin*†

1. INTRODUCTION

Prolonged exposures to hypoxia such as travel to altitude produce adaptations in ventilatory responses[1-3]. Repeated bouts of hypoxia such as those produced by obstructive sleep apnoea[4] also induce adaptations. Several investigators have observed an increased hypoxic responsiveness produced by repeated bouts of hypoxia[4-7] as have we. These adaptations are likely to be mediated by alterations in respiratory chemoreflex characteristics. We measured the changes in chemoreflex characteristics during repeated daily 20-minute exposures to isocapnic hypoxia and propose a graphical model and mechanism to explain the increased hypoxic responsiveness.

2. METHODS

Iso-oxic, modified rebreathing tests were used to assess changes in peripheral and central chemoreflex characteristics during hypoxia and hyperoxia[8]. These included the chemoreflex sensitivities and thresholds for carbon dioxide as well as basal ventilation independent of chemoreflexes. They were measured before and after exposure to 20 minutes of isocapnic hypoxia, daily for 2 weeks. Subjects repeated the identical protocol using room air as a control.

3. RESULTS

By the 14th day 6 of 7 subjects exhibited a hypoxic ventilatory response and hypoxic ventilatory decline during the exposures to isocapnic hypoxia. After each hypoxic episode, basal ventilation and chemoreflex sensitivity to carbon dioxide were not

* University of Toronto, Department of Physiology, Toronto, Ontario, Canada M5S 1A8.
† University of Toronto, Department of Anaesthesia, Toronto, Ontario, Canada M5S 1A8.

Frontiers in Modeling and Control of Breathing, edited by Poon and Kazemi.
Kluwer Academic/Plenum Publishers, New York, 2001.

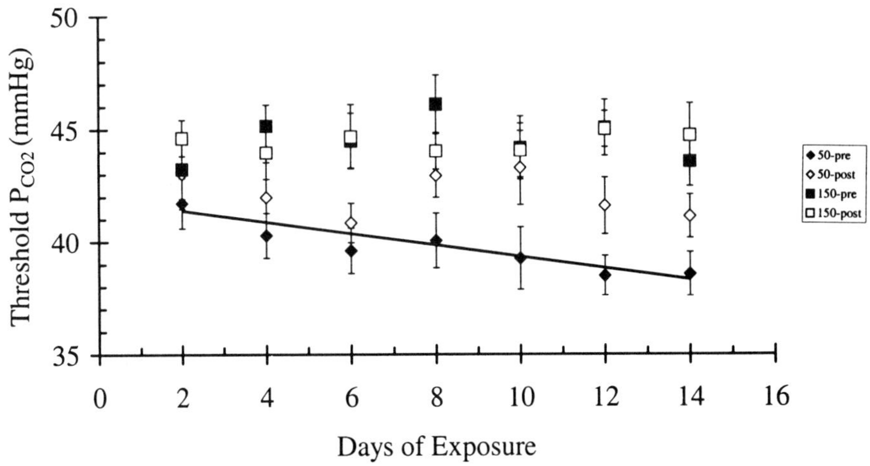

Figure 1. Average (SEM) chemoreflex thresholds for carbon dioxide of all subjects before (pre, filled) and after (post, open) exposure to isocapnic eucapnic hypoxia for rebreathing tests performed at an iso-oxic level of 50 mmHg (diamonds) and 150 mmHg (squares). There is a significant correlation between days of exposure and 50-pre thresholds ($R^2 = 0.855$, $p < 0.05$).

significantly altered in either the hypoxic or hyperoxic rebreathing tests. By contrast, there was an increase in the chemoreflex threshold for carbon dioxide for the hypoxic but not the hyperoxic rebreathing tests. Thus, single exposures to isocapnic hypoxia produced a significant increase in chemoreflex threshold. Over the course of the 14-day period, regression analysis demonstrated a progressively decreasing chemoreflex threshold for carbon dioxide for the hypoxic but not the hyperoxic rebreathing tests. (Figure 1). Thus, multiple exposures to isocapnic hypoxia produced a significant decrease in the chemoreflex threshold.

4. DISCUSSION

4.1. General

These changes in the chemoreflex threshold for carbon dioxide measured during hypoxic but not hyperoxic rebreathing tests suggest that the mechanism to account for the alterations in peripheral chemoreflex responsiveness not only in response to exposure to a single bout of isocapnic hypoxia but also in response to repeated bouts is change in peripheral chemoreflex threshold for carbon dioxide. We offer a graphical model to account for the changes observed in the hypoxic ventilatory response.

4.2. Proposed Model

Our model utilises the peripheral chemoreflex adaptations to suggest how they affect the hypoxic ventilatory response in humans. The model parameters are taken from data that has been previously published[8]. At the start of isocapnic hypoxia (end tidal partial pressures of oxygen and carbon dioxide of 40 mmHg), the chemoreflex threshold for carbon dioxide decreases and the chemoreflex sensitivity to carbon dioxide increases (Figure 2, moving along the isocapnic line from 1 to 2). As the hypoxic exposure progresses the chemoreflex threshold for carbon dioxide increases, causing the observed hypoxic ventilatory decline (Figure 2, moving along the isocapnic line from 2 to 3). We suggest that this threshold decrease might be due to a decrease in the effectiveness of synaptic transmission along the peripheral chemoreflex arc.

The second part of our model considers the effects of repeated bouts of hypoxia. Based on our findings, repeated daily exposures to isocapnic hypoxia produce a decrease in the peripheral chemoreflex threshold for carbon dioxide. Thus, at the onset of isocapnic hypoxia, the ventilatory response is markedly enhanced (Compare Figures 2 and 3 where ventilation progresses along the isocapnic line from 1 to 2). Nevertheless, as the hypoxic exposure continues, the chemoreflex threshold increases to the same value as before enhancement, so that the magnitude of the hypoxic ventilatory decline is

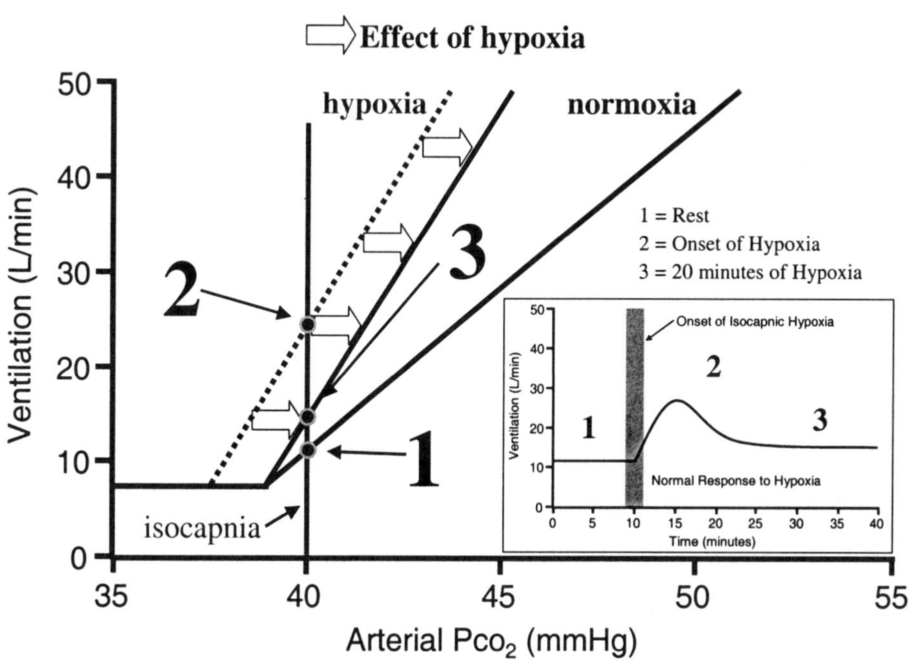

Figure 2. Modelling the response to a single exposure to isocapnic hypoxia. Using the normoxic chemoreflex response line (black line, 100 mmHg O_2) and that of hypoxia (dashed line, 40 mmHg PO_2) the acute ventilatory response to hypoxia is shown as ventilation progresses from 1 to 2 along the isocapnic line. The effect of 30 minutes of hypoxia, indicated as a shifting chemoreflex threshold (arrows), produces hypoxic ventilatory decline as ventilation progresses from 2 to 3 along the isocapnic line. An example of the normal hypoxic ventilatory response as predicted by the model is also shown (inset).

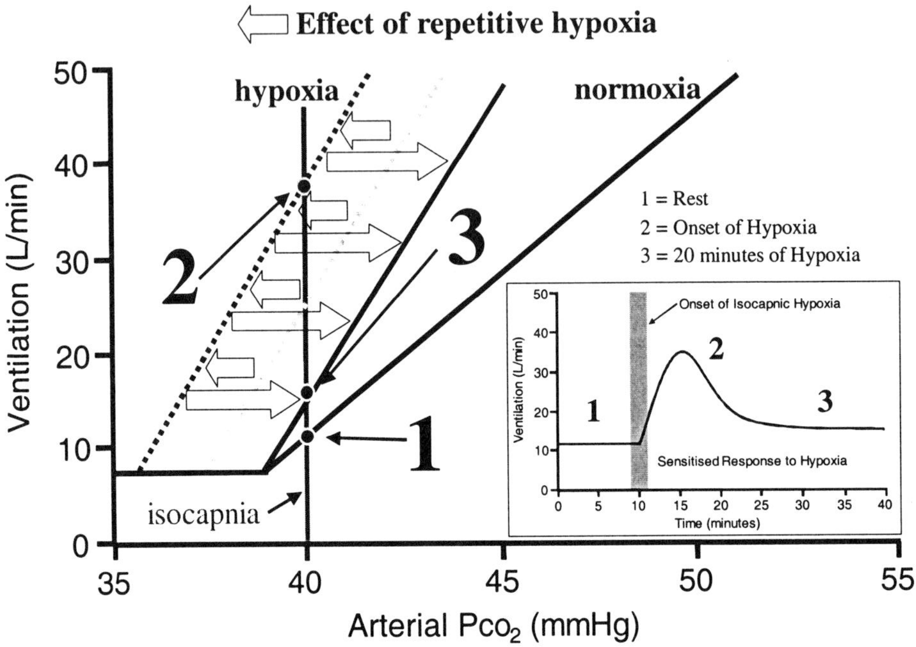

Figure 3. Modelling the response to a single exposure to isocapnic hypoxia after repeated daily exposures. This figure is similar to the previous figure (Figure 2) except that the initial response to hypoxia is enhanced after repeated daily exposures; the hypoxic chemoreflex response line has changed from the grey dashed line to the black dashed line. The result is a larger acute response (1 to 2); nevertheless, the value toward which ventilation approaches after 30 minutes of exposure is identical (2 to 3). An example of the sensitized hypoxic ventilatory response as predicted by the model is also shown (inset).

proportional to the initial response (Figure 3, ventilation progresses along the isocapnic line from 2 to 3). As previously, we suggest that the increased chemoreflex threshold might results from an increase in the effectiveness of synaptic transmission along the peripheral chemoreflex arc.

5. ACKNOWLEDGMENTS

We acknowledge the expert advice of the Respiratory Research Grouping at the University of Toronto. This research was approved by the University of Toronto Ethical Research Board and supported by a grant from the Ontario Thoracic Society.

6. REFERENCES

1. J. G. Tansley, M. Fatemian, L. S. Howard, M. J. Poulin, and P. A. Robbins, Changes in respiratory control during and after 48 h of isocapnic and poikilocapnic hypoxia in humans. *J. Appl. Physiol.* **85**, 2125-34 (1998).

2. L. S. Howard, and P. A. Robbins, Alterations in respiratory control during 8 h of isocapnic and poikilocapnic hypoxia in humans. *J. Appl. Physiol.* **78**, 1098-107 (1995).
3. M. Fatemian, and P. A. Robbins, Human ventilatory response to CO2 after 8 h of isocapnic or poikilocapnic hypoxia. *J. Appl. Physiol.* **85**, 1922-8 (1998).
4. K. Narkiewicz, P. J. van de Borne, C. A. Pesek, M. E. Dyken, N. Montano, and V. K. Somers, Selective potentiation of peripheral chemoreflex sensitivity in obstructive sleep apnea. *Circulation* **99**, 1183-9 (1999).
5. K. Katayama, Y. Sato, Y. Morotome, N. Shima, K. Ishida, S. Mori, and M. Miyamura, Ventilatory chemosensitive adaptations to intermittent hypoxic exposure with endurance training and detraining. *J. Appl. Physiol.* **86**, 1805-11 (1999).
6. K. Katayama, Y. Sato, K. Ishida, S. Mori, and M. Miyamura, The effects of intermittent exposure to hypoxia during endurance exercise training on the ventilatory responses to hypoxia and hypercapnia in humans. *Eur J. Appl. Physiol.* **78**, 189-94 (1998).
7. T. V. Serebrovskaya, I. N. Karaban, E. E. Kolesnikova, T. M. Mishunina, L. A. Kuzminskaya, A. N. Serbrovsky, and R. J. Swanson, Human hypoxic ventilatory response with blood dopamine content under intermittent hypoxic training. *Can. J. Physiol. Pharmacol.* **77**, 967-73 (1999).
8. J. Duffin, R. M. Mohan, P. Vasiliou, R. Stephenson, and S. Mahamed, A model of the chemoreflex control of breathing in humans: model parameters measurement. *Respir. Physiol.* **120**, 13-26 (2000).

POST-HYPOXIC FREQUENCY DECLINE CHARACTERIZED IN THE RAT WORKING HEART BRAINSTEM PREPARATION

Thomas E. Dick, Mathias Dutschmann, and Julian F.R. Paton*

1. INTRODUCTION

The respiratory response to brief episodes of hypoxia is dynamic, i.e., the breathing pattern changes progressively both during and following an hypoxic episode[1]. The initial response in breathing with hypoxic exposure includes a sharp and acute increase in breathing frequency (F). Subsequently, F decreases as exposure to hypoxia continues. This decrease in F is a form of short-term depression that is referred to as "hypoxic ventilatory decline." Immediately after the cessation of hypoxia and return to hyperoxic gas, F decreases further below the baseline frequency; the latter may be a form of short-term depression which is called post-hypoxic frequency decline (PHFD)[1,2].

The neural mechanisms underlying changes in frequency during and following hypoxia have been investigated by numerous investigators[2-5]. During hypoxia, decreases in frequency were associated with both disfacilitatory and inhibitory mechanisms[3,5]. The mechanism of the decrease in frequency after hypoxia is unknown but it does depend on the expiratory-facilitatory areas in the ventrolateral (vl) pons. Inactivating neurons in the vl pons blocks PFHD selectively[2] sparing the initial response to hypoxia. The physiologic consequence of the decrease in breathing frequency following hypoxia are also unknown but it occurs in humans during NREM sleep[6]. Thus, PHFD may represent a short-term modulation of motor patterns formed by the ponto-medullary pattern generator.

In this study, we utilized the working heart-brainstem preparation[7] (WHBP), a reduced in-situ preparation with advantages over in-vivo and in-vitro approaches[7,8]. First, we verified that the WHBP expressed PHFD and that this depended on activity in the vl pons. Second, we tested the hypotheses that both PHFD and stimulation of the vl pons are associated with increases in post-inspiratory activity (stage-I expiratory activity).

* Thomas E. Dick, Division of Pulmonary and Critical Care Medicine, Department of Medicine, Case Western Reserve University, Cleveland, Ohio 44106-4941. Mathius Dutschmann and Julian F.R. Paton, Department of Physiology, University of Bristol, Bristol, England.

2. METHODS

All experimental procedures were approved by the Home Office and Animals Ethics Committee at the Bristol University. Rats (125-180 g) were anesthetized in halothane and transected below the diaphragm. The head and thorax were placed in ice-cold saline, skinned and decerebrated. Part of the dorsal ribcage was removed and the left thoracic phrenic nerve isolated. In addition, in 5 of 14 preparations, the central end of the cervical vagus nerve was also isolated for recording efferent activity. The preparation was placed in the recording chamber and the skull stabilized.

The descending aorta was cannulated and the preparation perfused with a Ringer's solution containing ficoll (1.25%) that was gassed with 95% O_2 and 5% CO_2. Perfusate was warmed to 31°C and filtered. Perfusion pressure was measured from the aorta and adjusted by changing pump rate until an eupneic pattern of phrenic nerve discharge was obtained (typically 75-95 mmHg). Phrenic and vagal nerves were recorded with suction electrodes as indices of the respiratory cycle.

2.1. Protocols

The first protocol characterized the respiratory response to hypoxia before and after blocking activity in the vl pons and then after recovery from the blockade. The respiratory response was determined from phrenic nerve activity and in a subset of

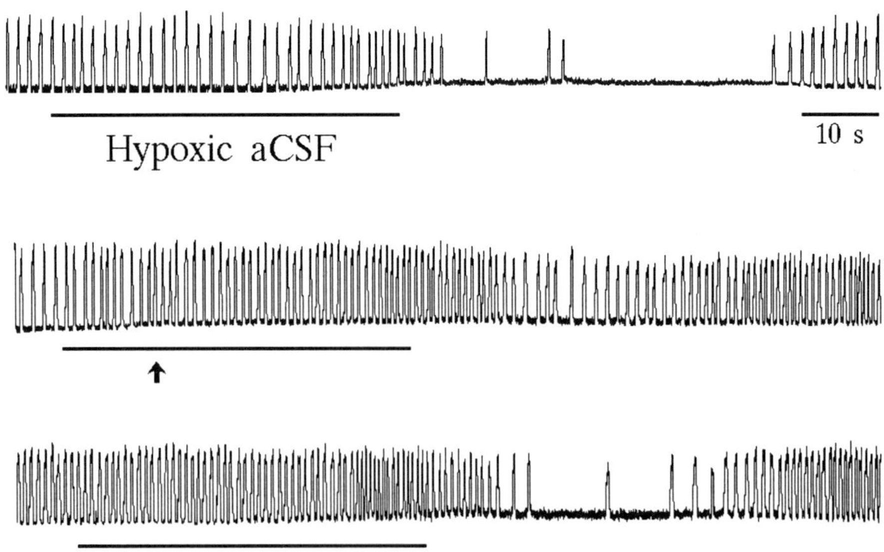

Figure 1. Hypoxic ventilatory response (HVR) in the working heart brainstem preparation (WHBP) before and after isoguvacine injection in the vl pons. (Tracings) Phrenic nerve activity was integrated to measure HVR (8% O_2, 92% N_2, Line indicates hypoxic exposure period). (Top) Before isoguvacine, respiratory frequency increased during hypoxia and decreased following hypoxia, i.e., post-hypoxia frequency decline. (Middle) In the presence of isoguvacine (injected bilaterally into vl pons, second injection indicated by arrow), HVR remained but PHFD was attenuated. (Bottom) After isoguvacine washout, PHFD returned.

animals from both phrenic and efferent vagal nerve activities. We blocked activity in the vl pons by injecting a $GABA_A$ receptor agonist, isoguvacine (100 mM, 40-60 nl), bilaterally. During the protocol, vl pontine sites were identified by bilateral microinjections of L-glutamate (100mM, 10-50 nl) to evoke an increase in duration of expiration (TE). Injection sites were marked with pontamine blue for histological identification. In the first seven experiments, hypoxic exposure was effected by switching the gas bubbling the perfusate from $95\%O_2/5\%CO_2$ to $8\%O_2/92\%N_2$ for 60-120 s. Even though the duration of hypoxia varied across preparations, the duration of exposures was the same within a preparation. In the remaining seven experiments, hypoxic exposure was caused by switching perfusates from one gassed with $95\%O_2/5\%CO_2$ to one gassed with $5\%O_2/5\%CO_2/90\%N_2$ for 60 s. This latter method produced a more consistent response and maintained isocapnia but the results from both methods were combined because PHFD was elicited with both techniques for making the tissue hypoxic.

The second experimental protocol tested the hypothesis that medullary expiratory activity was activated by both vl pontine stimulation and during PHFD. A glass micropipette was placed in the vl pons at a location that evoked a prolongation of TE with L-glutamate microinjections (100 mM). A glass microelectrode filled with NaCl (4 mM) was placed in the vl medulla to record respiratory-modulated activity. We recorded single-unit activity in response to vl-pontine stimulation and during and following hypoxia (60 s). Histological analysis was performed to verify electrode placement.

2.2. Analysis

For the first protocol, we measured cycle lengths as defined by the onset of phrenic burst at the following time points: 1) immediately before exposure, 2) at peak frequency during hypoxia and 3) immediately after post-hypoxic apnea. We also measured the duration of the apnea following hypoxia. For baseline and the post-hypoxic pattern, we averaged three consecutive cycles and tested for differences using a repeated-measures, one-way ANOVA followed with Students-Newman-Kuels post-hoc analysis. We accepted significance at $p<0.05$ level. For the second protocol, we determined discharge frequency of respiratory neurons while perfusing with hyperoxic, isocapnic perfusate and activating the vl pons with L-glutamate, and then during and following hypoxia. We compared the duration and frequency of neuronal firing.

3. RESULTS

Similar to previous results in vivo, brief hypoxia in the WHBP resulted in short-term changes in the phrenic nerve activity (Fig. 1): there was an initial increase followed by a decrease in frequency during exposure. Respiratory frequency showed a further decline after hypoxia (i.e. PHFD). The PHFD was due to a prolongation of TE and was transient with the pattern progressively returning to baseline frequency. For the group of animals (n=7), F during PHFD was significantly slower than baseline ($p<0.05$). Further, the decrease in breathing frequency depended on the vl pons. Bilateral injections of isoguvacine in the vl blocked the PHFD. Thus, there was no statistical difference in F at baseline and at PHFD following isoguvacine injections. After allowing time (45-90 min) for recovery and isoguvacine to be cleared, the difference between pre- and post-hypoxic F was significant ($p < 0.05$). This inactivation of the vl pons caused no significant

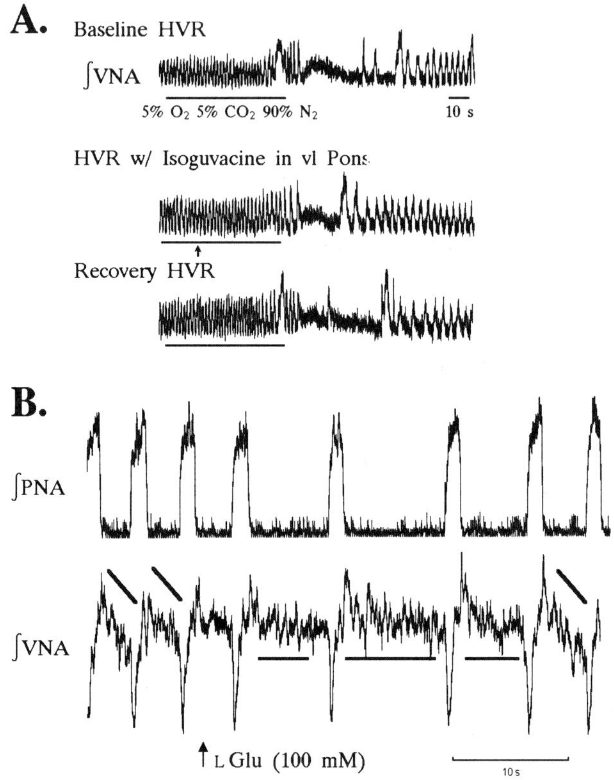

Figure 2. Vagal motor activity during hypoxia and PHFD (A) and after L-Glutamate injection in the vl pons (B). Vagal Nerve (efferent) Activity (VNA) served as an index of post-inspiratory (PI) activity. In **A**, vagal efferent activity, before (top tracing), with (middle) and after recovery from (bottom) isoguvacine injections in the vl pons. During PHFD, TE prolongation was coupled with sustained VNA. In **B**, phrenic (top) and vagal (bottom) motor activities. Similar to the effects evident in VNA during PHFD (see A), L-Glutamate activation of the vl pons increased duration rather than amplitude of PI activity.

disturbances in either the base line breathing pattern or frequency. Histological analysis confirmed that microinjection sites were in the vl pons. Thus, the initial results replicate previous findings obtained in vivo.

During PHFD, vagal efferent activity extended during the prolonged TE (Fig. 2A). This is in contradistinction to prolonged TE during the short-term depression of F that occurs during hypoxia at which time vagal efferent activity not only is prolonged but also activated (Fig. 2A). After bilateral injections of isoguvacine in the vl pons, TE is shortened during PHFD with vagal motor activity remaining correlated to the duration of TE. Following recovery, vagal efferent activity returned to the pre-drug PHFD pattern (Fig. 2A). Thus, during PHFD, PI activity was sustained rather than evoked was evident in the animals (n=5). Sustained rather than evoked vagal efferent activity was also recorded in response to vl pontine stimulation (Fig. 2B). In response to vl pontine stimulation, TE prolonged and vagal efferent activity persisted at a constant level rather than decrementing as in baseline. (Fig. 2B; Compare the decrementing slope of activity

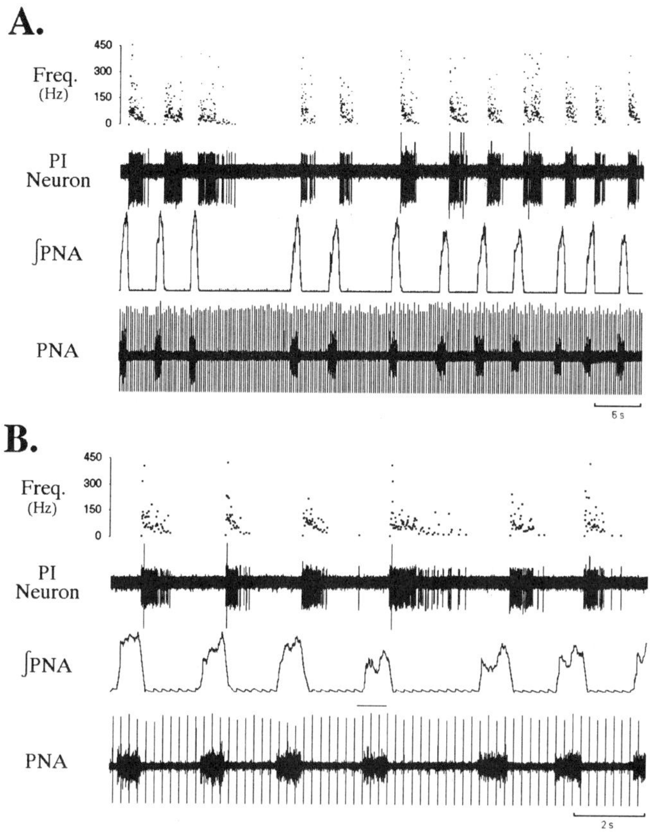

Figure 3. Medullary post-inspiratory (PI) activity after a brief hypoxic episode (A) and after L-Glutamate injection in the vl pons (B). Tracings: A and B: Instantaneous discharge frequency of a medullary PI neuron (top), raw recording of neural activity (second) integrated phrenic nerve activity (third) and raw phrenic nerve activity (bottom). A and B same neuronal recording. A) Switch from hypoxic to hyperoxic perfussate occurs before tracings begin. Sustained PI activity occurs during the prolongation of TE during PHFD. B) Similarly with activation of the vl pons with L-Glutamate, duration rather than frequency of PI activity increased.

with the horizontal slope associated with L-Glutamate injection). This effect was transient and PI activity returned to baseline within a three respiratory cycles.

A similar pattern of sustained activity was evident in medullary PI neurons (Fig. 3). During PHFD, PI neuronal activity increased its duration of discharge in 8 of 9 PI neurons (Fig. 3A). For this population, the duration of activity increased 36±20% during the expiratory phase. In contrast to vagal efferent activity, 5 of these 8 neurons also increased their peak discharge frequency by 120±85% (range of 245%-25% increase).

The increase in the discharge duration of PI neurons was the most consistent effect among the different types of neurons. Across the population of recorded medullary neurons, 3 Tonic E (n=5), no E Aug (n=7), 0 Pre-I (n=4), 2 I Aug (n=8), 1 early I (n=2) and 0 late I (n=2) increased their duration of activity and this was not necessarily coupled with increases in their discharge frequency.

L-Glutamate activation of the vl pons also evoked increases in both the duration (n=8

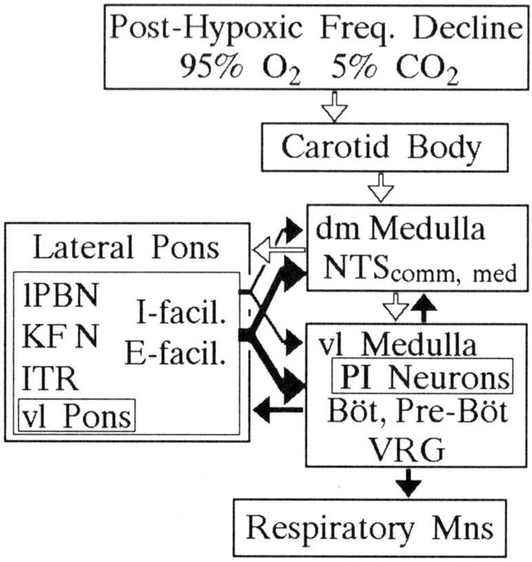

Figure 4. Our model depicting the elements that determine the respiratory frequency (F) after brief exposures of hypoxia in the WHBP. The key elements are: 1) the medial (med) and commissural (comm) nuclei of the solitary tract (NTS) where the afferent fibers from the carotid body terminate in the dorsomedial (dm) medulla, 2) post-inspiratory (PI) neurons in the ventral respiratory group (VRG), the Bötzinger and pre-Bötzinger complexes in the ventrolateral (vl) medulla, 3) the lateral parabrachial (lPBN) and Kölliker-Fuse (KF N) nuclei of the dorsolateral pons, the intertrigeminal regions (ITR) in the lateral pons, and the ventrolateral (vl) pons. These pontine areas form a column in the lateral pons such that dorsal aspect (lPBN and dorsal KF N) is inspiratory-facilitatory (I-facil), whereas the ventral aspect (ventral KF N, ITR, and the vl pons) are expiratory-facilitatory (E-facil). We propose that the respiratory pattern is generated by a pontomedullary network that possesses dynamic properties. After a brief hypoxic event, post-hypoxic frequency decline is a consequence of the absence of excitatory afferent activity (open arrow), and the increased influence of E-facil neurons (large solid arrow between pontine and medullary respiratory areas). Thus, pontine E-facil input results in the persistence of PI activity.

of 9) and discharge frequency (n=6 of 9) of the recorded PI neurons. Increased duration of activity was evoked from the vl pons in 7 of the 8 neurons that increased their discharge frequency during PHFD (Fig. 3B) and relative to the baseline cycles, these neurons increased their duration of discharge by 40±24 percentage. Similarly, 3 of the 5 neurons in which discharge frequency increased during PHFD also increased their discharge frequency in response to glutamate in the vl pons (113±1.00 %).

4. DISCUSSION

Our results show that PHFD can be studied in the WHBP and that it depends on the integrity of the vl pons. Further, activation of the vl pons can prolong the post-inspiratory activity of both vl medullary neurons and that recorded from the vagal motor outflow. A qualitatively similar effect was obtained during PHFD in that the duration of discharge of vl medullary post-inspiratory neurons and the post-inspiratory motor activity recorded from vagal efferents both increased. A prolongation of expiration associated with PHFD has been reported in reduced[2, 4] as well as unanesthetized rodents[9, 10]. Thus, we conclude that the WHBP reflects the respiratory behavior of PHFD and further, that this behavior is associated with prolongation of post inspiratory activities.

Activity of post-inspiratory neurons following vl pontine activation correlated with that during post-hypoxic frequency decline. In particular, the duration of activity was prolonged with minimal effect on frequency of discharge. We propose that vl pons becomes activated at the onset of PHFD and this provides an excitatory synaptic drive to post-inspiratory neurons thereby extending expiration. That we found extension of post-inspiratory activity in the vagus efferent activity indicates that expiratory motor neurons and respiratory muscles are also activated.

We speculate that a pontomedullary network is essential for generating the breathing rhythm and pattern during eupnoea as well as the respiratory response to peripheral chemoreceptor stimulation (Fig. 3). Previous studies support roles for both the dl and vl pons in the maintenance of a normal respiratory pattern by influencing frequency[11-18]. Although the dl pons has a major influence on the generation of the resting breathing pattern and, thus, is an element of the respiratory pattern generator[17, 18], we focused on the role of the vl pons in the present study. Whilst the dl pons contains inspiratory- and expiratory-facilitatory loci[19, 20], the vl pons is primarily an expiratory-facilitatory area[13, 14]. Possibly both intrapontine and pontomedullary interactions contribute to the generation of respiratory activity which may involve an active contribution from the dl pons during eupnoea and involve the vl pons after exposure to hypoxia.

The pontomedullary circuit has a dynamic interaction[16, 21]. As we showed here in the WHBP blockade of pontine activity in expiratory-facilitatory areas of the vl pons blocked PHFD. Further, activity of vl medullary post-inspiratory neurons is prolonged after chemoreceptor and/or vl pontine activation. Thus, we speculate that pontine activity affects vl medullary respiratory neurons (especially post-inspiratory neurons) after hypoxia (Fig. 4). Thus 'memory' or 'plasticity' in the response may result from a dynamic reconfiguration of the connectivity and the relative strength of the influence of the different components of the pontomedullary respiratory pattern generator. In particular, our working hypothesis is that following hypoxia, expiratory facilitatory neurons in the vl pons modulates the breathing pattern by increasing the duration and frequency of the medullary PI neuronal discharge. This type of plasticity is considered to be "use-dependent" plasticity paired with the hypoxic stimulus[21]. In models of respiratory rhythmogenesis, increasing PI activity prolongs TE in as much as it inhibits inspiratory activity[22].

Even though the physiological relevance for PHFD is unknown, we can speculate that PHFD may be related to decreasing metabolism in the period following hypoxia, i.e., as a mechanism to decrease sympathetic activity which is respiratory-modulated. Ontogenetically, the lateral pons inhibits fetal breathing movements when the ewe is exposed to hypoxia[23]. The purpose for this inhibition is to reduce motor activity in the fetus and to conserve energy because fetal breathing movements do not assist in gas exchange. Further, the increase activity of laryngeal adductors during PHFD would restrict airflow maintaining lung volume and facilitating gas exchange following hypoxia.

In conclusion, our model for the control of breathing consists of a pontomedullary neuronal network that generates the respiratory rhythm[8, 17, 18]. Under hypoxic conditions, the generalized level of excitation increases and specific elements of the network are recruited to integrate chemoreceptor afferent information. These elements, essential for central processing of hypoxic effects, possess dynamic properties and are called into play at specific times.

5. ACKNOWLEDGMENTS

The authors gratefully acknowledge the support by the British Heart Foundation to JFRP, Deutsche Forschungsgemeinschaft to MD, and National Heart, Lung and Blood Institute (HL 25830) to TED.

6. REFERENCES

1. F. L. Powell, W. K. Milsom, and G. S. Mitchell, Time domains of the hypoxic ventilatory response. *Respir. Physiol.* **112**, 123-134 (1998).
2. S. K Coles and T. E. Dick, Neurones in the ventrolateral pons are required for post-hypoxic frequency decline in rats. *J. Physiol.* (London) **497**, 79-94 (1996).
3. J. E. Melton, J. A. Neubauer, and N. H. Edelman, GABA antagonism reverses hypoxic respiratory depression in the cat. *J. Appl. Physiol.*, **69**:1296-1301 (1990).
4. F. Hayashi, S. K. Coles, K. B. Bach, G. S., Mitchell, and D. R. McCrimmon, Time-dependent phrenic nerve responses to carotid afferent activation: intact vs. decerebellate rats. *Amer. J. Physiol.* **265**, R811-R819 (1993).
5. D.W. Richter, P. Schmidt-Garcon, O. Pierrefiche, A. M. Bischoff, and P. M. Lalley, Neurotransmitters and neuromodulators controlling the hypoxic respiratory response in anaesthetized cats. *J. Physiol.* (London) **514**, 567-578 (1999).
6. K. Gleeson, Ventilatory pattern following hypoxic stimulation in nonrapid eye movement and rapid eye movement sleep. *Sleep* **16**: S146-S147 (1993).
7. J. F. R. Paton, A working heart-brainstem preparation of the mouse. *J. Neurosci. Meth.* **65**, 63-68 (1996).
8. M. Dutschmann, R.J.A. Wilson, and J.F.R Paton, Respiratory activity in neonatal rats. Autonomic *Neurosci.: Basic and Clinical*, **84**: 19-29. (2000).
9. A. Mizusawa, H. Ogawa, Y. Kikuchi, W. Hida, K. Shirato, Role of the parabrachial nucleus in ventilatory responses of awake rats. *J. Physiol.* **489**, 877-884 (1995).
10. S. Subramanian, F. Han, B. O. Erokwu, T. E. Dick, K. P. Strohl, Do genetic factors influence the Dejours phenomenon? *Proceedings of the VIII Oxford Conference* (This volume).
11. W. Wang, M.-L. Fung, and W. M. St. John, Pontile regulation of ventilatory activity in the adult rat. *J. Appl. Physiol.* **74**, 2801-2811 (1993).
12. S. F. Morrison, S. L. Cravo, and H. M. Wilfehrt, Pontine lesions produce apneusis in the rat. *Brain Res.* **652**, 83-86 (1994).
13. J. S. Jodkowski, S. K. Coles, and T. E. Dick, A 'pneumotaxic centre' in rats. *Neurosci. Lett.* **172**, 67-72 (1994).
14. J. S. Jodkowski, S. K. Coles, and T. E. Dick, Prolongation in expiration evoked from ventrolateral pons of adult rats. *J. Appl. Physiol.* **82**, 377-81 (1997).
15. T. E. Dick, S. K. Coles, and J. S. Jodkowski, A 'pneumotaxic centre' in the ventrolateral pons of rats. in: *Ventral Brainstem Mechanisms and Control Functions*, ed. O. Trouth, R. Millis, H. Kiwell-Schöne, and M. Schläfke, pp.723-737. (Marcel Dekker, New York, 1995).
16. T. E. Dick and S. K. Coles, Ventrolateral pons mediates short-term depression of respiratory frequency after brief hypoxia. *Respir. Physiol.* **121**, 87-100 (2000).
17. W. M. St. John, Neurogenesis of patterns of automatic ventilatory activity. *Prog. Neurobiol.* **56**, 97-117 (1998).
18. R. St. Jacques and W. M. St. John, Transient, reversible apnoea following ablation of the pre-Botzinger complex in rats. *J. Physiol.* (London) **520**, 303-314 (1999).
19. N. L. Chamberlin and C. B. Saper,. Topographic organization of respiratory responses to glutamate microstimulation of the parabrachial nucleus in the rat. *J. Neurosci.* **14**, 6500-6510 (1994).
20. N. L. Chamberlin and C. B. Saper, A brainstem network mediating apneic reflexes in the rat. *J. Neurosci.* **18**, 6048-6056 (1998).
21. C.-S. Poon and M.S. Siniaia, Plasticity of the cardiorespiratory neural processing: classification and computational functions. *Respir. Physiol.* **122**, 83-109 (2000).
22. D. W. Richter, D. Ballantyne, and J. E. Remmers, How is the respiratory rhythm generated? A model. *NIPS* **1**, 109-112 (1986).
23. P. D. Gluckman and B. M. Johnson Lesions in the upper lateral pons abolish the hypoxic depression of breathing in unanaesthetized fetal lambs in utero. *J. Physiol.* (London) **382**: 373-83 (1987).

MODEL OF NONASSOCIATIVE LEARNING IN VAGAL-PONTINE MODULATION OF THE RESPIRATORY RHYTHM

Daniel L. Young, Marina S. Siniaia, and Chi-Sang Poon*

1. INTRODUCTION

Recent studies[1-5] have revealed several time-adaptive components in the classic Hering-Breuer inflation reflex (HBIR)[6,7] modulation of the respiratory rhythm elicited by lung inflation or electrical vagal stimulation. The primary adaptive component features a slow exponential decay that evidences a gradual habituation of HBIR, possibly due to time-dependent adaptations in the primary central pathway for neurotransmission of vagal-volume feedback (VVF) to the respiratory central pattern generator (RCPG). A secondary adaptive component is characterized by an abrupt post-stimulus rebound and exponential recovery of the respiratory rhythm following the cessation of the vagal input, an effect that has been linked to NMDA receptor-dependent desensitization of neurotransmission in a secondary vagally-modulated pathway via the pontine pneumotaxic region.[3,4] The combined habituation-desensitization of HBIR via the primary and secondary central pathways conform with the classic dual-process model of nonassociative learning.[4,8] These observations provide the first indication that nonassociative learning may play a role in the vagal control of the respiratory rhythm.

This paper explores the physiological implications of such nonassociative learning in respiratory control under varying pathophysiological states. Novel concepts under this new paradigm were derived from computer simulation of a model of the RCPG, the respiratory pump and the vagal-pontine feedback pathways connected in a closed loop.

2. METHODS

A model of the RCPG and its feedback pathways was implemented on the GENESIS

* DLYoung, MS Siniaia, and C-S Poon, Harvard-MIT Division of Health Sciences and Technology, Massachusetts Institute of Technology, Cambridge, Massachusetts, 02139, U.S.A. Correspondence: DL Young, Bldg. 56-030, Massachusetts Institute of Technology, Cambridge, Massachusetts, 02139, U.S.A. Phone: (617) 253-2131; Fax: (617) 253-7648; email: dlyoung@mit.edu

Frontiers in Modeling and Control of Breathing, edited by Poon and Kazemi.
Kluwer Academic/Plenum Publishers, New York, 2001.

software package.[9] Previous studies[10-12] have established a basic RCPG model with VVF that demonstrates a three-phase respiratory pattern. Here we extended this model by including nonassociative learning mechanisms in the vagal-pontine pathways. This model however does not include the recently reported inspiratory promoting mechanisms[13,14] or network reverberation found after NMDA receptor blockade.[14]

2.1. RCPG Network Model

The core RCPG network is composed of seven Hodgkin-Huxley type neurons of either incrementing or decrementing spike frequency profiles[10] (Fig. 1). Modulating the basic respiratory pattern are the pons neuron and pulmonary stretch receptors (PSR) which are implemented as simple neurons without spike frequency adaptation or ramp activities. Common to all cell types are the following ion-channel types: the K delayed rectifier current, the fast Na current, the transient A-current, and the K leakage current. See ref. 10 for further details on the neuron model.

Incrementing (ramp) spiking activities result from the interaction between $K_{AHP}(Ca)$ and Ca_T low threshold channels while decrementing activities result from the interaction between $K_{AHP}(Ca)$ and Ca_L high threshold channels. A simple model is utilized to account for Ca^{2+} intracellular concentration dynamics:[9]

$$\frac{d[Ca^{2+}]}{dt} = B\, I_{Ca} - \frac{[Ca^{2+}]}{\tau}$$

where $[Ca^{2+}]$ is the calcium concentration in moles/m^3. On the right hand side of this equation, the first term accounts for the total inward calcium current I_{Ca} while the second term accounts for Ca^{2+} depletion with a single decay time constant, τ (for incrementing and decrementing neurons τ= 0.003 and .0133 s, respectively). The constant, *B*, converts the inward calcium current into concentration.

Synaptic connections are either excitatory (resting ionic potential, E_k = 0.045V) or inhibitory (E_k = -0.082V). The synaptic conductances ($g_{syn}(t)$) are modeled by an alpha function[15,16] with a time constant of 0.01 s and a maximal conductance of 0.8 S/m^2. The efficacy of synaptic transmission is controlled by a weight factor (*W*) and the effective conductance is the product of *W* and $g_{syn}(t)$.

2.2. Model of Respiratory Pump

Following Rybak et al.[11] the integrated phrenic nerve activity and the lung volume (LV) are modeled as leaky integrators connected in cascade, with the former being driven by ramp-I and post-I activities. A DC component superimposed in LV controls the end expiratory volume. Scaled lung volume (LV_S=LV*g, where gain, g = 0.6*e^{-12} A/m^3 in control) is used as the current input to the PSR. Hence changing the gain changes the phrenic drive to tidal volume ratio.

2.3. Model of Vagal-Pontine Feedback Pathways

Previous studies[3-5] have shown that pontine activity increases expiratory time but does not affect inspiratory time. These experimental observations were modeled by a hypothetical positive feedback pathway between con-E2 and the pons neuron, although other network configurations are also possible.

Activity-dependent plasticity was incorporated into the primary and secondary vagal-pontine pathways to simulate the observed nonassociative learning in HBIR[4,5]. In the primary pathway homosynaptic plasticity was included at the PSR projections to post-I and early-I neurons or in the neurons in the solitary nucleus (not shown). In the secondary pathway, heterosynaptic plasticity was included to modify the excitatory input from con-E2 to the pons neuron. These synaptic weight changes were modeled as follows:

$$\Delta W = (PRA - T_{PRA}) * R_W * dt$$

where ΔW is the dimensionless discrete weight change, *PRA* is a dimensionless measure of presynaptic activity, T_{PRA} is the presynaptic activity threshold, R_W is a constant rate of synaptic weight change (s^{-1}), and *dt* is the time step in seconds. To prodůce synaptic accommodation (for desensitization and habituation) at suprathreshold presynapatic activities, R_W was set less than zero (-0.0075 s^{-1}). This negative rate caused *W* to decrease when *PRA* was above T_{PRA} and increase when *PRA* was below T_{PRA}. Weights were bounded by maximum and minimum values ($W_{max} = 2*W_0$ and $W_{min} = W_0/4$, where W_0 is the initial synaptic weight). The synaptic weights were updated as follows:

$$W(n+1) = W(n) + \Delta W * S$$

where *n* is the time index, the scale, $S = N^{1/2}$ or $(1\text{-}N)^{1/2}$ when the synapses are decreasing or increasing, respectively, and the normalized weight, $N = (W\text{-}W_{min})/(W_{max}\text{-}W_{min})$).

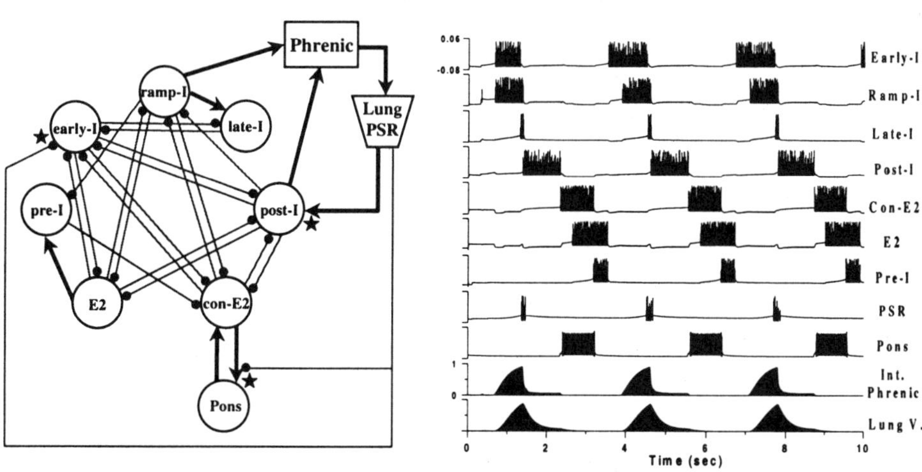

Figure 1. *Left*: Schematic of the RCPG consisting of seven core interconnected neurons, mechanical feedback, and the pons. Filled stars indicate hypothetical sites of nonassociative learning. Excitatory and inhibitory synaptic connections are depicted with arrows and circles, respectively. *Right*: neuronal activities of each cell type, integrated phrenic, and the lung volume during eupneic breathing. Membrane voltages are plotted in Volts while integrated phrenic activity and lung volume are plotted in normalized units.

3. RESULTS AND DISCUSSION

Simulation of the model under control conditions closely mimicked the reported three-phase respiratory pattern (Fig. 1). In eupneic breathing, the pons should have minimal influence on the breathing pattern since the pons/con-E2 pathway is desensitized by vagal feedback. This eupneic pattern set the baseline for subsequent simulations studying the effects of the nonassociative learning in respiratory pattern generation.

3.1. Vagal Stimulation

To test this model we simulated the response to vagal stimulation at 80Hz after bivagotomy. The results demonstrated the salient features found experimentally, namely initial reflex responses followed by habituation and post-stimulus rebound/memory in breathing frequency and expiration time. Inspiratory time and phrenic amplitude also showed accommodation during stimulation but no post-stimulus memory.

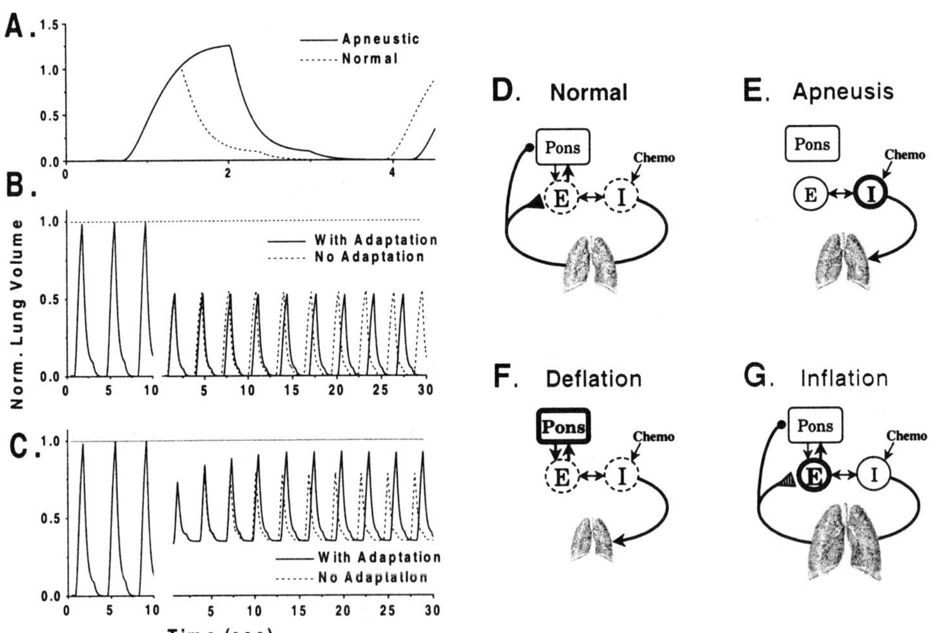

Figure 2. On the left-hand side three simulations of different physiologic states are shown: A: apneustic breathing resulted from coincident pons blockade and vagotomy; B: lung restriction; and C: lung inflation. Panels B and C illustrate that nonassociative learning dynamically modulated the breathing rhythm helping to nullify the effect of altered VVF. On the right-hand side four scenarios are shown: D: eupneic breathing when VVF predominately controls I-offswitch while pons is densensitived by VVF activity; E: apneusis occurs when both the pons and VVF are removed; F: during lung deflation, VVF is reduced/abolished and the pons becomes sensitized and maintains a stable rhythm; G: during lung inflation, VVF is enhanced, the primary pathway is habituated, and the pons is desensitized.

3.2. Apneustic Breathing

Removing both the vagal-volume feedback and the pontine neuron from the model produced an apneustic breathing pattern characterized by prolonged inspiration (+100%), shortened expiration (-10%), decreased breathing frequency (-11%), and increased peak phrenic amplitude (+7%) (Fig. 2A). This breathing pattern is similar to that found experimentally after vagotomy and pontine lesions[4,17] or NMDA receptor blockade.[4,18-20] In contrast, abolishing VVF alone did not result in apneustic breathing in the model, in agreement with experimental observations. These findings suggest that the pons acts as a fail-safe mechanism preventing apneusis when VVF is inactive.

3.3. Lung Restriction

End-expiratory lung volume is reduced in certain restrictive lung diseases such as pulmonary fibrosis, bronchitis and respiratory distress syndrome. In such cases, the lung restriction diminishes overall ventilation as well as vagal-volume feedback.

Decreasing the lung volume gain, g, by 50% simulated lung restriction (Fig. 2B). Without any adaptation after lung restriction, inspiratory time increased by about 15% while expiratory time decreased by 12.5%. With adaptation, however, the synaptic weights increased via nonassociative learning in the primary and secondary pathways due to reduced PSR activities. As a result, the normal respiratory pattern was partially restored, namely inspiratory time was only 10% larger than control while expiratory time was only 2% smaller than control. These adaptations reflected beneficial compensatory changes in the vagal-volume feedback pathway in response to chronic lung restriction that served to preserve the inspiratory off-switch.

3.4. Lung Inflation

Lung distension occurs in certain abnormalities such as emphysema and pneumothorax. Over-inflation of the lungs results in sustained elevation of functional residual capacity and increase in vagal volume feedback.

To investigate the compensatory changes in response to such lung distension, the DC component in the lung volume model was increased so that the end-expiratory volume (EEV) was nearly 30% of the normalized end-tidal volume (Fig. 2C). This change caused PSR to fire more frequently throughout ventilation. Without adaptation (R_W = 0), increased EEV reduced inspiratory time by 65%, increased expiratory time by 17%, and decreased peak lung volume by 20%, together reducing overall ventilation. However, with $R_W < 0$, adaptations in the primary and secondary pathways (decreasing synaptic weights) tended to partially restore normal lung volume (90% of control) as well as inspiratory (90%) and expiratory times (109%). Therefore, the adaptations effectively suppressed the HBIR-related apneic effects of lung inflation.

4. CONCLUSION

In this chapter it was shown via simulation of a RCPG model that nonassociative learning in vagally mediated pathways could effectively enhance the dynamic range of the respiratory rhythm under varying physiological states by fine-tuning synaptic weights in vagal-pontine feedback pathways. The pons acted as a back up to vagal-volume feedback whenever the latter was inactive or reduced.

5. ACKNOWLEDGMENTS

D. L. Young is recipient of a National Institute of Mental Health individual predoctoral fellowship award 1F31MH12697. The work was supported in part by National Institute of Health grant HL60064.

6. REFERENCES

1. D. L. Young, M. S. Siniaia, and C. -S. Poon, Adaptive modulation of respiratory rhythm by vagal input: effect of gating, *Soc. Neurosci. Abstr.* 378, (1998).
2. C.-S. Poon and M. S. Siniaia, Plasticity of cardiorespiratory neural processing: classification and computational function, *Respir. Physiol.* **122**, 83-109 (2000).
3. C. -S. Poon, D. L. Young, and M. S. Siniaia, High-pass filtering of carotid-vagal influences on expiration in rat: role of N-methyl-D-aspartate receptors, *Neurosci. Lett.* **284**(1-2), 5-8 (2000).
4. M. S. Siniaia, D. L. Young, and C. -S. Poon, Habituation and desensitization of the Hering-Breuer reflex in rat, *J. Physiol.* (London) **523**, 479-91(2000).
5. D. L. Young, M. S. Siniaia, and C. -S. Poon, A gated neural differentiator-integrator linked to nonassociative learning in the mammalian respiratory central pattern generator, (submitted).
6. J. Breuer, Self-steering of respiration through the nerves vagus (circa 1868), in: *Breathing: Hering-Breuer Centenary Symposium*, edited by R. Porter (Churchill, London, 1970), pp. 365-397.
7. E. Hering, Self-steering of respiration through the nerves vagus (circa 1868), in: *Breathing: Hering-Breuer Centenary Symposium*, edited by R. Porter (Churchill, London, 1970), pp. 359-364.
8. P. M. Groves and R.F. Thompson, Habituation: a dual-process theory, *Phys. Rev.* **77**, 419-450 (1970).
9. J. M. Bower and D. Beeman, *The book of Genesis: exploring realistic neural models with the General Neural Simulation System*, (Springer-Verlag New York, New York, 1995).
10. I. A. Rybak, J. F. Paton, and J. S. Schwaber, Modeling neural mechanisms for genesis of respiratory rhythm and pattern. I. Models of respiratory neurons, *J. Neurophysiol.* **77**, 1994-2006 (1997)
11. I. A. Rybak, J. F. Paton, and J. S. Schwaber, Modeling neural mechanisms for genesis of respiratory rhythm and pattern. II. Network models of the central respiratory pattern generator, *J. Neurophysiol.* **77**, 2007-26 (1997).
12. I. A. Rybak, J. F. Paton, and J. S. Schwaber, Modeling neural mechanisms for genesis of respiratory rhythm and pattern. III. Comparison of model performances during afferent nerve stimulation, *J. Neurophysiol.* **77**, 2027-39 (1997).
13. K. Takano and F. Kato, Inspiration-promoting vagal reflex under NMDA receptor blockade in anaesthetized rabbits, *J. Physiol. (London)* **516**, 571-82 (1999).
14. D. L. Young, M. S. Siniaia, and C. -S. Poon, NMDA receptor blockade unmasks novel gating and memory mechanisms in vagal control of respiratory rhythm, (this volume).
15. W. Rall, Distinguishing theoretical synaptic potentials computed for different soma-dendritic distribution of synaptic inputs, *J. Neurophysiol.*, **30**, 1138-1168 (1967).
16. J. J. B. Jack, D. Noble, and R. W. Tsien, *Electric Current Flow in Excitable Cells* (Calderon Press, Oxford, 1975).
17. W. Wang, M. L. Fung, and W. M. St John, Pontile regulation of ventilatory activity in the adult rat, *J. Appl. Physiol.* **74**, 2801-11 (1993).
18. A. S. Foutz AS, J. Champagnat, and M. Denavit-Saubie, Involvement of N-methyl-D-aspartate (NMDA) receptors in respiratory rhythmogenesis, *Brain Res.* **500**, 199-208 (1989).
19. R. Monteau, P. Gauthier, P. Rega, and G. Hilaire, Effects of N-methyl-D-aspartate (NMDA) antagonist MK-801 on breathing pattern in rats, *Neurosci. Lett.* **109**, 134-9 (1990).
20. C. A. Connelly, M. R. Otto-Smith MR, and J. L. Feldman, Blockade of NMDA receptor-channels by MK-801 alters breathing in adult rats, *Brain Res.* **596**, 99-110 (1992).

NMDA RECEPTOR BLOCKADE UNMASKS NOVEL GATING AND MEMORY MECHANISMS IN VAGAL CONTROL OF RESPIRATORY RHYTHM

Daniel L. Young, Marina S. Siniaia, and Chi-Sang Poon*

1. INTRODUCTION

Afferent feedback from lung receptors plays a key role in the vagal self-steering control of the respiratory rhythm.[1, 2] Although this control process is generally considered to be purely reflexive, there is increasing recognition that the central processing of these afferent inputs may involve more complex mechanisms such as memory and gating in the respiratory controller.[3] For example, it was recently shown[4] that the Hering-Breuer reflex may be adapted centrally through dual-process nonassociative learning in the form of habituation and desensitization in parallel vagal-pontine pathways to the respiratory central pattern generator (RCPG).

In this chapter we report novel central processes that are manifested in vagal-mediated pathways after NMDA receptor (NMDA-R) blockade, including quantized inspiratory-promoting effects, threshold gating, and associative learning. Based on these observations, several models of central processing of vagal afferent inputs are proposed.

2. METHODS AND MATERIALS

2.1. Animal Preparation and Electrophysiology

Procedures for animal preparation were as described previously.[4] Experiments were performed on 9 adult urethane-anesthetized, paralyzed, and artificially ventilated Sprague-Dawley rats. Expired CO_2 was monitored and maintained between 4.6-4.8% end-tidal by a servo-ventilator.

A phrenic nerve was dissected at the C5 level and mounted for recording. Both vagus nerves were sectioned and the central end of the contralateral nerve was stimulated with a

* DLYoung, MS Siniaia, and C-S Poon, Harvard-MIT Division of Health Sciences and Technology, Massachusetts Institute of Technology, Cambridge, Massachusetts, 02139, U.S.A. Correspondence: DL Young, Bldg. 56-030, Massachusetts Institute of Technology, Cambridge, Massachusetts, 02139, U.S.A. Phone: (617) 253-2131; Fax: (617) 253-7648; email: dlyoung@mit.edu

bipolar electrode. The stimulus current threshold (T; 10-20μA) was assessed as the minimal current necessary to produce a visible response in phrenic activity. Stimulus currents (1-2.5*T) were applied at frequencies between 20-100Hz (0.1 ms pulse duration). Current intensities were well below the threshold necessary to activate C-fibers.[5,6]

The non-competitive NMDA receptor (NMDA-R) antagonist dizocilpine maleate (MK-801; 1mg/kg) was administered i.v. to block NMDA-Rs.[7-9] Drug effects were studied after a stabilization period of 20 minutes.

2.2. Stroboscopic Stimulation

Stroboscopic stimulation is a periodic probing method that allows separation of coactive direct, memory, and gated signals. This technique has traditionally been used to temporally filter signals of varying frequencies in engineering problems[10] and has also been used in some respiratory studies.[3] Recently we utilized this method to study vagally induced memory responses in the RCPG.[11,12]

Stroboscopic stimuli were applied repeatedly over an eight-minute period during either inspiratory (S_I) or expiratory (S_E) cycles. S_I was initiated at the start of phrenic activity with its duration matched to control inspiratory duration value (T_I) (0.15-0.50 sec duration). Conversely, S_E was triggered on the falling edge of phrenic activity with its duration matched to control expiratory duration value (T_E) (0.4-0.7 sec duration).

2.3. RCPG Nonassociative Learning Model

A model (Fig. 1) derived from previous studies[4,11-13] sets the stage for this report. In these previous studies, a dual adaptive process in response to continuous and phasic vagus nerve stimulations was demonstrated. The primary adaptive component resulted from nonassociative habituation in the primary vagal afferent-NTS projection to the E neuron, a pathway that is thought to mediate the classic Hering-Breuer inflation reflex (HBIR).[14,15] Secondary nonassociative memory-type adaptation occurred in the pontine pneumotaxic region, a mechanism that was abolished by pontine lesioning or NMDA-R blockade.[4] Moreover, it was shown that the parallel secondary pathway influenced only the expiratory phase of respiration,[11,13] suggesting that the pontine-E modulation may be gated by phasic I activity. Fig. 1 illustrates one possible configuration of the pontine-E pathway that is compatible with the experimental data.[4,11-13]

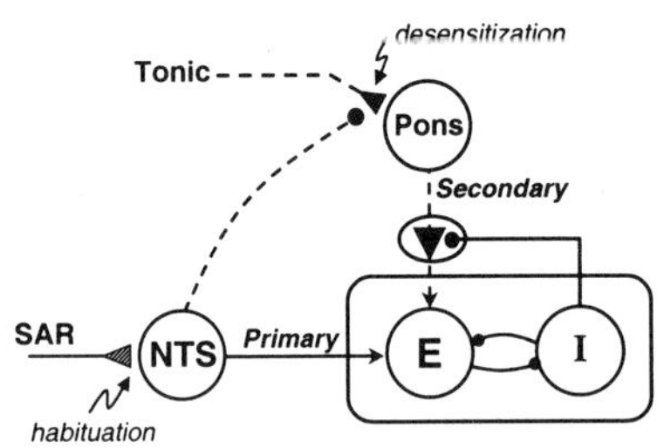

Figure 1. A schematic model of RCPG processing of vagal-afferent feedback in control condition.

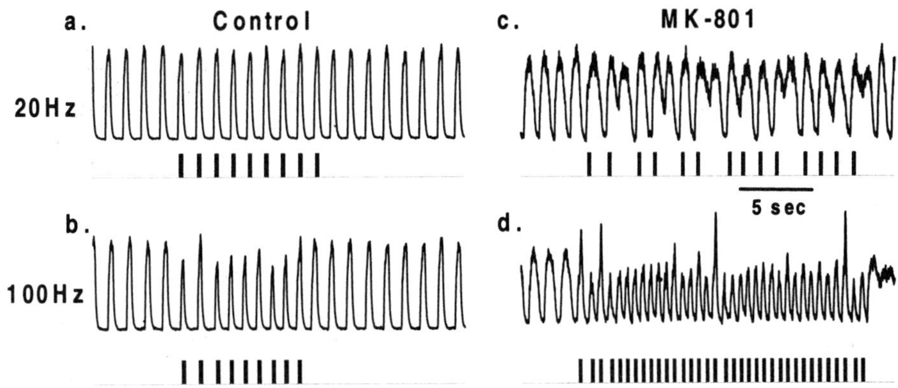

Figure 2. Examples of inspiratory promotion (I-promotion) during S_I are shown. (a,b) show control responses and (c,d) show responses after MK-801 administration at 20 and 100Hz (2*T). Note reversal of the S_I response from I-promotion to I-inhibition as the frequency was increased and the quantal nature of I-promotion after MK-801. A singularity occurs at an intermediate frequency (~80Hz) when both responses can be observed. Upper traces: integrated phrenic; lower traces: stimulus marker.

The respiratory central pattern generator (RCPG) in the model is considered to be a simplified flip-flop oscillator. This simple bi-phasic oscillator is utilized because the detailed wave shapes of the inspiratory and expiratory patterns are not being studied, but rather qualitative features such as the duration of inspiration and expiration are of interest. This model reproduces many of the salient features of a bi-phasic oscillator provided a means for phase switching is present.[16, 17] Here E and I neurons are presumed to oscillate via intrinsic relaxation (e.g. decrementing levels of activity) and mutual inhibition. Vagal and pontine activities modify this basic oscillatory pattern.

3. RESULTS AND DISCUSSION

Since NMDA-R blockade abolishes the pontine input to the RCPG,[4, 13] the present results after MK-801 administration revealed the structure of the vagal-afferent pathway in the model of Fig. 1.

3.1. Quantized Inspiratory-Promoting Responses

NMDA-R blockade uncovered an inspiratory-promoting (I-promoting) tertiary pathway. S_I activated this pathway when applied at low intensities (i.e. at low currents (<2*T) and/or low frequencies (<80Hz)) (Fig. 2c). Eight-minute data showed that this unique pathway exhibited no memory effects. Hence S_I input exerted direct influence on inspiration via the tertiary pathway at low amplitudes and frequencies.

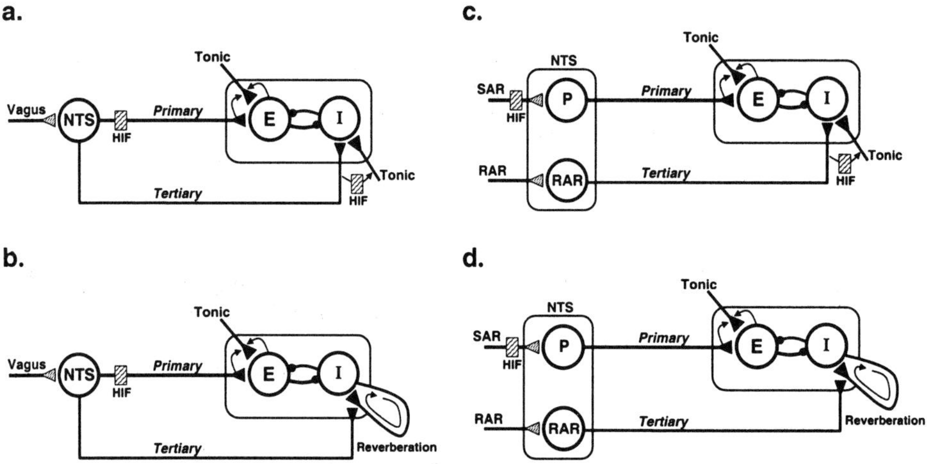

Figure 3. Four schematic models of RCPG processing of vagal-afferent feedback post MK-801 are illustrated. (a,b) depict a single vagal input with central filtering in the primary pathway; (c,d) depict peripheral filtering of separate vagal projections from SAR and RAR; (a,c) depict high intensity STP at I; and (b,d) depict a reverberating mechanism acting on I. All models show an associative STP involving a tonic input at E. E: expiratory neuron; I: inspiratory neuron; P: pump cell; RAR, RAR cell; HIF: high intensity filter.

This I-promoting pathway, however, was obscured during high intensity S_I (Fig. 2d). At such high intensities (2.5*T and/or 100Hz), T_I may be shortened via the primary HBIR pathway. Moreover, high intensity S_I also indirectly shortened T_E indicating activation of a novel high-threshold memory component in the tertiary pathway.

S_E at high intensities also caused I-promotion. In this case the response was indirect because the stimulus and I-promotion were out of phase.

The I-promoting responses resembled the I-promoting frequency dependent effect found in rabbits after NMDA-R blockade[18] and may be related to the Hering-Breuer deflation response (HBDR).[19] I-promotion may result from either excitation of I neurons or inhibition of E neurons. However, because we found no shortening of T_E during low intensity S_E, I-promotion likely resulted from a direct excitatory projection to I-neurons (Fig. 3).

As shown in Fig. 2c, I-promotion outlasted the stimulus suggesting a memory process. One mechanism that could account for this memory is short-term potentiation (STP) of a tonic I drive (Fig. 3a,c). However, integrated phrenic activity often displayed a multi-apex profile when S_I and S_E lengthened T_I. Prolonged T_I's appeared quantized in multiples of one, two or three times control T_I. The quantum nature of this apneustic response suggests a summation of multiple time-delayed inspiratory discharges rather than a simple prolongation of the initial inspiration. An alternative model that would explain the quantal nature of the apneustic-like breaths is therefore a reverberating process activated by the tertiary pathway and potentiated by high intensity stimulations (Fig. 3b,d).

3.2. Associative Vagal Short-term Potentiation

Our previous data showed that administration of NMDA-R antagonist MK-801 abolished the dominant pontine influence on the respiratory rhythm unmasking STP in the primary pathway.[13] The time constant associated with this STP was short during control and long after NMDA-R blockade.[13]

Analysis here of S_E eight-minute stimulations also revealed STP in the primary pathway after NMDA-R blockade. S_E caused a sudden lengthening of T_E followed by a secondary exponential increase in T_E ($\tau = 2.43 \pm 0.44$ sec). This bi-phasic response is indicative of a direct reflex component combined with STP in the primary pathway. Moreover, T_E gradually decayed towards control upon withdrawal of the stimulus suggesting relaxation of the STP ($\tau = 5.71 \pm 0.96$ sec). However, acting via the same primary pathway, S_I at high intensities shortened T_I directly without an STP component.

This difference between S_I and S_E responses suggests that STP in the primary pathway post MK-801 is codependent on vagal and E neuron activities - a novel form of paired or associative potentiation[20, 21] (Fig. 3). Namely while both stimuli caused a direct response via the primary pathway, only S_E produced STP and afterdischarge. Because the potentiation remained active post-stimulation, a persistent or tonic drive is likely involved as shown in the models. This unique central pairing mechanism may also be operative in control conditions although it may be masked by counteractive pontine adaptations.

3.3. Threshold Gating

Two instances of threshold gating were observed after NMDA-R blockade. The first was the occurrence of a stimulus memory in the tertiary pathway for high but not low intensity S_I and S_E. This result suggests that STP or the reverberating process may be threshold gated (Fig. 3).

The second was the dramatic switching of I-promotion to I-inhibition as the stimulus amplitude and/or frequency was increased (Fig. 2c,d). We propose two alternate models to explain these data. The first (Fig. 3a,b) suggests that the gating occurs centrally. Hence low intensity stimuli activate the tertiary pathway while only high intensity stimuli are able to activate the primary, I-inhibiting pathway. The second scenario (Fig. 3c,d) depicts peripheral gating. Here two different fiber types are activated: low threshold I-promoting fibers (e.g. RAR fibers) and high threshold I-inhibiting fibers (e.g. SAR fibers). In both proposed models, gating prevents low threshold stimuli (< 2*T or < 80Hz) from activating the primary vagal-afferent pathway that is involved in the HBIR. Above the high gating threshold, both pathways are activated and compete to control the rhythm. The data suggest that the high-threshold HBIR pathway overshadows the I-promoting pathway when co-activated at high intensities.

4. CONCLUSION

The Hering-Breuer reflex has been shown to exhibit memory and gating processes in vagal driven pathways. The present study demonstrated several novel mechanisms that were manifested in these pathways after NMDA-R blockade. Such mechanisms included quantized I-promoting responses, threshold gating, and associative learning.

5. ACKNOWLEDGMENTS

D. L. Young is recipient of a National Institute of Mental Health individual predoctoral fellowship award 1F31MH12697. The work was supported in part by National Institute of Health grant HL60064.

6. REFERENCES

1. J. Breuer, Self-steering of respiration through the nerves vagus (circa 1868), in: *Breathing: Hering-Breuer Centenary Symposium*, edited by R. Porter (Churchill, London, 1970), pp. 365-397.
2. E. Hering, Self-steering of respiration through the nerves vagus (circa 1868), in: *Breathing: Hering-Breuer Centenary Symposium*, edited by R. Porter (Churchill, London, 1970), pp. 359-364.
3. F. L. Eldridge and D. E. Millhorn, Oscillation, gating, and memory in the respiratory control system, in: *Handbook of Physiology*, Section 3: The Respiratory System, Vol. II: Control of Breathing, part 1, edited by A. P. Fishman, N. S. Cherniack, J. G. Widdicombe (American Physiological Society, Bethesda, 1986), pp. 93-114.
4. M. S. Siniaia, D. L. Young, and C. -S. Poon, Habituation and desensitization of the Hering-Breuer reflex in rat, *J. Physiol.* **523**, 479-91(2000).
5. A. S. Paintel, Vagal sensory receptors and their reflex effects, *Physiol. Rev.* **53**, 159-227 (1973).
6. D. R. Bergren and D. F. Peterson, Identification of vagal sensory receptors in the rat lung: are there subtypes of slowly adapting receptors? *J. Physiol.* (London) **464**, 681-698 (1993).
7. C. A. Connelly, M. R. Otto-Smith, and J. L. Feldman, Blockade of NMDA receptor-channels by MK-801 alters breathing in adult rat, *Brain Res.* **20**, 99-110 (1992).
8. C. -S. Poon, M. S. Siniaia, D. L. Young, and F. L. Eldridge, Short-term potentiation of carotid chemoreflex: an NMDAR-dependent neural integrator, *Neuroreport.* **10**(11), 2261-5 (1999).
9. C. -S. Poon, Z. Zhou, and J. Champagnat, NMDA receptor activity in utero averts respiratory depression and anomalous long-term depression in newborn mice, *J. Neurosci.* **20**(9), RC73 (2000).
10. P. Hariharan, *Basics of Interferometry*, (Academic Press, San Diego, 1992).
11. D. L. Young, M. S. Siniaia, and C. -S. Poon, Adaptive modulation of respiratory rhythm by vagal input: effect of gating, *Soc. Neurosci. Abstr.* 378, (1998).
12. D. L. Young, M. S. Siniaia, and C. -S. Poon, A gated neural differentiator-integrator linked to nonassociative learning in the mammalian respiratory central pattern generator, (submitted).
13. C. -S. Poon, D. L. Young, and M. S. Siniaia, High-pass filtering of carotid-vagal influences on expiration in rat: role of N-methyl-D-aspartate receptors, *Neurosci. Lett.* **284**(1-2), 5-8 (2000).
14. K. Ezure, Synaptic connections between medullary respiratory neurons and considerations on the genesis of respiratory rhythm, *Prog. Neurobiol.* **35**; 429-450 (1990).
15. F. Hayashi, S. K. Coles, and D. R. McCrimmon, Respiratory neurons mediating the Breuer-Hering reflex prolongation of expiration in rat, *J. Neurosci.* **16**(20), 6526-36 (1996).
16. J. Duffin, A model of respiratory rhythm generation, *NeuroReport* **4**, 1215-1218 (1991).
17. M. Matsugu, J. Duffin, and C. -S. Poon, Entrainment, instability, quasi-periodicity, and chaos in a compound neural oscillator, *J. Comput. Neurosci.* **5**(1), 35-51 (1998).
18. K. Takano and F. Kato, Inspiration-promoting vagal reflex under NMDA receptor blockade in anaesthetized rabbits, *J. Physiol.* **516** (Pt 2), 571-82 (1999).
19. E. A. Koller and P. Ferrer, Studies on the role of the lung deflation reflex, *Resp. Physiol.* **10**, 172-183 (1970).
20. C.-S. Poon, Synaptic plasticity and respiratory control. in: *Bioengineering Approaches to Pulmonary Physiology and Medicine*, edited by M.C.K. Khoo, (Plenum Press, New York, 1996), pp. 93-113.
21. C.-S. Poon and M. S. Siniaia, Plasticity of cardiorespiratory neural processing: classification and computational function, *Respir. Physiol.* **122**, 83-109 (2000).

BEYOND CHEMOREFLEX: PLASTICITY, REDUNDANCY AND SELF-ORGANIZATION IN RESPIRATORY CONTROL

A workshop summary

Workshop Chairs: Susan A. Ward* and Chi-Sang Poon*

1. PLASTICITY AND REDUNDANCY ARE KEY TO SELF-ORGANIZATION (C.-S. Poon*)

This Workshop addressed emerging issues of central integration of chemoreceptor inputs during challenges such as hypoxia, hypercapnia and exercise. Classical chemoreflex models assume that respiratory output increases in direct proportion to the sum of chemoreceptor inputs via a hard-wired neural controller, with additive interactions between afferent and efferent pathways (Figure 1A). While this structure has proved satisfactory in explaining the chemical control of breathing, it seems to fail in other respects such as exercise[1]. Increasing evidence now points to the respiratory controller probably being *plastic-wired* rather than hard-wired, with inputs and outputs demonstrating redundancy[2]. These recent findings have ushered in a gradual paradigm shift toward an adaptive respiratory controller model structure (see Figure 1B).

Why such structural redundancy and plasticity, which would certainly add to system complexity? This intriguing question is closely related to an important concept in system modeling — *self-organization* — which is a fundamental attribute of intelligent systems, and for which plasticity and redundancy represent two key ingredients[2]. Structural redundancy provides for the necessary "spare parts" or back-up in case of a failure in the primary pathway(s), while plasticity makes it possible to either turn on or turn off each spare part as necessary. Such a compensatory action is not possible with a hard-wired system.

* Susan A. Ward, University of Glasgow, Glasgow, UK G12 8QQ (Co-Chair); Chi-Sang Poon, Massachusetts Institute of Technology, Cambridge, MA (Co-Chair). Discussants: Thomas E. Dick, Case Western Reserve University, Cleveland, OH; Kenneth Lukowiak, University of Calgary, Calgary, Canada T2N 4N1; Donald R. McCrimmon, Northwestern University Medical School, Chicago, IL 60611; Frank L. Powell, University of California at San Diego, La Jolla, CA; Kingman P. Strohl, Case Western Reserve University, Cleveland, OH.

Frontiers in Modeling and Control of Breathing, edited by Poon and Kazemi.
Kluwer Academic/Plenum Publishers, New York, 2001.

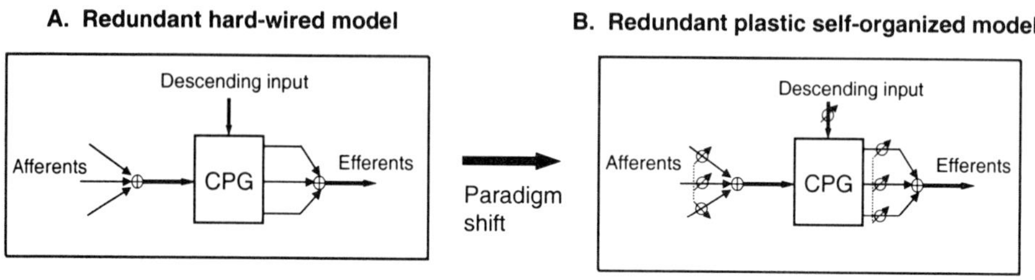

Figure 1. Respiratory control structures

For example, the central and peripheral chemoreflex pathways manifest redundancy and plasticity under certain conditions, and therefore their interaction may not be simply additive. Our recent studies have revealed both phasic and long-term depression (LTD) of synaptic transmission in the nucleus tractus solitarius[3,4], the first relay station of carotid chemoreceptor and other primary afferents. Such plasticity-dependent self-organization of the peripheral chemoreceptor pathway may contribute to the stability of the respiratory system. Paradoxically, we recently also found that an over-expression of LTD in the same brainstem area in some mutant animals may have deleterious effects on breathing, as these mutant animals all died of respiratory failure shortly after birth[5]. Apparently, in some circumstances plasticity could result in *self-disorganization* also.

2. PLASTICITY AND REDUNDANCY IN PONTO-MEDULLARY INTERACTIONS DURING HYPOXIA (T.E. Dick[*])

We are addressing plasticity and redundancy in pontine peripheral chemoreceptor inputs. For example, the ventrolateral (vl) pons contributes to *use/disuse-dependent, pairing-specific* plasticity (c.f., ref. 2) in the context of *posthypoxic frequency decline* (PHFD) — the undershoot in respiratory frequency (through prolongation of expiration) that follows brief hypoxic exposures[6]. This plasticity may depend on an increase in expiratory-modulated activity in the vl pons[7] (possibly the dorsolateral (dl) pons also) acting through medullary post-inspiratory neurons[8]. Use/disuse-dependent plasticity is an attractive concept to us because we theorize that the pons is a part of a reverberating neural network generating breathing pattern. And as Poon has indicated (see 1., above), the teleological significance of use/disuse-dependent, pairing-specific plasticity in ventilatory control could be beneficial or detrimental. Indeed, this plasticity may compromise the respiratory response to recurrent hypoxic episodes.

The dl pons is generally considered to interact with vagal slowly-adapting pulmonary stretch receptor (PSR) input to control inspiratory duration. This network may function as a Hebbian reverberating circuit with intrinsic feedback loops and redundancy. Whether the vl (and dl) pons provides a similar interaction with chemoreceptor inputs is unclear. While the hypoxic ventilatory response depends on intact peripheral chemoreceptor input, we have preliminary evidence that PHFD is still present (but attenuated) after carotid sinus denervation and, further, that this plasticity depends on the pons. These data support a potential for chemoreflex redundancy in the central nervous system[9] which appears to involve the pons as a component or reside in the pons itself.

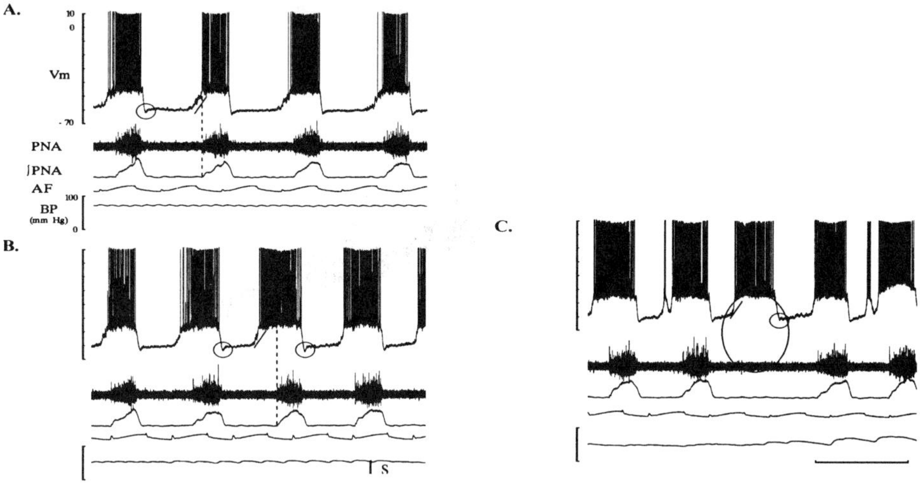

Figure 2. Inspiratory neuron recorded intracellularly in dl pons before (A) and during (B and C, continuous record with the last few ms in B overlapping with C) hypoxia. Vm, membrane potential; PNA, ∫PNA, raw and integrated phrenic nerve activity; AF, airflow; BP, blood pressure. During baseline conditions (A), the neuron depolarized before PNA and hyperpolarized (circle, reversed with Cl^- injection) at the end of PNA. During hypoxia (B and C), the depolarization began earlier (dashed line), the slope increased (trajectory away from diagonal line), and the neuron discharged in the absence of PNA (oval). The post-burst hyperpolarization was absent after the neuron's discharge, indicating its mediation by extrinsic synaptic input. Further, this neuron's activity pattern is dissociated from the motor behavior during hypoxia, indicating that "efference copy" is not driving the activity of this particular neuron. (Preliminary work of Jodkowski, Coles and Dick).

Finally (see Figure 2), we have observed a pontine "inspiratory" cell (i.e. normally discharging with phrenic nerve activity) whose discharge characteristics were substantially altered during hypoxic exposures. That is, the onset of discharge advanced progressively to commence well before that of the phrenic nerve even though expiration, i.e. the inter-burst interval was shortening. Then, the cell discharged independently of phrenic nerve activity, i.e., the total burst occurred in a lengthened expiratory period and in the absence of phrenic nerve activity. This behavior would be consistent with a Hebbian feedback co-variance model, in that this pontine cell would be part of a reverberating pontomedullary network integrating afferent feedback from peripheral chemoreceptor input and the efference copy from the central pattern generator.

3. PLASTICITY IN THE VENTILATORY RESPONSE TO HYPOXIA (F.L. Powell*)

A major issue in this field has been a lack of agreement in definition of the terms plasticity and redundancy and, therefore, confusion in their application. Poon and Siniaia[2] have provided definitions that all Workshop participants appear to agree upon. Our recent studies defined multiple neural mechanisms meeting the definition of plasticity for the hypoxic ventilatory response during acclimatization to hypoxia[13]. These mechanisms operate on different time scales and it is interesting that they span the time course of change in stimuli to different groups of CO_2-sensitive chemoreceptors, which was previously thought to explain the complex time course of the hypercapnic ventilatory response. This raises fundamental questions about the relationship between models and their physiological basis. For example, are there "equivalent circuits" for plasticity in control system models that are independent of the precise physiological mechanisms involved? Are there universal model characteristics that allows us to test for different plasticity mechanisms? Answers to such questions may be critical for identifying respiratory plasticity in humans. Comparative studies reveal striking similarities in the HVR change during chronic hypoxia amongst species, but it is impossible to confirm certain physiological mechanisms in humans with current technology[13]. Models describing universal properties of plasticity may obviate the need for invasive experiments to determine the exact mechanism in every case. Addressing this issue offers a promising area for productive collaborations between modelers and experimentalists.

4. GAIN-MODULATION OF BULBOSPINAL PREMOTOR RESPIRATORY NEURONS: A POTENTIAL CELLULAR MECHANISM FOR ADAPTIVE CONTROL (D.R. McCrimmon*)

Synaptic inhibition has long been proposed to contribute to respiratory phase transitions and sculpting of the detailed pattern of respiratory motor output. Recent work of Zuperku et al.[11,12] also implicates a tonic GABAergic mechanism in constraining the control activity of bulbospinal (BS) premotor respiratory neurons and providing a potent gain control mechanism that multiplicatively modulates their response to synaptic inputs. Application of the competitive GABAA receptor antagonist bicuculline (BIC) yielded amplified replicas of this control discharge pattern (i.e., by a constant multiple). This response differs qualitatively from that elicited by application of an excitatory amino acid to the same neuron. Continuous application of NMDA or AMPA increases the discharge rate by a constant amount throughout the entire respiratory cycle. Thus, in contrast to their GABAergic gain modulation, BS neurons can linearly combine an excitatory input with the prevailing pattern over a wide output range.

The impact of the gain modulation was further evaluated with respect to sensory inputs. Inputs from PSRs produce a bidirectional response in expiratory BS neurons. PSR discharge frequency is directly related to lung inflation and transpulmonary pressure (PTP). Below a PTP threshold, these neurons increase their discharge rate as PTP increases. Increases in PTP above the threshold level cause graded decreases in discharge frequency. BIC amplified both the low-PTP excitatory and higher-PTP inhibitory components of the inflation response to the same degree.

These findings suggest that a $GABA_A$ receptor-mediated mechanism can gain-modulate both the amplitude of excitatory and inhibitory inputs and the prevailing spontaneous neuronal activity. Gain modulation is a powerful means to control the BS neuronal output pattern without loss or distortion of the information content derived from the waxing or waning of discharge frequency within a respiratory cycle. This mechanism may be fundamentally important in adaptive behaviors such as the precision tuning of multiple respiratory motor output patterns, the dynamic reconfiguration of underlying neural circuits due to changes in brain state (e.g., sleep/waking), and differential control of excitability in neurons which participate in many different motor behaviors.

5. THE RESPIRATORY CENTRAL PATTERN GENERATOR OF *LYMNAEA*: IT IS HARD-WIRED, BUT IT IS NOT (K. Lukowiak*)

There are times when a neuron and its connections must be precisely specified. One such time is the formation of a specific neuronal circuit that mediates an important homeostatic behavior - aerial respiratory behavior in the pond snail *Lymnaea*. When given to opportunity (as in cell culture experiments) to make inappropriate synaptic connections (i.e., synapses *not* seen in the animal) with neurons that possess the necessary postsynaptic receptors for the transmitter released by the potential presynaptic neuron, we found that the CPG neurons only synapsed with the cells that they should (*should* being defined here as what is seen in the animal). Thus, at this level of analysis, there is no plasticity. Similarly, transplantation studies (where a single neuron is transplanted into a host central nervous system) have similarly shown *specificity*. Moreover, cell killing and cell transplantation studies demonstrate that other neurons do not take over the function of the CPG neurons. It appears that there is no redundancy.

However, there is great plasticity within the formed CPG. Although tightly constrained in how it is put together, the formed CPG's functionality can be altered, to persist for long periods of time (up to months). We know this because we can train snails not to perform certain CPG-controlled behavior that should normally be occurring[10]. The sites of plastic change in the CPG involve changes in both intrinsic membrane properties and synaptic connectivity. We believe we have shown, for the first time, that a *single* neuron in this circuit is a site for long-term memory storage. Thus, at one level there is little or no plasticity, while at another level and time there is great plasticity and change.

6. GENETIC VARIANCE AND MODELING VENTILATORY CONTROL (K.P. Strohl*)

Respiratory frequency and tidal volume are output variables used in many models designed to explain or predict ventilatory behavior in health and disease. This behavior stands at the interface between psychologic and physiologic processes and the environment. There are a number of structural and functional elements involved in the production of ventilatory behavior, one of which is the genetic make-up of the organism. There is substantial variation among otherwise healthy individuals in ventilatory behavior at rest, during exercise and with chemosensory challenges, some of which result from inheritance[14]. Hence, evolutionary advantage and probability estimates for survival could be key elements to incorporate into ventilatory behavior models. This is exemplified by Tibetan, Andean and Ethiopian Highlanders, who evidence significant differences in key respiratory control variables, e.g., resting ventilation, O_2 saturation and isocapnic HVR.[15] These group differences exist in healthy men and women who are part of culturally and socially developed and expanding (i.e., successful) human populations, who happen to live about 4,000m above sea level in three geographically separated locations. These populations also differ in hemoglobin concentration, a key determinant of tissue O_2 delivery. These natural experiments suggest that humans, as well as other animals indigenous at higher altitudes, adapt using ventilatory behavior as only one of many potential physiologic functions that preserve cellular respiration in hypobaric hypoxia. Indeed, the actions of apparent and latent gene polymorphisms in the founding populations in these three continents to a large degree have determined the different strategies for evolutionary fitness in each of these current populations.

There are two major conclusions that I reach from this literature. First, ventilatory behavior models based solely on one population clearly are not generally applicable and additional features beyond breathing pattern must be incorporated. Second, modeling approaches incorporating genetic mechanisms and the plasticity arising out of genetic variance may be useful in explaining, predicting, and designing alternative strategies for successful adaptations to hypoxic exposure occurring in the course of human disease.

7. CONCLUDING REMARKS (S.A. Ward*)

The potential for an important role for plasticity, redundancy and self-organization in ventilatory control is appreciable. However, the ineluctable challenge remains to robustly discriminate their putative physiological equivalents and establish their proportional contributions to the overall control mechanism.

8. REFERENCES

1. S.A. Ward, Control of exercise hyperpnea in humans: a modeling perspective, *Respir. Physiol.* **122,** 149-166 (2000).
2. C.-S. Poon, and M.S. Siniaia, Plasticity of cardiorespiratory neural processing: classification and computational functions. *Respir. Physiol.* **122,** 83-109 (2000).
3. Z. Zhou, J. Champagnat, and C.-S. Poon, Phasic and long-term depression in brainstem nucleus tractus solitarius neurons: differing roles of AMPA receptor desensitization. *J. Neurosci.* **17,** 5349-5356 (1997).
4. Z. Zhou, and C.-S. Poon, Field potential analysis of synaptic transmission in spiking neurons in a sparse *and irregular neuronal structure in vitro.* J. Neurosci. Methods **94,** *193-203 (2000).*
5. C.-S. Poon, Z. Zhou, and J. Champagnat, NMDA receptor activity in utero averts respiratory depression and anomalous long-term depression in newborn mice, *J. Neurosci.* **20,** (RC73), 1-6 (2000).
6. S.K. Coles, and T.E. Dick, Neurones in ventrolateral pons are required for post-hypoxic frequency decline in rats. *J Physiol. (London)* **497,** 79-94 (1996).
7. T.E. Dick, and S.K. Coles. Ventrolateral pons mediates short-term depression of respiratory frequency after brief hypoxia, *Respir. Physiol.* **121**, 87-100 (2000).
8. T.E. Dick, M. Dutschmann, and J.F.R. Paton, Post-hypoxic frequency decline characterized in the rat working heart brainstem preparation. *This volume* (2001).
9. W.M. St. John, and S.C. Wang, Integration of chemoreceptor stimuli by caudal pontile and rostral medullary sites. *J. Appl. Physiol.* **41,** 612-622 (1976).
10. K. Lukowiak, and N. Syed, Learning, memory and a respiratory central pattern generator. *Comp. Biochem. Physiol. Part A* **124,** 265-274 (1999).
11. Z. Dogas, M. Krolo, E.A. Stuth, M. Tonkovic Capin, F.A. Hopp, D.R. McCrimmon, and E.J. Zuperku, Differential effects of GABAA receptor antagonists in the control of respiratory neuronal discharge patterns. *J.Neurophysiol.* **80,** 2368-2377 (1998).
12. D.R. McCrimmon, E.J. Zuperku, F. Hayashi, Z. Dogas, C.F.L. Hinrichsen, E.A. Stuth, M. Tonkovic-Capin, M. Krolo, and F.A. Hopp, Modulation of the synaptic drive to respiratory premotor and motor neurons. *Respir.Physiol.* **110,** 161-176 (1997).
13. F.L. Powell, M.R. Dwinell, and E.A. Aaron, Measuring ventilatory acclimatization to hypoxia: comparative aspects, *Respir. Physiol.* **122**, 271-286 (2000).
14. F. Han, and K.P. Strohl, Inheritance of ventilatory behaviour in rodent models. *Respir. Physiol.* **121,** 247-256 (2000).
15. C.M. Beall, Tibetan and Andean contrasts in adaptation to high-altitude hypoxia. *Adv. Exp. Med. Biol.* **475,** 63-74 (2000).

Chemoreflex and Exercise Control of Breathing

MODELING THE VENTILATORY RESPONSE TO VARIATIONS IN END-TIDAL PCO_2 IN PATIENTS WHO HAVE UNDERGONE BILATERAL CAROTID BODY RESECTION

M. Fatemian, A. Dahan, S. Meinesz, A. van der Mey, and P. A. Robbins*

1. INTRODUCTION

Studies on patients who have had bilateral carotid body resection for the relief of asthmatic symptoms[1,2] or bilateral carotid endarterectomy for transient cerebral ischemia[3] suggest that the fast component of ventilatory response to CO_2 arises from carotid body activity. However it is also possible that the effect on the ventilatory response to CO_2 was related to their underlying respiratory or vascular disease. The current study sought to examine the ventilatory responses to CO_2 of adult human subjects who had undergone bilateral carotid body resections for glomus tumors. These subjects have normal lung and vascular function.

2. METHODS

Four such subjects were recruited. Each subject was exposed to a multi-frequency binary sequence (MFBS) in end-tidal P_{CO_2} ($P_{ET_{CO_2}}$). This sequence was as designed by Pedersen *et al.* [4], and consisted of a series of steps between two different levels of $P_{ET_{CO_2}}$ which were 1-2 Torr and 10 Torr above the subject's normal air-breathing value for $P_{ET_{CO_2}}$, respectively. The total duration of the sequence was 1408s. Two models were fitted to the data. The first was a one-compartment model, with a central gain term (Gc), time constant (Tc), pure delay (Dc), bias term (B), and trend term. The second was a two-compartment model, which in addition to all of the terms of the one-compartment model above, had a peripheral gain term (Gp), time constant (Tp), and pure delay (Dp). In both models, a parallel model of the noise processes was used.

* M. Fatemian, and P.A. Robbins, University Laboratory of Physiology, University of Oxford, Oxford OX1 3PT, UK. A. Dahan, Departments of Anesthesiology and Physiology, S. Meinesz, Department of Otorynolaryngology, A. van der Mey, Department of Physiology, Leiden University Medical Ctr., Leiden, The Netherlands.

Frontiers in Modeling and Control of Breathing, edited by Poon and Kazemi.
Kluwer Academic/Plenum Publishers, New York, 2001.

3. RESULTS

The results from this study were compared with those for normal subjects exposed to the MFBS in PET_{CO_2} under hyperoxic conditions obtained by Pedersen *et al*[4]. In none of the carotid body resected subjects did the two-compartment model fit significantly better than the one-compartment model, whereas, for intact subjects, the two-component model produced significantly better fit on 6 out of 9 occasions (chi-squared = 4.95, $p<0.05$). Average parameter values for these subjects together with those obtained from the study by Pedersen *et al*[4] are given in table 1.

Table 1. Average parameter values (mean±SD) from model fits

	1-compartment model		2-compartment model	
	CBR	Control[4](hyperoxia)	CBR	Control[4](hyperoxia)
Gc (l/min/Torr)	1.42±0.2†	2.10±0.4	1.2±0.2*	2.0±0.3
Tc (s)	71±12*	42±6	119±44	133±32
Dc (s)	7.2±3.9	7.2±1.6	5.6±1.8**	14.7±1.8
B (Torr)	35.8±0.9*	31.2±1.7	36.4±0.6†	33.4±1.4
Gp (l/min/Torr)			0.33±0.08*	0.75±0.16
Tp (s)			20.8±4.6	12.0±2.8
Dp(s)			17.3±1.6**	6.4±0.7

† $0.05<p<0.1$; * $p<0.05$; ** $p<0.01$

4. CONCLUSIONS

This study provides additional data on human subjects who have undergone CBR. However, in contrast with previous studies, these subjects did not have co-existent lung or vascular disease. The findings are in the general direction predicted from previous results in humans and from experimental studies in animals, namely: 1) a rightwards shift of the ventilation-PCO_2 response relation; 2) a reduction in the ventilatory sensitivity to CO_2; and 3) a loss of rapid ventilatory sensitivity to changes in PCO_2. The absence of a significant improvement in fit of the two-compartment model when compared with the one-compartment model in all subjects who had undergone CBR supports the notion that, when a significant improvement in fit does occur, it can be associated with presence of a peripheral chemoreflex response. Furthermore, in relation to the parameters of the fast compartment, it is of note that the mean value for Gp for CBR subjects is under half that associated with normal subjects in hyperoxia, and that the values obtained for Dp for CBR subjects would more commonly be associated with the slow (central chemoreflex) loop.

5. REFERENCES

1. Y. Honda, S. Watanabe, I. Hashizume, Y. Satomura, N. Hata, Y. Sakakibara, and J. W. Severinghaus, Hypoxic chemosensitivity in asthmatic patients two decades after carotid body resection, *J. Appl. Physiol.* **46**, 632-638 (1979).
2. J. W. Bellville, B. J. Whipp, R. D. Kaufman, G. D. Swanson, K. A. Aqleh, and D. M. Wiberg, Central and peripheral chemoreflex loop gain in normal and carotid body-resected subjects, *J. Appl. Physiol.* **46**, 843-853 (1979).
3. J. G. Wade, C. P. Larson Jr., R. F. Hickey, W. K. Ehrenfeld, and J. W. Severinghaus, Effect of carotid endarterectomy on carotid chemoreceptor and baroreceptor function in man, *N. Engl. J. Med.* **282**, 823-829 (1970).
4. M. E. F. Pedersen, M. Fatemian, and P. A. Robbins, Identification of fast and slow ventilatory responses to carbon dioxide under hypoxic and hyperoxic conditions in humans, *J. Physiol.* **521.1**, 273-287 (1999).

CEREBRAL BLOOD FLOW SENSITIVITIES TO CO_2 WITH THE STEADY-STATE METHOD AND READ'S REBREATHING METHOD

Jaideep J. Pandit[*], Ravi M. Mohan[†], Nicole D. Paterson[†], and Marc J. Poulin[†]

1. INTRODUCTION

In 1967, Read described a rebreathing technique to estimate the central chemoreflex ventilatory sensitivity to CO_2[1] Subsequent studies have reported that the ventilatory sensitivity measured during rebreathing exceeds the sensitivity measured by the steady-state method by about 60%.[2-4] It is well-accepted that brain tissue P_{CO2} (Pt_{CO2}) is the ultimate determinant of the central ventilatory response. A gradient always exists between end-tidal P_{CO2} (PET_{CO2}) and Pt_{CO2} in the steady state, but during rebreathing these variables are presumed to be equal. Theoretically, for a unit rise in PET_{CO2} the corresponding rise in Pt_{CO2} is the same during rebreathing, but is smaller in the steady state [5]. Therefore, when ventilation is plotted against PET_{CO2} for these two methods, the rebreathing method yields a greater slope.

It is known that CO_2 has a profound influence on cerebral blood flow (CBF). Cerebral blood flow in turn might affect the central ventilatory response, since flow might affect the degree to which CO_2 is washed out from brain tissue. Therefore, it is possible that the differences between the ventilatory sensitivity as measured by the two methods might be related to differences in CBF which might occur using the two methods. The purpose of this study was to compare the CBF response to CO_2 during rebreathing with that measured by the steady-state method. If Pt_{CO2} is also the ultimate determinant of CBF, we would expect that, as with the ventilatory response, the CBF response to CO_2 is greater with rebreathing than with the steady-state method.

* Nuffield Department of Anaesthetics, John Radcliffe Hospital, Oxford OX3 9DU, UK.
† University Laboratory of Physiology, Parks Road, Oxford OX1 3PT, UK

Frontiers in Modeling and Control of Breathing, edited by Poon and Kazemi.
Kluwer Academic/Plenum Publishers, New York, 2001.

2. METHODS

2.1. Protocols

This study had the approval of the Central Oxford Research Ethics Committee. Six healthy volunteers undertook each of two protocols in random order. In the steady-state protocol, a dynamic end-tidal forcing system was used to control the subjects' end-tidal gases[6]. PET_{CO2} was initially held at 1.5 Torr above the subject's normal values for 8 min. Subjects were then exposed to three steps of hypercapnia with PET_{CO2} held at 6.5 Torr, 11.5 Torr and 16.5 Torr above the subject's normal values for 4 min, each step separated by a 4 min period in which PET_{CO2} was returned to 1.5 Torr above normal for 4 min. End-tidal P_{O2} (PET_{O2}) was held at 300 Torr throughout the protocol.

In the rebreathing protocol, subjects also initially underwent a 10 min period in which the end-tidal gases were controlled by dynamic end-tidal forcing (PET_{CO2} held at 1.5 Torr above normal values) in order to measure the baseline value of CBF. Subjects were then switched to breathing from a 6 L bag containing a mixture of 93.5% O_2 and 6.5% CO_2. Rebreathing from the bag lasted at least 3 min.

CBF was measured using a 2 MHz pulsed Doppler system to insonicate the proximal segment of the middle cerebral artery. The intensity-weighted mean velocity (V_{IWM}) was used as the index of CBF[7].

Subjects wore noseclips and breathed through a mouthpiece for both protocols. Expired gas at the mouth was measured by a mass spectrometer and analysed for PET_{CO2} and PET_{O2}. Respiratory volumes were measured by a turbine measuring device and flows by a pneumotachograph in series with the mouthpiece. Data concerning experimental time, respiratory data, CBF, gas flow and composition were stored every 20 ms in a data acquisition computer. An ECG was used to monitor heart rate and finger pulse oximeter for oxygen saturation.

2.2. Data Analysis

The 20 ms data was averaged into 15 sec periods.

The ventilatory sensitivity to CO_2 was obtained by least-squares linear regression analysis of the plot of ventilation against PET_{CO2}. The steady-state protocol yielded four points in this plot: one at eucapnia and one taken from the last min of each of the three steps of hypercapnia. The rebreathing protocol yielded a variable number of points, dependent upon the length of time the subject was able to rebreathe. A mean ventilatory sensitivity to CO_2 was obtained for each subject, and the mean for all subjects combined was used as the group mean for each protocol.

Similarly, the CBF sensitivity to CO_2 was obtained in a similar manner by least-squares linear regression analysis of the plots of CBF against PET_{CO2} for each of the protocols.

Student's t-test was used to compare the means of ventilatory CO_2 sensitivity between the protocols, and the CBF CO_2 sensitivity between the two protocols.

3. RESULTS

Figure 1 shows an example of the gas profile for each of the two protocols. Figure 2 shows an example of the resulting ventilatory response, and Figure 3 an example of the resulting CBF response, for each of the two protocols.

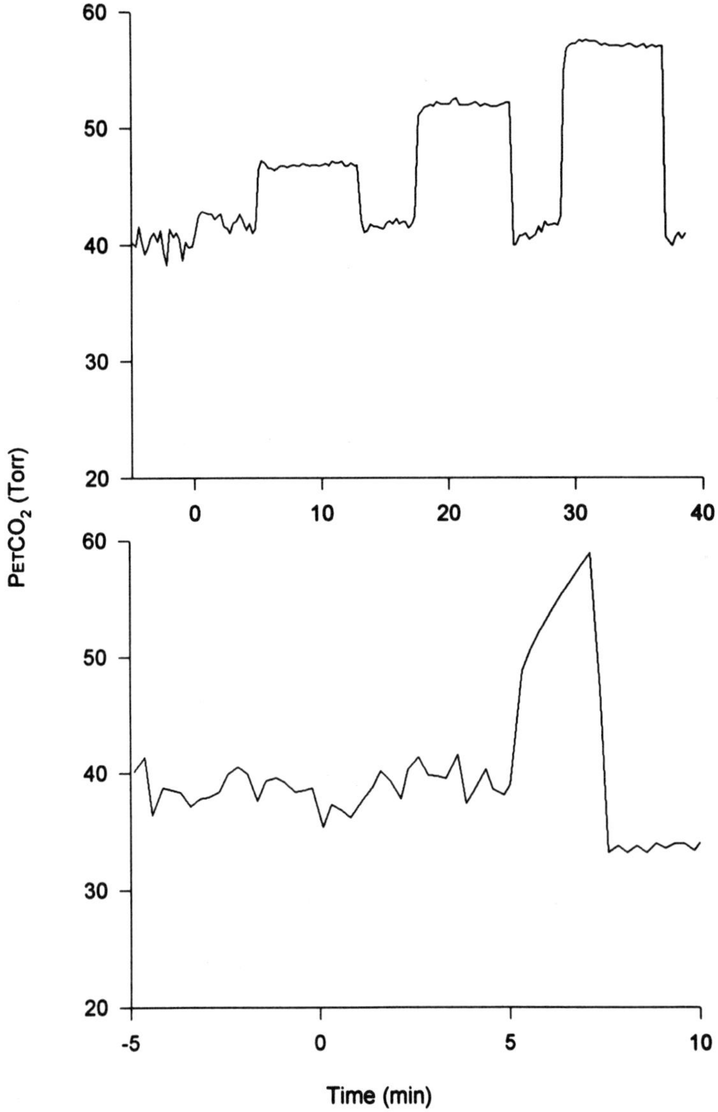

Figure 1. End-tidal P_{CO_2} against time for (top) steady-state protocol and (bottom) rebreathing protocol (breath-by-breath data from one example subject). The period of rebreathing is when the $P_{ET_{CO_2}}$ rises; the point at which $P_{ET_{CO_2}}$ begins to decline again is when rebreathing was stopped. Note the different timescales of the two protocols.

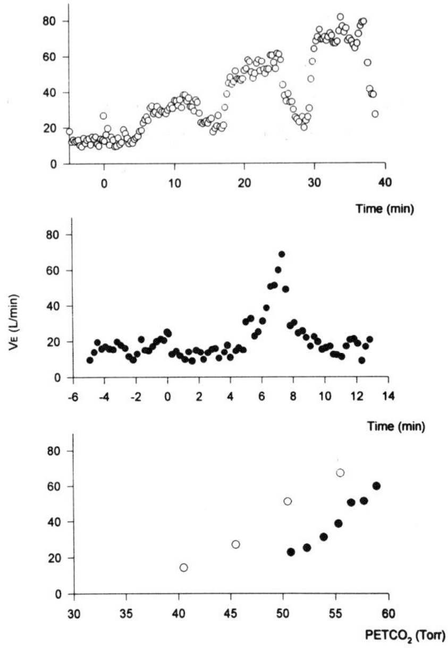

Figure 2. Ventilation against time for (top) steady-state protocol and (middle) rebreathing protocol (15 sec averages for one example subject). Bottom panel shows the plots of VE against $P_{ET_{CO_2}}$ for steady state (○) and for rebreathing (•). Linear regression of the points in the bottom panel yielded the ventilatory responses reported in Table 1.

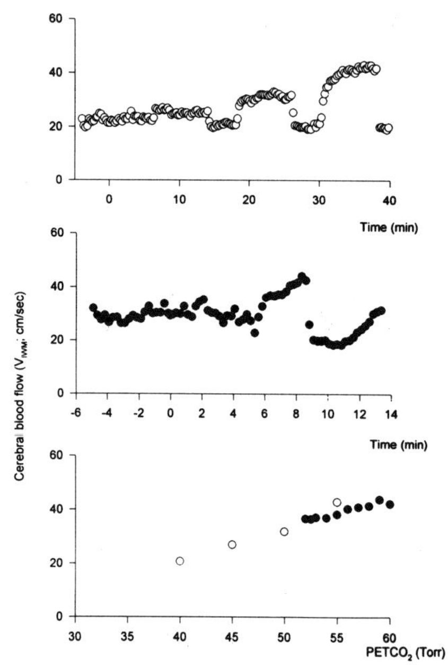

Figure 3. CBF (V_{IWM}) against time for (top) steady-state protocol and (middle) rebreathing protocol (15 sec averages for one example subject). Bottom panel shows the plots of V_{IWM} against $P_{ET_{CO_2}}$ for steady state (○) and for rebreathing (•). Linear regression of the points in the bottom panel yielded the CBF responses reported in Table 1.

Table 1. Ventilatory and CBF CO_2 sensitivities;* denotes $P < 0.02$, comparisons between rebreathing and steady state

Subject	Ventilatory sensitivity to CO_2 (l/min/Torr)		CBF sensitivity to CO_2 (V_{IWM}, % increase/Torr)	
	Rebreathing	Steady-state	Rebreathing	Steady-state
101	4.25	3.30	3.7	5.6
107	6.82	1.96	3.7	4.1
111	0.94	1.19	2.0	3.7
796	2.18	1.50	1.1	2.8
951	1.75	1.96	2.4	2.5
971	1.38	2.28	2.3	4.6
Mean	2.89	2.03	2.5	3.8*
(SD)	(2.24)	(0.73)	(1.0)	(1.2)

Table 1 shows the calculated ventilatory and CBF CO_2 sensitivities for each subject, and for the group as a whole. The ventilatory sensitivity to CO_2 during rebreathing exceeded the steady-state sensitivity by 42% but this was not statistically significant. The sensitivity of CBF to CO_2 was 34% lower during rebreathing than in the steady state ($P < 0.02$, Student's t-test).

4. DISCUSSION

Whereas Read's rebreathing method yields a higher value for ventilatory CO_2 sensitivity higher than does the steady-state method it yields a value for CBF CO_2 sensitivities significantly lower than does the steady-state method.

4.1. Comparison of Ventilatory Data with Other Studies

The mean of previous studies shows that the ventilatory CO_2 sensitivity measured by the rebreathing method exceeds that measured by the steady-state method by 58% (95% confidence interval 16-200%)[2-4]. In our study, the ventilatory CO_2 sensitivity measured by rebreathing exceeded that measured by steady state by a similar amount (42%). However, this was not statistically significantly different from the mean value measured by the steady-state method, largely due to the great variability in results produced by rebreathing. It is interesting that Honda and Miyamura have shown that rebreathing can produce very variable results[2].

4.2. Cerebral Blood Flow Data

This is a novel result and we are not aware of any previous study which has assessed the CBF CO_2 sensitivity using the Read technique. It is important to note some of the potential limitations of our methodology. First, measurement of CBF by flow in the middle cerebral artery may not be representative of total CBF. However, this method has been well-validated in a number of settings and there is no strong reason to suppose why it should influence the results of this study. Second, there has been some controversy as to which index is the best measure of CBF using the Doppler system: V_{IWM} may not be the most appropriate index[5, 7, 8]. However, using peak velocity (V_P) does not alter our results. Third, it could be argued that since the duration of the two protocols differed, experimental conditions of necessity could not be matched between the protocols. In particular, in the steady-state method, high levels of oxygen were inspired for more longer periods than during rebreathing. However, there is little evidence that supranormal levels of oxygen (in contrast to the effects of hypoxia) affect CBF[9].

4.3. Physiological Interpretation of Results

If the rebreathing method produces an equality of PET_{CO2} and Pt_{CO2}, and if Pt_{CO2} determines CBF CO_2 sensitivity, then we can predict that the rebreathing method will yield higher values for CBF CO_2 sensitivity than will the steady-state method. In fact, we have found the opposite to be the case. There are two possible interpretations, each of which have important implications:

One possibility is that Pt_{CO2} is not the main determinant of the CBF CO_2 response. There is some support for this notion from the study of Severinghaus & Lassen (1967)[10], who argued that blood vessel wall (intramural) P_{CO2} was important.

Another possibility is that during rebreathing, PET_{CO2} and Pt_{CO2} are not, in fact, equal. And if this is the case, then it would undermine the central principle which underlies the rebreathing technique and would suggest that Read's method is not appropriate for either respiratory or cerebral blood flow studies. There appears to be some independent evidence from animal studies to support this notion[4]. This study does not provide further insight into which of these two possibilities is the more likely.

5. REFERENCES

1. D.J.C. Read and J. Leigh, Blood-brain tissue P_{CO2} relationships and ventilation during rebreathing, *J. Appl. Physiol.* **23**, 53-70 (1967).
2. Y. Honda and M. Miyamura, Increased ventilatory response to CO_2 by rebreathing in consecutive daily trials, *Jpn. J. Physiol.* **22**, 13-23 (1972).
3. M.S. Jacobi, C.P. Patil, and K.B. Saunders, Comparison of transient, steady state and rebreathing methods measuring the ventilatory response to carbon dioxide in man, *J. Physiol.* **394**, 58P, (1987).
4. Berkenbosch, J.G. Bovill, A. Dahan, J. DeGoede, and I.C.W. Olievier, The ventilatory CO2 sensitivities from Read's rebreathing method ans the steady-state method are not equal in man, *J. Physiol.* **411**, 367-377 (1989).
5. M.J. Poulin and P.A. Robbins, Influence of cerebral blood flow on the ventilatory response to hypoxia in humans, *Exp. Physiol.* **83**, 95-106 (1998).
6. P.A. Robbins, G.D. Swanson, and M.G. Howson, A prediction-correction scheme for forcing alveolar gases along certain time courses, *J. Appl. Physiol.* **52**, 1353-1357 (1982).
7. M.J. Poulin, P-J. Liang, and P.A. Robbins, Dynamics of the cerebral blood flow response to step changes in end-tidal P_{CO2} and P_{O2} in humans, *J. Appl. Physiol.* **81**, 1084-1095 (1996).
8. M.J. Poulin and P.A. Robbins, Indexes of flow and cross-sectional area of the middle cerebral artery using Doppler ultrasound during hypoxia and hypercapnia in humans, *Stroke* **27**, 2244-2250 (1996).
9. S.A. Bew, L.M. Field, D.W. Droste, and P. Razis, The effect of high concentrations of inspired oxygen on middle cerebral artery blood velocity measured by transcranial Doppler, *Exp. Physiol.* **79**, 593-596 (1994).
10. J.W. Severinghaus and N. Lassen, Step hypocapnia to separate arterial from tissue P_{CO2} in the regulation of cerebral blood flow, *Circ. Res.* **20**, 272-279 (1967).

MODELING THE INTERACTION BETWEEN PERFUSION PRESSURE AND CO_2 ON CEREBRAL BLOOD FLOW

Michael R. Edwards, D.C. Lin, and Richard L. Hughson*

1. INTRODUCTION

It is well recognized that steady state changes in cerebral perfusion pressure (CPP) and arterial CO_2 (PCO_2) alter cerebral blood flow (CBF). With the advent of Doppler ultrasound, a more dynamic understanding of how these variables affect CBF can now be investigated. Recently, several authors have characterized spontaneous, beat-by-beat changes in CPP on CBF using frequency domain analyses.[1, 2] However, since PCO_2 shows its own pattern of variability, these authors have failed in their analysis to consider the impact of spontaneous, breath-by-breath oscillations in PCO_2 on CBF.

Understanding how CBF is controlled has important considerations for ventilation since blood flow will ultimately determine the delivery of CO_2 to the central chemoreceptors. In this study, we have explored how breath-by-breath changes in PCO_2 and beat-by-beat changes during spontaneous breathing at rest affect CBF, independently, using spectral analysis. We have also introduced a new model to provide a possible solution for two inputs, CPP and PCO_2, on the regulation of CBF.

2. METHODS

Eight healthy young subjects (6 men, 2 women) were tested at rest in a seated position. End-tidal CO_2 ($P_{ET}CO_2$) was measured by mass spectrometry, CBF was estimated from mean cerebral blood flow velocity (MFV) by Doppler ultrasound, and CPP was determined from noninvasive arterial tonometry corrected for the vertical displacement from the brain. Cerebrovascular resistance index (CVRi) was calculated as CPP/MFV. Data recording began only after a stable end-tidal CO_2 ($P_{ET}CO_2$) was observed.

* Michael R. Edwards and Richard L. Hughson, Cardiorespiratory and Vascular Dynamics Lab, Department of Kinesiology, University of Waterloo, Waterloo, Ontario, N2L 3G1, CANADA. D.C. Lin, Ryerson Polytechnical University, Toronto, Ontario, M5B 2K3, CANADA.

A continuous estimate of $PaCO_2$ was calculated from the expired PCO_2 profile as described by Whipp et al.[3]. Cross-spectral analysis was performed between the input variables CPP and $PaCO_2$ and the output variables MFV and CVRi. A linear relationship between the input and output variables was assumed when squared coherence exceeded 0.5 permitting evaluation of transfer function gain and phase relationships. Autoregressive moving average with exogenous input (ARX) modeling was performed with inputs PCO_2 and CPP and outputs MFV or CVRi. ARX modeling is an adaptation of methods used previously to study the arterial baroreflex[4]. Step responses were obtained by integration of the impulse response of the ARX model.

3. RESULTS

Our model incorporated the continuous estimate of PCO_2 derived from the time series data of expired PCO_2[3] and allowed us to report the cross-spectral relationship between PCO_2 and MFV or CVRi. This method reflects within breath and between breath oscillations in PCO_2.

3.1. Group Data

All subjects showed high frequency gains for CPP→MFV and seven of eight subjects showed no coherence at very low frequencies (<0.07 Hz). The phase shift between CPP and MFV was positive indicating MFV preceded changes in CPP. For five of eight subjects, there was a high coherence (>0.5) between CPP and CVRi at the very low frequencies (<0.07 Hz). The phase shift was negative indicating that CVRi followed changes in CPP. We observed high coherence between PCO_2 and MFV in seven of eight subjects in the high frequency region and in four of eight subjects in the low frequency region. Spectral analysis between PCO_2 and CPP showed correlation in three subjects in the low frequency region and all subjects showed correlation in the high frequency region. Analysis with ARX modeling indicates that the magnitude of the gain for PCO_2 to MFV is approximately eight fold smaller than that for CPP to MFV expressed as (cm/s)/mmHg PCO_2 and (cm/s)/mmHg CPP respectively.

3.2. Individual Data

Two subjects are contrasted in terms of spontaneous variability in PCO_2 (Figures 1 and 2). Visually, subject CA shows much more variability in PCO_2 (Figure 2, top). The cross-spectra between PCO_2 → MFV (figure 2, bottom) and cross-spectra between PCO_2 → CVRi indicate much higher gains for PCO_2 in the low frequency for subject CA compared with JL. Figures 3 and 4 show results from ARX modeling for the same individuals (JL, figure 3 and CA, figure 4). The gain of the relationships between the inputs and output are expressed independently from the impulse functions for CPP (A) and PCO_2 (B) and from the step functions obtained as the integral of the impulse responses for CPP (C) and PCO_2 (D). Despite similar gains for CPP, subject CA has a ten fold greater gain for PCO_2. In addition, figures 3 and 4 indicate the return of MFV to baseline in the step response to CPP (C) and the offset of MFV for the step response to PCO_2 (D).

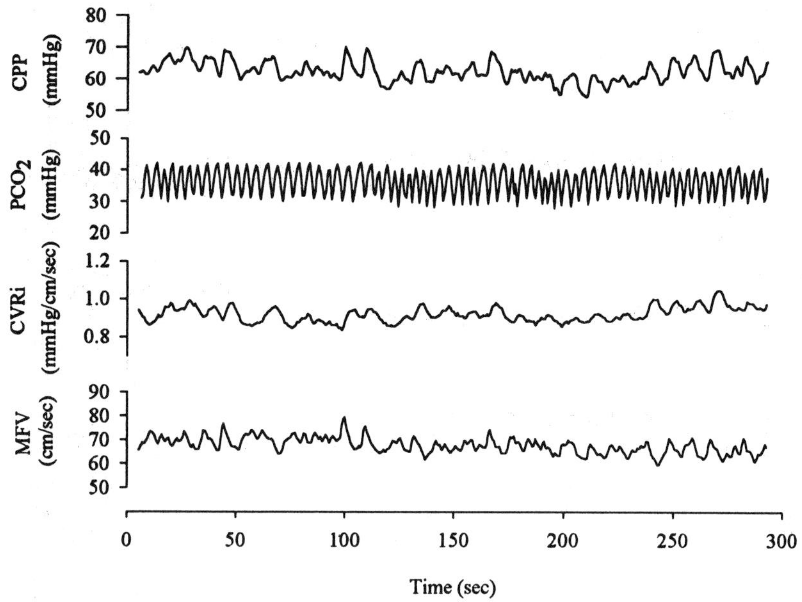

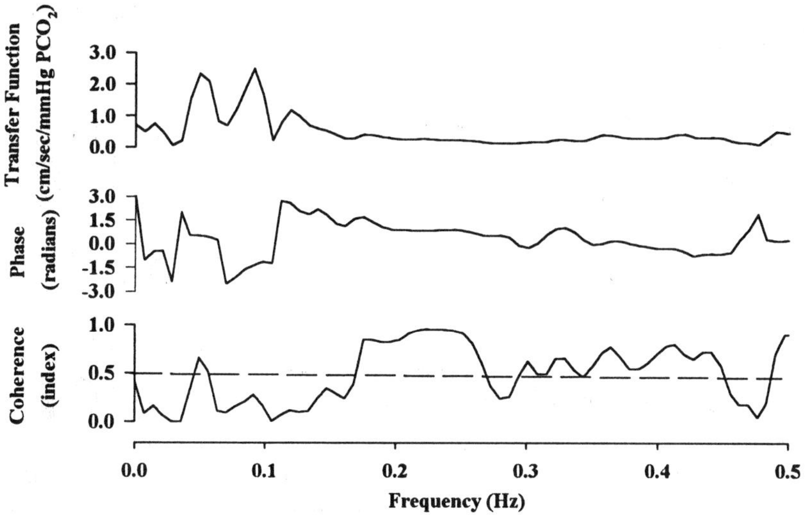

Figure 1. Time series data (top) for subject JL are processed to reveal the transfer function, phase and coherence for the spectral relationship between $PCO_2 \rightarrow$ MFV (bottom). Given the relatively small amplitude of oscillation in PCO_2 (top), it is not surprising that there is little coherence in transfer functions.

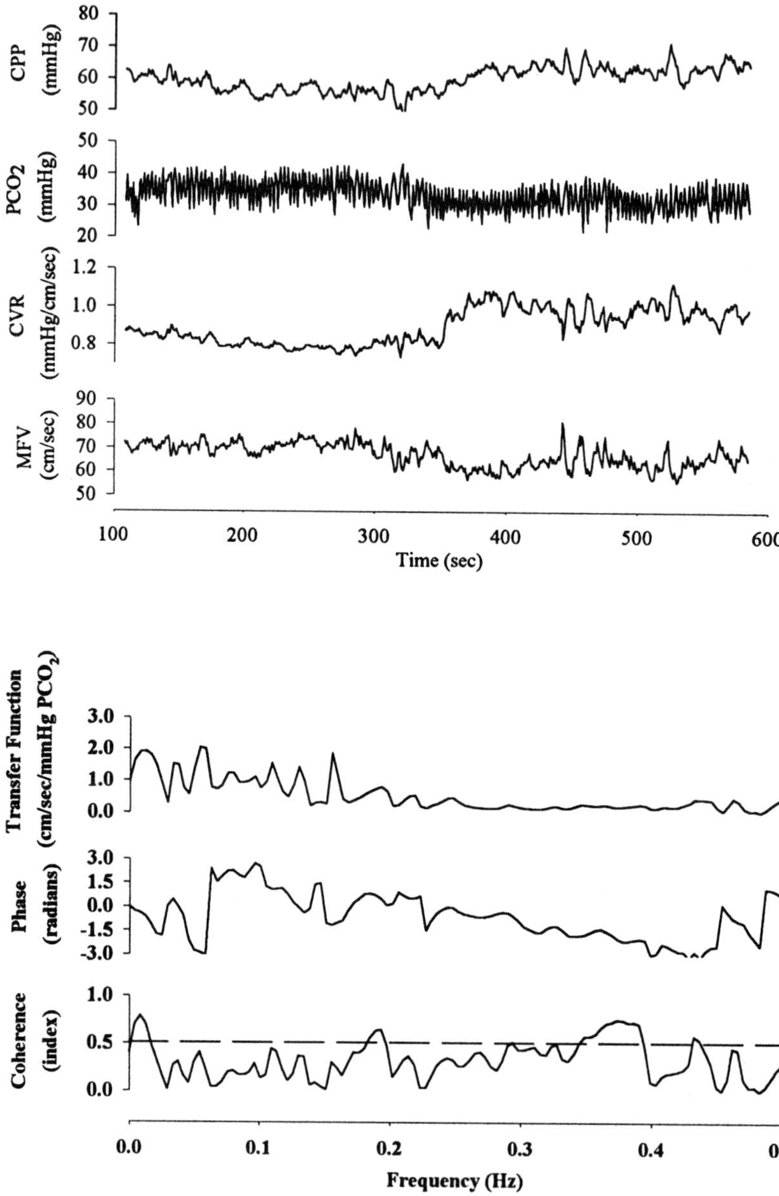

Figure 2. Time series data (top) for subject CA are processed to reveal the transfer function, phase and coherence for the relationship between $PCO_2 \rightarrow$ MFV (bottom). Unlike Fig. 1 (Subject JL), there was a large amplitude, low frequency swing in PCO_2 causing changes in MFV due to changes in CVR.

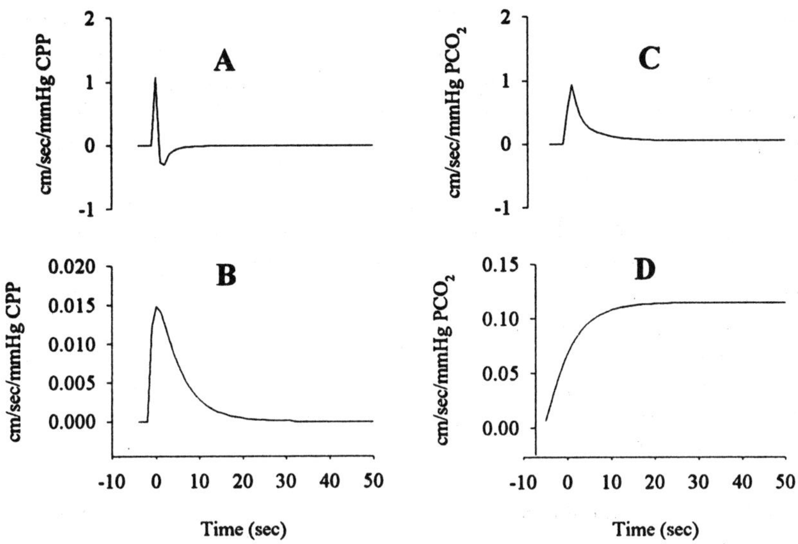

Figure 3. ARX modeling was used to determine the simultaneous effects of CPP and PCO_2 on MFV for subject JL. The gain of the relationships between the inputs and output are expressed independently from the impulse functions for CPP (A) and PCO_2 (B) and from the step functions obtained as the integral of the impulse responses for CPP (C) and PCO_2 (D).

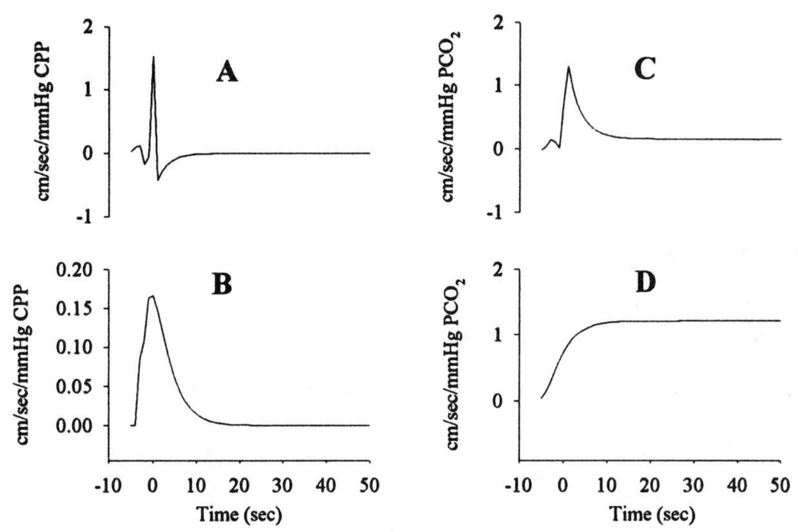

Figure 4. ARX modeling of the simultaneous inputs of CPP and PCO_2 on the output response of MFV was applied to subject CA. The individual panels show impulse responses for CPP (A) and PCO_2 (B), and step responses for CPP (C) and PCO_2 (D). Note the return of MFV to baseline in the step response to CPP (C) and the offset of MFV for the step response to PCO2 (D). Also note in Fig. 3 and 4 the same scales for panels A and B, but different scales for C and D.

4. DISCUSSION

The strong coherence between CPP and MFV within the frequency range of 0.07-0.30 Hz and a poorer relationship within the low frequency range (<0.07 Hz) is consistent with the concept of cerebral autoregulation and reflects the high pass filter nature of this system. In contrast, the high coherence between CPP and CVRi in both low and high frequency regions suggests that CVR is the regulated variable. Previous researchers have pointed to the phase lead of MFV ahead of CPP as another aspect of cerebral autoregulation. It is illogical that the output variable (MFV) should precede the input variable. A more physiological conclusion is reached for phase relationship when CVRi, which is the manipulated variable in the system regulating CBF, lags changes in CPP.

Although we observed high coherence between PCO_2 and MFV in both the high and low frequency region, it is the low frequency response that appeared to be linked to the magnitude of the variation in PCO_2 in this frequency domain. This is clearly seen in figures 1 and 2 and emphasizes the need to measure PCO_2 when modeling CBF.

The high coherence between PCO_2 and CPP indicates that it is important to consider the interactions of CPP and PCO_2 in the regulation of CBF. Furthermore, there are serious limitations with cross-spectral analysis that appear to have been ignored in previous studies that explored only CPP and MFV. Cross-spectral methods are not causal. That is, they do not have the ability to discriminate between multiple input-output relationships. An alternative approach is autoregressive moving average with exogenous inputs (ARX) modeling that allows for simultaneous solution of two inputs (PCO_2 and CPP) and one output. ARX modeling appears to yield satisfactory solutions in some cases of spontaneous breathing. However, in other cases, the input amplitude and frequency content might need to be broadened to provide a valid solution. This conclusion is similar to that reached recently by Panerai et al.[5].

5. ACKNOWLEDGMENTS

This research was supported by grants from the Natural Sciences and Engineering Research Council of Canada.

6. REFERENCES

1. P. Blaber, R. L. Bondar, F. Stein, P. T. Dunphy, P. Moradshahi, M. S. Kassam, and R. Freeman, Transfer function analysis of cerebral autoregulation dynamics in autonomic failure patients, *Stroke* **28**, 1686-1692 (1998).
2. R. Zhang, J. H. Zuckerman, and B. D. Levine, Deterioration of cerebral autoregulation during orthostatic stress: insights from the frequency domain, *J. Appl. Physiol.* **85**(3), 1113-1122 (1998).
3. B. J. Whipp, N. Lamarra, S. A. Ward, J. A. Davis, and K. Wasserman, Estimating arterial PCO_2 from flow-weighted and time-averaged alveolar PCO_2 during exercise, in: *Respiratory Control: A Modeling Perspective*, edited by G. D. Swanson, F. S. Grodins, and R. L Hughson (Plenum Press, New York, 1991), pp. 91-99.
4. D. D. O'Leary, D. C. Lin, and R. L. Hughson, Determination of baroreflex gain using auto-regressive moving-average analysis during spontaneous breathing, *Clin. Physiol.* **19**(5), 369-377 (1999).
5. R. B. Panerai, D. M. Simpson, S. T. Deverson, P. Mahoney, P. Hayes, and D. H. Evans. Multivariate dynamic analysis of cerebral blood flow regulation in humans. *IEEE Trans. Biomed. Engin.* **47**, 419-423 (2000).

CENTRAL AMINO ACID NEUROTRANSMITTERS, VENTILATORY OUTPUT AND METABOLISM DURING ACUTE HYPOXIA IN ANESTHETIZED RATS

M.R. Dwinell, H. Kazemi, J.T. Lam and F.L.Powell*

1. INTRODUCTION

Acute hypoxia decreases $\dot{V}_{O_2}$ in small mammals.[1] The magnitude of the fall in $\dot{V}_{O_2}$ is related to the age of the animal, the level of hypoxia and the degree of development at birth.[1] In addition to the fall in $\dot{V}_{O_2}$ with exposure to acute hypoxia, ventilation changes by first increasing from peripheral chemoreceptor stimulation and then decreasing in a time-dependent manner. The mechanism of such "roll-off", or hypoxic ventilatory decline (HVD), is not fully understood.[2] Although the carotid bodies are required for the increase in ventilation during the hypoxic ventilatory response, they are not necessary for the decrease in metabolism that occurs with hypoxia.[1] We propose calling this the hypoxic metabolic response (HMR), by analogy to the hypoxic ventilatory response (HVR).

Amino acid neurotransmitters in the ventral medulla surface (VMS), both excitatory and inhibitory, may be involved in the biphasic ventilatory response to acute hypoxia.[3,4] Application of the glutamate receptor (NMDA) antagonists, MK-801 and AP-5, to the VMS blocked the increase in the phrenic nerve response to hypoxia in anesthetized rats.[3] GABA, the inhibitory neurotransmitter, may be involved in the HVD as demonstrated in neonatal piglets.[5] Application of a GABA antagonist to the VMS prevented the HVD in anesthetized rats.[3] Brainstem amino acid concentrations measured in anesthetized rats during exposure to hypoxia change in parallel with changes in ventilation.[4] Glutamate concentrations in the region just below the ventral medullary surface rapidly increased with hypoxia, while the concentration of GABA decreased slowly, corresponding with the time course of HVD.

* M.R. Dwinell, Dept. of Medicine, University of California, San Diego, La Jolla, CA 92093. H. Kazemi, Dept. of Medicine, Massachusetts General Hospital, Boston, MA 02114. J.T. Lam, Dept. of Medicine, University of California, San Diego, La Jolla, CA 92093. F.L. Powell, Dept. of Medicine and White Mountain Research Station, University of California, San Diego, La Jolla, CA 92093.

Frontiers in Modeling and Control of Breathing, edited by Poon and Kazemi.
Kluwer Academic/Plenum Publishers, New York, 2001.

During studies investigating central chemoreceptors and the control of breathing, a change in metabolism was observed under certain conditions, suggesting that central respiratory centers may also be involved in the depression of $\dot{V}_{O_2}$ measured during exposure to acute hypoxia. Application of GABA by ventriculocisternal perfusion to chemosensitive areas of the ventral medulla depresses $\dot{V}_{O_2}$ and $\dot{V}_{CO_2}$ in anesthetized dogs.[6] We hypothesized that a common mechanism in the VMS may contribute to the neural integration of changes in O_2 supply ($\dot{V}_I$) and O_2 demand ($\dot{V}_{O_2}$) as well as the HVR. Therefore, the goal of this experiment was to determine whether the fall in metabolic rate during hypoxia depends on amino acid neurotransmission within the ventrolateral medulla in the CNS.

2. METHODS

Three groups of adult male Charles River Sprague-Dawley rats were used. Group 1 (control, n=5, 462 ± 11.7 g) had the minimal amount of surgery (*i.e.* the ventral surface of the medulla was not exposed). The ventral medullary surface was exposed in Group 2 (sham, n=4, 365 ± 16.0 g) and Group 3 (experimental drug, n=7, 352.1 ± 13.7 g).

2.1. Animal Preparation

All rats (Groups 1, 2 and 3) were initially anesthetized with isoflurane while breathing 100% O_2. A tracheal cannula was inserted and the animals were artificially ventilated (model 680 rodent respiratory, Harvard) with 50% O_2-balance N_2. Femoral arterial and venous catheters were inserted for arterial blood pressure measurement (model P23 ID pressure transducer, Stratham), arterial blood gas sampling (Instrumentation Laboratory), and intravenous fluid injection. After the catheterization was complete, the animals were gradually switched from isoflurane to urethan (1.6 g/kg iv). Additional anesthetic was administered (urethan) as needed as judged by a change in arterial blood pressure following a toe pinch. End-tidal P_{CO_2} (PET_{CO_2}) was measured with a flow-through capnograph (model 1265 Capnoguard, Novametrix). Rectal temperature was measured, and body temperature was maintained close to 37° degrees with a circulating-water pad and a deltaphase isothermal pad.

A ventral approach was used to expose the ventral medullary surface (VMS) for drug application in Groups 2 and 3 as previously described.[4] $\dot{V}_{O_2}$ was determined using a mass spectrometer (Perkin-Elmer, 1100 Medical Gas Analyzer) to measure the difference between F_{O_2} and F_{CO_2}, the inspired and mixed-expired gases.

Hypoglossal nerve activity was measured in some of the animals in Groups 2 (n=2) and 3 (n=4). In these animals, a bilateral vagotomy was performed. The hypoglossal nerve was exposed on the right side, cut, desheathed, and placed on a bipolar platinum hook electrode immersed in mineral oil. The nerve signal was preamplified (P5 series, Grass), filtered, rectified, and integrated using a moving time averager (model MA-821, CWE; time constant = 20 ms) to acquire a moving average of peak nerve activity. The raw and integrated hypoglossal signals were displayed on a chart recorder (Gould) and stored on a computer by use of the BIOPAC data collection program (MP100A, BIOPAC Systems). These rats were paralyzed with pancuronium bromide (2.5 mg/kg iv).

2.2. Experimental Protocol

The animal was allowed to stabilize for 40-60 minutes after surgery. An arterial blood gas sample was collected and the ventilator settings were adjusted if Pa_{CO_2} was not in the normal range. In those rats in which nerve activity was measured, the ventilator was adjusted to maintain P_{ETCO_2} at 3 Torr above the CO_2 threshold for hypoglossal nerve activity. Once the appropriate ventilator settings were determined, the ventilator settings were not changed throughout the course of the experiment.

For Group 1 (no surgical exposure of the VMS), blood pressure, $\dot{V}_{O_2}$, $\dot{V}_{CO_2}$, and arterial blood gas measurements were made during a 15 min control period (F_{IO_2} = 0.3). The rats were then exposed to 20-30 minutes of hypoxia (F_{IO_2} = 0.1) with measurements taken throughout the exposure to hypoxia. Measurements were taken during a 15 min recovery period (F_{IO_2} = 0.3).

The second group of rats (Group 2, VMS exposed and sham applied to surface). Blood pressure, $\dot{V}_{O_2}$, $\dot{V}_{CO_2}$, arterial blood gases were measured during a 15 min control period (F_{IO_2} = 0.25), 10 min following the application of saline, during 20-30 min of hypoxia (F_{IO_2} = 0.10) and during 15 min of recovery (F_{IO_2} = 0.25). Group 3 was studied using the same protocol with drug application rather than saline application to the VMS. MK801 (8 mM; n=7) and bicuculline (10 μM; n=3) were applied to the surface of the VMS using small cotton pledgets.

2.3. Data Analysis

A two-factor multivariate repeated-measures analysis was used to determine significant differences in the metabolic and blood gas data between rats in Group 2 and 3. $P<0.05$ was considered significant. Values are means ± SE.

3. RESULTS

3.1. Metabolic Rate

In Group 1 (control), $\dot{V}_{O_2}$ was 1.625 ± 0.1 ml/min/100g min during 30% O_2 and fell rapidly to 1.043 ± 0.1 ml/min/100g within 2 min after exposure to hypoxia (F_{IO_2} = 0.10) ($P<0.05$). After 20 min of hypoxia, $\dot{V}_{O_2}$ remained low at 1.086 ± 0.1 ml/min/100g and returned to control levels (1.785 ± 0.1 ml/min/100g) immediately when normoxia was restored (Fig. 1). $\dot{V}_{CO_2}$ did not fall as rapidly as $\dot{V}_{O_2}$ upon exposure to hypoxia, but continued to fall throughout the hypoxic exposure, resulting in a transient increase in R ($\dot{V}_{CO_2}/\dot{V}_{O_2}$). In Group 2 (sham), $\dot{V}_{O_2}$ dropped significantly from 1.261 ± 0.04 ml/min/100g in normoxia to 0.914±0.08 ml/min/100g after 3 min of hypoxia ($P<0.05$). $\dot{V}_{O_2}$ was 0.891 ± 0.1 ml/min/100g after 20 min of hypoxia and rose to 1.477 ± 0.09 ml/min/100g 15 min after return to normoxia (Fig. 2A). The changes in $\dot{V}_{O_2}$ upon exposure to hypoxia were not different in Group 3 as compared to Group 2 regardless of which drug was applied (Fig. 2A).

The time course of changes in $\dot{V}_{CO_2}$ were similar among the rats with the VMS exposed (Group 2 and 3). $\dot{V}_{O_2}$ decreased immediately upon exposure to acute hypoxia while $\dot{V}_{CO_2}$ decreased more slowly (Fig. 2B).

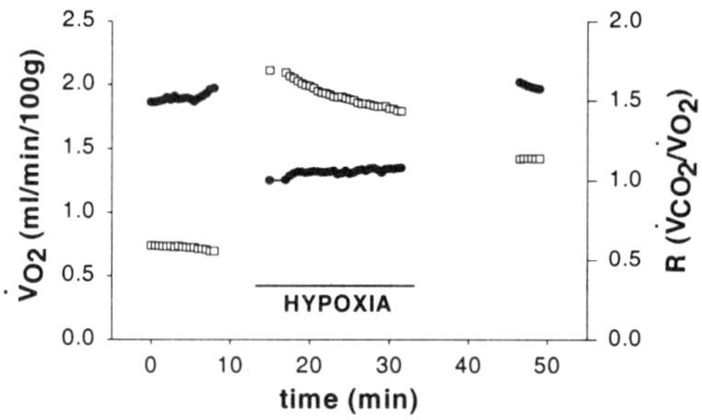

Figure 1. $\dot{V}_{O_2}$ and R before, during and after 20 min of hypoxia in one anesthetized rat from Group 1 (control). •, $\dot{V}_{O_2}$ (ml/min/100g); □, R ($\dot{V}_{CO_2}/\dot{V}_{O_2}$).

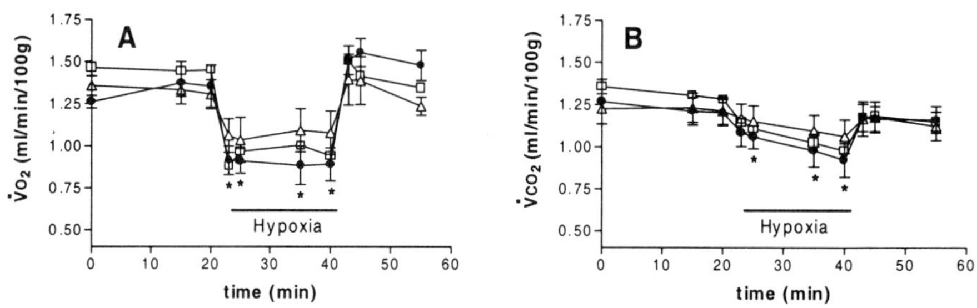

Figure 2. $\dot{V}_{O_2}$ and $\dot{V}_{CO_2}$ in Groups 2 & 3 before, during and after exposure to acute hypoxia. •, saline; □, MK801 (8 mM); Δ, bicuculline (10 μM). *, significantly different than control, normoxia.

3.2. Arterial Blood Gases

Arterial blood gas values changed similarly during acute hypoxia in all groups of rats. Values for rats in Group 1 (control) are given in Table 1. Pa_{O_2} decreased rapidly and Pa_{CO_2} decreased more slowly, in phase with the changes in $\dot{V}_{O_2}$ and $\dot{V}_{CO_2}$, respectively.

In Group 2 (sham), Pa_{O_2} decreased from 88.8 ± 3.9 mmHg (normoxia) to 38.0 ± 0.4 mmHg after 3 min of hypoxia and was 37.2 ± 1.6 mmHg after 20 min of hypoxia. Pa_{CO_2} was 31.2 ± 1.7 mmHg during normoxia, 32.8 ± 1.6 mmHg after 3 min of hypoxia and fell to 26.8 ± 1.2 mmHg after 20 min of hypoxia. In Group 3 (experimental drug), Pa_{O_2} was 106.0 ± 7.2 mmHg during normoxia, 39.9 ± 2.4 mmHg after 3 min of hypoxia and 39.0 ± 3.2 mmHg after 20 min of hypoxia. Pa_{CO_2} was 34.1 ± 2.6 mmHg during normoxia, 30.0 ± 2.4 mmHg after 3 min of hypoxia and 24.4 ± 2.0 mmHg following 20 min of hypoxia.

Table 1. Pa_{CO_2} (mmHg), Pa_{O_2} (mmHg) and pH values from rats in Group 1

Group 1	TB (°C)	Pa_{CO_2}	Pa_{O_2}	pH
Control	35.9 ± 0.2	31.6 ± 2.2	155.6 ± 8.9	7.360 ± 0.02
5 min hypoxia	36.0 ± 0.2	27.0 ± 2.0	35.8 ± 2.6	7.348 ± 0.03
20 min hypoxia	35.8 ± 0.3	25.5 ± 0.3	35.7 ± 2.7	7.314 ± 0.03
5 min normoxia	35.8 ± 0.2	30.8 ± 2.3	156.1 ± 13.5	7.290 ± 0.03

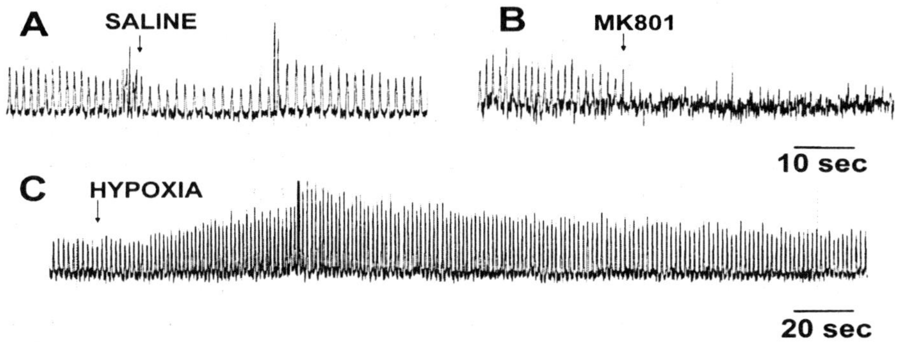

Figure 3. Hypoglossal nerve activity recorded during application of saline (A) and MK801 (B) to the VMS. In C, hypoglossal nerve activity during acute hypoxia recorded in sham rat.

3.3. Hypoglossal Nerve Activity

Hypoglossal nerve activity recorded in sham rats (n=2) or MK801 treated rats (n=4) was significantly different (Fig. 3). In the sham rats, nerve activity increased initially during hypoxia and rolled-off as Pa_{CO_2} began to fall during the 20 min of hypoxia. No attempt was made to hold Pa_{CO_2} constant in any of the experiments. After application of MK801 to the VMS, hypoglossal nerve activity was reduced, often silenced. Activity did not return during the hypoxic exposure in any rat with MK801 application to the VMS. In one MK801 rat, hypoglossal nerve activity returned weakly during exposure to $F_{I}CO_2$ = 0.10 at the end of the protocol.

4. DISCUSSION

These results demonstrate that the hypoxic metabolic response (HMR, or fall in $\dot{V}_{O_2}$ upon acute exposure to hypoxia) is extremely fast. $\dot{V}_{O_2}$ decreases within the first 2 min necessary for us to wash out the ventilator system and measure stable changes in mixed expired gas concentrations after changing inspired gases. Such rapid direct effects of systemic hypoxia on cellular metabolic rates have not been reported, suggesting this may be a neurally mediated regulatory response. The arterial chemoreceptors are not necessary for this response[1] and our present data indicate that glutamate and GABA neurotransmission in the VMS is not involved. Initially, the decrease in $\dot{V}_{O_2}$ cannot be explained by a passive Q_{10} effect either because it occurs during constant body temperature. However, the HMR may involve a change in the neural set point for body temperature because it does decrease during prolonged hypoxia. A neural pathway for this hypothetical reflex remains to be defined.

These experiments were stimulated by the initial observation that ventriculocisternal perfusion of GABA in the anesthetized dog decreased metabolism.[6] Furthermore, the VMS is involved in integration of respiratory responses to diverse stimuli such as CO_2 and metabolic changes, such as those during exercise.[7] However, application of GABA (and glutamate) antagonists to the VMS of anesthetized rats did not alter the HMR. We know that the drugs reached respiratory control structures in the VMS because the ventilatory motor output during hypoxia was changed in Group 3 rats. It is possible that amino acid neurotransmission in structures outside the VMS are involved in the HMR or that only exogenously induced changes in GABA are effective at changing metabolism in these experimental preparations. Further experiments are necessary to test these ideas.

The results may also have implications for studying the time domains of the HVR in rat preparations. The time course of the decrease in $\dot{V}_{CO_2}$ and Pa_{CO_2} during acute hypoxia was similar to that for ventilatory motor output during HVD. HVD occurs in cats and humans even if Pa_{CO_2} is held constant during hypoxia[8] but it is not universally observed in rats.[9] Hence, HVD in the rat may deserve further investigation. Also, the potential effects of a change in metabolic rate during acute hypoxia in the anesthetized rat preparation may have other effects on ventilation besides those mediated by Pa_{CO_2}, for example through a change in state of the ventilatory control system.

In conclusion, the HMR, *i.e.* fall in $\dot{V}_{O_2}$ during acute hypoxia, occurs immediately in the anesthetized rat and is independent of changes in ventilatory output. Furthermore, this fall in $\dot{V}_{O_2}$ is not dependent on NMDA or GABA receptors on the VMS.

5. ACKNOWLEDGMENTS

Supported by National Heart, Lung and Blood Institute Grants HL-17731 and HL-07212. M.R. Dwinell is a Parker B. Francis Fellow in Pulmonary Research.

6. REFERENCES

1. J.P. Mortola and H. Gautier, Interaction between metabolism and ventilation: Effects of respiratory gases and temperature, in: *Regulation of Breathing*, edited by J.A. Dempsey and A.I. Pack (Marcel Dekker, New York, 1995), pp. 1011-1064.
2. G.E. Bisgard and J.A. Neubauer, Peripheral and central effects of hypoxia, in: *Regulation of Breathing*, edited by J.A. Dempsey and A.I. Pack (Marcel Dekker, New York, 1995), pp. 617-668.
3. I. Soto-Arape, M.D. Burton, and H. Kazemi, Central amino acid neurotransmitters and the hypoxic ventilatory response, *Am. J. Respir. Crit. Care Med.* **151**:1113-1120 (1995).
4. B. Hoop, J.L. Beagle, T.J. Maher, and H. Kazemi, Brainstem amino acid neurotransmitters and hypoxic ventilatory response, *Respir. Physiol.* **118**:117-129 (1999).
5. J. Huang, C. Suguihara, D. Hehre, J. Lin, and E. Bancalari, Effects of GABA receptor blockade on the respiratory response to hypoxia in sedated newborn piglets, *J. Appl. Physiol.*, **77**:1006-1010 (1994).
6. M.P. Kneussl, P. Pappagianopoulus, B. Hoop, and H. Kazemi, Effect of centrally administered γ-aminobutyric acid on metabolic function, *J. Appl. Physiol.* **61**:472-476 (1986).
7. H.V. Forster, P.J. Ohtake, L.G. Pan, and T.F. Lowry, and P. Martino, Effect of breathing of surface ventrolateral medullary cooling in awake, anesthetized and asleep goats, *Respir. Physiol.* **110**:187-197 (1997).
8. J.A. Neubauer, J.E. Melton, and N.H. Edelman, Modulation of respiration during brain hypoxia, *J. Appl. Physiol.* **68**:441-451 (1990).
9. F.L. Powell, W.K. Milsom, and G.S. Mitchell, Time domains of the hypoxic ventilatory response, *Respir. Physiol.* **112**:123-134 (1998).

GENE REGULATION DURING INTERMITTENT HYPOXIA: EVIDENCE FOR THE INVOLVEMENT OF REACTIVE OXYGEN SPECIES

Gautam Adhikary, David Kline, Guoxiang Yuan, Ganesh K. Kumar, Michel S. Simonson, Neil S. Cherniack, and Nanduri R.Prabhakar*

1. INTRODUCTION

Intermittent hypoxia occurs in many breathing disorders including sleep apneas, asthma, chronic lung diseases and can have cardio-respiratory consequences including hypertension and altered ventilatory responses to hypoxia and hypercapnia [1,2]. Despite its clinical significance, little is known about the molecular mechanisms associated with recurrent episodes of hypoxia. It is being increasingly recognized that induction of genes is one of the fundamental molecular mechanism that underlies physiological adaptations to chronic hypoxia [3]. Recent studies have shown that the protooncogene *c-fos* is one such gene that is induced during hypoxia, and it regulates the expression of downstream genes (e.g., tyrosine hydroxylase) via activator protein-1 transcription factor [4]. To begin to understand the molecular mechanisms associated with intermittent hypoxia, we asked whether recurrent episodes of hypoxia induces *c-fos* expression, if so, by what mechanisms. For this purpose, we exposed rat pheochromocytoma cells (PC 12) to intermittent hypoxia, and found that *c-fos* is up-regulated by transcriptional activation which involves reactive oxygen species (ROS). Further, intermittent hypoxia is more potent in inducing *c-fos* expression than sustained hypoxia of comparable duration.

2. METHODS

2.1. Cell Culture

PC12 cells (original clone from Dr. L. Green) were grown in a humidified incubator at 37^{0}C circulated with 10% CO_2 , 21% O_2) as described earlier[4]. In the experiments

* Department o Physiology and Biophysics, Case Western Reserve University, Cleveland, OH 44106.
Tel: 216-368-8636, Fax: 216-368-1693, email: nrp@po.cwru.edu

Frontiers in Modeling and Control of Breathing, edited by Poon and Kazemi.
Kluwer Academic/Plenum Publishers, New York, 2001.

involving treatment with drugs, cells were pre-incubated for 30 min with desired concentration of the drug or vehicle and then were subjected to normoxia or to intermittent hypoxia.

2.2. Exposure of Cell Culture to Intermittent Hypoxia

We designed a humidified chamber in which cell cultures can be exposed to alternating cycles of brief hypoxia and re-oxygenation. The duration of hypoxia and re-oxygenation can be adjusted by a timed solenoid valve. The temperature in the chamber is maintained at 37^0C. Ambient oxygen levels within the chamber are continuously monitored by an oxygen analyzer (Beckman LB2) and recorded on a strip chart recorder. Cells were subjected to 15 sec of hypoxia (1% O_2, 10% CO_2) followed by 3 min of normoxia (21% O_2 , 10% CO_2) for 60 cycles. Cells exposed to 21% O_2 and 10% CO_2 acts as control.

2.3. Northern Blot Assay

The level of c-*fos* and β-actin mRNA levels were analyzed by northen blot analysis as described previously [5]. The loading of RNA was normalized by β-actin transcript level. Quantification of *c-fos* mRNA levels and β-actin levels was accomplished by optical density measurements using phosphoimager.

2.4. Repoter Gene Assay

Transcriptional activation of c-*fos* mRNA was analyzed by reporter gene assay using c-*fos*-luciferase plasmid as describe previously [4]. Briefly, cells were plated in 60-mm tissue culture plates at a density $0.5X10^6$ cells/plate. In a growth medium containing serum. For transfection, DNA liposome mixture was prepared using lipofectamine (10 μg), 1 μg of *fos*-LUC and 0.5 μg of pRSV β-galactosidase (internal control to determine transfection efficiency) in a 2 ml serum free medium. Cells were incubated in DNA liposome mixture (2 ml/plate) for 4 hr followed by addition of 2 ml of serum containing medium. After 36 hr, cells were exposed to intermittent hypoxia for desired duration. Total amount of plasmid to be transfected per plate was maintained equal by adding pUC19 DNA. Control experiments were run in parallel by exposing cells to normoxia. Luciferase activity was measured by Luminometer using luciferin as substrate. For measurements of β-galactosidase activity, 10 μl of lysate was incubated with 200 μl of the reaction mixture containing β-galactam ™ as substrate for 30 min at room temperature. Following incubation, 300 μl of luminence enhancer was added to the reaction mixture and the resulting luminance was measured with the luminometer. Reporter gene assays were in the linear range.

2.5. Measurements of Aconitase Activity

Following exposure to either normoxia or to intermittent or to sustained hypoxia, cells were lysed and the cell lysate was centrifuged for 2 min at 14,000x *g*. Aconitase activity was measured in the supernatant fractions. The enzymatic activity was monitored at 25^0 C in a spectrophotometric cells in 300 μl reaction medium containing 50 mM Tris-

HCl, pH 7.4, 5 mM sodium citrate, 0.6 mM $MnCl_2$, 0.2 mM NADP, 1-2 units of isocitrate dehydrogenase, and 10 μg of protein equivalent of the cell extract. Enzyme reaction was monitored by spectrophotometer at 340 nm.

2.6. Data Analysis

The data are expressed as mean±SEM from 3-5 individual experiments each run duplicate. Statistical analysis was performed by analysis of variance (ANOVA) or by paired 't' test where appropriate and p values less than 0.05 were considered significant.

3. RESULTS

3.1. Intermittent Hypoxia Induces *c-fos* mRNA via Transcriptional Activation of the c-*fos* Gene

An example depicting the effect of episodic hypoxia (60 cycles; 15 s 1% O_2 – 3 min 21% O_2) on *c-fos* mRNA expression is shown in Figure 1.

As can be seen in left panel for figure1, *c-fos* mRNA is elevated in cells exposed to intermittent hypoxia, whereas β-actin mRNA levels, which served as controls were unaffected. Average results from 5 individual experiments showed that *c-fos* mRNA levels increased by 3.5 fold compared to normoxic controls ($p<0.01$; n =5). As can be seen from right panel, 15 min of sustained hypoxia had no discernable effect on *c-fos* mRNA expression.

To determine whether transcriptional activation of the *c-fos* gene by IH contributes to increased c-*fos* mRNA expression, we performed reporter gene assay using luciferase expression vector. Intermittent hypoxia (60 cycles) increased *c-fos* promoter activity by 2.5 fold, where as comparable duration of sustained hypoxia has no effect. Neither intermittent nor sustained hypoxia had any effect on β-galactosidase activity, which served as a control for assessing the transfection efficiency.

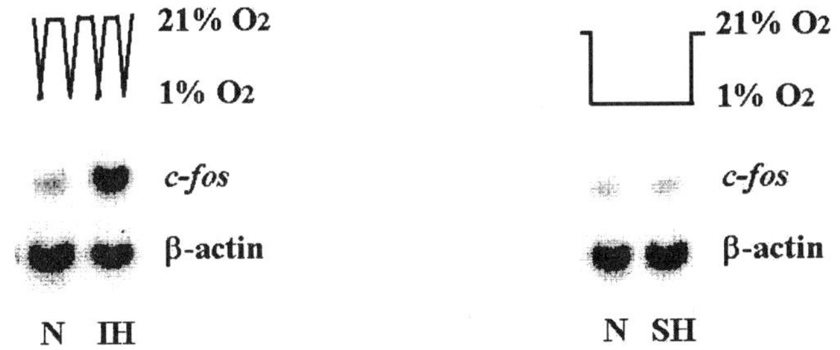

Figure 1. Induction of the c-*fos* gene expression by intermittent hypoxia in PC12 cells. Cells were subjected to normoxia (N) 21% O_2, intermittent hypoxia (IH 1% O_2, 15 sec H+3 min N for 60 cycles, which equals to 15 min. of hypoxia) and sustained hypoxia(SH;1% O_2) for 15 min. Total RNA was isolated and Northern hybridization was done with c-*fos* and β-actin c-DNA probes. Intermittent hypoxia (Left panel) induced c-*fos* expression, whereas 15 min sustained hypoxia (Right panel) had little effect on c-*fos* expression.

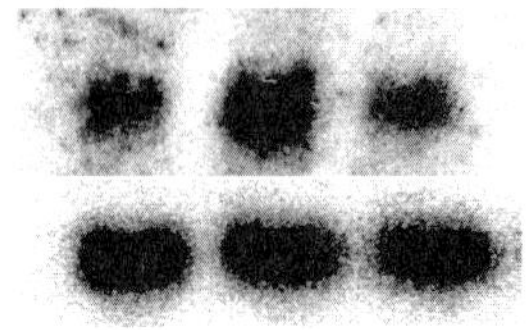

Figure 2. Reactive oxygen species (ROS) are involved in IH induced *c-fos* gene expression. Northern hybridization of RNA from cells exposed to normoxia (21% O2), intermittent hypoxia (IH; same paradigm as in figure 1) with and without SOD mimic (Mn(III) TMPyP, 25 μM). SOD mimic prevented *c-fos* mRNA expression by intermittent hypoxia.

3.2. Evidence for the Involvement of Superoxide Ions in *c-fos* Induction by Intermittent Hypoxia

Several studies have documented that Reactive Oxygen Species (ROS) stimulate immediate early gene expression including *c-fos* [6]. We tested the idea that superoxide anion ($O_2^{\cdot-}$), a member of ROS family, is released during re-oxygenation phase of intermittent hypoxia, and contributes to *c-fos* induction. Superoxide ions inhibit the enzyme activity of aconitase. This inhibition of aconitase activity is often used as an index of superoxide ion levels. We monitored aconitase activity in cells exposed to intermittent as well as sustained hypoxia. Aconitase activity was 74$\pm$6% less in cells exposed to recurrent hypoxia compared to normoxic controls ($p<0.01$), indicating the release of superoxide ions; whereas aconitase activity was unchanged in cells exposed to sustained hypoxia. These results indicate that superoxide ion levels are increased during intermittent but not sustained hypoxia. To test whether $O_2^{\cdot-}$ plays a role in induction of *c-fos*, cells were pre-treated with superoxide dismutase mimic (SOD mimic; Mn(III) TMPyP; 25 μM), potent scavenger of superoxide ions, and then were exposed to intermittent hypoxia. As shown in Figure 2, *c-fos* was no longer induced by intermittent hypoxia in presence of the SOD mimic.

4. DISCUSSION

The present study used rat PC12 cells and the *c-fos* gene as models to examine the effects of episodic hypoxia on gene expression. It is clear from our results that episodic hypoxia stimulates *c-fos* expression and at least in part by transcriptional activation, as evidenced by reporter gene assay. Increased stability of mRNA also contributes to increased mRNA levels. Whether intermittent hypoxia affects *c-fos* mRNA stability, however, remain to be studied. Greenberg *et. al.* [7] reported increased c-Fos protein expression in the central nervous system of rats exposed to 30 days of intermittent hypoxia. However, from this study it is not clear whether elevated c-Fos protein were due to increased stability of the protein (post-translational modification) and/or due to increased transcriptional activation of the *c-fos* gene. The results from the present study provide evidence for transcriptional activation of *c-fos* that may potentially contribute to elevated c-Fos protein by intermittent hypoxia . It has been well established that sustained hypoxia stimulates *c-fos* gene expression in isolated cells [8]. It is interesting that when we compared the effects of sustained hypoxia of comparable duration and intensity with intermittent hypoxia, we found significant induction of *c-fos* gene only with intermittent hypoxia. These observations imply that intermittent hypoxia is more potent in activating gene expression than sustained hypoxia. What makes intermittent hypoxia more effective than sustained hypoxia? The feature that distinguishes episodic from sustained hypoxia is the intermittent re-oxygenation that occurs in the former but not in the later. By analogy, intermittent hypoxia seems to resemble ischemic-re-perfusion. Since ROS levels, especially superoxide ions are increased during re-perfusion [9], and they stimulate gene expression, we reasoned that the effects of intermittent hypoxia are mediated by ROS. Supporting such a notion are the findings: a) intermittent hypoxia decreased aconitase activity, indicating increased superoxide generation; and b) Superoxide scavengers ablated *c-fos* induction by intermittent hypoxia. These observations suggest that gene induction by intermittent hypoxia not only involves ROS, but may critically depend on episodic re-oxygenation more than hypoxia itself (Figure 3).

It is interesting to note that sustained hypoxia had little effect on aconitase, indicating that superoxide levels were unaffected. Therefore, the absence of superoxide generation might explain, why comparable levels of sustained hypoxia is less effective in inducing *c-fos*. These observations then suggest that ROS might be acting synergistically with hypoxia making it more potent in regulating gene expression.

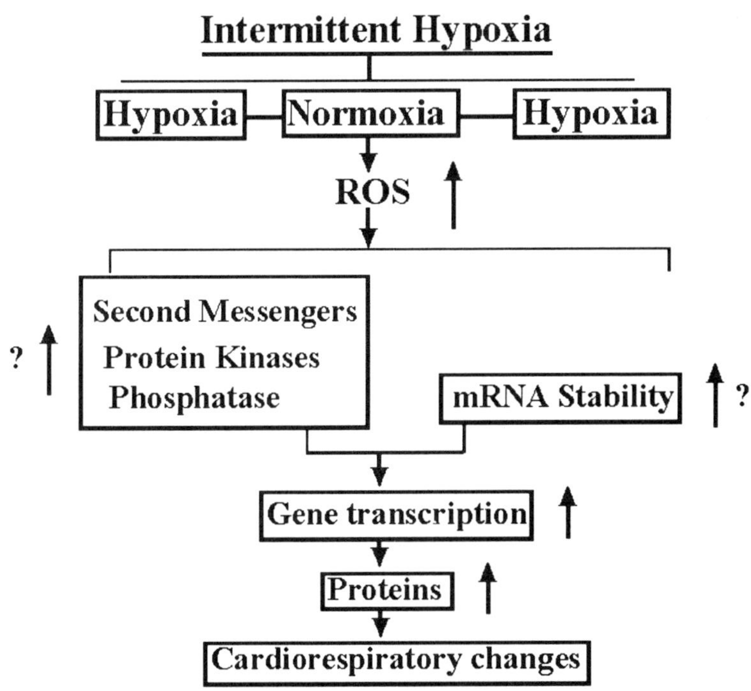

Figure 3. Schematic presentation of the importance of episodic re-oxygenation and the involvement of reactive oxygen species (ROS) in gene regulation.

What might be the physiological significance of *c-fos* induction by intermittent hypoxia? c-Fos, the protein encoded by the *c-fos* gene contributes to the formation of activator protein-1 (AP-1) transcription factor by forming a hetrodimer with Jun protein, encoded by another member of the immediate early gene family. While *c- Jun* is constitutively expressed, *c- fos* is induced in response to a variety of stimuli including hypoxia. AP-1 in turn regulates transcriptional activation of target genes by binding to DNA at AP-1 consensus sites [10]. Tyrosine hydroxylase (TH) gene, which encodes the rate limiting enzyme in catecholamine synthesis, is one such gene that has AP-1 consensus binding sites in the hypoxic responsive element (HRE) in its promoter[11]. Studies from us [4] and others [11] demonstrated that stimulation of the TH gene by hypoxia requires AP-1 transcription factor and involves induction of the *c-fos* gene. Given that catecholamine levels are elevated during intermittent hypoxia[12], it is possible that stimulation of the c-*fos* gene might potentially contribute to increased catecholamine synthesis, via AP-1 transcription factor.

5. ACKNOWLEDGMENTS

This work was supported by National Heart, Lung, and Blood Institute Grant HL-25830.

6. REFERENCES

1. Fletcher, E. C., G. Bao, and C. C. Miller, III. Effect of recurrent episodic hypocapnic, eucapnic and hypercapnic hypoxia on systemic blood pressure. *J. Appl. Physiol.* **78**: 1516-1521, 1995..
2. Fletcher, E. C., R. D. DeBehnke, M. S. lovoi, and A. B. Gorin. Undiagnosed sleep apnea in patients with essential hypertension. Ann. Intern. Med. **103**: 109-195, 1985
3. Bunn H. F., R. O. Pyton. Oxygen sensing and molecular adaptation to hypoxia. *Respir. Physiol.* **95**: 1-10, 1996..
4. Mishra, R., G. Adhikary, M. S. Simonson, N. S. Cherniack, and N. R. Prabhakar Role of *c-fos* in hypoxia-induced AP-1 cis-element activity and tyrosine hydroxylase gene expression. *Mol. Brain. Res.* **59**: 47-83, 1998.
5. Adhikary G., Premkumar, D.R. and N. R. Prabhakar, Dual influence of nitricoxide on gene regulation during hypoxia. *Adv. Exp. Med. Biol.,* **475**: 285-292, 2000.
6. Dalton T. P., Shertzer, G. H. and A. Puga. Regulation of gene expression by reactive oxygen. *Annu. Rev. Pharmacolol. Toxicol.* **39**: 67-101, 1999..
7. Greenberge, H. E., A. L. Sica, S. M. Scharf, and D. A. Ruggiero. Expression of *c-fos* in the rat brainstem after chronic intermittent hypoxia. *Brain Res.* **816**: 638-645, 1999.
8. Prabhakar N. R., B. C. Shenoy, M. S. Simonson, and N. S. Cherniack. Cell selective induction and transcriptional activation of immediate early genes by hypoxia. *Brain Res.* **697**: 266-270, 1995.
9. Semenza G. L. Cellular and Molecular dissection of reperfusion injury ROS within and without. *Circ. Res.* **86**: 117-118, 2000.
10. Angel, P. and M. Karin. The role of Jun and AP-1 complex in cell-proliferation and transformation. *Biochemica et Biophysica Acta*, **1072**: 129-157, 1991.
11. Norris, M.L., D. E., Millhorn. Hypoxia induced protein binding to O2-responsive sequencs on tyrosine hydroxylase *gene. J. Biol. Chem.*, **270**: 23774-23779, 1995.
12. Serebrovskya, T.V., Karaban, I. N., Kolesnikova, E. E., Mishunina, T. M., Kuzminskaya, L. A., Serebrovsky, A. N. and R. J. Swanson. Human hypoxic ventilatory response with blood dopamine content under intermittent hypoxia training. *Can. J. Physiol. Pharmacol.* **77**: 967-973, 1999

RESPONSE SURFACE MODELING OF DRUG INTERACTIONS ON CARDIORESPIRATORY CONTROL

Erik Olofsen, Diederik J.F. Nieuwenhuijs, Elise Y. Sarton, Luc J. Teppema, and Albert Dahan*

1. INTRODUCTION

Causes for perioperative respiratory complications include anesthetic- and opioid-induced reduced ventilatory drive.[1] Previous studies predominantly described the influence of single agents on the performance of the controller. Since in clinical practice an anesthetic is combined with an opioid, we designed a study on the interaction of an inhalational anesthetic, sevoflurane, and an opioid, alfentanil, on ventilation and heart rate during normoxia and acute isocapnic hypoxia. In the next section, we first review possible methodologies to quantify pharmacodynamic interactions. Subsequently we describe the analysis of the experimental data.

2. PHARMACODYNAMIC INTERACTION MODELS

Pharmacodynamic interaction models attempt to describe the relationship between the concentrations C_i of n simultaneously administered drugs and a resulting measurable effect parameter E under study:

$$E = f(C_1, C_2, \cdots, C_n) + \varepsilon \tag{1}$$

where the function f represents the knowledge postulated about the relationship and ε denotes unexplained variability in measurements of the effect. The function f, visualized in n+1–dimensional space, will be referred to as a "response surface".

In the following subsections, we briefly review various possible forms of f for one and two simultaneously present drugs. With a little loss of generality, we assume that drugs have the effect of depression, *i.e.,* they decrease the value E, and that complete depression is possible. Often, and also for the data we will discuss later, it is impossible to take into

*E. Olofsen, Department of Anesthesiology, Leiden University Medical Center, P.O. Box 9600, 2300 RC Leiden, The Netherlands. E-mail: e.olofsen@lumc.nl

Frontiers in Modeling and Control of Breathing, edited by Poon and Kazemi.
Kluwer Academic/Plenum Publishers, New York, 2001.

account all known mechanisms of action, and only empirical modeling is feasible. Even then, especially for the higher–dimensional case, finding a simple and appropriate form of f is not a trivial task. For thorough discussions of the subject, see Refs. 2 and 3.

2.1. Pharmacodynamic Models for a Single Drug

The simplest model to describe the depression from the baseline effect parameter E_0 is the linear one, given by

$$E = E_0 \cdot \left(1 - \left(\frac{C}{C_{50}}\right)/2\right) \tag{2}$$

where C_{50} is the concentration at which 50% depression occurs.

The inhibitory sigmoid E_{max} model is widely used in pharmacology. It originates from the binding of n drug molecules to a receptor molecule given by the reaction $n\mathrm{D}+\mathrm{R} \rightleftharpoons \mathrm{DR}$. Assuming that depression is directly dependent on the amount of bound receptors with respect to the total, one obtains

$$E = E_0 \cdot \left(1 - \frac{[\mathrm{DR}]}{[\mathrm{R}]+[\mathrm{DR}]}\right) = E_0 \cdot \left(1 - \frac{[\mathrm{D}]^n/k}{1+[\mathrm{D}]^n/k}\right) = E_0 \cdot \left(1 - \frac{\left(\frac{C}{C_{50}}\right)^\gamma}{1+\left(\frac{C}{C50}\right)^\gamma}\right) \tag{3}$$

where $k = C_{50}^\gamma$ is the equilibration constant; we replaced n by γ to emphasize that for empirical use n is allowed to be non-integer.

The function discussed by Richards in Ref. 4 attempts to model reaction rates rather than the equilibrium situation, but may be of empirical value. It can be written as

$$E = E_0 \cdot \left(1 - \left(\frac{C}{C_{50}}\right)^\gamma \cdot \left(1 - \left(\frac{1}{2}\right)^\delta\right)\right)^{1/\delta} \tag{4}$$

This function has the interesting property that, with only one parameter extra, it is a generalization of both the linear (when $\delta = 1$ and $\gamma = 1$) and the above sigmoid model (when $\delta = -1$). Note that when $\delta < 0$, E becomes only zero asymptotically when $C \to \infty$. See Figure 1 for examples of concentration–effect relationships.

2.2. Pharmacodynamic Models for Two Drugs

The simplest way to incorporate a second drug is through the assumption that the interaction is additive, *i.e.*, their mechanisms of action are identical but only their potencies are different. For example, the Richards model would generalize to

$$E = E_0 \cdot \left(1 - \left(\frac{C_1}{C_{50,1}} + \frac{C_2}{C_{50,2}}\right)^\gamma \cdot \left(1 - \left(\frac{1}{2}\right)^\delta\right)\right)^{1/\delta} \tag{5}$$

The question now is how to proceed in case we might be dealing with antagonistic or synergistic drug interactions. The criteria an "ideal" pharmacodynamic interaction model should meet are listed in Ref. 5, and a nice review of common drug interaction models is given in its web supplement. Three main approaches are outlined below.

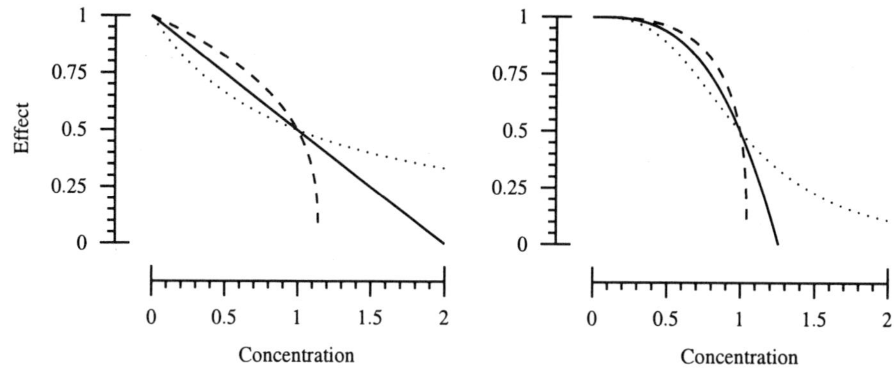

Figure 1. Examples of concentration–effect relationships given by the Richards model (Eq. 4) with $C_{50} = 1$, $\delta = 3$ (dashed), 1 (solid), -1 (dotted) and $\gamma = 1$ (left panel) and $\gamma = 3$ (right panel).

2.2.1. *Logistic Regression Model*

For the case $\delta = -1$ (refer to Eq. (3)), the model is equivalent to the logistic regression model usually describing the probability of the occurrence of an event as a function of the logarithm of concentration. Interaction could then be modeled by adding a second concentration term and an interaction term as is common in logistic regression. Unfortunately, this model predicts profound synergism when the interaction term is zero.

2.2.2. *Isobole Methods*

By generalizing mutually exclusive and mutually non-exclusive binding of two drugs on a receptor[6], the following equation can be derived:

$$E = E_0 \cdot \left(1 - \frac{\left(\frac{C_1}{C_{50,1}}\right)^{\gamma_1} + \left(\frac{C_2}{C_{50,2}}\right)^{\gamma_2} + \alpha \cdot \left(\frac{C_1}{C_{50,1}}\right)^{\gamma_1} \cdot \left(\frac{C_2}{C_{50,2}}\right)^{\gamma_2}}{1 + \left(\frac{C_1}{C_{50,1}}\right)^{\gamma_1} + \left(\frac{C_2}{C_{50,2}}\right)^{\gamma_2} + \alpha \cdot \left(\frac{C_1}{C_{50,1}}\right)^{\gamma_1} \cdot \left(\frac{C_2}{C_{50,2}}\right)^{\gamma_2}} \right) \tag{6}$$

where α is an interaction parameter. Note that isoboles are determined concentration combinations that yield fixed values of the numerator, for example 50%. One problem is that for $\gamma_1 \neq 1$ or $\gamma_2 \neq 1$ the interaction is not purely additive when $\alpha = 0$. Also a solution by Greco *et al.*[7] provides insufficient flexibility.

2.2.3. *Empirical Response Surface Modeling*

In respiration physiology, concentration-effect relationships are often remarkably linear. Here we therefore opt to generalize the Richards model (which is linear when $\delta = \gamma = 1$). To do this, two ideas, discussed in Ref. 5, are fundamental. First, the combination of two drugs is considered to act like a single drug with a certain concentration–effect relationship, and second, its properties, so the parameters of the relationship (here δ and γ) are only dependent on the ratio of the concentrations of the two drugs, given by $Q = U_1/(U_1 + U_2)$, where $U_1 = C_1/C_{50,1}$ and $U_2 = C_2/C_{50,2}$. The generalized Richards model is then given by Eq. (5) with appropriate definitions of the functions $\delta(Q)$ and $\gamma(Q)$ provided.

3. ALFENTANIL–SEVOFLURANE INTERACTION ON CARDIORESPIRATORY CONTROL

This section describes the analysis of alfentanil–sevoflurane interaction data, which has been published in greater detail elsewhere.[8]

3.1. Methods

Nine healthy male volunteers participated after approval of the protocol by the local Human Ethics Committee. The effect of sevoflurane and alfentanil, given separately and in combination, was assessed on resting normoxic ventilation and the changes in ventilation due to acute isocapnic hypoxia (duration of hypoxia = 3 min, $P_{ET}O_2 = 48$ mmHg). The end-tidal concentrations of sevoflurane (C_S) used were 0, 0.1, 0.2 and 0.3%; the target alfentanil plasma concentrations (C_A) used were 0, 10, 20, 30, 40 and 50 ng/ml. The nature of the interaction was assessed for, amongst others, the parameters ventilation ($\dot{V}_i$), heart rate (*HR*) and their dependence on the arterial blood saturation, $\Delta\dot{V}_i/\Delta S_pO_2$ and $\Delta HR/\Delta S_pO_2$. We used the generalized Richards model with free $\gamma(Q) = \gamma$, and an interaction term on δ under the constraint that $\delta(0) = \delta(1) = 1$ to allow for both linearity and nonlinearity, given by

$$E = E_0 \cdot \left(1 - \left(\frac{C_A}{C_{50,A}} + \frac{C_S}{C_{50,S}}\right)^{\gamma} /2 \cdot I(Q)\right) \tag{7}$$

where $I(Q)$ is a spline function with parameters $I_{\max}$ and $Q_{\max}$ which denote the maximum value of the interaction and the value of Q at which $I(Q_{\max}) = I_{\max}$, respectively. The population model parameters and their interindividual variabilities were estimated using NONMEM.[9] The likelihood-ratio test was used to determine statistical significance ($P < 0.01$) of nonlinearity ($\gamma \neq 1$) and synergism ($I_{\max} > 1$).

3.2. Results

Pure additive interaction was found for the depression of $\Delta\dot{V}_i/\Delta S_pO_2$ and $\Delta HR/\Delta S_pO_2$. Depression of $\Delta\dot{V}_i/\Delta S_pO_2$ by 50% occurred at an alfentanil concentration of 31.3 ± 2.8 ng/ml and a sevoflurane concentration of 0.27 ± 0.05 ET% (see Figure 2). Synergistic interaction was found for normoxic $\dot{V}_i$ and *HR*. Depression of $\dot{V}_i$ by 25% occurred at an alfentanil concentration of 37.7 ± 11 ng/ml and a sevoflurane concentration of 0.73 ± 0.44 ET% with interaction parameters $I_{\max} = 1.92 \pm 0.28$ at $Q_{\max} = 0.68 \pm 0.11$ (see Figure 3). Nonlinearities were not present.

3.3. Sensitivity Analysis

Sensitivity analysis can be employed to determine how much a model parameter contributes to the fit. A change of more than 6.64 points in the value of -2 log likelihood

(-2LL) corresponds to a statistical significance of $P < 0.01$.[9] In Figures 2 and 3, -2LL is plotted as a function of γ to assess how well it is established that the concentration–effect relationships are linear.

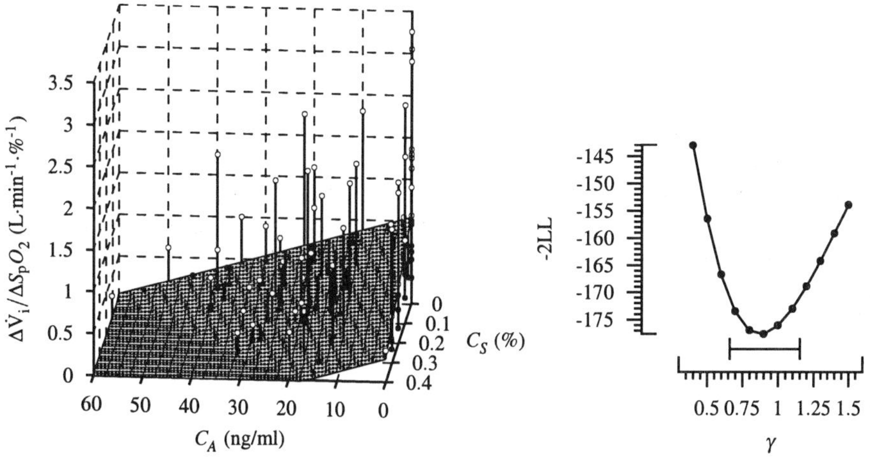

Figure 2. Left panel: response surface model of depression of hypoxic sensitivity as a function of alfentanil and sevoflurane concentration; open and closed circles denote data points above and below the surface respectively. Right panel: sensitivity analysis for γ; see text for explanation.

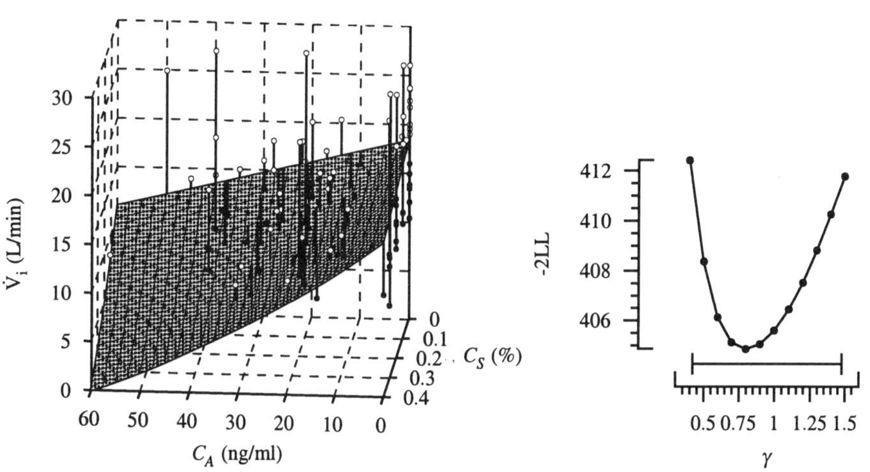

Figure 3. Left panel: Response surface model of ventilatory depression; see legend of Fig. 2 for explanation. Right Panel: sensitivity analysis for γ; see text for explanation.

The depicted confidence intervals determine families of shapes of the concentration–effect relationship (Eq. 7) that are indistinguishable from the optimal ones: for hypoxic sensitivity, the minimum value of -2LL is -177.6, at $\gamma = 1$ it is -176.1, and the confidence interval is approximately (0.66,1.15); for ventilation, the minimum value of -2LL is 404.9, at $\gamma = 1$ it is 405.6 and the confidence interval is approximately (0.42,1.84).

4. CONCLUSIONS

Response surface modeling seems a valuable technique to analyze the interaction of drugs on cardiorespiratory control. In the concentration ranges studied, alfentanil and sevoflurane depress the studied parameters in a dose-dependent linear fashion. The interaction on hypoxic drive is additive, while the interaction on ventilation is synergistic; the interaction on heart rate is of the same nature. The parameters representing responses to hypoxia show greater drug sensitivity than the parameters representing their baseline values.

5. REFERENCES

1. A. Dahan and L. Teppema, Influence of Low-Dose Anaesthetic Agents on Ventilatory Control: Where Do We Stand?, *Br. J. Anaesth.*, **83**(2), 199–201 (1999).
2. G. E. P. Box and N. R. Draper, *Empirical Model-Building and Response Surfaces* (Wiley, New York, 1987).
3. M. C. Berenbaum, What is Synergy?, *Pharmocol. Rev.*, **41**(2), 93–141 (1989).
4. F. J. Richards, A Flexible Growth Function for Empirical Use, *J. Exp. Botany*, **10**(29), 290–300 (1959).
5. C. F. Minto, T. W. Schnider, T. G. Short, K. M. Gregg, A. Gentilini, and S. L. Shafer, Response Surface Model for Anesthetic Drug Interactions, *Anesthesiology*, **92**(6), 1603–1616 (2000).
6. T.-C. Chou and P. Talalay, Generalized Equations for the Analysis of Inhibitions of Michaelis-Menten and Higher-Order Kinetic Systems with Two or More Mutually Exclusive and Nonexclusive Inhibitors, *Eur. J. Biochem.*, **115**(1), 207–216 (1981).
7. W. R. Greco, H. S. Park, and Y. M. Rustum, Application of a New Approach for the Quantification of Drug Synergism to the Combination of Cis-Diamminedichloroplatinum and 1-β-D-Arabinofuranosylcytosine, *Cancer Res.*, **50**(17), 5318-5327 (1990).
8. A. Dahan, D. J. F. Nieuwenhuijs, E. Olofsen, E. Y. Sarton, R. Romberg, and L. J. Teppema, Response Surface Modeling of Alfentanil–Sevoflurane Interaction on Cardiorespiratory Control and Bispectral Index, *Anesthesiology* **94**(6), 982–991 (2001).
9. S. L. Beal and L. B. Sheiner, *NONMEM User's Guides* (NONMEM Project Group, University of California at San Francisco, San Francisco, 1999)

A META-ANALYSIS OF THE EFFECT OF LOW-DOSE VOLATILE ANAESTHETICS ON THE VENTILATORY RESPONSE TO HYPOXIA

Analysis for sources of heterogeneity

Jaideep J. Pandit[*]

1. INTRODUCTION

Knill *et al.* found that most volatile agents at subanaesthetic doses (< 0.2 MAC) depress the acute ventilatory response to hypoxia (AHVR)[7-13]. Most subsequent studies have agreed with this for most agents, but there has been controversy with respect to isoflurane and sevoflurane. With isoflurane, there has been inconsistency between results of different groups. Knill *et al.* found that 0.1 MAC isoflurane depressed AHVR by 50%[13], but Temp *et al.* found no effect[21]. Since it is known that sustained hypoxia can lead to a decline in the ventilatory response (hypoxic ventilatory decline or HVD), the ramp method of inducing hypoxia[13] may have caused an underestimation of the true hypoxic response. When Temp *et al.* measured AHVR using the slow "ramp" and rapid "step" tests in the same subjects, they confirmed that AHVR was lower in the "ramp" tests[22]. Van den Elsen *et al.*[23] suggested that the state of arousal of subjects might also contribute to different results. When external stimuli to subjects were absent (as in Knill's experiments), a depressant effect of isoflurane was observed. When extraneous audiovisual input was used (as in Temp's studies), no such effect was seen[23]. It is unclear, however, why these differences under different experimental conditions have been observed only with isoflurane and not, for example, with enflurane or halothane where different research groups have reported largely consistent results[3,4,7,9-12,14-16]. Pandit *et al.* have argued that the subject's state of arousal alone is unlikely to explain such marked effects on the AHVR[17]. Watching television appears to abolish the marked depression by low-dose anaesthetic[23] yet painful electric shock has no effect[18]. The issues with sevoflurane are different: groups generally agree on the degree of depression of AHVR but this is much less (20 – 36%) than occurs with other agents[17-19]. Different anaesthetic agents may therefore not affect AHVR by a common mechanism.

* Nuffield Department of Anaesthetics John Radcliffe Hospital, Oxford OX3 9DU, United Kingdom.

In summary, the issue is whether the variability in results of different research groups arises as a result of different agents, or as result of different study conditions. The factors which have probably received most attention as potential sources of variability are: ramp vs step input of hypoxia; the level of background CO_2; and the presence or absence of stimulation of subjects. Anaesthetic dose also differs between subjects.

The purpose of this study was twofold. The first was to conduct a meta-analysis of the effect of sub-anaesthetic doses of volatile agents on AHVR, with a view to establishing an overall (or "average") effect. The second aim was to examine whether the group of published reports was statistically homogenous. If not, it was planned to assess whether sources of heterogeneity arose mainly from the use of different agents or of different study conditions, concentrating on the speed of induction of hypoxia, the level of background CO_2, the presence or absence of subject stimulation, and anaesthetic dose. This appears to be the first attempt at meta-analysis of the literature in this field, and the attempt to describe potential sources of heterogeneity of the data is also novel.

2. METHODS

2.1. Data Acquisition

Published articles in English were obtained by MEDLINE and manual search (1966-2000). Inclusion criteria were: (1) study in healthy human subjects; (2) anaesthetic dose < 0.2 MAC; (3) AHVR (litre.min^{-1}) the end-point for both control and test group). Anaesthetic agents studied in only one or two articles were excluded. Abstracts were not included. In total, 21 articles were included, incorporating 205 subjects. In all articles, subjects served as their own controls. Some articles presented the results of one agent studied at different doses. To maximise the data available, each dose was counted as if it were a different study. Thus, one published *article* might yield one *study* at 0.05 MAC and one *study* at 0.1 MAC. In this manner, 21 published *articles* yielded a total of 37 separate *studies* for analysis[3,4,6-25]. In addition to the anaesthetic agent used and its dose, three other study conditions were noted: (a) if there was a ramp or rapid step hypoxic stimulus; (b) whether studies were conducted at isocapnia or hypercapnia; and (c) if there was stimulation of the subjects (either by audio-visual input or pain). A study was marked a "ramp" study either if the method used was that of progressive hypoxia (eg, induced by rebreathing) or used a method in which the minute ventilation failed to rise within one min of the hypoxic challenge. A study was taken to be at isocapnia if the mean end-tidal PCO_2 was held at no greater than 0.26 kPa above natural values.

2.3. Data Analysis

2.3.1. Meta-Analysis of Treatment Effect

The data obtained was in the form of a continuous variable and therefore the method of D'Agostino and Weintraub was used to pool the results[1]. For each of the 37 studies the mean change and SD were computed and from these the "treatment effect", g, and its 95% confidence intervals were calculated. The value g is the standardised difference of the treatment means (with values of the order of 0.2, 0.5, 0.8 considered small, medium and large effects respectively). This procedure yielded 37 values of *g*, one for each study. These values were then combined to obtain a value of *g* (and 95% confidence interval) for the 37 studies as a whole. The *z* statistic was performed on this value. Homogeneity of the data was assessed by (a) an inspection of the graph of results obtained; (b) a formal test of homogeneity which involves computing a $\chi 2$ statistic: if this value is less than the degrees of freedom for the group, this suggests homogeneity within the group[1].

2.3.2. *Analysis for Sources of Heterogeneity*

The standardised treatment effect *g* is the appropriate measure to obtain the overall statistical effect of the meta-analysis. A more intuitive measure and one which is therefore more amenable to examine sources of heterogeneity, is the AHVR in the presence of anaesthetic (AHVR2) expressed as a fraction of the AHVR in controls (AHVR1). AHVR2/AHVR1 ratios were calculated for each study; and the means calculated for each of the four anaesthetic agents; for each of the two speeds of hypoxic stimulus; for each of the two levels of background CO_2; for each of the two levels of subject stimulation and for each of twelve levels of anaesthetic dose. Comparisons between the means were made using analysis of variance (ANOVA) with the fixed factors being: anaesthetic agent (four levels), speed of hypoxic input (two levels), background CO_2 (two levels), stimulation (two levels), and dose (twelve levels). A value of $P < .05$ taken as significant (Minitab for Windows version 9.2; General Linear Model). Significant influences of a factor would indicate that it contributed to heterogeneity.

3. RESULTS

Figure 1 shows the treatment effect, *g*, plotted along with its 95% confidence intervals for each of the 37 studies. We can pose four questions: (1) what is the mean effect of anaesthetic? (2) is there homogeneity? (3) what are the main sources of variation for the results as a whole? (4) for any given agent, what are the sources of variation in results? (1) For all studies taken together, *g* is 0.83 (95% CI = 0.65-1.01; *z* statistic = 9.04, $P < 0.05$). By averaging the AHVR2/AHVR1 across all studies, this corresponds to a significant mean reduction in the control AHVR by anaesthetic of 43% (CI 35% - 51%). (2) The confidence intervals of 26 of 37 studies include a *g* value of zero, which suggests the data is not homogenous. The test of homogeneity yielded a χ^2 statistic of 46.3. This is larger than the degrees of freedom for the group (36) and provides no statistical support for homogeneity. (3) Figure 2 shows the analysis for sources of heterogeneity using the ratios AHVR2/AHVR1. This figure suggests that the greatest variation arises when data are grouped according to agent. ANOVA yielded three significant sources of variability for the group as a whole: anaesthetic agent ($P = 0.001$), stimulation ($P = 0.008$) and dose ($P = 0.009$). (4) However, subject stimulation remained a significant source of variation only for the isoflurane results ($P=0.006$) and not for the halothane, enflurane or sevoflurane results. Dose remained significant only for the halothane and enflurane results but insufficient studies were available to assess its effect with isoflurane or sevoflurane.

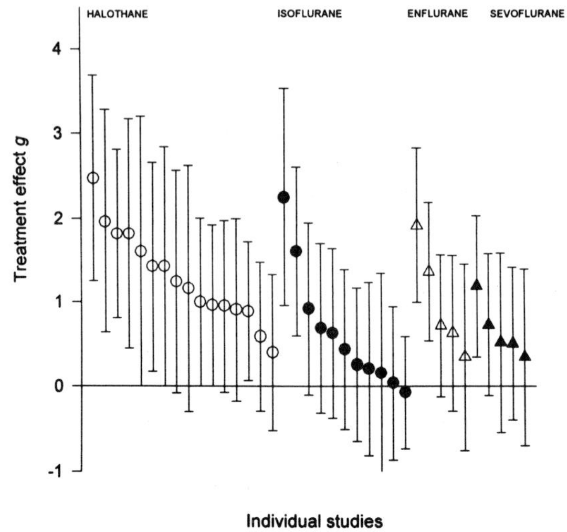

Figure 1. Treatment effect, $g \pm 95\%$ confidence intervals plotted for each of the 37 studies in the meta-analysis. Data are grouped in order of effect for each agent: halothane (○); isoflurane (●); enflurane (△); sevoflurane (▲).

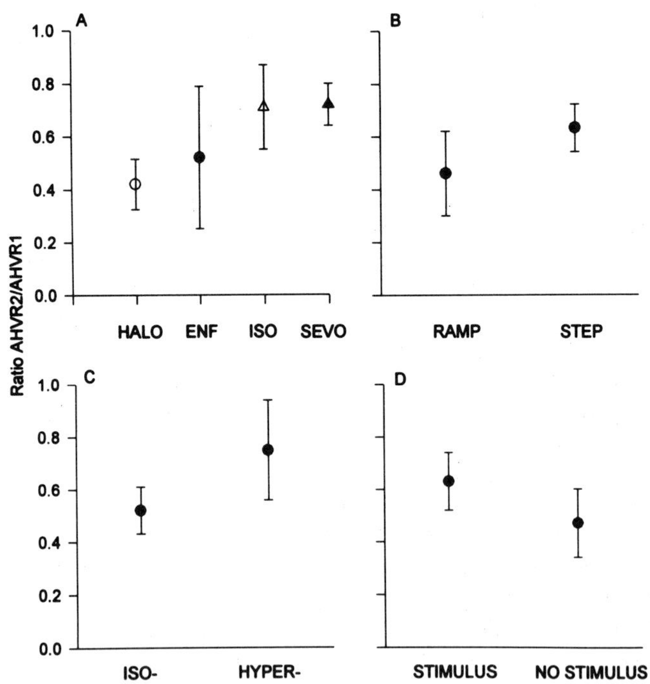

Figure 2. Examining sources of heterogeneity. A: the ratio AHVR2/AHVR1 ± 95% CI , is calculated separately for each of the four anaesthetic agents (symbols as in Figure 1); B: AHVR2/AHVR1 ± 95% CI is calculated for studies using either ramp or step hypoxia; C: AHVR2/AHVR1 ± 95% CI is calculated for studies using either isocapnia or hypercapnia; D AHVR2/AHVR1 ± 95% CI is calculated for studies using either subject stimulation or no stimulation.

4. DISCUSSION

The factor most strongly contributing to variable results across all studies is the anaesthetic agent. Speed of hypoxic input and level of background CO_2 do not influence the results. Subject stimulation has a significant overall effect, but its influence depends on the agent studies. Similarly, dose has a significant effect but its influence is dependent upon the agent.

4.1. Critique of the Methods

This study is subject to the general criticisms of all meta-analyses, which have been debated fully elsewhere[5]. In the main, the technique is used by epidemiologists and public health doctors to determine the effectiveness of a treatment at the population level. The question being asked, therefore, is usually a clinical one. This study differs in that the same technique has been used to address a scientific, rather than a clinical question. Furthermore, the "average effect" (which is the main concern of most meta-analyses) is of less importance here than is the analysis for sources of heterogeneity. In these respects, meta-analysis has been used in a very novel way.

The number of studies was maximised by counting as a separate study each occasion on which a different dose of anaesthetic was used within an article. This might be criticised because such "studies" might not be truly independent with some subjects being included in the analysis more than once, yet the meta-analysis treats them statistically as independent studies. This potential problem is universal to respiratory physiology: 8 of van den Elsen *et al*'s[23] subjects participated in 2 previous studies on halothane[3,4]; and the same subjects appear in papers by Nagyova and colleagues[14-16]. Therefore, regarding studies within a published article as separate is little different from regarding the different articles themselves as independent. The impact of this on the results was assessed: For each article, one study was selected at random and the analysis repeated using only these studies. The results were the same.

Finally, some study conditions have been perhaps artificially grouped under one heading. It is accepted that subject stimulation is not an all-or-nothing phenomenon, but in the absence of an objective quantitative measure or arousal, this seems reasonable. The effect of the exact hypercapnic stimulus on the results was assessed graphically and no relationship between the AHVR2/AHVR1 ratio with anaesthetic and no relationship was found.

4.2. Physiological Significance of the Results

It is perhaps not surprising that the results suggest that different anaesthetic agents have different effects. Individual anaesthetics have different effects on many body systems and their effects are well-documented. Indeed, it might be curious if we supposed that different agents were to have a common mechanism of action on the chemoreflex control of breathing alone.

5. REFERENCES

1. R.B. D'Agostino and M. Weintraub. Meta-analysis: a method for synthesizing research. *Clin Pharm Ther* **58**: 605-16 (1995).
2. A. Dahan and L. Teppema. Editorial 1. Influence of low-dose anaesthetic agents on ventilatory control: where do we stand? *Br J Anaesth* **83:** 199-201 (1999).
3. A. Dahan, M.J.L.J. van den Elsen, A. Berkenbosch, J. DeGoede, I.C.W. Olievier, J.W. van Kleef and J. Bovill. Effects of subanesthetic halothane on the ventilatory responses to hypercapnia and hypoxia in healthy volunteers. *Anesthesiology* **80:** 727-738 (1994).
4. A. Dahan, M.J.L.J. van den Elsen, A. Berkenbosch, J. DeGoede, I.C.W. Olievier, A.G.L. Burm and J.W. van Kleef. Influence of a subanesthetic concentration of halothane on the ventilatory response to step changes into and out of sustained isocapnic hypoxia in healthy volunteers. *Anesthesiology* **81:** 850-859 (1994).
5. A.R. Feinstein. Meta-analysis: statistical alchemy for the 21st century. *J Clin Epid* **48**: 71-79 (1995).
6. I.T. Foo, S.E. Martin, R.J. Lee, G.B. Drummond and P.M. Warren. Interaction of the effects of domperidone and sub-anaesthetic concentrations of isoflurane on the immediate and sustained hypoxic ventilatory response in humans. *Br J Anaesth* **74**: 134-140 (1995).
7. A.W. Gelb and R.L. Knill. Subanesthetic halothane: its effect on regulation of ventilation and relevance to the recovery room. *Can Anesth Soc J* **25**: 488-494 (1978).
8. R.L. Knill and J.L. Clement. Variable effect of anaesthetics on the ventilatory response to hypoxaemia in man. *Can Anaesth Soc J* **29:** 93-99 (1982).
9. Knill RL, Clement JL. Site of selective action of halothane on the peripheral chemoreflex pathway in humans. *Anesthesiology* 1984; **61:** 121-126.
10. R.L. Knill and J.L. Clement. Ventilatory responses to acute metabolic acidemia in humans awake, sedated and anesthetized with halothane. *Anesthesiology* **62:** 745-753 (1985).
11. R.L. Knill and A.W. Gelb. Ventilatory responses to hypoxia and hypercapnia during halothane sedation and anesthesia in man. *Anesthesiology* **49:** 244-251 (1978).
12. R.L. Knill, P.H. Manninen and J.L. Clement. Ventilation and chemoreflexes during enflurane sedation and anaesthesia in man. *Can Anaesth Soc J* **26:** 353-360 (1979).
13. R.L. Knill, H.T. Kieraszewicz and B.G. Dodgson. Chemical regulation of ventilation during isoflurane sedation and anaesthesia in humans. *Can Anaesth Soc J* **30:** 607-614 (1983).
14. B. Nagyova, K.L. Dorrington and P.A. Robbins. Effect of low-dose enflurane on the ventilatory response to hypoxia in humans. *Br J Anaesth* **72:** 509-514 (1994).
15. B. Nagyova, K.L. Dorrington, E.W. Gill and P.A. Robbins. Comparison of the effects of sub-hypnotic concentrations of propofol and halothane on the acute ventilatory response to hypoxia. *Br J Anaesth* **75:** 713-718 (1995).
16. B. Nagyova, K.L. Dorrington, M.J. Poulin and P.A. Robbins. Influence of 0.2 minimum alveolar concentration of enflurane on the ventilatory response to sustained hypoxia in humans. *Br J Anaesth* **78:** 707-713 (1997).
17. J.J. Pandit, J. Manning-Fox, K.L. Dorrington and P.A. Robbins. Effects of subanaesthetic sevoflurane on ventilation. 2: Response to acute and sustained hypoxia in humans. *Br J Anaesth* **83:** 210-6 (1999).
18. E. Sarton, A. Dahan, L. Teppema, M. van den Elsen, E. Olofsen, A. Berkenbosch A and J. van Kleef. Acute pain and central nervous system arousal do not restore impaired hypoxic ventilatory response during sevoflurane sedation. *Anesthesiology* **85:** 295-303 (1996).
19. E. Sarton, M. van der Wal, D. Nieuwenhuijs, L. Teppema, J.L. Robotham and A. Dahan. Sevoflurane-induced reduction of hypoxic drive is sex-independent. *Anesthesiology* **90:** 1288-93 (1999).
20. A. Sollevi and S.G.E. Lindahl. Hypoxic and hypercapnic ventilatory responses during isoflurane sedation and anaesthesia in women. *Acta Anaesthesiol Scand* **39**: 931-938 (1995).
21. J.A. Temp, L.C. Henson and D.S. Ward. Does a subanesthetic concentration of isoflurane blunt the ventilatory response to hypoxia? *Anesthesiology* **77:** 1116-1124 (1992).
22. J.A. Temp, L.C. Henson and D.S. Ward. Effect of a subanesthetic minimum alveolar concentration of isoflurane on two tests of the hypoxic ventilatory response. *Anesthesiology* **80:** 739-750 (1994).
23. M.J.L.J. van den Elsen, A. Dahan, A. Berkenbosch, J. DeGoede, J.W. van Kleef and I.C.W. Olievier. Does subanesthetic isoflurane affect the ventilatory response to acute isocapnic hypoxia in healthy volunteers? *Anesthesiology* **81:** 860-867 (1994).
24. M. van den Elsen, A. Dahan, J. DeGoede, A. Berkenbosch and J. van Kleef. Influence of subanesthetic isoflurane on ventilatory control in humans. *Anesthesiology* **83**: 478-490 (1995).
25. C.H. Young, G.B. Drummond and P.M. Warren. Effect of sub-anaesthetic concentration of halothane on the ventilatory response to sustained hypoxia in healthy humans. *Br J Anaesth* **71:** 642-647 (1993).

SIMULATION OF THE RESPIRATORY CONTROL SYSTEM

James Duffin[*] [†]

1. INTRODUCTION

I describe a simulation of the chemoreflex control of breathing implemented on a personal computer. The simulation is intended as an educational aid to illustrate complex respiratory behaviours and as a test bed for hypotheses concerning chemoreflex control.

The computer display (figure 1) takes the form of a chart recorder, plotting key respiratory variables, with accompanying digital readouts. Once activated, it therefore acts like an accelerated version of a laboratory monitoring system, displaying 20 minutes of simulated time in about 2 minutes. In addition to the chart recorder, the computer display features mouse and keyboard operated controls that allow a number of interventions to be applied as the simulation continues so that the consequences can be observed. These include alterations in the fraction of inspired gases, end-tidal forcing and rebreathing. In addition, several switches may be set to simulate exercise, sleep, hyperventilation and breath holding.

The simulation program was implemented with the LabVIEW programming language (National Instruments) on a personal computer (Microsoft Windows 9x operating system), and may be downloaded from www.utoronto.ca/respgrp/sim.htm. The source code is available from the author on request.

2. SIMULATION OVERVIEW

Operation of the simulation program is simple; once started the simulation displays the changes in the respiratory variables for a simulated time of 20 minutes, or less if the stop button is operated. If the "save file" button has been activated, the user is prompted for a file name to store the data for export to other programs for analysis and display.

* University of Toronto, Department of Physiology, Toronto, Ontario, Canada M5S 1A8.
† University of Toronto, Department of Anaesthesia, Toronto, Ontario, Canada M5S 1A8.

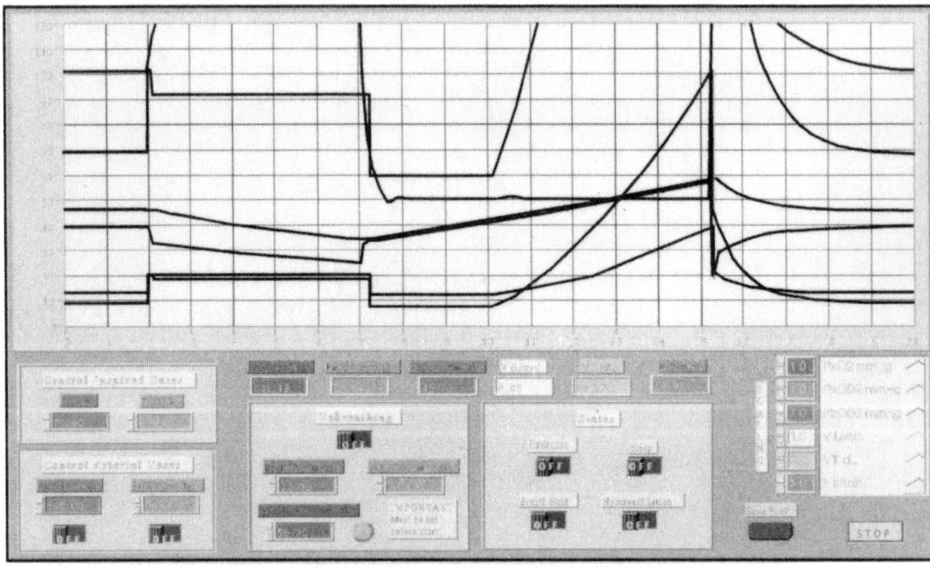

Figure 1. The computer simulation display

As the simulation proceeds, control values for the interventions are entered using either the mouse or keyboard. The digital controls may be altered either by clicking on the arrows beside them or by clicking within the digital control and then using the keyboard to enter new values; a click outside the control or the "enter" key enters the new value so that the user can choose the time that the change should occur. The mouse operates the switches.

Several scenarios can be implemented. The inspired fractions of oxygen and carbon dioxide gases can be altered so that changes in ambient conditions such as ascent to altitude may be simulated. End-tidal forcing protocols may also be simulated, by adjusting the arterial partial pressures of oxygen and carbon dioxide. These particular controls become effective and maintain the arterial gases at the set values when the switches beside them are activated with a click of the mouse. Rebreathing protocols can also be implemented. When the rebreathing switch is operated, ventilation is from a bag containing a pre-set gas mixture, with oxygen added to maintain a pre-set iso-oxic partial pressure. The controls for rebreathing bag initial Po_2 and Pco_2, as well as the iso-oxic Po_2 to be maintained during the rebreathing, must be set before starting the simulation.

The four mouse-operated "state" switches act to produce the following conditions:

- Exercise - a 1 L/min oxygen uptake exercise.
- Sleep - the basal ventilatory drive is withdrawn.
- Breath Hold - pulmonary gas exchange ceases.
- Hyperventilation - basal ventilation is increased to 18.4 L/min.

A number of respiratory variables are displayed on the chart recorder as the simulation continues and their colour and style can be adjusted to suit as well as their scaling. They include:

Pa_{O2}	- Arterial partial pressure of oxygen
Pa_{CO2}	- Arterial partial pressure of carbon dioxide
Pb_{CO2}	- Brain tissue (central) partial pressure of carbon dioxide
V	- Pulmonary ventilation in L/min
VT	- Tidal Volume in dL (these units fit the general scales)
F	- Breathing frequency in breaths/min.

Figure 2 shows a selection of simulation variables exported to a graphing and display program; they illustrate the instability resulting when sleep occurs under hypoxic conditions.

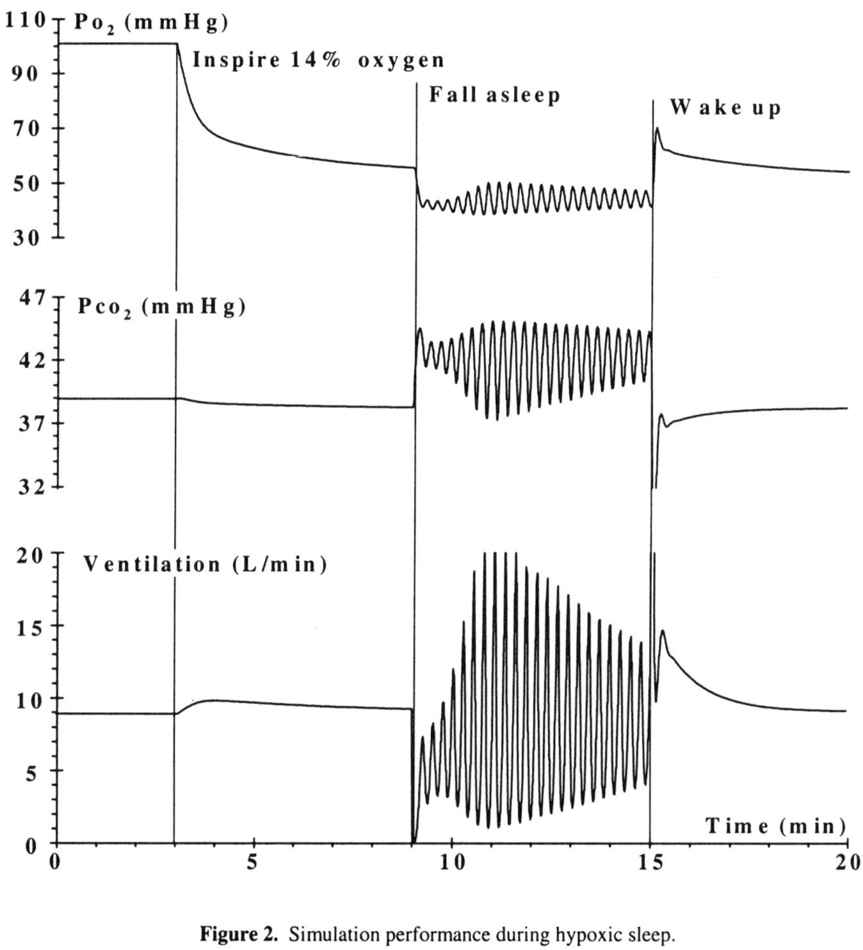

Figure 2. Simulation performance during hypoxic sleep.

3. MODEL OVERVIEW

The controlled system of the model consists of three main compartments, lung, brain tissue and other tissue, each with its own material balance equation for O_2 and CO_2, as well as a rebreathing bag compartment. Dissociation curves for O_2 and CO_2 are modelled with empirically derived equations. Blood flow is distributed between compartments with transit delays and flows dependent on cardiac output (proportional to metabolism), as well as Sa_{O2}. Normal brain blood flow is set proportional to resting cardiac output and the ratio of brain volume to tissue volume, and also depends on Pa_{CO2} and Sa_{O2}. The chemoreflex control equations[1] determine the ventilation, tidal volume and breathing frequency in response to changes in central and peripheral chemoreceptor stimuli. The central chemoreceptor responds to brain compartment $Pbco_2$, and the peripheral chemoreceptors respond to $Paco_2$ and Pao_2 after their own transit delay.

4. DISCUSSION

The model is relatively straightforward with respect to the controlled part of the system, and like others before it adopts a simple few-compartment approach with details like time delays and other refinements limited only by computing power[2-6]. But the more interesting and controversial aspects of the simulation are in the model for the controller part of the loop; that is the chemoreflexes. I recently published the details describing how the chemoreflex model was developed[1], so I will focus here on some of the major facets of its implementation because they highlight gaps in current knowledge.

To model the chemoreflexes I used data from iso-oxic modified rebreathing tests gathered from a large number of subjects. This technique provides a more detailed view of the ventilatory responses than steady state techniques and shows that, particularly during iso-oxic hypoxia, the response is not a single linear one but composed of two portions[7]. It also shows that in the hypocapnic range, the sub-threshold or "basal ventilation" is independent of P_{CO2} and P_{O2}.

The latter observation led to one of the major conclusions from these rebreathing experiments that was incorporated into the model of the chemoreflexes; that hypoxia is not an independent ventilatory stimulus but acts only by changing the peripheral chemoreflex sensitivity to CO_2. This finding contradicts conventional wisdom, but I am convinced by my own experiments that it is the case, basing my conclusion upon observations of a lack of ventilatory response to hypoxia in the hypocapnic range when P_{CO2} is below the chemoreflex threshold[7, 8].

The concept is important because it challenges the idea that a specific "hypoxic ventilatory response" can be measured. The isocapnic P_{CO2} chosen for the hypoxic challenge will determine the magnitude of the response; not only because the response is proportional to the amount by which the isocapnic P_{CO2} chosen exceeds the chemoreflex CO_2 threshold, but also because it is proportional to the chemoreflex sensitivity to CO_2. Thus, for comparisons of hypoxic ventilatory responses in the same subjects to be valid, the isocapnic PCO2 must be the same; differences may then be due to either a change in chemoreflex threshold or a change in the chemoreflex sensitivity, and which of these factors is responsible cannot be determined from the hypoxic ventilatory response. Similarly, comparisons of hypoxic ventilatory responses between different populations such as lowlanders vs. highlanders will be affected by differences in their respective chemoreflex thresholds and sensitivities to CO_2.

To incorporate the model of the chemoreflexes into the simulation, I needed to make several assumptions. First, in order to separate the peripheral and central chemoreflex drives, I assumed that they had the same chemoreceptor threshold, and combine additively to alter ventilation when a drive threshold is exceeded. Information to confirm such assumptions in humans is scant. The two segment ventilatory responses obtained using the modified rebreathing technique were not, as was first hypothesised[7], due to a separation of the peripheral and central chemoreflex responses, but due to a change in the pattern of ventilation where frequency begins to takes over as tidal volume approaches its limits[1]. This assumption appears to be correct for most subjects but perhaps there are separate thresholds in some subjects? Also, whether these two drives add or combine in a more complex way is also as yet undetermined, although the model prediction works correctly for an average subject.

The second assumption, that the peripheral chemoreflex is silenced by hyperoxia, enabled me to use the hyperoxic measured responses as a measure of the central chemoreflex mediated response and determine the peripheral chemoreflex mediated responses by subtraction. This view was held as an established one in the past[9], but has recently come into question. Because of the arterial location of the peripheral chemoreceptors they respond quickly by comparison with the central chemoreceptors located in the medulla, and methods to separate the ventilatory responses mediated by each chemoreceptor rely on temporal separation. One way is to use a two-component model incorporating two time delays and two exponential rises to fit the ventilatory response to a step end-tidal forcing of P_{CO2}; this method finds two responses even in hyperoxia[10]. Another way is to observe the ventilatory response to a brief small increase in inspired P_{CO2} and compare the ventilatory response delays in hypoxia with those of hyperoxia[11]; this method finds that hyperoxia silences the peripheral chemoreceptors. The two methods therefore provide contradictory evidence, perhaps because of the different stimuli used.

The final assumption concerned the interaction of chemoreflex and behavioural ventilatory drives; I assumed they are separate and additive. Each has their own thresholds for affecting ventilation and they do not combine to produce a single ventilatory drive that affects ventilation when a threshold is exceeded, but add to the ventilation independently. Thus, ventilation is constant at a basal value until the chemoreflex drive exceeds a threshold. This assumption is based on observations of the effect of exercise on chemoreflex threshold[12, 13], but it may not apply to ventilatory drives other than exercise such as those related to “wakefulness”.

I conclude by observing that, as always, the modelling exercise demonstrates its value by focussing on the specific assumptions necessary for its implementation, and thereby challenges us to find the missing information experimentally.

5. ACKNOWLEDGMENTS

I acknowledge the expert advice of the Respiratory Research Grouping at the University of Toronto. This research was supported by a grant from the Ontario Thoracic Society.

6. REFERENCES

1. J. Duffin, R. M. Mohan, P. Vasiliou, R. Stephenson, and S. Mahamed, A Model of the Chemoreflex Control of Breathing in Humans: Model Parameters Measurement. *Respir Physiol* **120**, 13-26 (2000).
2. F. S. Grodins, J. Buell, and A. J. Bart, Mathematical analysis and digital simulation of the respiratory control system. *J Appl Physiol* **22**, 260-76 (1967).
3. W. S. Yamamoto, Computer simulation of ventilatory control by both neural and humoral CO2 signals. *Am J Physiol* **238**, R28-35 (1980).
4. G. S. Longobardo, N. S. Cherniack, and B. Gothe, Factors affecting respiratory system stability. *Ann Biomed Eng* **17**, 377-96 (1989).
5. M. C. Khoo, A. Gottschalk, and A. I. Pack, Sleep-induced periodic breathing and apnea: a theoretical study. *J Appl Physiol* **70**, 2014-24 (1991).
6. L. Chiari, G. Avanzolini, and M. Ursino, A comprehensive simulator of the human respiratory system: validation with experimental and simulated data. *Ann Biomed Engng* **25**, 985-99 (1997).
7. R. Mohan, and J. Duffin, The effect of hypoxia on the ventilatory response to carbon dioxide in man. *Respir Physiol* **108**, 101-15 (1997).
8. T. Rapanos, and J. Duffin, The ventilatory response to hypoxia below the carbon dioxide threshold. *Can J Appl Physiol* **22**, 23-36 (1997).
9. D. J. C. Cunningham, Review lecture: studies on arterial chemoreceptors in man. *Journal of Physiology* **384**, 1-26 (1987).
10. M. E. Pedersen, M. Fatemian, and P. A. Robbins, Identification of fast and slow ventilatory responses to carbon dioxide under hypoxic and hyperoxic conditions in humans. *J Physiol (Lond)* **521 Pt 1**, 273-87 (1999).
11. J. G. Whitwam, J. Duffin, and A. Triscott, The use of sodium bicarbonate to test peripheral chemoreceptor activity in man. *Brit J Anaesth* **43**, 1198-1199 (1971).
12. J. Duffin, and G. V. McAvoy, The peripheral-chemoreceptor threshold to carbon dioxide in man. *Journal of Physiology* **406**, 15-26 (1988).
13. K. Casey, J. Duffin, and G. V. McAvoy, The effect of exercise on the central-chemoreceptor threshold in man. *Journal of Physiology* **383**, 9-18 (1987).

LONG-HAUL FLIGHTS MAY INDUCE RESPIRATORY CHANGES SIMILAR TO VENTILATORY ACCLIMATISATION TO ALTITUDE

Marzieh Fatemian, David Y. Kim, Marc J. Poulin, and Peter A. Robbins*

1. INTRODUCTION

Ventilatory acclimatisation to altitude is associated with a progressive increase in ventilation, a progressive decrease in end-tidal P_{CO_2} (PET_{CO_2})[1] and a progressive increase in the acute ventilatory sensitivity to hypoxia (AHVR)[2]. The duration and severity of hypoxia generally associated with these changes are both substantially greater than those associated with commercial airline flight. Nevertheless, commercial airline flight normally involves some reduction in cabin pressure, and this study sought to determine whether the associated reduction in inspired P_{O_2} could also induce acclimatisation.

2. METHODS

Twelve healthy subjects were exposed in a purpose-built chamber[3] where the composition of the inspired gas could be altered and maintained at a desired level. Two protocols were employed for these chamber exposures: protocol H, 8 hours of mild hypoxia (inspired P_{O_2} of 127 Torr corresponding to a median altitude value of 1894m during commercial airline flights[4]) and protocol C, 8 hours of air breathing as a control. Subjects were exposed to each of the two protocols on different days in a random order. In order to detect any effect of the long exposure on the respiratory control, the air-breathing PET_{CO_2} and the ventilatory response to acute hypoxia (AHVR) were measured before and after the chamber exposure. Measurements of AHVR were carried out by imposing a square-wave variation in end-tidal P_{O_2} (6 cycles, each with a period of 2 min, with end-tidal P_{O_2} varying between 100 Torr and 50 Torr) at constant PET_{CO_2}, using an end-tidal forcing system[5, 6]. A dynamic model, relating end-tidal P_{O_2} to ventilation [7], was fitted to the ventilatory data to obtain numerical values for hypoxic sensitivity (Gp) and

* University Laboratory of Physiology, University of Oxford, Oxford OX1 3PT, U.K.

Frontiers in Modeling and Control of Breathing, edited by Poon and Kazemi.
Kluwer Academic/Plenum Publishers, New York, 2001.

euoxic ventilation (Vc). Analysis of variance was used to compare the changes in $P_{ET}CO_2$, Gp and Vc after the chamber exposures.

3. RESULTS

The oxygen saturation of the blood, as determined by pulse oximetry, was 98.0±0.4 % (mean±SD) during the final period of the air-breathing exposure and 95.3±1.1 % during the final period of the exposure to hypoxia. The end-tidal PO_2 was 106.2±3.6 Torr during the final period of the air-breathing exposure and 79.8±2.3 Torr during the final period of the exposure to hypoxia.

An example of the ventilatory responses to the square-wave variation in end-tidal PO_2 imposed before and half-an-hour after the chamber exposures is shown for one subject in Figure 1.

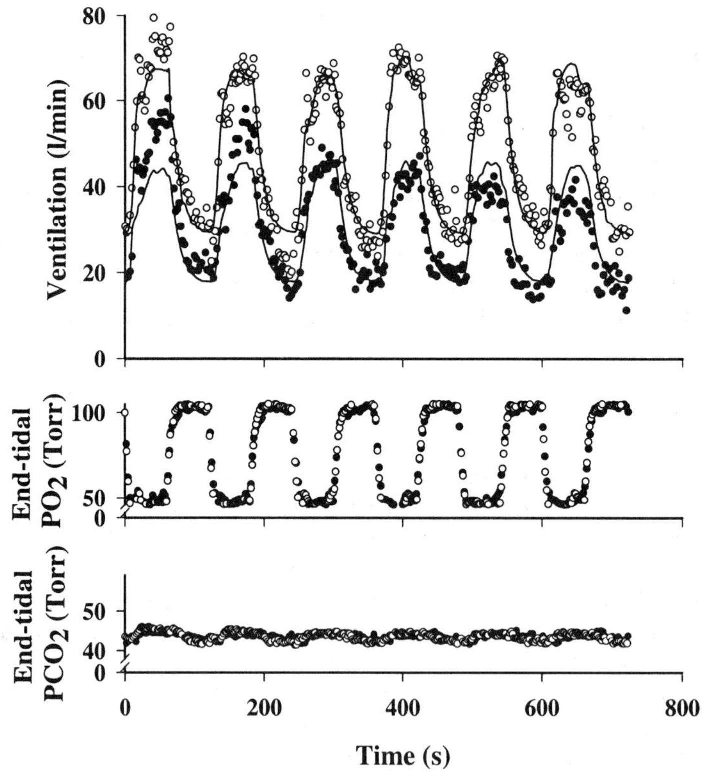

Figure 1. The ventilatory response to acute hypoxia before and after the 8-h hypoxic exposure for one subject. Breath-by-breath ventilation (top), end-tidal PO_2 (middle) and end-tidal PCO_2 (bottom) before exposures (closed symbols) and half-an-hour after exposures (open symbols). Full lines (upper panels) show fit of respiratory model used to estimate ventilatory sensitivity to hypoxia (Gp) and ventilation in absence of hypoxia at fixed end-tidal PCO_2 ($\dot{V}c$).

Following the hypoxic exposure (protocol H), there was a significant reduction in air-breathing $P_{ET_{CO_2}}$ (from 39.2±1.8 Torr to 38.1±1.5 Torr, mean±SD, $p<0.05$), a significant increase in hypoxic sensitivity, Gp, (from 0.84±0.54 l/min/% to 1.13±0.66 l/min/%, $p<0.05$) and a significant increase in Vc (from 11.1±2.4 l/min to 14.7±5.2 l/min, $p<0.05$). There were no significant changes in any of these measurements, following the air-breathing exposure (protocol C), ($P_{ET_{CO_2}}$: 39.2±1.7 Torr vs 39.0±1.7 Torr; Gp: 0.95±0.40 l/min/% vs 0.99±0.45 l/min/%; Vc: 11.6±3.3 l/min vs 11.6±2.5 l/min).

4. CONCLUSIONS

We conclude that the mild hypoxia associated with long-haul airline flights does induce significant acclimatisation within the respiratory control system.

5. ACKNOWLEDGMENT

This study was approved by the Central Oxford Research Ethics Committee and supported by the Wellcome Trust.

6. REFERENCES

1. C. G. Douglas, J. S. Haldane, Y. Henderson, and E. C. Schneider, Physiological observations made on Pike's Peak, Colorado, with special reference to adaptation to low barometric pressure, *Phil. Trans. Roy. Soc.* **B203**, 185-318 (1913).
2. H. V. Forster, J. A. Dempsey, M. L. Birnbaum, W.G. Reddan, J. Thoden, R.F. Grover, and J. Rankin, Effect of chronic exposure to hypoxia on ventilatory response to CO_2 and hypoxia, *J. Appl. Physiol.* **31**, 586-592 (1971).
3. L. S. G. E. Howard, R. A. Barson, B. P. A. Howse, T.R. McGill, M.E. McIntyre, D.F. O'Connor, and P.A. Robbins, A chamber for controlling the end-tidal gas tensions over sustained periods in humans, *J. Appl. Physiol.* **78**, 1088-1091 (1995).
4. J. J. Cottrell, Altitude exposures during aircraft flight, *Chest* **92**, 81-84 (1988).
5. P. A. Robbins, G. D. Swanson, and M. G. Howson, A prediction-correction scheme for forcing alveolar gases along certain time courses, *J. Appl. Physiol.* **52**, 1353-1357 (1982).
6. M. G. Howson, S. Khamnei, M. E. McIntyre, D. F. O'Connor, and P. A. Robbins, A rapid computer-controlled binary gas-mixing system for studies in respiratory control, *J. Physiol.* **394**, 7P (1987).
7. I. D. Clement and P. A. Robbins, Dynamics of the ventilatory response to hypoxia in humans, *Respir. Physiol.* **92**, 253-275 (1993).

SUSTAINED MODERATE HYPEROXIA AUGMENTS THE ACUTE HYPOXIC RESPONSE IN AWAKE GOATS

Jay K. Herman*, Ken D. O'Halloran,† and Gerald E. Bisgard*

1. INTRODUCTION

Sustained exposure to low oxygen induces a time-dependent increase in pulmonary ventilation that is termed ventilatory acclimatization to hypoxia (VAH). The time it takes for this increase in ventilation to reach a stable plateau varies greatly across species but in the goat it occurs in about 4-6 h[1]. Previous reports in the goat have indicated that this process is not only dependent on intact carotid bodies (CB)[2], but also on the progressive increase in CB output that occurs with prolonged hypoxic but not hypercapnic stimulation[3, 4]. It is unclear, however, whether or not acclimatization occurs over an oxygen continuum or whether it is strictly a function of a hypoxic stimulus.

Prolonged exposure to high oxygen levels (≥ 1 atm) results in respiratory impairment as evident by labored breathing[5], blunted responses to hypoxia[6-9] and cyanide[10]. Much of this respiratory impairment is thought to stem from the toxic effects of high oxygen on the ultrastructure and function of both the lung[11] and carotid body[8, 12]. Moderate levels of hyperoxia, on the other hand, have been used clinically to improve oxygenation[13] and in respiratory research to suppress the CB output, without the apparent problems associated with oxygen toxicity. Therefore the purpose of the present study was to use moderate hyperoxia as a means of physiologically suppressing the CB to test the hypothesis that sustained moderate hyperoxia would induce a time-dependent decrease in ventilation that would exhibit a similar time-dependent profile as VAH in the conscious goat model.

2. METHODS

2.1. Animal Preparation

Adult female goats (mean body weight 54 ± 7 kg (± SD)) were used in this study. The goats were prepared, trained and data collected as previously described[14].

---

* Department of Comparative Biosciences, University of Wisconsin, Madison, WI 53706 USA.
Correspondence: JKH, Ph: +1 608 262-2962; fax: +1 608 263-3926; e-mail: HermanJ@svm.vetmed.wisc.edu
† Department of Physiology, University College, Dublin, Dublin2, Ireland.

Frontiers in Modeling and Control of Breathing, edited by Poon and Kazemi.
Kluwer Academic/Plenum Publishers, New York, 2001.

2.2. Protocol

The goats were allowed approximately 20-30 min to adapt to the breathing circuit. Once stable baseline variables were established, 30 min of control respiratory, cardiovascular and blood gas variables were collected. At the end of this period, the inspired gas mixture was adjusted until a stable level of hypoxia was achieved (arterial PO_2 (PaO_2) ≈ 40 torr for all groups). Inspired CO_2 was added and continuously adjusted to maintain arterial PCO_2 ($PaCO_2$) isocapnic to the normoxic baseline $PaCO_2$. The goats were maintained at this isocapnic PaO_2 level for 30 min. Arterial blood samples were taken at a minimum of every 5-min to ensure minimal arterial blood gases fluctuations. The goats were then returned to room air for an additional 30 min to collect a second control period. At this time point the goats were exposed to 1 of 4 protocols consisting of 4 h of either normoxia (room air), hyperoxia (PaO_2 ≈ 245 torr), poikilocapnic hypoxia, or isocapnic hypoxia (PaO_2 ≈ 40 torr for both hypoxic exposures). The protocol is demonstrated graphically on figure 1C. Following this 4 h period, the goats breathed

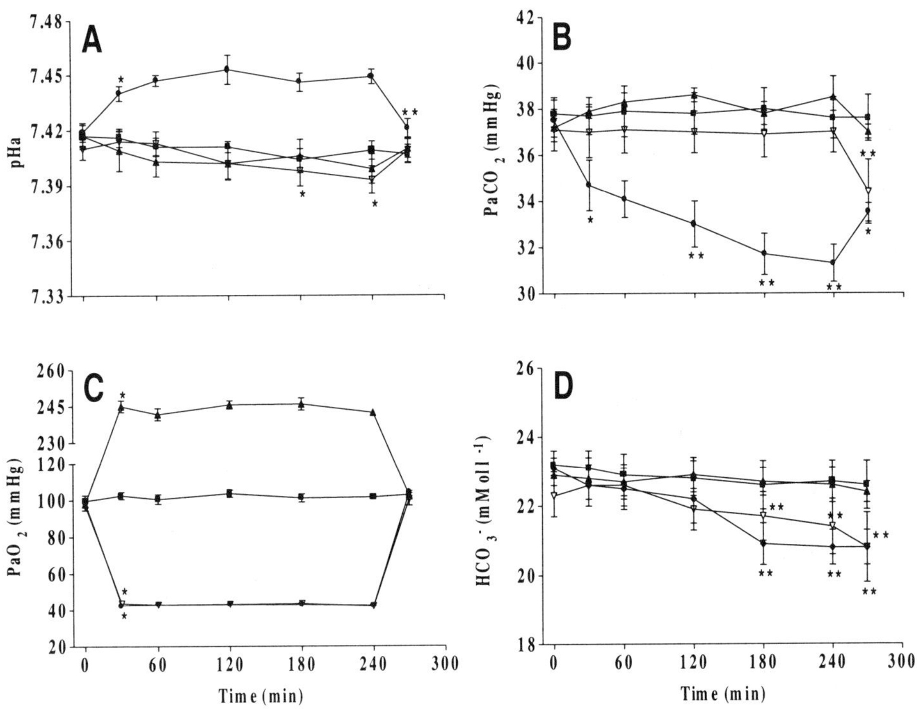

Figure 1. Arterial blood gas and pH before, during and after 4 h of normoxia (■), moderate hyperoxia (▲), poikilocapnic (•) and isocapnic (∇) hypoxia. A) arterial pH. B) arterial PCO_2 C) arterial PO_2 D) HCO_3^- concentration (mMol l^{-1}). * significantly different from baseline control ($p < 0.05$) ** significantly different from acute (30 min) response. Data presented as mean ± sem.

room air for an additional 30 min before being exposed to an additional 30 min of isocapnic hypoxia. Each goat served as its own control and a minimum of two weeks was allowed before the goat was re-tested with a different protocol. All goats were used in each protocol with the exception of one goat that was only used in the normoxic and hyperoxic protocols.

3. RESULTS

Arterial blood gas variables are shown in figure 1. There was a tendency for $PaCO_2$ to increase with sustained hyperoxia but this increase never achieved significance (Fig 1B). Poikilocapnic hypoxia resulted in a significant increase in pHa (Fig 1A) and a time-dependent decrease in $PaCO_2$ (Fig. 1B) and HCO_3^- concentration (mMol· l^{-1}; Fig 1D). Upon return to normoxic conditions there was a $PaCO_2$ increased in the poikilocapnic group although it remained significantly lower than the initial control level. Isocapnic hypoxia resulted in a tendency for pHa to decrease with time (Fig 1A) and this was associated with a fall in the HCO_3^- concentration (Fig 1D). Following the 4 h isocapnic hypoxic protocol, on return to normoxia, $PaCO_2$ fell significantly and was no longer significantly different from the $PaCO_2$ of the poikilocapnic hypoxic group at this time point.

Neither sustained hyperoxia nor normoxia induce any directional change in expired minute ventilation (V_E) throughout the 4 h protocol (Fig. 2). Poikilocapnic hypoxia

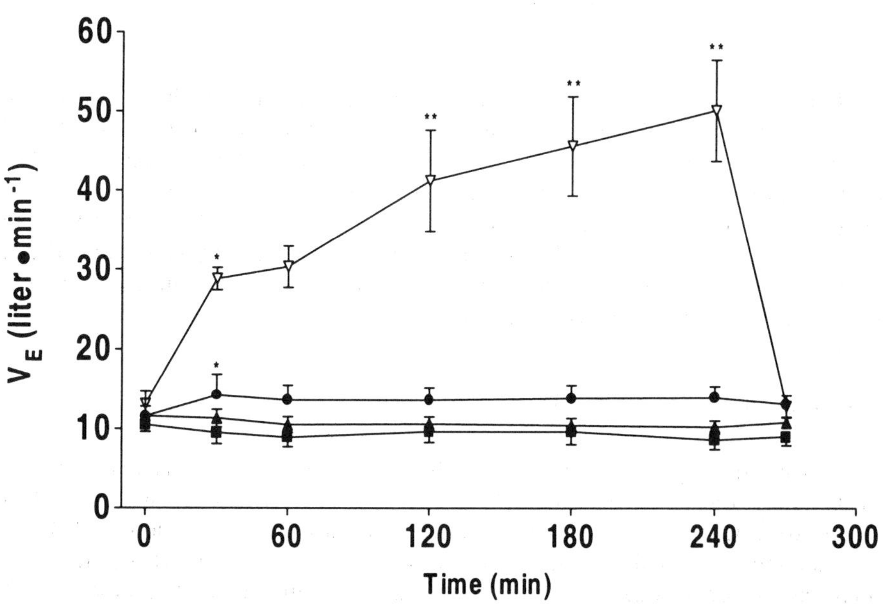

Figure 2. Expired minute ventilation (V_E) before, during and after 4 h of normoxia (■), moderate hyperoxia (▲), poikilocapnic (•) and isocapnic (∇) hypoxia. * significantly different from baseline control ($p < 0.05$) ** significantly different from acute (30 min) response. Data presented as mean ± sem.

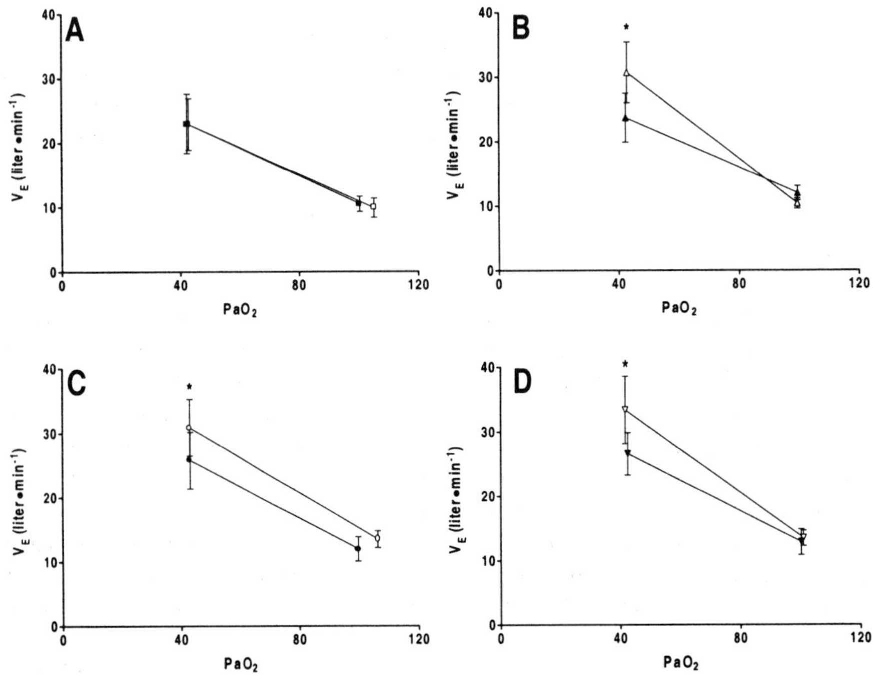

Figure 3. Acute isocapnic hypoxic ventilatory responses before (closed symbol) and after (open symbol) the 4 h protocols. A) Normoxia B) Hyperoxia C) Poikilocapnic hypoxia D) Isocapnic hypoxia. * significantly different from initial response ($p < 0.05$). Data presented as means ± sem.

resulted in a slight elevation in V_E that was maintained during the protocol while isocapnic hypoxia resulted in a time dependent increase in V_E such that by the 2nd – 4th h of hypoxia V_E was significantly elevated above the initial acute hypoxic response. V_E in each protocol returned to the control level following the return to room air.

Four hours of normoxia did not alter the acute hypoxic ventilatory response (Fig. 3A). Similar to what was found following both poikilocapnic hypoxia (Fig 3C) and isocapnic hypoxia (Fig. 3D) sustained hyperoxia resulted in a significant elevation in the acute hypoxic ventilatory response (Fig. 3B).

4. DISCUSSION

The main findings of the present study are as follows: 1) Sustained moderate hyperoxia did not alter V_E but does induce a slight increase in $PaCO_2$ suggesting that a time-dependent decrease in pulmonary ventilation did occur. 2) Baseline $PaCO_2$ following the isocapnic and poikilocapnic hypoxia protocols was similar suggesting that the mechanisms of acclimatization/deacclimatization are similar to both protocols. This conclusion is further supported by the observation that the HCO_3^- concentration decreased similarly in both hypoxic protocols. 3) Sustained moderate hyperoxia results in an augmentation of the acute hypoxic ventilatory response.

The finding that the acute hypoxic ventilatory response was elevated following sustained hyperoxia differs from that reported in humans in which hyperoxia resulted in a blunted acute hypoxic response[15]. It is not clear why these two studies resulted in opposing responses but may reflect either species difference or the slightly higher oxygen levels and longer duration of exposure used by Ren et al[15]. This last scenario seems most probable as our findings are similar to the reports in both humans[16] and rats[17] in which pre-conditioning by breathing 100% oxygen for 10 min augments the acute hypoxic ventilatory response. Gozal[17] further showed that in rats, neuronal nitric oxide synthesis inhibitors could block this augmentation. It is not currently possible to ascertain if reactive oxygen or nitrogen species were involved in augmenting the acute hypoxic response in the goat following the hyperoxic protocol. However, these studies would indicate that moderate levels or short durations of hyperoxic exposure may modulate the ventilatory control system making the chemoreceptive reflex more sensitive to hypoxia.

5. ACKNOWLEDGMENTS

We would like to thank Gordon Johnson for his excellent technical assistance. (Supported by NIH Grants HL 15473 and HL 10069).

6. REFERENCES

1. G. E. Bisgard, and J. A. Neubauer, Peripheral and central effects of hypoxia on control of ventilation, in: *Regulation of Breathing* edited by J. A. Dempsey and A. I. Pack (Dekker, New York, 1995), Second edition. Vol. 79. pp. 617-668.
2. C. A. Smith, G. E. Bisgard, A. M. Nielsen, L. Daristotle, N. S. Kressin, H. V. Forster, and J. A. Dempsey, Carotid bodies are required for ventilatory acclimatization to chronic hypoxia, *J. Appl. Physiol.* **60,** 1003-1010 (1986).
3. G. E. Bisgard, M. A. Busch, L. Daristotle, A. Bressenbrugge, and H. V. Forster, Carotid body hypercapnia does not elicit ventilatory acclimatization in goats, *Respir. Physiol.* **65**, 113-125 (1986).
4. A. M. Nielsen, G. E. Bisgard, and E. H. Vidruk, Carotid chemoreceptor activity during acute and sustained hypoxia in goats, *J. Appl. Physiol.* **65**, 1796-1802 (1988).
5. D. Torbati, A. Mokashi, and S. Lahiri, Effects of acute hyperbaric oxygenation on respiratory control in cats, *J. Appl. Physiol.* **67**, 2351-2356 (1989).
6. R. Arieli, D. Kerem, and Y. Melamed, Hyperoxic exposure affects the ventilatory response to hypoxia in awake rats, *J. Appl. Physiol.* **64**, 181-186 (1988).
7. I. Liberzon, R. Arieli, and D. Kerem, Attenuation of hypoxic ventilation by hyperbaric O2; effects of pressure and exposure time, *J. Appl. Physiol.* **66**, 851-856 (1989).
8. D. Torbati, A. K. Sherpa, S. Lahiri, A. Mokashi, K. H. Albertine, and C. Digiulio, Hyperbaric oxygenation alters carotid body ultrastructure and function, *Respir. Physiol.* **92**, 183-196 (1993).
9. R. Arieli, Normoxic, hyperoxic, and hypoxic ventilation in rats continuously exposed for 60 h. to 1ATA O2, *Aviat. Space Environ. Med.* **65**, 1122-1127 (1994).
10. S. Lahiri, A. Mokashi, M. Shirahata, S. Andronikou, Chemical respiratory control in chronically hyperoxic cats, *Respir. Physiol.* **82**, 201-215 (1990).
11. L. E. Otterbein, L. L. Mantell, and A. M. Choi, Carbon monoxide provides protection against hyperoxic lung injury, *Am. J. Phyiol.* **276**, L688-L694 (1999).
12. C. Di Giulio, M. Di Muzio, G. Sabatino, L. Spoletini, F. Amicarelli, C. Di Ilio, and A. Modesti, Effect of chronic hyperoxia on young and old rat carotid body ultrastructure, *Exp. Gerontol*, **33**, 319-29.

13. R. Greif, O. Akca, E.-P. Horn, A. Kurz, and D. I. Sessler, Supplemental perioperative oxygen to reduce the incidence of surgical-wound infection, *N. Engl. J. Med.* **342**, 161-167 (2000).
14. J. K. Herman, K. D. O'Halloran, G. S. Mitchell, and G. E. Bisgard, Methysergide augments the acute, but not the sustained, hypoxic ventilatory response in goats. *Respir. Physiol.* **118**, 25-37 (1999).
15. X. Ren, M. Fatemian, and P. A. Robbins, Changes in respiratory control in humans induced by 8 h of hyperoxia, *J. Appl. Physiol.* **89**, 655-662 (2000).
16. Y. Honda, H. Tani, A. Masuda, T. Kobayashi, T. Nishino, H. Kimura, S. Masuyama, and T. Kuriyama, Effect of prior O2 breathing on ventilatory response to sustained isocapnic hypoxia in adult humans, *J. Appl. Physiol.* **81**, 1627-1632 (1996).
17. D. Gozal, Potentiation of hypoxic ventilatory response by prior O2 breathing is modulated by nNOS activity in the conscious rat, *J. Appl. Physiol.* **85**, 129-132 (1998).

RESPIRATORY EFFECTS OF BREATHING HIGH OXYGEN DURING INCREMENTAL EXERCISE IN HUMANS

Jaideep J Pandit* and Peter A. Robbins†

1. INTRODUCTION

During incremental exercise to exhaustion, minute ventilation ($\dot{V}_E$) increases in a linear manner with respect to work rate, oxygen consumption ($\dot{V}_{O2}$) and carbon dioxide production ($\dot{V}_{CO2}$), until a threshold level of work rate. After this point, the slope of the $\dot{V}_E$/work rate relationship is steeper than it was at lower work rates. This point is known as the ventilatory anaerobic threshold (VAT). At approximately the same threshold work rate, blood lactate concentration begins to rise significantly from its resting level (the lactate anaerobic threshold, LAT), and this may or may not be causally related to the VAT. Wasserman *et al.*[1] have argued that lactic acidosis contributes to an additional component of the ventilatory response via stimulation of the peripheral chemoreceptors, and that this is an adaptive response in which the ensuing respiratory alkalosis limits the fall in pH generated by metabolic acidosis. Consistent with this hypothesis, it has been observed that subjects who have undergone carotid body resection do not show the characteristic slope of the $\dot{V}_E$/work rate relationship above the anaerobic threshold and lose the inflection point of the VAT[1]. However, subjects with McArdle's syndrome (myophosphorylase deficiency) produce no lactic acid and hence lack a lactate anaerobic threshold, but they nonetheless demonstrate a ventilatory anaerobic threshold[2]. Paterson *et al.*[3] have argued that arterial plasma potassium (K^+), rather than lactic acid, contributes to the hyperventilation above VAT in these (and perhaps also in normal) subjects. Regardless of whether K^+ or lactate is the chemical stimulus to the extra ventilation above VAT, there is general agreement that both these stimuli act via the final common pathway of the carotid bodies.

Henson *et al.*[4] attempted to silence the carotid bodies using an intravenous infusion of dopamine. Their prediction was that hyperventilation above the VAT would be abolished and the ventilatory response would resemble that seen in the carotid body-resected subjects of Wasserman *et al.*[1]. In fact, they found that the shape of the $\dot{V}_E$/work

* Nuffield Department of Anaesthetics, John Radcliffe Hospital, Oxford OX3 9DU, UK
† University Laboratory of Physiology, Parks Road, Oxford OX1 3PT, UK

Frontiers in Modeling and Control of Breathing, edited by Poon and Kazemi.
Kluwer Academic/Plenum Publishers, New York, 2001.

rate relationship was unaffected by dopamine and hyperventilation above VAT still occurred. This would suggest that peripheral chemoreceptor drive is not essential for the characteristic response during incremental exercise. One problem with their study was that in two out of their six subjects dopamine had no discernible effect on the hypoxic response at rest, so that overall, the ventilatory response to hypoxia at rest was only reduced, but not abolished, by dopamine infusion. The carotid bodies were not truly "silenced" by the protocol and may, therefore, have been sufficiently active during incremental exercise to generate the normal response.

The purpose of our study was to repeat the general approach of Henson *et al.*[4]. We planned to use high oxygen (O_2) to overcome some of these problems with dopamine, and to employ repeated studies in the same subjects. We selected an inspired fraction of 66% O_2 (rather than 100% O_2) for two reasons. First, 100% O_2 is known to stimulate ventilation independently, possibly due to the Haldane effect raising CO_2 levels in the central chemoreceptors[5]. Second, Bannister & Cunningham reported better exercise performance using 66% rather than 100% O_2[6]. Our predictions were: (1) if the $\dot{V}_E$/work rate relationship was altered, then this would support the interpretation of Wasserman and colleagues concerning the role of carotid bodies in incremental exercise; (2) if the $\dot{V}_E$/work rate relationship was unaltered, then this would suggest either that mechanisms other than carotid bodies are sufficient to generate the normal ventilatory response to incremental exercise, or that the carotid bodies are not fully silenced by high inspired O_2.

2. METHODS

Six healthy male subjects participated in the study. Their mean age, height and weight (± SD) were: 20 ± 1 years; 1.8 ± 0.6 m; 72 ± 11 kg. They were asked to refrain from drinking coffee or alcohol prior to study. This study had the approval of the Central Oxford Research Ethics Committee. Subjects wore noseclips and breathed through a mouthpiece which was connected to the gas supply via low resistance tubing. The total gas flow was maintained at a level high enough to prevent rebreathing (up to 180 L/min). Expired gas at the mouth was sampled for end-tidal partial pressures of O_2 and CO_2 (PET_{O2}, PET_{CO2}) by a mass spectrometer. Respiratory volumes were measured by a turbine volume measuring device and flows by a pneumotachograph in series with the mouthpiece. Data concerning experimental time, gas flow and gas composition were stored every 20 msec in a data acquisition computer. Oxygen consumption ($\dot{V}_{O2}$) and carbon dioxide production ($\dot{V}_{CO2}$) were calculated on a breath-by-breath basis (STPD) as described previously[7]. Subjects exercised on a Mijnhardt KEM-3 electromagnetically-braked cycle ergometer. A heparinised venous blood sample was taken before and at the end of each period of exercise for determination of plasma lactate

Each subject undertook the protocol breathing air four times and breathing 66% O_2 four times, yielding a total of eight experimental periods for each subject. Exercise testing occurred on different days over a period of two months and the order of air breathing and O_2 breathing tests was random. At the start of each experimental period, the subject sat on the cycle ergometer and, for the relevant protocol, oxygen was introduced. After 90 sec, the subject undertook a 30 sec period of unloaded pedalling. Then, the work rate was increased in ramps of 20 W every min. Subjects were asked to keep the pedal rate between 50 and 70 per min.

A bilinear relationship was fitted to the breath by breath data of ventilation against time for each experimental period, to obtain values for (a) the inflection point (ie, VAT), (b) the slope below the VAT (slope 1) and (c) the slope above the VAT (slope 2). An adaptation of the method of Beaver *et al.*[8] was used. Briefly, a guess was made "by eye" of the VAT inflection point in the $\dot{V}_E$-time plot and a linear relationship fitted to the data above and below this point. The mean sum of squared error of the fit was calculated. Then, different inflection points were selected at intervals either side of the first selected point, and the process repeated. The final VAT (expressed as a point on the time axis in min) and slopes accepted were those which yielded the smallest mean sum of squared error. We used a standard subroutine for minimising the sum of squared residuals (subroutine E04FDF, Numerical Algorithms Group, Oxford, UK). The values so obtained for VAT, slope 1 and slope 2 were averaged to obtain the mean for each subject, and these subject values used to obtain the mean for the group for each of the two protocols. The significance of the differences between the group mean values of the air and O_2 protocols was assessed using Student's t-test for the following variables: VAT; slope 1; slope 2; peak work rate achieved; change in peak lactate. The null hypothesis was that O_2 does not affect these variables. Statistical significance was accepted at the $P < 0.05$ level.

3. RESULTS

Figure 1 shows the result from one subject (845). Table 1 shows the results for the group for all variables.

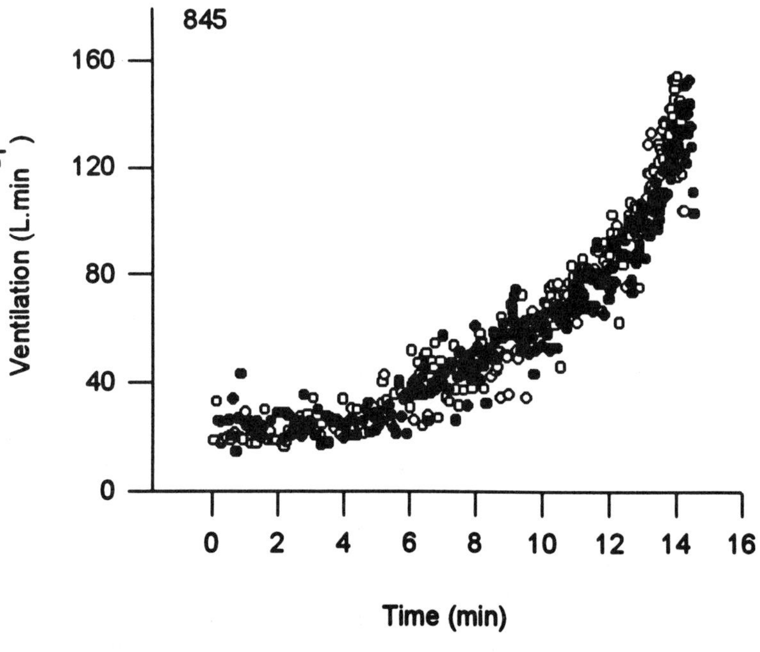

Figure 1. Plot of ventilation against time for one example subject for air breathing (○) and O_2 breathing (•).

Table 1. Estimates using bilinear model fit of ventilatory anaerobic threshold (VAT), slopes, peak work rate and change in lactate (ΔLactate) for air and O_2. None of the comparisons between values for air and O_2 for each variable was significant (Student's t-test)

Subject	Slope 1 ($L.min^{-2}$)		Slope 2 (min)		AT (W)		Peak work rate (mmol)		Δ Lactate ($L.min^{-2}$)	
	Air	O_2	Air	O_2	Air	O_2	Air	O_2	Air	O_2
835	2.51	3.19	15.0	14.7	9.00	9.52	335	330	12.7	11.9
842	4.23	3.55	16.5	18.9	10.6	12.2	315	345	9.45	10.5
845	4.79	3.94	17.2	18.1	9.73	11.1	275	285	9.17	10.3
848	3.15	2.74	16.8	13.5	8.70	8.81	260	270	8.30	8.90
849	6.39	5.68	17.2	17.4	10.1	10.3	285	285	10.9	8.88
853	4.98	4.66	14.9	13.1	10.6	11.1	290	305	9.47	8.20
Mean (SD)	4.34 (1.39)	3.96 (1.07)	16.3 (1.05)	16.0 (2.49)	9.79 (0.80)	10.5 (1.22)	293 (27)	303 (29)	10.0 (1.56)	9.78 (1.37)

4. DISCUSSION

The main result of this study is that breathing 66% O_2 does not change the nature of the ventilatory response during incremental exercise to exhaustion. Oxygen did not affect peak work rate or lactate production, so we can conclude that the results did not arise due to differences in degree of exercise intensity between the protocols.

Our result is largely consistent with that of Henson *et al.*[4]. However, it differs from the result predicted from the observations of Wasserman *et al.*[1] on carotid body-resected subjects. Two possible interpretations of our finding are either (a) other mechanisms not involving the carotid bodies ensure that a normal ventilatory response to incremental exercise occurs; (b) the carotid bodies are not silenced by 66% O_2.

It is surprising that there are relatively few studies which have examined the ventilatory responses to incremental exercise under conditions of altered carotid body drive. Hughes *et al.*[9] assessed the effect of 31% O_2 during incremental exercise and could find no significant effect, but this might have been because 31% O_2 is not sufficient to silence the carotid bodies. Hogan *et al.*[10] investigated the effect of 61% O_2 on the lactate anaerobic threshold in six subjects and found it to be unchanged. They did not discuss the implications of their finding for ventilatory control, but it is possible to calculate from their data that VAT was also unchanged by O_2. In a brief communication, Henson *et al.*[11] have reported that breathing 100% O_2 did not affect the anaerobic threshold or maximum ventilation during incremental exercise, which is in agreement with our results, but the full details of their experiment are unpublished (D.S. Ward, personal communication).

Taken together with our result, these findings seem to be at odds with the prediction from carotid body-resected subjects. One interpretation is that there are mechanisms other than peripheral chemoreceptors, which can still generate the normal ventilatory response to incremental exercise, though it is unclear which precise mechanisms might be involved.

Another interpretation is that high O_2 does not sufficiently reduce the activity of the carotid bodies and there is increasing evidence that this might be the case. High O_2 might abolish the carotid body response to acidosis (H^+)[12] and to arterial potassium (K^+)[13], but there are other putative stimuli to breathe in exercise, such as cathecholamines, adenosine and oscillations (rather than mean values) of arterial pH and P_{CO2}. It is not established whether O_2 abolishes the carotid body response to these stimuli. There is also evidence that high O_2 may not abolish the response to CO_2. Animal studies clearly demonstrate a residual CO_2 response of afferent carotid body discharge[14]. Recently, Pedersen *et al.*[15] have employed multi-frequency binary sequence (MFBS) exposures of hypercapnia in humans and observed that a rapid (presumably peripheral) component of the ventilatory response to CO_2 persists against background hyperoxia. This, second, interpretation of our result would have an important implication for methodology in respiratory physiology studies. The magnitude of the decline in ventilation on sudden introduction of O_2 during steady exercise has long been taken to represent the activity of the carotid bodies[16]. If, in fact, O_2 at best only reduces the carotid body response to certain stimuli, but does not abolish it to all stimuli, it is unclear precisely which aspect of carotid body function this "Dejours test" is actually measuring.

5. REFERENCES

1. K. Wasserman, B.J. Whipp, S.N. Koyal, and M.G. Cleary, Effect of carotid body resection on ventilatory and acid-base control during exercise, *J. Appl. Physiol* **39**, 354-358 (1975).
2. J.M. Hagberg, E.F. Coyle, J.E. Carroll, J.M. Miller, W.H. Martin, and M.H. Brooke, Exercise hyperventilation in patients with McArdle's disease, *J. Appl. Physiol* **52,** 991- 994 (1982).
3. D.J. Paterson, J.S. Friedland, D.A. Bascom, I.D. Clement, D.A. Cunningham, R. Painter, and P.A. Robbins, Changes in arterial K+ and ventilation during exercise in normal subjects and subjects with McArdle's syndrome., *J. Physiol.* **429**, 339-348 (1990).
4. L.C. Henson, D.S. Ward, and B.J. Whipp. Effect of dopamine on ventilatory response to incremental exercise in man, *Respir. Physiol.* **89**, 209-224 (1992).
5. M.J. Miller, and S.M. Tenney, Hyperoxic hyperventilation in carotid-deafferented cats, *Respir. Physiol.* **23**:, 23-30 (1975).
6. R.G. Bannister, and D.J.C. Cunningham, The effects on the respiration and performance during exercise of adding oxygen to the inspired air, *J. Physiol.* **125**, 118-137 (1954).
7. J.J. Pandit, and P.A. Robbins, Ventilation and gas exchange during sustained exercise at normal and raised CO_2 in man, *Resp. Physiol.* **88**, 101-112 (1992).
8. W.L. Beaver, K. Wasserman, and B.J. Whipp, A new method for detecting anaerobic threshold by gas exchange, *J. Appl. Physiol* **60**, 2020-2027 (1986).
9. R.L. Hughes, M. Clode, R.H.T. Edwards, T.J. Goodwin, and N.L. Jones, The effect of inspired O2 on cardiopulmonary and metabolic responses to exercise in man, *J. Appl. Physiol* **24**, 336-347 (1968).
10. M.C. Hogan, R.H. Cox, and H.G. Welch. Lactate accumulation during incremental exercise with varied inspired oxygen fractions, *J. Appl. Physiol.* 55: 1134-1140 (1983).
11. L.C. Henson, T.T. Nguyen, B.J. Whipp, and D.S. Ward. Effect of hyperoxia on ventilation above lactate threshold, *FASEB J.* **6**, 1370 (1992).

12. M. Pokorski, and S. Lahiri, Aortic and carotid chemoreceptor responses to metabolic acidosis in the cat, *Am. J. Physiol.* **244**, R652-R658 (1983).
13. R.E. Burger, J.A. Estavillo, P. Kumar, P.C.G. Nye, and D.J. Paterson, Effects of potassium, oxygen, and carbon dioxide on steady-state discharge of cat carotid body receptors, *J. Physiol.* **401**, 519-531 (1988).
14. J.H.G.M. Van Beek, A. Berkenbosch, J. DeGoede, and C.N. Olievier, Influence of peripheral O_2 tension on the ventilatory response to CO_2 in cats, *Respir. Physiol.* **51**, 379-390 (1983).
15. M.E.F. Pedersen, M. Fatemian, and P.A. Robbins. Identification of fast and slow ventilatory responses to carbon dioxide under hypoxic and hyperoxic conditions in humans, *J. Physiol* **521**, 273-287 (1999).
16. P. Dejours, Intérêt méthodologique de l'étude d'un organisme vivant à la phase initiale de rupture d'un équilibre physiologique. *Comptes rendus des Séances de l'Académie des Sciences Paris* **245**, 1946-1948 (1957).

INTRA-BREATH PROFILES OF ALVEOLAR GAS EXCHANGE AND THE INITIAL LIMB-TO-LUNG TRANSIT DELAY DURING EXERCISE IN HUMANS

Relevance for modeling the exercise hyperpnea

H. B. Rossiter*, N. P. Marshall†, S. A. Ward‡, and B. J. Whipp*

1. INTRODUCTION

Following the onset of constant load exercise, pulmonary ventilation ($\dot{V}_E$), O_2 uptake ($\dot{V}O_2$) and CO_2 output ($\dot{V}CO_2$) all increase exponentially as a function of time, subsequent to the limb-to-lung transit time which is termed phase I (ϕI)[1]. The onset of the exponential phase (ϕII) for $\dot{V}O_2$ and $\dot{V}CO_2$ is "consequent" to the arrival of altered mixed venous gas tensions at the pulmonary capillary bed resulting from increased muscle metabolic rate[1]. It is not known, however, whether the onset of the ϕII $\dot{V}_E$ coincides with that of $\dot{V}O_2$ and $\dot{V}CO_2$, suggestive of a 'pulmonary' mechanism of mediation, or whether its onset is delayed i.e. consistent with a transit delay to downstream sites of chemoreception (e.g. carotid bodies) - evidence that is crucial for assembling a mechanistic control model of the transient.

It has been suggested that the time after exercise onset when ϕI ends and ϕII begins (the limb-to-lung transit delay) may be determined from the onset of the fall in respiratory exchange ratio (R) after the start of exercise. However, the digital nature of the breath (e.g. figure 1) and the breath-to-breath noise[2] reduces the fidelity of the transit time determination: there is, to date, no sufficiently good estimate of this transit time.

In order to investigate the immediacy of the onset of the cardiodynamic increases in gas exchange, Weissman *et al.*[3] used a prolonged exhalation that spanned the exercise onset. They found that increases in gas exchange consistent with increases in pulmonary perfusion could be detected after a short delay of ~1.5 s.

Weissman *et al.*[3] measured this increased gas exchange as increased rates of change of PO_2 and PCO_2 at the mouth during the 'alveolar' phase of the intra-breath gas profile.

* Department of Physiology, St. George's Hospital Medical School, Tooting, London, SW17 ORE, UK
† Sport and Exercise Science Research Centre, South Bank University, London SE1 0AA, UK
‡ Centre for Exercise Science and Medicine, IBLS, University of Glasgow, Glasgow, G12 8QQ

Frontiers in Modeling and Control of Breathing, edited by Poon and Kazemi.
Kluwer Academic/Plenum Publishers, New York, 2001.

However, a subsequent change in the composition of mixed venous blood entering the pulmonary capillary bed should also be detected in the same fashion i.e. the result of blood from the muscle reaching the lung having undergone gas exchange at the onset of exercise. Alterations in the intra-breath gas profiles of exhaled gas should, therefore, be detectable during a prolonged exhalation timed to span the expected limb-to-lung transit delay - ensuring that the closing volume does not confound the interpretation. The closing volume is manifest towards the end of an exhalation, approaching residual volume, and results in a greater contribution to the expirate from the apical areas of the lung with a high $\dot{V}_A / \dot{Q}$. The closing volume, therefore, causes a deflection in the expired gas tensions opposite to those expected from altered mixed venous gas tensions due to exercise, i.e. closing volume causes the alveolar profile of PO_2 to increase and PCO_2 to decrease, whereas increased muscle metabolism causes the PO_2 profile to decrease and PCO_2 to increase.

2. METHODS

In order to characterise the duration of ϕI more precisely from the intra-breath profiles of P_AO_2 and P_ACO_2, six healthy male subjects (aged 22 to 59) performed repeated prolonged exhalations after the onset of exercise (under informed consent as approved by the Local Research Ethics Committee). The prolonged exhalation began on the second breath after exercise onset, ending at residual volume (RV). Figure 1 presents a schematic of the protocol employed, where the alveolar slope ($P_AO_2.s^{-1}$) was measured at rest (*slope a*) and during exercise (*slope b*). The transit time was determined by the interval between exercise onset and the deflection in $P_AO_2.s^{-1}$. The subjects were given visual feedback to aid the maintenance of a constant expiratory flow. Each subject also performed the prolonged exhalations at rest and during the exercise steady state, where mixed venous gas tensions were not expected to be changing. Each subject subsequently performed a single-breath nitrogen washout test as an index of the normalcy of airflow distribution.

The profiles of CO_2, O_2 and N_2 partial pressures in the alveolar phase of the exhalations were determined by mass spectrometry (*QP9000*, CaSE, Gillingham, Kent) and the inspired and expired volumes were measured by a turbine volume sensor (Interface Associates, CA, USA). Exercise was initiated on command from rest on an electromagnetically braked cycle ergometer (*Excaliber*, Lode, NL), with the flywheel 'pre-spun' and rotating at 60 rpm. Other variables recorded were six-lead ECG (*Q5000*, Quinton, USA) and oxygen saturation (*Biox 3740*, Ohmeda, CO, USA).

3. RESULTS

During the prolonged exhalation an acceleration in the rate of the P_AO_2 decline was evident beginning at an average of 11.7s (S.D. ± 1.35) and 10.0s (S.D. ± 2.01) after the onset of 70W and 200W cycling respectively, with no apparent change in the profile of P_ACO_2 increase (Figure 2). Surprisingly, however, an acceleration in the P_AO_2 decline was also evident during the control breaths at rest and during 70W steady-state exercise, i.e. when mixed-venous gas tensions were presumably constant (Figure 3).

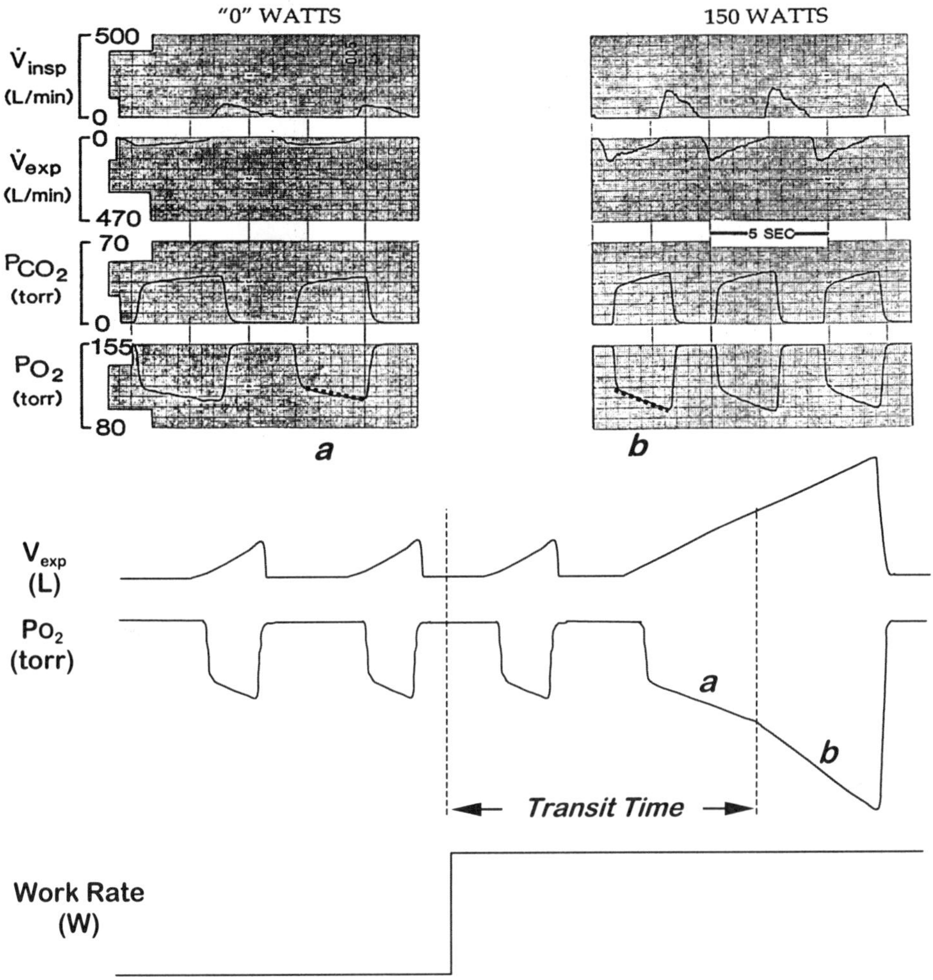

Figure 1. On-line data (top) show the typical breathing pattern during the steady-state of unloaded and loaded cycling, the slope of the alveolar gas phase of exhalation increasing from ***a*** to ***b***. Schematic illustration of the methods used for measurement of the limb-to-lung transit delay (bottom). The prolonged exhalation was performed after the onset of exercise to span the likely transit delay at which point the slope of the P_AO_2 change was expected to increase from slope ***a*** towards slope ***b***.

The steepening of the alveolar Po_2 slope "mid expiration" was independent of metabolic rate, averaging a 67% (S.D. ± 13.7) increase in the rate of decline ($\Delta P_AO_2.s^{-1}$). These findings were highly reproducible, with each of four repeats for each subject typically producing the same result. Indeed, an acceleration in $\Delta P_AO_2.s^{-1}$ was observed in 68 of the 75 prolonged exhalations performed.

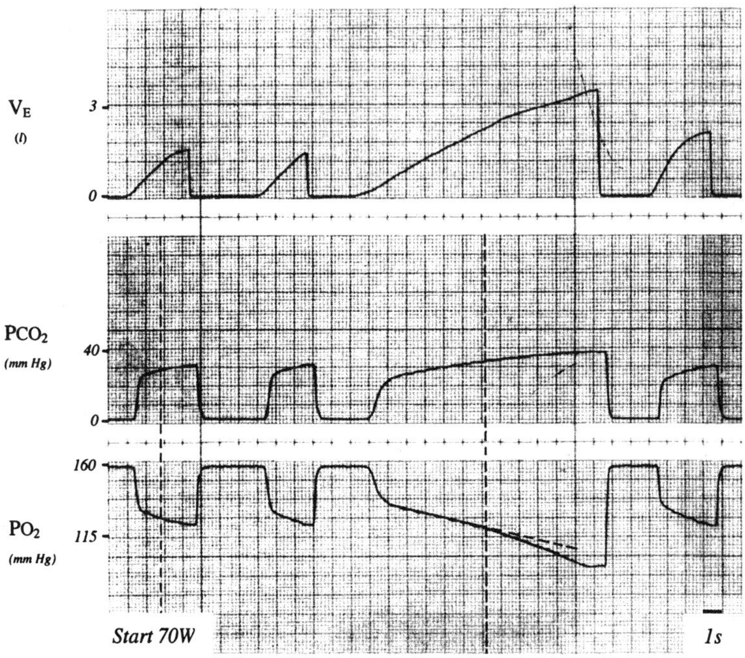

Figure 2. A representative example of a prolonged exhalation performed on the second breath after the onset of exercise (70 W). Note that the P_AO_2 slope in the alveolar phase of the prolonged exhalation steepens (indicated by the vertical dotted line) such that the rate of fall is greater resulting in a more rapid decrease of intra-breath R (Kim *et al.*[4]). Also note the effect of the closing volume on P_AO_2 during the last second of prolonged exhalation.

Consequently, R during the prolonged exhalation (computed by the method of Kim *et al.*[4]) began to decline more rapidly at approximately mid-exhalation for both the exercise-transition breaths and the control steady-state breaths (Table 1). This occurred at an R value which was not significantly different among the conditions and which averaged 0.5 (± 0.14); i.e. a value appreciably greater than that of 0.32 used to estimate "true" mixed venous PCO_2[4, 5].

Table 1. Acceleration of intra-breath R ($R.l^{-1}$) during prolonged exhalation[a] (pre- and post-deflection) at rest, at the onset and during steady-state exercise (70 W) in six subjects: values are subject mean[b] (± S.D.)

	Rest		Rest to 70 W		70 W steady-state	
Subj.	Pre	Post	Pre	Post	Pre	Post
1	0.252 (0.1)	0.414 (0.2)	0.192 (0.1)	0.473 (0.2)	0.152 (0.1)	0.556 (0.4)
2	0.260 (0.0)	0.550 (0.0)	0.162 (0.0)	0.430 (0.2)	0.245 (0.1)	0.320 (0.1)
3	0.122 (0.0)	0.227 (0.1)	0.179 (0.1)	0.458 (0.2)	0.164 (0.0)	0.352 (0.0)
4	0.210 (0.0)	0.640 (0.0)	0.178 (0.1)	0.405 (0.1)	0.163 (0.1)	0.303 (0.0)
5	0.367 (0.1)	0.757 (02)	0.225 (0.1)	0.395 (0.0)	0.245 (0.1)	0.320 (0.1)
6	0.341 (0.2)	0.724 (0.4)	0.307 (0.1)	0.588 (0.2)	0.414 (0.0)	0.596 (0.1)
Mean	0.257 (0.1)	0.552 (0.2)	0.207 (0.1)	0.458 (0.2)	0.231 (0.1)	0.409 (0.1)

[a] $R = s - (F_IO_2 * s) - F_ICO_2 / 1 - (F_IO_2 * s) - F_ICO_2$, where s is the slope of the R-line at particular values of P_AO_2 and P_ACO_2 from Kim *et al.*[4]. [b] mean of *n* repeats for each subject where $n = 3$ (± 1.1).

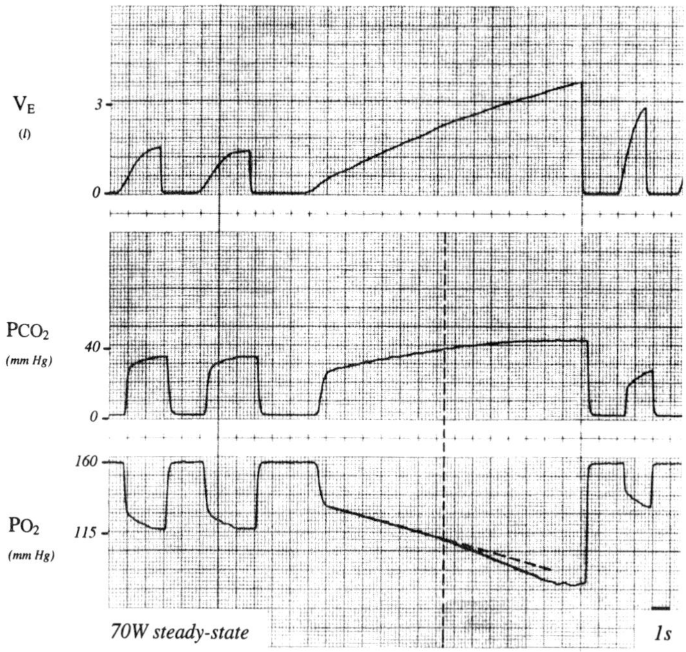

Figure 3. A representative example of a prolonged exhalation performed during the steady state of exercise (70 W). The P_AO_2 slope in the alveolar phase of the prolonged exhalation steepens (indicated by the vertical dotted line) even during a region where pulmonary blood flow and mixed venous gas tension are not expected to be changing. Also note the effect of the closing volume on the P_AO_2 during the last second of prolonged exhalation.

4. DISCUSSION

Without the recognition that the accelerated decline in P_AO_2 and R also occurs during the prolonged exhalation when mixed venous gas tensions are presumably not changing (i.e. on control breaths), one would be tempted to ascribe the changes following the exercise onset to accelerated intra-breath gas exchange resulting from increased muscle metabolic rate being manifest at the limb-to-lung transit delay. Clearly, this is not the case. The observed changes are also not consistent with a closing volume effect consequent to altered topographical basal-to-apical alveolar emptying patterns. The effect of the closing volume can be clearly seen during the last second of the prolonged exhalation in figure 2 as an upwards (decelerative) turn in the P_AO_2 slope. Rather, our results are more consistent with a diminishing contribution to the expirate of widely-dispersed regions having high $\dot{V}_A/\dot{Q}$ and rapid-time-constants for emptying, the consequences of which confound the attempted measurement of the muscle to lung transit time. Although it has been suggested that a deep inspiration may cause an increase in convection dependent inhomogeneity in the lung and influence inert gas mixing and gas diffusion[6], our subjects, in these studies, were instructed not to make a deep inspiration prior to the prolonged exhalation and all expressed a 'normal' phase III slope during single-breath nitrogen washout testing.

Interestingly one of the first records of the alveolar gas profiles by mass spectrometry during a prolonged exhalation at rest presented by Rahn and Farhi[7] – courtesy of Fowler and Hugh-Jones[8] - also reveals a clear deflection in the slope of $P_{A}O_2$.

One of the most striking findings is that changes in R as a function of P_{CO_2} during the prolonged exhalation did not conform to a linear function, an assumption underlying the estimation of "true" mixed venous P_{CO_2} proposed to occur at an R of 0.32.

Importantly, however, as the acceleration of R decline is also manifest in the steady-state of both rest and exercise it undermines the use of prolonged exhalations to estimate the limb-to-lung transit time in order to establish a frame of reference for the onset of the ϕII exercise hyperpnoea. This important determination awaits resolution.

5. REFERENCES

1. Whipp, B. J., Ward, S. A., Lamarra, N., Davis, J. A., and Wasserman, K., Parameters of ventilatory and gas exchange dynamics during exercise, *J. Appl. Physiol.* **52**: 1506-1513 (1982).
2. Lamarra, N., Whipp, B. J., Ward, S. A., and Wasserman, K., Effect of interbreath fluctuations on characterising exercise gas exchange kinetics, *J. Appl. Physiol.* **62**: 2003-2012 (1987).
3. Weissman, M. L., Jones, P. W., Oren, A., Lammara, N., Whipp, B. J. and Wasserman, K., Cardiac output increase and gas exchange at start of exercise *J. Appl. Physiol.* **52:** 236-244 (1982).
4. Kim, T. S., Rahn, H. and Farhi, L. E., Estimation of true venous and arterial PCO_2 by gas analysis of a single breath, *J. Appl. Physiol.* **21:** 1338-1344 (1966).
5. Rahn, H. and Fenn, W. O., *A graphical analysis of the respiratory gas exchange: The O_2-CO_2 diagram*, Published by the (American Physiological Society, Washington, D.C. 1955).
6. Cotton, D. J., Prabhu, M. B., Mink, J. T. and Graham, B. L., Effect of ventilation inhomogeneity on "intrabreath" measurements of diffusing capacity in normal subjects, *J. Appl. Physiol.* **75:** 927-932 (1993).
7. Rahn, H., and Farhi, L. E., Ventilation, perfusion, and gas exchange – the $\dot{V}_A/\dot{Q}$ concept, In: *Handbook of Physiology, Section 3: Respiration, Volume I* Eds: Fenn, W. O., and Rahn, H., (American Physiological Society, Washington, D.C. 1964).
8. Fowler, K. T., and Hugh-Jones, P., Mass spectrometry applied to clinical practice and research, *B.M.J.*, **1**:1205-1211 (1957).

SKELETAL MUSCLE CHEMOREFLEX IN EXERCISE VENTILATORY CONTROL

David M. Systrom, Mirko Hrovat, David Oelberg and Homayoun Kazemi*

1. INTRODUCTION

The mechanisms by which minute ventilation (VE) is tightly linked to the metabolic demands of exercising skeletal muscle remain uncertain[1]. The idea that by-products of exercising muscle metabolism could stimulate peripheral afferent nerves communicating with regions of the central nervous system important in cardiorespiratory regulation (the skeletal muscle chemoreflex) is most compelling for the pressor response to exercise[2], but a growing body of literature suggests it is important in the control of ventilation as well [3-7].

2. ANIMAL STUDIES

The skeletal muscle chemoreflex is mediated by free endings of small myelinated (group III) and unmyelinated (group IV) fibers [6, 8-12]. Spinal cord transmission for the skeletal muscle chemoreflex appears to be glutamatergic and can be potentiated by substance P [13]. The CNS projections of such afferents are clustered in or near the ventrolateral medulla[14], where their glutamatergic neurotransmission[15] can be attenuated by muscarinic activity [16]. Several metabolites are capable of evoking the skeletal muscle chemoreflex [6, 11, 17]. For instance, Rotto et al [11] demonstrated that lactic acid, but not equimolar sodium lactate at neutral pH, injected into the femoral artery of anesthetized cats increases activity of group IV neural afferents. A subsequent study by these investigators showed that lactic acid was more potent than equimolar HCl [6].

Recent ^{31}P magnetic resonance spectroscopy (MRS) animal work from this laboratory also supports a role for skeletal muscle acidosis as a metaboreflex stimulus [8].

---

* David M. Systrom, Homayoun Kazemi, Pulmonary and Critical Care Unit, Massachusetts General Hospital and Harvard Medical School, Boston, Massachusetts 02114. Mirko Hrovat, Department of Radiology, Brigham & Women's Hospital, Boston, MA 02115. David Oelberg, Pulmonary & Critical Care Division, Danbury Hospital, Yale University School of Medicine, Danbury, CT 06810.

Frontiers in Modeling and Control of Breathing, edited by Poon and Kazemi.
Kluwer Academic/Plenum Publishers, New York, 2001.

As a function of decreasing pHi, monoprotonated Pi becomes diprotonated and the net Pi peak resonates closer to that of PCr in a graded fashion[19]. pHi by MRS is nearly identical to that measured by micro-elctrodes from a single cell[20, 21]. pHe can be measured by the chemical shift vs PCr of a phosphorylated, pH-sensitive marker which distributes to the ECF (phenylphosphonic acid) .

It was reasoned that because the endings of groups III and IV are bathed in skeletal muscle interstitial fluid, the ECF pH (pHe) should be of critical importance to the reflex. Seven lightly anesthetized rats underwent unilateral continuous 5 Hz sciatic nerve stimulation in an 8.45 T NMR magnet that resulted in a mixed lactic acidosis and respiratory alkalosis, with no net change in arterial pH. Gastrocnemius intracellular pH fell from 7.30 ± 0.03 at rest to 6.72 ± 0.05 at 2.4 min of stimulation, and then rose to 7.05 ± 0.01 ($p < 0.05$), despite ongoing stimulation and muscle contraction. Despite arterial hypocapnia, pHe showed an immediate drop from its resting baseline of 7.40 ± 0.01 to 7.16 ± 0.04 units ($p < 0.05$) and remained acidic throughout the stimulation protocol. During the on- and off-transients for 5 Hz stimulation, changes in the pH gradient between intracellular and extracellular compartments suggested recruitment of saturable sarcolemmal ion transport mechanisms. (Figure 1). It was concluded that extracellular fluid pH of exercising skeletal muscle meets temporal and qualitative criteria necessary for a ventilatory metaboreflex mediator in a setting where arterial pH does not.

Interestingly, the metaboreflex is probably operative during low-level dynamic exercise that does not result in net lactic acid production[8]. The nature of the stimulus under these conditions is unknown, but it is possible that oxidatively produced carbonic acid (read below) or other metabolites such as potassium[17, 22] and cyclooxygenase pathway products are responsible[11].

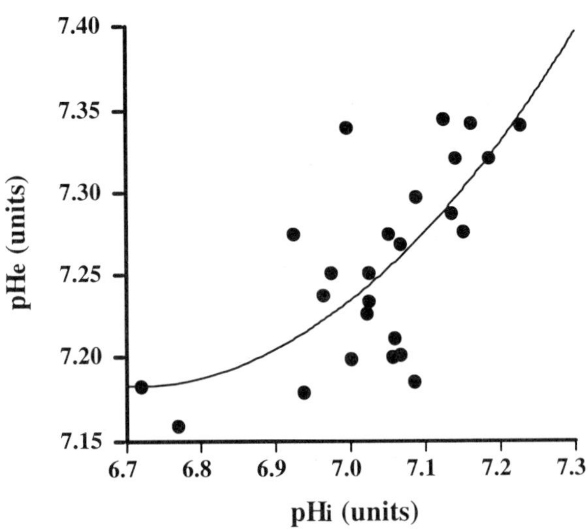

Figure 1. Simultaneously measured intra- (pHi) and extracellular fluid pH (pHe) of rat gastrocnemius muscle during 5 Hz sciatic nerve stimulation. pHe, a putative ventilatory chemoreflex stimulant, falls immediately despite arterial hypocapnia (not shown).

3. HUMAN STUDIES-INTENSE EXERCISE

Recent work suggests the skeletal muscle chemoreflex plays a role in the human during intense exercise. The most commonly utilized tool has been the application of limb positive pressure to trap metabolites in the exercising extremity, thereby accentuating the neural and attenuating any arterial regulatory pathways[4, 5, 7, 23-25]. Hyperventilation has been found if the exercise is sufficiently intense to produce lactic acidemia [4, 5, 7, 23-26].

None of these studies directly measured candidate stimuli to the muscle chemoreflex, however. We have recently adapted a metabolic cart for use in the fringe field of a 1.5 T whole body magnet, allowing simultaneous measurement of large muscle pHi and breath-by-breath ventilation[18, 27, 28] in the human. Using these techniques, a prospective, randomized investigation asked if skeletal muscle pHi mediates the ventilatory chemoreflex in man [29]. Twelve healthy subjects performed three bouts of isotonic submaximal quadriceps exercise on each of two days. Bilateral circumferential lower-extremity positive pressure of 45 mmHg was applied to the proximal thighs during exercise (BLPPex) or recovery (BLPPrec), in a randomized order. With BLPPex, end-exercise VE was 43 % higher (53.3 ± 3.8 vs 37.3 ± 2.2 L/min, mean ± SEM, $p < 0.0001$), and arterialized P_{CO_2} lower (33 ± 1 vs 36 ± 1 mmHg, $p = 0.0009$). Relative hyperventilation with BLPPex persisted for the first minute of recovery. Although blood lactate concentration was slightly higher at end-exercise with BLPPex, arterialized pH was not different. Skeletal muscle pHi was more acidic at end-exercise with BLPPex (6.44 ± 0.07 vs 6.62 ± 0.07, $p = 0.004$) and the difference persisted into early recovery (Figure 2). The pattern of pHi fall and recovery mirrored in a reciprocal fashion the ventilatory response. VE was linearly related to pHi, but not to pHart, during exercise for individual subjects.

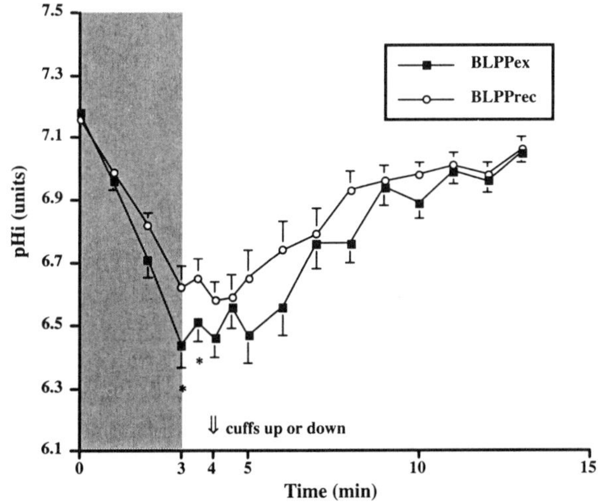

Figure 2. Human vastus medialis intracellular pH (pHi) during 3 minutes of constant load exercise and recovery, with proximal thigh cuff occlusion applied during exercise (BLPPex) or recovery (BLPPrec). BLPPex was associated with a more acid pHi and hyperventilation.

If BLPP results in increased perception of muscular effort, central motor command[30] could have driven ventilation independently of the muscle chemoreflex. In our study BLPP applied during exercise did in fact decrease PCr/Pi, a marker of bioenergetic stress, and leg discomfort was higher. Conversely, BLPP applied in the absence of effort (recovery) has also resulted in hyperventilation[23, 29]. Although such recovery data would seem to militate against feedforward regulation, it is possible that centrally-mediated "after-discharge"[1] continues to drive ventilation under these conditions. It is difficult to invoke central command as a cause of hyperventilation, however, when BLPP is applied only well into recovery [29].

4. HUMAN STUDIES-SUBMAXIMAL EXERCISE

To determine if acid change in muscle contributes to *submaximal* exercise ventilatory drive (before changes in arterial pH and lactate are seen), we inhibited CO_2 clearance from muscle with BLPP to 45 mmHg and/or an acute infusion of acetazolamide (ACTZ). It was reasoned that during low-level exercise, ACTZ would slow the removal of CO_2 produced from oxidative metabolism and alter the classically described relation between VE and VCO_2. It was hypothesized that if muscle pH is an important stimulus for the ventilatory chemoreflex, the previously described VE/pHi relationship would remain unchanged. Nine healthy volunteers underwent maximum incremental quadriceps exercise in a 1.5 T, 60 cm bore NMR magnet on two days separated by one week. Subjects received ACTZ 500 mg intravenously over 1 h before exercise, in a double-blind, saline placebo (PLB) controlled fashion. During submaximal exercise ACTZ was associated with an increased slope of the VE/VCO_2 relationship (41 (36-45, 95% CI) vs 28 (24-32), but no change in the VE/pHi relation (overlapping 95% CI's for slope and intercept).

The data are consistent with a skeletal muscle ECF hydrogen ion stimulus to ventilation throughout the entire range of incremental exercise. They also suggest that cuff-induced vascular distention is not necessary for the VE/pHi relationship.

5. CONCLUSIONS

Our studies confirm that hyperventilation can be evoked by partial circulatory occlusion and inhibition of carbonic anhydrase, during both low level and intense exercise, as well as recovery. In these studies, the ventilatory response cannot be fully explained by stimulation by arterialized blood pH, nociception, venous distension or central command, but seems better related to acid changes in exercising skeletal muscle itself.

6. ACKNOWLEDGMENTS

David M. Systrom is supported by NIH K24 HL04022-02 and Mirko Hrovat by NIH R21 CA87734.

7. REFERENCES

1. Dempsey, J.A., Exercise hyperpnea. *Adv. Exp. Med. Biol.* **393**(133): 133-136 (1995).
2. Coote, J.H., Hilton, S.M., and Perez-Gonzalez, J.F., The reflex nature of the pressor response to muscular exercise. *J. Physiol. (Lond.)* **215**: 789-804 (1971).
3. Bennett, F.M., A role for neural pathways in exercise hyperpnea. *J. Appl. Physiol.* **56**: 1559-1564 (1984).
4. Clark, A.L., Piepoli, M., and Coats, A.J.S., Skeletal muscle and the control of ventilation on exercise: evidence for metabolic receptors. *Eur. J. Clin. Invest.* **25**: 299-305 (1995).
5. Eiken, O. and Bjurstedt, H., Dynamic exercise in man as influenced by experimental restriction of blood flow in the working muscles. *Acta Physiol. Scand.* **131**: 339-345 (1987).
6. Rotto, D.M., Stebbins, C.L., and Kaufman, M.P., Reflex cardiovascular and ventilatory responses to increasing H^+ activity in cat hindlimb muscle. *J. Appl. Physiol.* **67**: 256-263 (1989).
7. Williamson, J.W., Raven, P.B., Foresman, B.H., and Whipp, B.J., Evidence for an intramuscular ventilatory stimulus during dynamic exercise in man. *Respir. Physiol.* **94**(2): 121-135 (1993).
8. Adreani, C.M., Hill, J.M., and Kaufman, M.P., Responses of group III and IV muscle afferents to dynamic exercise. *J. Appl. Physiol.* **82**(6): 1811-1817 (1997).
9. Kaufman, M.P., Longhurst, J.C., Rybicki, K.J., Wallach, J.H., and Mitchell, J.H., Effects of static muscular contraction on impulse activity of groups III and IV afferents in cats. *J. Appl. Physiol.* **55**: 105-112 (1983).
10. Mitchell, J.H. and Schmidt, R.F., *Cardiovascular reflex control by afferent fibers from skeletal muscle receptors*, in *Handbook of Physiology. Cardiovascular System. Peripheral Circulation and Organ Blood Flow*, J.T. Shepherd, F.M. Abbound, and S.R. Geiger, Editors. 1983, American Physiological Society: Bethesda. p. 623-658.
11. Rotto, D.M. and Kaufman, M.P., Effects of metabolic products of muscular contraction on the discharge of group III and IV afferents. *J. Appl. Physiol.* **64**: 2306-2313 (1988).
12. Rybicki, K.J., Kaufman, M.P., Kenyon, J.L., and Mitchell, J.H., Arterial pressure responses to increasing interstitial potassium in hindlimb muscle of dogs. *Am. J. Physiol.* **247**: R717-R721 (1984).
13. Beyaert, C.A., Hill, J.M., and Kaufman, M.P., Substance P analogues potentiate the pressor response to microinjection of L-glutamate into laminas I and II of the cat dorsal horn. *Brain Res.* **759**(1): 175-179 (1997).
14. Li, J., Hand, G.A., Potts, J.T., Wilson, L.B., and Mitchell, J.H., c-Fos expression in the medulla induced by static muscle contraction in cats. *Am. J. Physiol.* **272**:H48-55 (1997).
15. Hand, G.A., Potts, J.T., Treuhaft, B.S., Wilson, L.B., Petty, F., and Mitchell, J.H., Static muscle contraction elicits a baroreflex-dependent increase in glutamate concentration in the ventrolateral medulla. *Brain Res* **748**(1-2): 211-218 (1997).
16. Ally, A., Hand, G.A., and Mitchell, J.H., Central injection of physostigmine attenuates exercise-induced pressor response in conscious cats. *Am. J. Physiol.* **273**:R393-399 (1997).
17. McCloskey, D.I. and Mitchell, J.H., Reflex cardiovascular and respiratory responses originating in exercising muscle. *J. Physiol. (Lond.)* **224**: 173-186 (1972).
18. Evans, A., Tsai, L.W., Oelberg, D.A., Kazemi, H., and Systrom, D.M., Skeletal muscle ECF pH error signal for exercise ventilatory control. *J. Appl. Physiol.* **84**: 90-96 (1998).
19. Moon, R.B. and Richards, J.H., Determination of intracellular pH by ^{31}P magnetic resonance. *J. Biol. Chem.* **248**: 7276-7278 (1973).
20. Bangsbo, J., Johansen, L., Quistorff, B., and Saltin, B., NMR and analytic biochemical evaluation of CrP and nucleotides in the human calf during muscle contraction. *J. Appl. Physiol.* **74**: 2034-2039 (1993).
21. Gadian, D.G., *Nuclear Magnetic Resonance And Its Applications To Living Systems*. 1982, New York: University Press.
22. Wildenthal, K., Mierzwiak, D.S., Skinner, N.S.J., and Mitchell, J.H., Potassium-induced cardiovascular and ventilatory reflexes from the dog hindlimb. *Am. J. Physiol.* **215**: 542-548 (1968).
23. Piepoli, M., A.L., C., and Coats, A.J.S., Muscle metaboreceptors in hemodynamic, autonomic, and ventilatory responses to exercise in men. *Am. J. Physiol.* **269** *(Heart Circ. Physiol. 38):* H1428-H1436 (1995).
24. Sargeant, A.J., Rouleau, M.Y., Sutton, J.R., and Jones, N.L., Ventilation in exercise studied with circulatory occlusion. *J. Appl. Physiol.* **50**: 718-723 (1981).
25. Stanley, W.C., Lee, W.R., and Brooks, G.A., Ventilation studied with circulatory occlusion during two intensities of exercise. *Eur. J. Appl. Physiol.* **54**: 269-277 (1985).

26. Comroe, J.H. and Schmidt, C.F., Reflexes from the limbs as a factor in the hyperpnea of muscular exercise. *Am. J. Physiol.* **138**: 536-547 (1943).
27. Systrom, D.M., Kanarek, D.J., Kohler, S.J., and Kazemi, H., ^{31}P nuclear magnetic resonance spectroscopy study of the anaerobic threshold in humans. *J. Appl. Physiol.* **68**(5): 2060-2066 (1990).
28. Larson, D.E., Hesslink, R.L., Hrovat, M.I., Fishman, R.S., and Systrom, D.M., Dietary effects on exercising muscle metabolism and performance by ^{31}P-MRS. *J. Appl. Physiol.* **77**: 1108-1115 (1994).
29. Oelberg, D.A., Evans, A.B., Hrovat, M.I., Pappagianopoulos, P.P., Patz, S., and Systrom, D.M., Skeletal muscle chemoreflex and pHi in exercise ventilatory control. *J. Appl. Physiol.* **84**: 676-682 (1998).
30. Eldridge, F.L., Central integration of mechanisms in exercise hyperpnea. *Med. Sci. Sports Exerc.* **26**: 319-327 (1994).
31. Eldridge, F.L. and Gill-Kumar, P., Central neural respiratory drive and afterdischarge. *Respir. Physiol.* **40**: 49-63 (1980).

DISTENTION OF VENOUS STRUCTURES IN MUSCLES AS A CONTROLLER OF RESPIRATION

Philippe Haouzi*, Bruno Chenuel, Bernard Chalon, and Andrew Huszczuk

1. INTRODUCTION

The search for a CO_2 flow-related control mechanism of ventilation ($\dot{V}E$) that could account for $PaCO_2$ homeostasis[1-5] remains rather disappointing. Indeed, although activation of chemoreceptors or mechanoreceptors located in specific sites of the cardiovascular system, i.e. the right heart[6, 7, 8] , the pulmonary circulation[9, 10], the carotid bifurcation or the ascending aorta[11] can trigger ventilatory reflexes, their role in adjusting ventilation to the rate of incoming blood flow, when metabolism varies, is far from being established[3, 12]. It remains therefore unclear what afferent information reaching the central nervous system could adjust the level of ventilation to the rate of CO_2 removed from the tissues and reaching the lungs, to maintain constant arterial PCO_2. The possible role of the peripheral circulation, mainly in the muscles, as a possible controller of respiration has been more recently proposed[13-16], since a large proportion of the mechanosensitive group III and IV muscle afferent fibers, which have long been shown to stimulate ventilation, are located in the adventitia of the venules[17-18]. Based on human[19-24] and animal[13, 14] studies in which the peripheral circulation was either impeded or increased, it was argued that during a peripheral vasodilatation, the change in volume of the post-capillary and/or venous network could be a source of $\dot{V}E$ stimulation[15]. Such a mechanism places therefore the circulation in the tissues as a possible site of mediation between peripheral gas exchange and ventilatory control.

In this paper, two aspects of this putative mechanism are reported. Firstly, if the degree of congestion of the venular or venous system in a muscle is sensed by the central nervous system and affect respiratory neurons, one should be able to identify an afferent traffic in the fibers belonging to group III and IV, originating in a skeletal muscle, which is influenced by the mechanical distention of the venous structures. Secondly, since it has long been recognized that total impediment of the muscular circulation following a muscular exercise reduces the normal ventilatory response[19-22], whereas partial occlusion during contraction seems to have a stimulatory effect [25, 26], we sought to manipulate the

* Philippe HAOUZI, Laboratoire de Physiologie- Faculté de Médecine, 9 avenue de la Forêt de Haye - 54505 Vandoeuvre-lès-Nancy, France. Tel: 33 3 83 59 27 37. Fax: 33 3 83 59 27 26. e-mail: p.haouzi@chu-nancy.fr

Frontiers in Modeling and Control of Breathing, edited by Poon and Kazemi.
Kluwer Academic/Plenum Publishers, New York, 2001.

muscular circulation in humans following an impulse change in WR (400 W for 12 seconds) in order to dissociate the ventilatory effects resulting from blocking the venous return from those related to a reduction in arterial supply to the skeletal muscles when their metabolic activity increases, but without the confounding effects of the contractions. Indeed, an impulse change in work rate (the first derivative of a constant work rate) has the unique "property" of triggering a metabolic, hemodynamic and ventilatory outcome 20-30 seconds after the cessation of the muscular contractions[2, 27, 28]. This allows the study of the ventilatory effects of non-painful manipulations of the peripheral circulation (total or partial occlusion, supine or sitting exercise) in hyperemic and metabolically active, but resting, muscles in humans.

2. MATERIAL AND METHODS

2.1. Responses of Group III and IV Muscle Afferent Fibers to Venous Congestion in the Cat

Afferent impulses arising from endings in the triceps surae muscles were recorded from the L7 and S1 dorsal roots in 29 anesthetized and paralyzed cats as previously described[15, 29]. A balloon tipped catheter (French 3, Arrows) was placed in the inferior vena cava above the iliac bifurcation via the right femoral vein. Afferent fibers conducting impulses between 2.5 and 30 m/s were classified as group III fibers and those conducting impulses at less than 2.5 m/s were classified as group IV fibers. The caval balloon was inflated for one minute. A fiber was regarded as responding if its activity increased by at least 100 % during the minute of venous occlusion.

2.2. Human Study

Six healthy subjects exercised on an electromagnetically braked hyperbolic cycle-ergometer (Lodde Excalibure), and breathed room air through a mouthpiece connected to a pneumotachograph (Medical Graphics System). Breath-by-breath minute ventilation and pulmonary gas exchange were determined as already described[22, 23]. Systemic arterial blood pressure was estimated non-invasively, measuring the finger arterial pressure (Finapres cuff placed around the first finger and Finapres Monitor, Ohmeda 2300, Ohmeda, Louisville USA). The finger was positioned at a level corresponding to that of the carotid bifurcation.

The impulse load consisted of a 400 W exercise applied for 12 seconds (4 tests in each subject). In four other tests, total vascular occlusion of the lower limb circulation (above the systolic pressure, 200-210 Torr) was performed by inflating for 90 seconds plastic cuffs placed around the upper thighs just at the cessation of the exercise bout. In four other tests, the cuffs were inflated at 90 Torr (partial occlusion) to produce incomplete impediment of the arterial supply while the venous return from the exercising muscle must have been significantly blocked. Four additional tests were performed in sitting and supine position at 280 Watts to trigger an increase in O_2 uptake similar to that obtained during the vascular obstruction procedures. In supine position, the legs were positioned 10 cm above the heart level.

3. RESULTS

3.1. Venous Occlusion and Muscle Group III and IV Fiber Activity

Seven of the 36 group IV fibers tested were excited by the obstruction of the inferior vena cava (from 7 ± 4.2 to 21.5 ± 22.9 imp/min). In all except one of these fibers, activity could be evoked by probing the triceps surae muscle. In four fibers the activity increased after a delay and remained elevated throughout the period of occlusion. The three remaining fibers adapted to venous occlusion (Figure 1). Two of the 15 group III fibers that were tested increased their activity from 0 imp/min to 7 - 8 imp/min.

In addition, we studied two of the fibers responding to venous occlusion, one belonging to the group III and one to the group IV, 5 to 7 minutes after an injection of isoproterenol at different levels of popliteal blood flow (Figure 1). The afferent activity triggered by the venous occlusion was directly proportional to the level of blood flow prior to the obstruction.

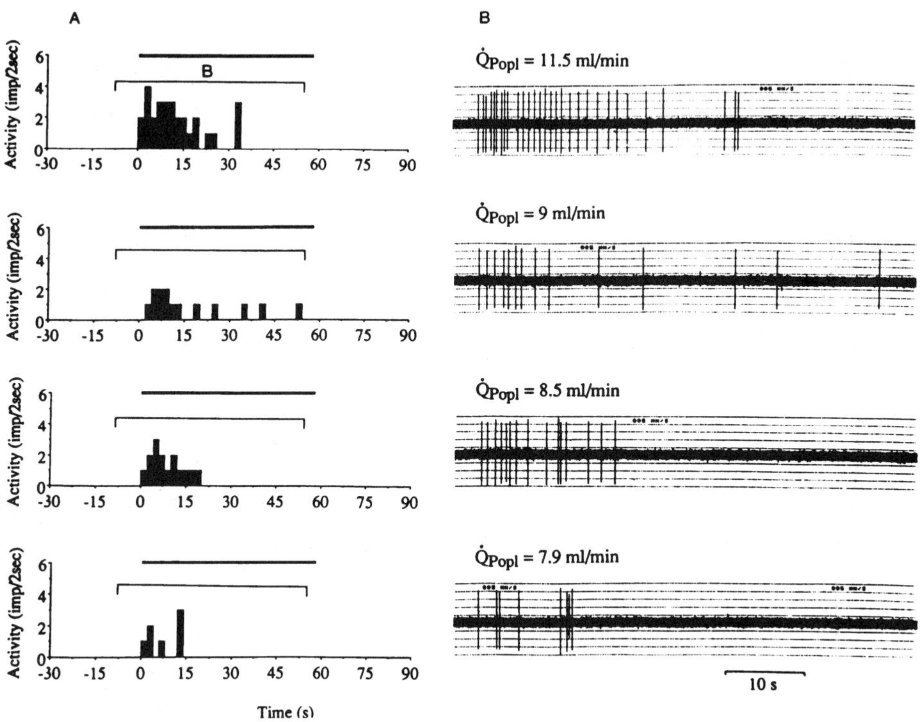

Figure 1. Effect of venous obstruction (thick horizontal bars) on the discharge of a group IV muscle afferent (conduction velocity 1.2 m/s) at four different levels of flow following an injection of isoproterenol. A. histograms of activity. B. recordings of the activity during time period depicted by bracket B. The higher the level of popliteal blood flow (Q_{popl}) prior to occlusion the greater the neural response to venous obstruction.

3.2. Human Study

Complete circulatory occlusion *reduced* significantly the normal delayed rise in ventilation (-29%) that normally occurred 20-25 seconds after the cessation of the contractions. In contrast, cuff inflation at 90 mmHg reduced the delayed rise in $\dot{V}O_2$ (25%) and $\dot{V}CO_2$ (-23%) in similar proportion to that during total obstruction but with a small or no reduction in ventilation. Finally, the $\dot{V}E$ response was significantly more blunted in supine (-19%) than in sitting position, despite similar $\dot{V}O_2$ as shown in Figure 2. Despite a similar rise in BP at the baroreceptor level (and similar $\dot{V}O_2$) during partial occlusion (+5 ± 2 Torr) and in supine (+7 ± 3 Torr), $\dot{V}E$ was dramatically depressed in supine but not during partial occlusion (see Figures 2, 3).

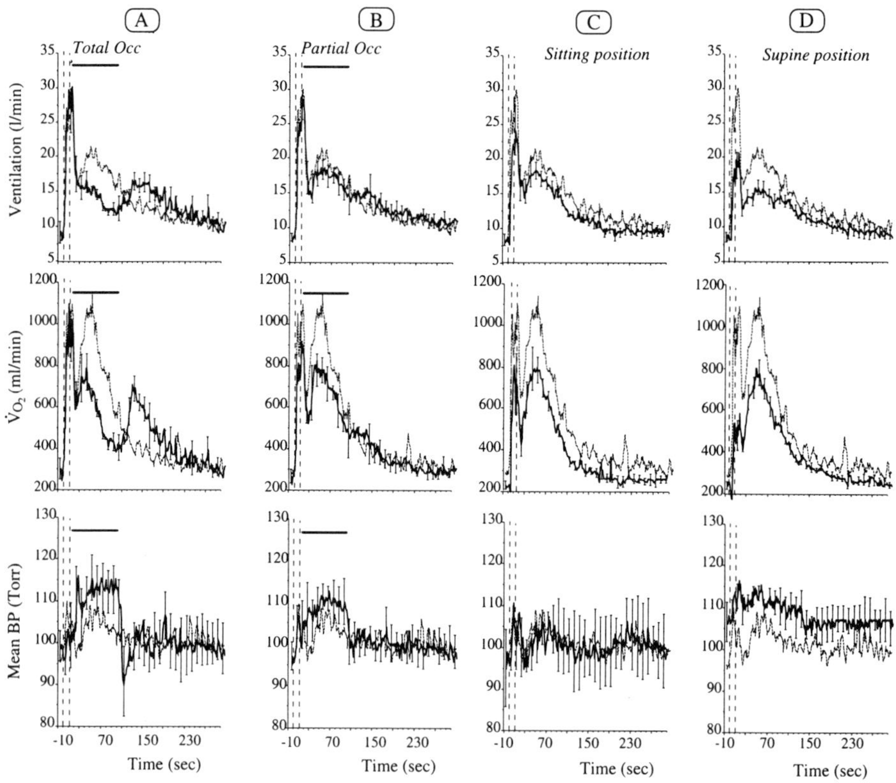

Figure 2. Effects on ventilation, pulmonary gas exchange, arterial blood pressure (BP) (mean ± sem) of suprasystolic vascular occlusion (A) and partial occlusion (B) for 90 seconds, following an impulse change in WR. The period of exercise is indicated by the two vertical dotted lines. The control tests are shown in dotted lines. The period of occlusion is depicted by the thick horizontal bar. Panels C and D show the responses to an impulse change in WR which triggers a delayed rise in $\dot{V}_{O_2}$ similar to that during the occlusions.

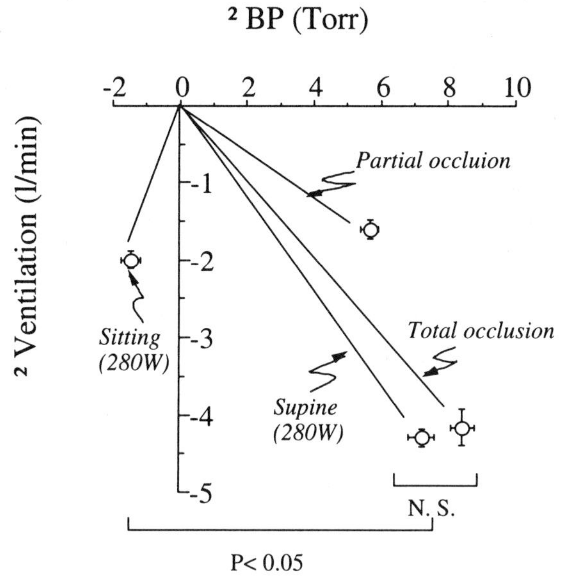

Figure 3. Averaged VE pic response as a function of BP changes measured at the level corresponding to that of the carotid baroreceptors. Data are mean ± sem differences from the control tests (400W). Whereas BP increased during partial occlusion as it did in supine or during total occlusion, VE was higher during partial occlusion than during total occlusion or in supine position.

4. DISCUSSION

The main finding of the present study was that total occlusion of the lower limb circulation depressed not only the expected $\dot{V}_{O2}$ and $\dot{V}_{CO2}$ response which followed the short burst of exercise but also the ventilatory response. The reduction in $\dot{V}_E$ was much larger than what would have been expected from a similar change in $\dot{V}_{O2}$, i.e. from an impulse of lower intensity, and perhaps more importantly from a partial occlusion. Finally, the $\dot{V}_E$ response was reduced in supine position. If preventing the rise in pulmonary gas exchange may *per se* prevent the rise in $\dot{V}_E$ during an occlusion, by reducing the stimulation of receptors located in the central circulation, a given reduction of O_2 uptake should lead to a similar reduction in $\dot{V}_E$, whether circulation is impeded totally or partially. This was obviously not the case. This implies that factors unrelated to the consequences of the reduction in venous flow to the central circulation, which includes the involvement of the arterial chemoreceptors, receptors located in the right heart or in the pulmonary circulation[4-11], dictate the changes in ventilation during the occlusion procedures.

We observed a systematic rise in BP during total obstruction. However, partial occlusion also increased BP, but with no real inhibitory effect on ventilation (see Figures 2 and 3). Indeed, despite higher BP during partial occlusion than in control, the $\dot{V}_E$

response was similar to or even higher than that triggered by a similar increase in pulmonary gas exchange. In other words, although the rise in arterial blood pressure may have contributed to the inhibition of $\dot{V}E$ during total occlusion, there must be another more powerful mechanism which counteracts the ventilatory inhibition mediated by the stimulation of the arterial baroreceptors[30] during partial occlusion, and /or conversely magnifies it during total obstruction. Furthermore, an almost identical increase in BP during partial occlusion did not depress $\dot{V}E$, whereas it reduced dramatically $\dot{V}E$ in supine, as already reported during constant work rate exercise[31]. Therefore, the contribution of the ventilatory component of the arterial baroreflex cannot solely explain the difference in $\dot{V}E$ response between positions. The nature of the signal that leads to a much lower $\dot{V}E$ in supine or during total occlusion than during partial occlusion or control, independently of the BP changes and $\dot{V}_{O2}$ reduction is still to be explained.

In humans, partial impediment of blood flow during exercise has been reported to stimulate ventilation[25]. In keeping with the concept of peripheral metaboreception[29, 32, 33, 34], it is hard to reconcile this observation with those showing that a depression in $\dot{V}E$ occurs when only the arterial supply is impeded during exercise[13, 23] or during total impediment of the circulation following a muscular exercise[19-22]. One should therefore examine the possibility that partial occlusion has affected not only the arterial supply, as total occlusion will do, but also the degree of distention of the muscle venules to a much larger extent than during total obstruction. Our study performed in the cat supports such a mechanism.

Group III and IV muscle afferent fibers have long been shown to stimulate ventilation[29, 32]. They respond to the deformation of their receptive fields and/or the accumulation of metabolic-by-products of exercise [33, 34]. The present study demonstrates that these endings can also respond to the mechanical distention of the peripheral vascular network, mainly from the venous and venular side in *resting* muscle giving the neurophysiological basis for the ventilatory and circulatory reflexes elicited during muscle congestion[12, 22, 35]. Such a property is the expected consequence of the anatomical location of many group IV endings - they are close to the veinules[17, 18] - and of their receptive properties - many of these endings are responding to mechanical stimuli[31].

The load applied on the venous return was much higher in sitting than in supine position. This constitutes a condition which will affect the degree of distention of the venular side of the muscle circulation, and thus will stimulate these endings which in turn may increase $\dot{V}E$ more vigorously than in supine at any given $\dot{V}_{O2}$. Similarly, during incomplete occlusion, the obstruction of the venous return may have distended the venous and venular side of the muscle circulation much more vigorously than during total occlusion, and thus may have stimulated thin muscle afferent fibers. During total occlusion, however, the various compartments of the peripheral circulation are to be affected differently: the pressure in the arterial side of the microcirculation will drop, whereas it may rise in the large capacitance vessels which will have to accommodate a large volume of blood to equilibrate pressures in this isolated compartment. The changes occurring in the venular side of the microcirculation in hyperemic muscles, are likely to be dictated by the transmission of the pressures from the arterial side, and may therefore drop during total occlusion, i.e. when the perfusion pressure decreases.

It is therefore proposed that the volume of blood in the peripheral venular system could mediate the ventilatory effects of vascular impediment through stimulation of the muscle group III and IV afferent fibers. This speculation is supported by the observation that a population of group III and IV afferent fibers do respond to the mechanical distention of the venular structures in the muscle. Such a system could inform the central nervous system of the circulatory status at the venular end of the muscular peripheral vascular network, matching ventilation to the degree of muscle hyperemia and thus coupling peripheral and pulmonary gas exchange systems.

5. REFERENCES

1. P. Dejours, Control of respiration in muscular exercise, in: *Handbook of Physiology: respiration*, vol I, edited by W.O. Fenn and H. Rahn (American Physiological Society, Washington, 1964), pp. 631-648.
2. B.J. Whipp, The control of exercise hyperpnea, in: *The regulation of breathing*, edited by T. Hornbein (New York, 1981), pp. 1069-1139.
3. B.J. Whipp, and S.A. Ward, Coupling of ventilation to pulmonary gas exchange during exercise, in: *Exercise pulmonary physiology and pathophysiology*, Vol. 52, chapt. 10, 1991, pp. 271-307.
4. A. Huszczuk, P.W. Jones, A. Oren, E.C. Schors, L.E. Nery, B.J. Whipp, and K. Wasserman, Venous return and ventilatory control, in: *Modeling and Control of Breathing*, edited by B.J. Whipp, D.M. Wiberg (Elsevier, New York, 1983), pp. 78-85.
5. J. Ponte, and M.J. Purves, Carbon dioxide and venous return and their interaction as stimuli to ventilation in the cat, *J. Physiol. Lond.* **274**, 455-475 (1978).
6. P.W. Jones, A. Huszczuk, and K. Wasserman, Cardiac output as a controller of ventilation through changes in right ventricular load, *J. Appl. Physiol.* **53**, 218-224 (1982).
7. D.R. Kostreva, E.J. Zuperku, R.V. Purtock, R.L. Coon, and J.P. Kampine, Sympathetic afferent nerve activity of right heart origin, *Am. J. Physiol.* **229**, 911-915 (1975).
8. T.C. Lloyd, Effect on breathing of acute pressure rise in pulmonary artery and right ventricle, *J. Appl. Physiol.* 57, 110-116 (1984).
9. R.W. Stremel, B.J. Whipp, R. Casaburi, D.J. Huntsman, and K. Wasserman, Hypopnea consequent to reduced pulmonary blood flow in the dog, *J. Appl. Physiol.* **46**, 1171-1177 (1979).
10. W.O. Kan, J.R. Ledsome, and C.P. Boulter, Pulmonary arterial distension and activity in phrenic nerve of anesthetized dogs, *J. Appl. Physiol.* 46, 625-631 (1979).
11. W.S. Yamamoto, and M.W. Edwards, Homeostasis of carbon dioxide during intravenous infusion of carbon dioxide, *J. Appl. Physiol.* **15**, 807-818 (1960).
12. A. Huszczuk, B.J. Whipp, A. Oren, E.C. Schors, M. Pokorski, I.E. Nery, and K. Wasserman, Ventilatory responses to partial cardiopulmonary bypass at rest and exercise in dogs, *J. Appl. Physiol.* **62**, 575-583 (1986).
13. A. Huszczuk, E. Yeh, J.A. Innes, I. Solarte, and K. Wasserman, Role of muscular perfusion and baroreception in the hyperpnea following muscle contraction in dog, *Respir. Physiol.* **91**, 207-226 (1993).
14. P. Haouzi, A. Huszczuk, J.P. Gille, B. Chalon, F. Marchal, J.P. Crance, and B.J. Whipp, Vascular distension in muscles contributes to respiratory control in sheep, *Respir. Physiol.* **99**, 41-50 (1995).
15. P. Haouzi, J.M. Hill, B.K. Lewis, and M.P. Kaufman, Responses of groups III and IV muscle afferents to distension of the peripheral vascular bed, *J. Appl. Physiol.* **87**, 545-553 (1999).
16 P. Haouzi, J.J. Hirsch, J.P. Gille, F. Marchal, J.P. Crance, and A. Huszczuk, Papaverine injection into the hindlimb circulation stimulates ventilation in sheep, *Respir. Physiol.* **105**, 143-153 (1996).
17. M.J. Stacey, Free nerve endings in skeletal muscle of the cat, *J. Anat.* **105**, 231-254 (1969).
18. M. Von Düring, and K.H. Andres, Topography and ultrastructure of group III and IV nerve terminals of cats gastrocnemius-soleus muscle, in: *The Primary Afferent Neuron: A Survey of Recent Morpho-Functional Aspects*, edited by W. Zenker and W.L. Neuhuber (Plenum Press, New York, 1990), pp.35-41.
19. P. Dejours, J.C. Mithoefer and A. Teillac, Essai de mise en évidence de chémorécepteurs veineux de ventilation, *J. Physiol. (Paris)* **47**, 160-163 (1955).
20. L.B. Rowell, L. Hermansen and J.R. Blackmon, Human cardiovascular and respiratory responses to graded muscle ischemia, *J. Appl. Physiol.* **41**, 693-701 (1976).
21. J.A. Innes, I. Solarte, A. Huszczuk, E. Yeh, B.J. Whipp and K. Wasserman, Respiration during recovery from exercise: effects of trapping and release of femoral blood flow, *J. Appl. Physiol.* **67**, 2608-2613 (1989).
22. P. Haouzi, A. Huszczuk, J. Porszasz, B. Chalon, K. Wasserman, and B.J. Whipp, Femoral vascular occlusion and ventilation during recovery from heavy exercise, *Respir. Physiol.* **94**, 137-150 (1993).

23. P. Haouzi, J.J. Hirsch, F. Marchal, and A. Huszczuk, Ventilatory and gas exchange response during walking in severe peripheral vascular disease, *Respir. Physiol.* **107**, 181-190 (1997).
24. P. Dejours, J.C. Mithoefer, and J. Raynaud, Evidence against the existence of specific ventilatory chemoreceptors in the legs, *J. Appl. Physiol.* **10**, 367-371 (1957).
25. O. Eiken, Responses to dynamic leg exercise in man as influenced by changes in muscle perfusion pressure, *Acta. Physiol. Scand.* 131, suppl. 566 (1987).
26. D.A. Oelberg, A.B.Evans, M.I.Hrovat, P.P. Pappagianopoulos, S. Patz, and D. Systrom, Skeletal muscle chemoreflex and pHi in exercise ventilatory control, *J. Appl. Physiol.* **84**, 676-683 (1998).
27. Y. Fujihara, J.R. Hildebrandt, and J. Hildebrandt, Cardiorespiratory transients in exercising man. I. Tests of superposition, *J. Appl. Physiol.* **35**, 58-67 (1973).
28. Y. Fujihara, J. Hildebrandt, and J.R. Hildebrandt, Cardiorespiratory transients in exercising man. II. Linear models, *J. Appl. Physiol.* **35**, 68-76 (1973).
29. M.P. Kaufman, Afferents from limb skeletal muscle, in: *Lung Biology in Health and disease, Regulation of breathing*, Vol. 79, edited by J. Dempsey and A.I. Pack (Basel, Dekker Inc., New-York, 1995), pp. 583-616.
30. M.J. Brunner, M.S. Sussman, A.S. Greene, C.H. Kallman and A.A. Shoukas, Carotid sinus baroreflex control of respiration, *Circ. Res.* **51**, 624-636 (1982).
31. D. Weiler-Ravell, D.M. Cooper, B.J. Whipp, and K. Wasserman, Control of breathing at the start of exercise as influenced by posture, *J. Appl. Physiol.* **55**, 1460-1466 (1983).
32. D.I. McCloskey, and J.H. Mitchell, Reflex cardiovascular and respiratory responses originating in exercising muscle, *J. Physiol. (London)* **224**, 173-186 (1972).
33. S. Mense, and M. Stahnke, Responses in muscle afferent fibers of slow conduction velocity to contractions and ischaemia in the cat, *J. Physiol. (London)* **342**, 383-397 (1983).
34. J.H. Mitchell, and R.F. Schmitt, Cardiovascular reflex control by afferent fibers from skeletal muscle receptors, in: *Handbook of Physiology. The cardiovascular System*, Sect. 2, Vol. III, Chapt. 17, American Physiological Society, (Bethesda, M.D., 1983), pp. 623-637.
35. J. McClain, C. Hardy, B. Enders, M. Smith, and L. Sinoway, Limb congestion and sympathoexcitation during exercise, *J. Clin. Invest.* **92**, 2353-2359 (1993).

VENTILATORY AND CIRCULATORY RESPONSES AT THE ONSET OF RAPID CHANGES IN POSTURE

Miharu Miyamura, Koji Ishida, Keisho Katayama, Yasutake Sato, and Norihiro Shima*

1. INTRODUCTION

The transition from rest to light or moderate intensity exercise is typically accompanied by an abrupt step-like increment in ventilation at the first exercise breath. This phase I response is observed during not only voluntary exercise and passive movement, but also during electrically induced muscle contraction. Many investigators have pursued the mechanisms of this phase I response; however, it is still a matter of dispute as to its nature. Since changes in ventilation are so rapid, a number of neural mechanisms, such as reflexogenic influences from the cerebral cortex and hypothalamic motor regions, afferent stimulus through the group III and IV fibers, and central and/or peripheral circulatory influences, are postulated to explain the phase I response.[1] Regarding the last factor, Huszczuk et al.[2] suggested that ventilation is controlled by a mechanism which acts rapidly, and which relies on the integrity of baroreception and/or vasoregulation; baroreception appears to be more important. Although it has been reported that hemodynamic responses are considerably affected by a sudden change in posture, it is likely that respiratory response is also modified by postural changes in man. To our knowledge, however, the initial ventilatory responses to rapid change in posture have not been studied in detail. The purpose of this study was to clarify the ventilatory and circulatory responses at the onset of sudden tilts from supine to upright, and vice versa.

* M Miyamura, K Ishida, K Katayama, Y Sato, and N Shima, Research Center of Health, Physical Fitness and Sports, Nagoya University, Nagoya 464-8601, Japan. Correspondence: M Miyamura. email: miyamura@htc.nagoya-u.ac.jp

Frontiers in Modeling and Control of Breathing, edited by Poon and Kazemi. Kluwer Academic/Plenum Publishers, New York, 2001.

2. METHODS

Seven healthy male students participated in this study as subjects with their informed consent. The average values (SE) of age, height, and body mass of the subjects were 27.0 (0.5) years , 173.3 (1.8) cm, and 65.9 (1.5) kg, respectively. All subjects were physically active but not participating in any regular training or organized sports. The subjects were instructed as to the experimental procedures, but not the purpose and results of this study until its completion. All subjects came to the laboratory twice on separate days. On the first day, each subject was familiarized only with the apparatus and testing procedures involved in the study. The actual experiment was conducted on a separate day at least a few days after that first familiarization day.

On the second day, the tilting tests were carried out on each subject using a tilt-board, i.e., the subject rested comfortably on the tilt-board in supine position for 10 min, and was asked to relax during the experimental period. The tilt- board was able to be rapidly tilted manually by the experimenters in about 1.5 s from horizontal (0°) to upright (70°), or the reverse. A tilt from upright to supine posture will be denoted as "down-tilt", and a tilt in the reverse direction as "up-tilt". These tilts were performed without any signal so as to prevent the subjects from being aware of when they started. The goniometer was attached to the axis of the tilt board in order to monitor the start of tilting. In addition, the down-tilt and up-tilt were initiated just before the start of the inspiratory phase; the experimenter checked this on an oscilloscope which connected to the hot-wire flowmeter. The up-tilt and the down-tilt were each repeated four times at approximately 5 min intervals to avoid the influence of prior tilting, and results of each replicate were averaged to determine the mathematical mean. The two tilt conditions were performed in a different order for each subject to avoid order effects.

Inspiratory minute ventilation (VI), tidal volume (VT), respiratory frequency Rf), end-tidal CO_2 and O_2 partial pressure ($PETCO_2$ and $PETO_2$) were determined breath-by-breath before, during and after tilting for a total of 90 s; the subjects breathed through a respiratory face mask attached to a hot-wire flowmeter that was calibrated prior to each experiment by means of a 2 - L calibration pump. The concentration of CO_2 and O_2 were analyzed by a gas analyzer; samples of gas were drawn continuously through the vinyl tube to obtain $PETCO_2$ and $PETO_2$. In addition, heart rate (HR) was measured beat-by-beat from R spike using an electrocardiogram. Blood pressure was also measured noninvasively beat-by-beat using a Finapres. A Finapres sensor was attached to the finger of the left hand and positioned at heart level. Systolic blood pressure (SBP) and diastolic blood pressure (DBP) were measured, and mean blood pressure (MBP) was calculated from SBP and DBP. All ventilatory and circulatory signals were continuously displayed on the screen of the oscilloscope and converted from analog to digital using an A-D converter at a sampling frequency of 100 Hz for 90 s, including 30 s before and 60 s after each tilt. These data were then stored in a hard disc unit and analysed by computer. Breath-by-breath and beat-by-beat data were aligned with the onset of each tilt, linearly interpolated between each breath or beat to yield a data point at each 1 s interval and ensemble averaging was done across all four repetitions.

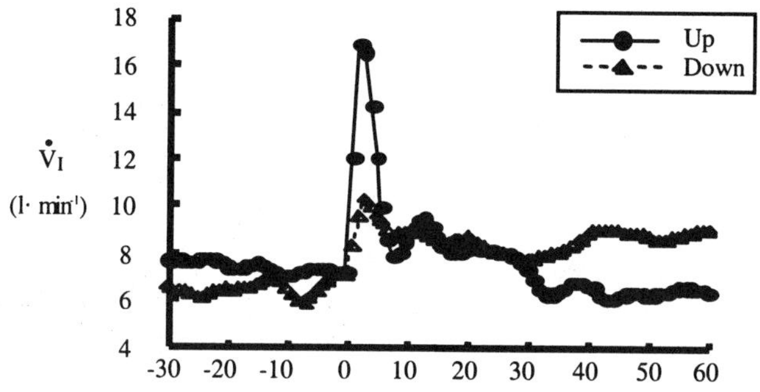

Fig. 1. An example of changes in ventilation ($\dot{V}_I$) before, during and after up-tilt (black circle) and down-tilt (black triangle).

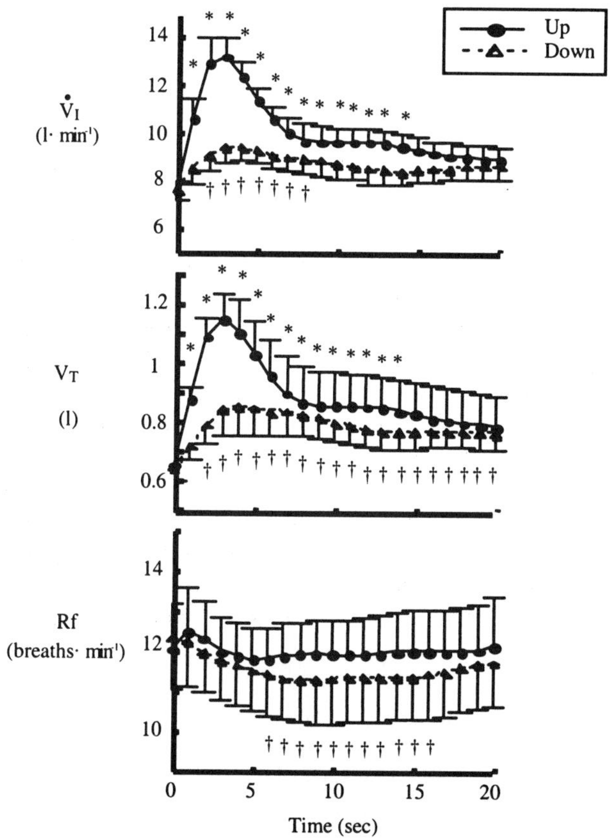

Fig. 2. Average values and standard errors of minute ventilation ($\dot{V}_I$: upper panel), tidal volume (V_T : middle panel) and respiratory frequency (Rf : lower panel) after up-tilt (black circle) and down-tilt (black triangle). * † indicate significant difference from resting condition.

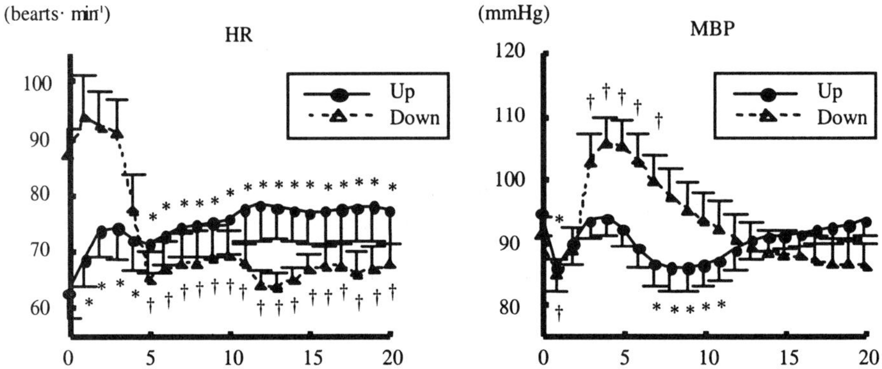

Fig. 3. Average values and standard errors of heart rate (HR : left panel) and mean blood pressure (MBP : right panel) after up-tilt (black circle) and down-tilt (black triangle). * † indicate significant difference from resting condition.

3. RESULTS

Figure 1 demonstrates the results of VI from one subject before, during and after the up-tilt and down-tilt. There was a rapid increase in ventilation both on transition from the supine to the upright and vice versa. Figure 2 shows the means and standard errors (SE) of ventilatory and circulatory responses for 20 s immediately after the up-tilt and down-tilt. As shown in Fig. 2, VI immediately after up-tilt and down-tilt increased significantly after 1 - 2 s and reached its peak value at 3 s. Average values of percentage changes of VI (ΔVI) at 3 s, which were estimated from rest values of 100%, were 177% in up-tilt and 123% in down-tilt. Also, VT increased significantly after both tilts, while Rf decreased significantly from 6 to 16 s in down-tilt, but did not change in up-tilt. Figure 3 shows the results of heart rate and blood pressure responses in both tilts. There were large differences in HR response between up-tilt and down-tilt, i.e., HR increased significantly at the first second in the tilt-up as compared with rest, while it abruptly decreased after 3 s in the down-tilt. These HR transients were larger in the down-tilt than in the up-tilt. On the other hand, MBP shows the biphasic change in both tiltings, i.e., MBP decreased significantly at first the second immediately after both tilts and reached its peak value at 4 s, and then decreased gradually in both tilting conditions until 9 s as shown in Fig. 3. SBP and DBP were followed almost the same tendency as MBP.

4. DISCUSSION

Our results show that the tilt from supine to upright and vice versa has large and immediate effects on ventilation, heart rate and blood pressure in a resting subject. First, VI increases significantly immediately after tilt-up and tilt-down which each took 1.5 s to be completed. This increase in VI was mainly due to an increase of VT (Fig. 2). These results basically coincided with data of Loeppky and Luft [3] who observed a change of pulmonary ventilation from supine to 60° upright or vice versa which took 15 s to be completed during resting condition. Haruna et al.[4] also found similar results with a tilt time of 2.5 s from supine to 80° upright or vice versa during ergometer cycling. It was found in this study that phase I response with respect to VI obtained in up-tilt was significantly greater than that in down-tilt. As described previously, neural mechanisms such as central command, afferent stimulus and circulatory influences have been postulated to explain the ventilatory responses at the onset of voluntary exercise and passive movement in humans until now. It is possible to assume that afferent impulse through group III and IV from lower limbs could be greater in up-tilt as compared with down tilt if the central command is the same. In other words, when compared with the down-tilt, the respiratory center seems to be more stimulated through afferent impulses by muscle contraction in the lower limbs in order to maintain upright posture. If so, ventilation may not be increased immediately after down-tilt because it seems that afferent impulses from lower limbs disappear within 1.5 s. However, it is of interest that, VI also increased significantly at the onset of down-tilt, though the degree of increase is relatively small, suggesting that an increase in VI in the down and up-tilt could be due to not only afferent impulses from the lower limbs, but also another stimulus such as vestibular reflex. However, further investigations are required to confirm this suggestion.

In the present study, HR increased significantly at the first second after up-tilt, while it decreased significantly from 3 - 4 s in down-tilt. More recently, Sundblad et al. [5] have observed the same latency during down-tilt within 3 s between 80° upright and horizontal (0°) during dynamic leg exercise at 50, 100 and 150W. These results indicate that the latency between the onset of tilt-induced changes in HR during down-tilt is almost the same, both at rest and during dynamic exercise, when the range of time and the angle of tilt are 1.5 - 2 s and 0°- 70 ~ 80°, respectively. On the other hand, in contrast to HR transients, MBP decreased significantly immediately after both tiltings, and then increased as shown in Fig. 3. These results coincide with the results of Pawelczyk and Raven [6] and Strange et al.[7] who have demonstrated that arterial pressure does change within the first second of neck suction in resting and exercising subjects, respectively. It is well known that the major arterial baroreceptor sites are located in the carotid sinus and the aortic arch. Recently Cui et al. [8] suggested that sinusoidal linear acceleration may suppress muscle sympathetic nerve activity in humans. Since there is also evidence that isolated carotid baroreceptor stimulation exerts important influences on the feed back control of BP and HR, MBP responses to an orthostatic change could be described as an adaptation of the carotid baroreflex function as suggested by Pawelczyk and Raven. [6]

In summary, our results suggest that rapid changes in VI, HR and MBP at the onset of up-tilt and down-tilt as applied here may be related to neural factors such as afferent impulses from the lower limbs, vestibular stimulation and baroreflexes.

5. REFERENCES

1. M. Miyamura, Control of ventilation during exercise in man with special reference to the feature at the onset. *Jpn. J. Physiol.* **44**:123-139 (1994).
2. Huszczuk, E. Yeh, J. A. Innes, L. Solarte, K. Wasserman, and B. J. Whipp, Role of muscle perfusion and baroreception in the hyperpnea following muscle contraction in dog. *Respir. Physiol.* **91**: 207-226 (1993).
3. J. A. Loeppky and U. C. Luft, Fluctuations in O_2 stores and gas exchange with passive changes in posture. *J. Appl. Physiol.* **39**: 47-53 (1975).
4. Y. Haruna, D. Linnarsson, and B. Tedner, Gas exchange and cardiovascular responses to tilting during graded exercise. *Physiologist,* **35**: S188-S189 (1992).
5. P. Sundblad, Y. Haruna, B. Tender, and D. Linnarsson, Short-term cardiovascular responses to rapid whole-body tilting during exercise. *Eur. J. Appl. Physiol.* **81**: 259-270 (2000).
6. J. A. Pawelczyk and R. B. Raven, Reduction in central venous pressure improve carotid baroreflex responses in conscious men. *Am. J. Physiol.* **257**: H1389-H1395 (1989).
7. S. Strange, L. B. Rowell, N. J. Christensen, and B. Saltin, Cardiovascular responses to carotid sinus baroreceptor stimulation during moderate to severe exercise in man. *Acta Physiol. Scand.* **138**:145-153 (1990).
8. J. Cui, S. Iwase, T. Mano, N. Katayama, and S. Mori, Muscle sympathetic nerve response to vestibular stimulation by sinusoidal linear acceleration in humans. *Neurosci. Lett.*, **267**:181-184 (1999).

CARDIORESPIRATORY RESPONSES TO WALKING AND RUNNING AT AN INCREMENTAL SPEED OF TREADMILL

Tatsuhisa Takahashi, Akiyoshi Okada, Toshiyo Tamura, and Yoshimi Miyamoto*

1. INTRODUCTION

During human locomotion, walking is the gait most commonly used at a low speed, and running is generally observed at a higher speed. A gait transition from walking to running occurs naturally as the speed of locomotion increases. To date, there have been many studies on human locomotion based on mechanical and metabolic analyses. In addition, in scientific and clinical studies using non-invasive monitoring techniques, the estimate of energy expenditure from ambulatory heart-rate monitoring has been gained a growing interest for the evaluation of physical activity and physiological stress.[1] However, cardiac responses associated with metabolic change during human locomotion have not been well understood. In the present study, therefore, we examined cardiorespiratory responses to walking and running on a treadmill at incremental speeds.

2. METHODS

Six healthy young male subjects with an average age (mean ± SD) of 21.7±1.2 years, an average height of 1.70±0.05 m and an average body mass of 65.3±5.8 kg, volunteered for this study. On a preliminary screening day each subject was informed of the procedures, risk, and benefits involved in the investigation, and each signed a written consent form. None of the subjects had either symptoms or history of cardiovascular disease, and they all had normal physical examinations and resting and exercise 12-lead electrocardiograms.

* T. Takahashi, Department of Biomedical Engineering, Faculty of Engineering, Yamagata University, Johnan 4-3-16, Yonzezawa 992-8510, Japan.
Phone: +81-238-26-3352, Fax: +81-238-26-2082, email: takahasi@eie.yz.yamagata-u.ac.jp

Frontiers in Modeling and Control of Breathing, edited by Poon and Kazemi.
Kluwer Academic/Plenum Publishers, New York, 2001.

The subjects performed walking or running on a motor-driven treadmill (AR-160A, Minato, Osaka, Japan) at a 0% gradient. The speed of the treadmill was increased from 1.0 km/h to 8.0 km/h by 0.2 km/h every minute, controlled by a personal computer (PC9801RX, NEC, Tokyo, Japan). Running was increased from 4.0 km/h to 8.0 km/h by 0.2 km/h every minute, because the subjects were not able to run naturally below the speed of 4.0 km/h.

The subjects were instructed not to consume caffeine for at least 24 h before the experiments. The subjects had eaten a light meal at least 3 h before exercise testing. Each subject was assigned to perform one exercise test at the same time of the day every week. All experiments were conducted at a room temperature of 19-22°C.

During treadmill testing, oxygen uptake ($\dot{V}O_2$) and heart rate (HR) were measured breath-by-breath using an on-line automated system.[2] Ventilatory airflow was monitored with a hot-wire-type pneumotachograph (RF-2, Minato, Osaka, Japan). The composition of expired gas was continuously analyzed with a medical mass spectrometer (WSMR-1400, Westron, Chiba, Japan) which was calibrated with a standard reference gas mixture before each measurement. HR was calculated from the R-R intervals, which were monitored by electrocardiography at the chest wall using bipolar electrodes. All data for cardiorespiratory variables were stored on diskettes for subsequent analysis by a personal computer (PC-9821Xa, NEC, Tokyo, Japan).

Cardiorespiratory data obtained from two repetitions of each test of walking and running for each subject were arranged separately with a 5-s interval time base using a Lagrange interpolation. These data were then averaged to yield a single data set for each subject. Where necessary, the two-tailed Student's t-test was used to test the difference between two means (± SD). Least-squares regression lines were fitted to HR versus $\dot{V}O_2$ plot in each of walking and running for each subject. Correlation between HR and $\dot{V}O_2$ was statistically examined using the Pearson correlation coefficient. For all statistical analyses, differences were considered significant at $P < 0.05$.

3. RESULTS

Figure 1A and 1B show the dynamic changes in $\dot{V}O_2$ and HR, respectively, during walking and during running at an increasing speed of 0.2 km/h every minute. $\dot{V}O_2$ and HR during walking increased nonlinearly with increasing speed, whereas $\dot{V}O_2$ and HR during running increased relatively linearly. Each of $\dot{V}O_2$ and HR was lower during walking than during running at the same speeds, while these variables crossed at 7.26± 0.68 and 7.40±0.65 km/h, respectively, for all the subjects.

The oxygen cost of walking to move a unit distance ($\dot{V}O_2$ divided by speed) decreased as speed rose up to 4.3 km/h, above which it increased (Fig. 2A). The mean minimum oxygen cost of walking for 6 subjects was 202.3±22.6 ml/kg/km at 4.2±0.2 km/h, whereas that of running was not detected in the range between 4.0 and 8.0 km/h.

The HR of both walking and running per one meter moved (HR cost) are also shown in Fig. 2B. The HR cost of walking decreased with increasing speed and then it increased slightly or leveled off. A minimum during walking was observed in 4 of 6 subjects, but was not observed during running. However, the ranges of the optimal

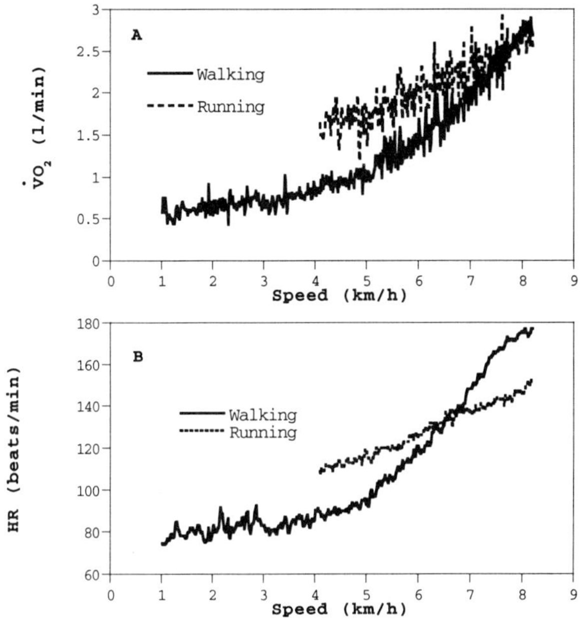

Figure 1. Oxygen uptake (A, $\dot{V}O_2$) and heart rate (B, HR) in walking and running at increasing treadmill speeds in a representative subject.

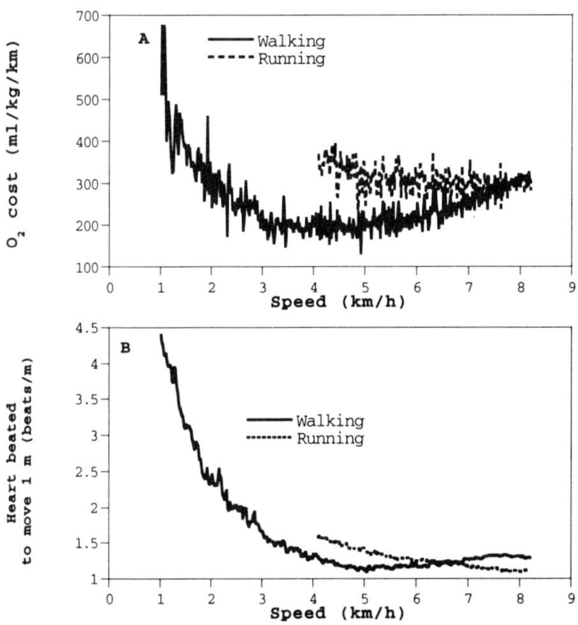

Figure 2. A: Oxygen uptake per unit weight per unit distance (oxygen cost, oxygen uptake divided by weight and speed, calculated from data Fig. 1). B: Heart beats to unit distance walked and run in same subject.

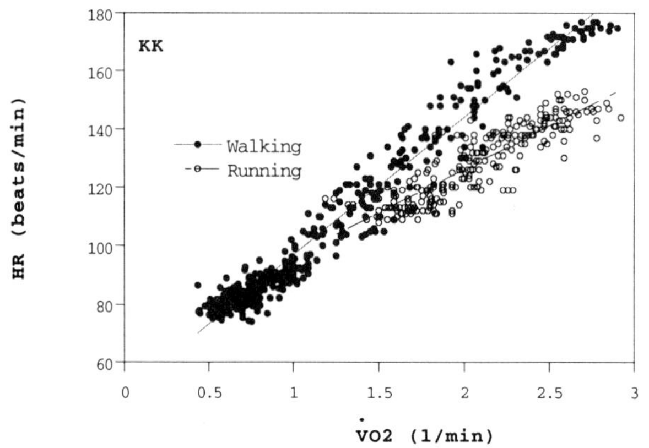

Figure 3. Heart rate (HR) against oxygen uptake ($\dot{V}O_2$) determined at increasing speed in walking and in running in the same subject.

walking speeds were higher and wider compared with the corresponding speeds for the oxygen cost.

To compare cardiac responses to walking and running, HR was replotted against $\dot{V}O_2$ determined at the increasing speeds (Fig. 3). The change in HR was proportional to that in $\dot{V}O_2$ under both walking and running conditions. A significant linear relationship between HR and $\dot{V}O_2$ was found for each subject in both walking and running. The mean slope of the HR-$\dot{V}O_2$ regression line was significantly greater ($P<0.05$) for walking than for running (37.3±8.0 vs. 25.3±7.6).

4. DISCUSSION

Many investigators have reported that the energy (metabolic) cost of walking can be expressed as oxygen uptake per a unit distance and that the speed requiring the lowest metabolic cost is considered to be the most efficient. Accordingly, the curve of oxygen uptake per body weight per kilometer, as a function of speed, was concave upward and exhibited a minimum value. The most efficient speed of level walking was 4.2 km/h, which is in good agreement with the reports of Cavagna et al.,[3] Robertson et al.,[4] and Zarrugh et al.[5]

The increased $\dot{V}O_2$ with speed was steeper in walking than in running, which is consistent with the results of Margaria et al.[6] Both $\dot{V}O_2$ and O_2 cost were lower for walking than for running at the same speeds below about 7.0 km/h, but vice versa at the higher speeds. However, the minimum oxygen cost of running seems to be similar to that of walking, because the oxygen cost of running decreased asymptotically to the level of the minimal value for walking. This is supported by the evidence that in the horses the minimum energy cost is identical between walking and running.[7]

Similarly, the curve of HR responses to move per unit distance (HR cost) decreased as walking speed increased and then it increased or leveled off. Moreover, the curves of HR cost of walking in 4 subjects were a "U" shape as a function of speed. Therefore, the optimal speed was not able to be defined as only a minimum oxygen cost.

In this study, we compared the responses of HR to walking and running. Since oxygen uptake for the two gaits was different at the same speeds, HR was plotted against $\dot{V}O_2$ for each gait. The more gentle slope of the HR-$\dot{V}O_2$ regression line for running compared to that for walking suggests that the efficiency of circulation may be higher during running than during walking, if it is assumed that the venous return is facilitated by the pumping action provided by the leg muscles.[8] Alternatively, it is possible that the increase in HR at a given level of $\dot{V}O_2$ was proportional to the size of muscle mass being used. It has been demonstrated that the slope of the linear relationship between HR and $\dot{V}O_2$ is increased for arm exercise compared with leg exercise.[9, 10] Despite their both being leg exercises, the quantity and/or quality of actually contracting muscles would be different between walking and running. The sense of effort for walking, as measured by a Rating of Perceived Exertion, while forced to maintain a gait over the preferred transition speed from walking to running was greater than that for running at the corresponding speed.[11] Our findings of the steeper slope of HR-$\dot{V}O_2$ relationship for walking compared to running are consistent with the study by Hreljac.[11]

The energy expenditure of walking in normal children and adults has been determined by measuring oxygen uptake.[1, 4, 6, 11, 12] Recently, heart rate has been used to estimate energy expenditure, not only because the measurement of heart rate is more straitforward and accurate and but also because a linear relationship between heart rate and oxygen uptake has been shown over wide range of treadmill walking speeds. The estimate of energy expenditure from ambulatory HR monitoring is important to assess physical activity and fitness levels for both scientific and clinical interest.[1] The slope of the HR-$\dot{V}O_2$ regression line for walking determined by ambulatory monitoring may be also used to assess fitness level. Åstrand and Rodahl[13] reported that with training, heart rate decreased for a given level of oxygen uptake and that the slope of the HR-$\dot{V}O_2$ relationship increased with bed rest and decreased with training.

Although a linear relationship was observed between HR and $\dot{V}O_2$ in each gait of walking and running, the slope of the HR-$\dot{V}O_2$ regression line was different between walking and running. Moreover, the gait speed results from the product of step rate and step length. The energy expenditures of walking and running depend on both step rate and step length.[12, 14] Therefore, the energy expenditure of both walking and running cannot be strictly estimated from speed alone, since a subject can be forced to walk or run at the same speed but with varying combinations of step rate and step length. Despite these limitations, the estimate of walking or running energy expenditure from a function of gait speed seems to be useful, according to the findings by Zarrugh and Radcliffe[12] and Cavanagh and Kram[14] who have demonstrated that the freely chosen step rate for walking and running was that requiring the least oxygen consumption at any given speed.

5. ACKNOWLEDGMENTS

This research was supported in part by a grant from the Ministry of Health and Welfare (no. H11-Longevity-039). We thank Mrs. Y. Kanezawa for assistance in preparation of the manuscript.

6. REFERENCES

1. J. Rose, J. G. Gamble, J. Lee, R. Lee, and W. L. Haskell, The energy expenditure Index: A method to quantitate and compare walking energy expenditure for children and adolescents, *J. Pediatr. Orthop.* **11**, 571-578 (1991).
2. T. Takahashi, and Y. Miyamoto, Influence of light physical activity on cardiac responses during recovery from exercise in humans, *Eur. J. Appl. Physiol.* **77**, 305-311 (1998).
3. G. A. Cavagna, F. P. Saibene, and R. Margaria, External work in walking, *J. Appl. Physiol.* **18**, 1-9 (1963).
4. R. J. Robertson, C. J. Caspersen, T. G. Allison, G. S. Skrinar, R. Anne Abbott, and K. F. Metz, Differentiated perception of exertion and energy cost of young women while carrying loads, *Eur. J. Appl. Physiol.* **49**, 69-78 (1982).
5. M. Y. Zarrugh, F. N. Todd, and H. J. Ralston, Optimization of energy expenditure during level walking, *Eur. J. Appl. Physiol.* **33**, 293-306 (1974).
6. R. Margaria, P. Cerretelli, P. Aghemo, and G. Sassi, Energy cost of running, *J. Appl. Physiol.* **18**, 367-370 (1963).
7. D. F. Hoyt, and C. R. Taylor, Gait and the energetics of locomotion in horses, *Nature* **292**, 239-240 (1981).
8. H. F. Stegall, Muscle pumping in the dependent leg, *Circ. Res.* **19**, 180-190 (1966).
9. E. Poulsen, and E. Asmussen, Energy requirements of practical jobs from pulse increase and ergometer test, *Ergonomics* **5**, 33-36 (1963).
10. Z. Vokac, H. Bell, E. Bautz-Holter, and K. Rodahl, Oxygen uptake/ heart rate relationship in leg and arm exercise, sitting and standing, *J. Appl. Physiol.* **39**, 54-59 (1975).
11. A. Hreljac, Preferred and energetically optimal gait transition speeds in human locomotion, *Med. Sci. Sports Exerc.* **25**, 1158-1162 (1993).
12. M. Y. Zarrugh, and C. W. Radcliffe, Predicting metabolic cost of level walking, *Eur. J. Appl. Physiol.* **38**, 215-223 (1978).
13. P.-O. Åstrand, and K. Rodahl, *Textbook of work physiology* (McGraw-Hill, Inc., New York, 1986).
14. P. R. Cavanagh, and R. Kram, Mechanical and muscular factors affecting the efficiency of human movement, *Med. Sci. Sports Exerc.* **17**, 326-331 (1985).

THE BREATH SOUND CHECK AND EXERCISE AT OR ABOUT THE VENTILATORY THRESHOLD

Robert W. Mertens, Harold J. Bell, and Robert C. Goode*

1. INTRODUCTION

There is a scarcity of information on the control of exercise intensity using ventilation and/or breathing sounds. We have reported[6, 7] that when a subject while jogging can "hear your breathing" that subject is at or near their Ventilatory Threshold (VT1). The VT1 has been described by Plowman and Smith[15] as "the first breakpoint in volume expired per minute." The use of heart rate and/or perceived exertion does not always indicate the metabolic intensity at which the subject is operating[1]. There is evidence that exercising at the VT1 is beneficial for cardiac transplant patients[12], asthmatics[16], diabetics[13], obese subjects[14], and young and older populations[5, 10, 18]. At the exercise intensity of the VT1 cardiovascular and respiratory improvements occur. Measurement of aerobic power (VO_2 max) is often employed as an indicator of the effectiveness of an exercise program, however, significant increases in VT1 can occur with minimal changes in VO_2 max and can result in health benefits[9,10].

VT1 is determined, in the laboratory by measuring ventilation, CO_2 production and O_2 consumption[17]. These procedures, while efficient, do not lend themselves to the field situation where activities such as jogging, or performing physical tasks occur. We have observed subjects on reaching an intensity of exercise such that they could "hear your breathing" that their heart rate was close to or at the threshold for a training effect[7, 11].

The experiments described in this paper are designed to examine: (A) the relationship between VT1 and the intensity of exercise at the point where a subject indicates that they can "hear your breathing" (the Breath Sound Check[7]) while exercising; and (B) the effects of exercise training at and about the VT1, on endurance performance indicators.

* Department of Physiology and Exercise Science Unit, Faculties of Medicine and Physical Education and Health, University of Toronto.

Frontiers in Modeling and Control of Breathing, edited by Poon and Kazemi.
Kluwer Academic/Plenum Publishers, New York, 2001.

2. THE 'BREATH SOUND CHECK' AND EXERCISE INTENSITY

2.1. Methods

In the first set of experiments, series I, seven male subjects were asked to indicate the point at which they "can *readily hear* your breathing" during a progressive submaximal exercise test. Series II, nine male subjects, was identical to series I except that the instruction "can *hear* your breathing" was employed. The 16 male subjects (series I and II combined) had the following characteristics: Age (18-30, M±SD = 22±3), BMI (18.5-31.5, M±SD = 23.8±3.3), Rest HR (60-93, M±SD = 70±8) Rest BP (90/50-165/70, M±SD = 120/68±21/±9), VO_2 max (31.0-57.5, M±SD = 47.5±7.7).

A progressive submaximal test was performed on an electrically braked cycle ergometer (Seimens-Elema Ergomed 920). Heart rate for all exercise tests was recorded with a Biopac MP100A data acquisition system using a three electrode setup with disposable electrodes.

Expired gases were collected continuously and analyzed every 20 seconds during a VO_2 max test for all subjects (SensorMedics 2900 Metabolic Measurement Cart). The CO_2 and O_2 analyzers were calibrated with previously chemically analyzed gases.

In series *(I)* and II subjects pedalled (60 rpm) for one minute at zero watts. After one minute the work rate was increased 25 watts per minute[3, 17] until the subjects indicated they could "*(readily)* hear your breathing" by raising their hand. At this time the work rate was held constant for five minutes. A second and third test were performed to assess the reliability of each phrase. Heart rate at the time when the subjects raised their hands was used to assess the reliability of the method (Table 1).

Table 1. Heart rate for reliability tests one through three

Subject		1st	2nd	3rd	M(1-3)	SD(1-3)
Series I	101	119	124	115	119	±5
	102	143	137	136	139	±4
	103	118	116	117	117	±1
	104	143	128	118	130	±13
	105	108	101	110	106	±5
	106	102	115	112	110	±7
	107	111	109	110	110	±1
Series II	108	100	101	98	100	±2
	109	103	105	114	107	±6
	110	117	124	142	128	±13
	111	115	112	112	113	±2
	112	110	124	128	121	±10
	113	144	127	144	138	±10
	114	96	93	92	94	±2
	115	166	146	149	154	±11
	116	137	124	133	131	±7

The test protocol for VO_2 max involved the subjects pedalling (60 rpm) for one min, at zero watts, then the work rate was increased 25 watts every minute until volitional exhaustion. The VT1 was determined as follows:

1) the first nonlinear increase or lowest point before an increase in the plot of ventilatory equivalent for oxygen ($VEQO_2$) and VO_2,
2) the first nonlinear increase in the plot of VE and VO_2, and
3) the first nonlinear increase in the plot of VCO_2 and VO_2 [4, 8].

When all three breakpoints occurred at a different VO_2, the plot with the clearest breakpoint was selected as the VT1. When there was no clear breakpoint, an average of the three breakpoints was calculated.

2.2. Results

A comparison was made for the two series ("readily hear your breathing and "hear your breathing") between the steady rate heart rate after the hand was raised (average heart rate of minutes two to five) and the heart rate at VT1 as determined from analysis of expired air (Table 2). The mean difference in heart rates between the two tests for all subjects combined is 6 beats per minute. Subjects 101, 103, 107, 108, and 109, had heart rates at the steady rate point that were some ~ 14% *less* than directly measured at the VT1. One subject (116) had a heart rate that was 41 beats above that observed at his VT1.

Table 2. Measure of agreement between BSC[a] steady rate[b] and VT1

Subject #		HR Max beats/min	BSC HR (2-5 min)		VTI		BSC-VTI	BSC-VTI
			beats	%	beats	%	beats/min	% HR max
Series I	101	182	128	70	149	82	-21	-12
	102	190	154	81	143	75	11	6
	103	196	126	64	148	76	-22	-11
	104	184	145	79	148	80	-3	-2
	105	193	118	61	123	64	-5	-3
	106	162	117	72	120	74	-3	-2
	107	182	119	65	144	79	-25	-14
Series II	108	180	106	59	133	74	-27	-15
	109	162	113	70	144	89	-31	-19
	110	171	137	80	144	84	-7	-4
	111	193	120	62	123	64	-3	-2
	112	187	145	78	130	70	15	8
	113	198	136	69	147	74	-11	-6
	114	178	98	55	104	58	-6	-3
	115	194	162	84	164	85	-2	-1
	116	180	154	86	113	63	41	23
	M:	183	130	71	136	74	-6	-4
	S.D.:	±11	±19	±10	±16	±9	±18	±10

[a] Breath Sound Check, [b] mean of HR two to five minutes after BSC indicated

2.3. Discussion

The results would suggest that the phrases "can readily hear your breathing " or "can hear your breathing" while exercising (the Breath Sound Check) can provide feedback that the active subjects are, within 15% of their VT1 and they will have a heart rate that is above the minimum for exercise benefit as recommended by the American College of Sports Medicine[2]. Recent evidence[5] and this paper (Part B) suggest that when adults exercise at their VT1 the intensity of exercise is sufficient such that subjects will achieve an increase in VT1 and VO_2 max as a result of the training. The Breath Sound Check can provide a cue for subjects that the activity is sufficiently demanding that they are likely to be receiving a "training effect". The "Breath Sound check" combined with the "Talk Test"[7] will ensure that the exercise is aerobic and provide an exercise guideline that will lead to improvement in cardiorespiratory endurance. The two guides, the "Breath Sound Check" and the "Talk Test" can result in each exercise subject controlling the intensity of aerobic exercise relative to their personal VT1 without the use of technical apparatus such as heart rate monitors.

3. CONSTANT AND INTERVAL LOAD TRAINING AT VT1

In the following experiment we wished to investigate the effects of exercising at the VT1 as compared to exercising at an intensity alternating 15% above and below VT1, as measured by heart rate. We wished to observe if there were changes in VT1 and VO_2 max and if the results were significantly dependent on the training protocol.

Table 3. Subject data at onset of training program

Subject	Sex (m/f)	Age (yrs.)	Weight (kg)	Training Protocol	VO_2max (ml/kg/min)	VT1 (ml/kg/min)
BG	m	59	83.2	CT	37.6	17.3
FH	f	30	79.0	CT	34.4	21.1
GS	f	44	75.0	CT	28.5	15.9
JS	f	29	70.8	CT	36.4	22.6
LV	f	32	76.5	CT	31.2	21.6
RB	f	22	70.4	CT	31.0	16.9
Mean± SD		31.2±11.3	75.8±4.9		33.2±3.5	19.2±2.9
CD	f	23	62.7	IT	27.9	18.5
LC	f	26	95.5	IT	24.5	15.7
MD	m	28	63.0	IT	36.7	17.9
PK	m	31	115.9	IT	24.5	15.1
RC	f	45	45.5	IT	25.3	11.2
SJ	f	20	52.2	IT	33.9	19.7
SM	m	23	84.5	IT	48.8	25.4
Mean± SD		28.6± 8.2	73.4±18.3		31.3±8.9	17.7±4.4

3.1. Methods

Thirteen subjects (five males and eight females) were assigned in a pseudo-random crossover design to either constant or interval load training (CT, IT), and trained with a brisk walk/jog for eight weeks using three, 30-minute sessions per week. Initial subject data is outlined in Table 3.

To collect data on subjects and determine their VO_2 max, VT1, and HR's, three maximal exercise tests were performed: one at the beginning of the study, one half-way through the training program (four weeks), and one at the end of the training program (eight weeks). Subjects performed VO_2 max tests on a motorized treadmill. Ventilatory gasses were collected from the mouth through a two-way non-rebreathing valve, while the subjects wore nose clips. Expired air flowed out through the valve into a wide-bore collection tube, connected to a dry gas meter. The expired gas was 'mixed' in this meter as its volume was measured. A sample of this mixed-expired gas was collected through a gas sample tube connected to a rapidly responding anaesthetic gas monitor. During the test, the subjects' HR was monitored using a finger probe pulse oximeter also connected to the anaesthetic gas monitor. Labview software (specially written) was used to acquire the data during these tests.

Subjects were trained using a walk/jog stepping pace that elicited a HR corresponding the appropriate workload (CT - at VT1 or IT - alternating ±15% about VT1). Workload was adjusted after four weeks to compensate for changes in the VT1. A polar HR monitor (Nissen, model PU-801) was used in conjunction with a Seiko digital metronome (model DM-33) in order to set and monitor the subjects' walk/jogging pace.

3.2. Results

Statistical analysis showed a significant increase in VO_2 max and VT1 between the pre, mid- and post-program exercise tests when expressed as either l/min or ml/kg/min values (P ranging from .004 to .035). The general treatment effects observed in the CT group, were also observed in the test data for the interval trained group. There was a statistically significant increase in VO_2 max and VT1 whether expressed in absolute or relative measures after four and eight weeks of training (P ranging from .001 to .026). Between group differences in improvement were determined to be statistically insignificant (P ranging from 0.073 to 0.876). See Table 4, below.

Table 4. Training induced changes in VO_2max and VT1 for both CT and IT groups after 4 and 8 weeks

Training	Pre		4-weeks		8-weeks	
Group	VO_2max	VT1	VO_2max	VT1	VO_2max	VT1
CT	33.2	19.2	34.6	21.5	40.1	24.7
IT	31.3	17.7	35.3	20.6	38.8	22.7

Units for measures are in (ml/kg/min).

3.3. Discussion

For sedentary persons beginning an exercise program at a training intensity corresponding to their VT1, significant improvements in endurance performance can be elicited through either CT or IT protocols. An individual using the Breath Sound Check method of targeting VT1 during exercise is working at an intensity approximating the threshold, which will likely fluctuate somewhat during the training session. Fluctuations in workload in the order of ±15% about VT1 appear to have insignificant effects on the training response, provided absolute workload for the session is equivalent to that provided by continuous exercise at VT1.

4. REFERENCES

1. Aellen, R., Hollmann, W., & Boutellier, U. (1993). Effects of aerobic and anaerobic training on plasma lipoproteins. *International Journal of Sports Medicine*, **14**, 396-400.
2. American College of Sports Medicine Position Stand. (1998). The Recommended Quantity and Quality of Exercise for Developing and Maintaining Cardiorespiratory and Muscular Fitness, and Flexibility in Healthy Adults. *Medicine and Science in Sports and Exercise*, **30**(6), 975-991.
3. Anderson, G. S., & Rhodes, E. C. A review of Blood Lactate and Ventilatory Methods of Detecting Transition Threshold. *Sports Medicine*, **8**(1), 43-55. (1989).
4. Beaver, W. L., Wasserman, K., & Whipp, B. J. (1986). A new method for detecting anaerobic threshold by gas exchange. *Journal of Applied Physiology*, **60**(6), 2020-2027.
5. Fabre, C., Masse-Biron, J., Ahmaidi, S., Adam, B., & Prefaut, C. (1997). Effectiveness of individualized aerobic training at the ventilatory threshold in the elderly. *Journal of Gerontology Series A Biological sciences and medical sciences*, **52**(5), B260-B266.
6. Goode, R. C., Mertens, R., Bell, H., & Piccinini, J. (1999). Voice, breathing sounds, just audible breathing, the ventilatory threshold and the control of exercise intensity. *Proceedings of Activity for Older Adults: From Research to Action*, 73-76.
7. Goode, R. C., Mertens, R., Shaiman, S., & Mertens, D. (1998). Voice, breathing, and the control of exercise intensity. *Advances in Experimental Medicine and Biology*, **450**, 223-229.
8. Gordon, N. F., & Scott, C. B. (1995). Exercise Intensity Prescription in Cardiovascular Disease: Theoretical Basis for Anaerobic Threshold Determination. *J. Cardiopulmonary Rehabilitation*. **15**(3), 193-196.
9. Haskell, W.L. (1994). Health consequences of physical activity: understanding and challenges regarding dose-response. *Medicine and Science in Sports and Exercise* **26**(6), 649-660.
10. Henritze, J., Weltman, A., Schurrer, R., & Barlow, K. (1985). Effects of training at and above the lactate threshold on the lactate threshold and maximal oxygen uptake. *European J. Appl. Physiol.*, **54**, 84-88.
11. Karvonen, M. J., Kentala, E., & Mustala, O. (1957). The Effects of Training on heart rate. *Acta Phsyiol. Scand.*, 307-315.
12. Kavanagh, T., Magdi, H. Y., Mertens, D. J., Kennedy, J., Campbell, R. B., & Sawyer, P. (1988). Cardiorespiratory responses to exercise training after orthotopic cardiac transplantation. *Circulation*, **77**(1), 162-171.
13. Kawaji, K., Fujita, Y., Yajima, Y., Shirataka, M., & Kubo, H. (1989). Usefulness of anaerobic threshold in estimating intensity of exercise for diabetics. *Diabetes Research and Clinical Practice*, **6**(4), 303-309.
14. Kim, H. S., Tanaka, K., & Maeda, K. (1991). Effects of exercise training at an intensity relative to lactate threshold in mildly obese women. *Annals of Physiological Anthropology*, **10**(4), 229-236.
15. Plowman, S. A. & Smith, D. L. (1997). *Exercise physiology: for health, fitness, and performance*. Boston: Allyn and Bacon.
16. Varray, A. L., Mercier, J. G., Terral, C. M., & Prefaut, C. G. (1991). Individualized aerobic and high intensity training for asthmatic children in an exercise readaptation program: Is training always helpful for better adaptation to exercise? *Chest* **99**(3), 579-586.
17. Wasserman, K., & Whipp, B. J. (1975). Exercise Physiology in Health and Disease. *American Review of Respiratory Disease*, **112**, 219-249.
18. Yamamoto, Y., Miyashita, M., Hughson, R. L., Tamura, S., Shinohara, M., & Mutoh, Y. (1991). The ventilatory threshold gives maximal lactate steady state. *European Journal of Applied Physiology*, **63**(1), 55-59.

IMPROVED O_2 TRANSPORT AND UTILIZATION CAPACITY FOLLOWING INTERMITTENT HYPOBARIC HYPOXIA IN RATS

Michiko Tanaka, Mou Nagasaka, Kiyoshi Sunaga, Atsuko Masuda, and Yoshiyuki Honda*

1. INTRODUCTION

Numerous studies have shown that exposure to hypobaric hypoxia induces altered respiration as well as circulatory activities. It has been suggested that acclimatization to high altitude leads to increased capillary density[1], red blood cell (RBC) counts and Hemoglobin (Hb) concentration[2,3], which may play a beneficial role in maintaining energy production of the muscles with low Po_2. Previous studies[4] have indicated that elevation of respiratory enzyme activity occurs in some muscles with a high oxidative capacity. Although many studies have reported changes of skeletal muscle metabolism from sea level to hypobaric hypoxia, there are few studies that address the opposite process, i.e. from hypobaric hypoxia to sea level[5]. Furthermore, respiratory alkalosis is initially induced by exposure to hypoxia, while the O_2 dissociation curve is shifted to the right due to an increased level of 2,3-diphosphoglycerate (DPG) in RBC resulting from the hypoxemia and alkalosis[6]. This right shift of the O_2 dissociation curve is assumed to facilitate a better O_2 unloading in venous blood at a given Po_2.

It remained to be studied whether or not O_2 transport as well as skeletal muscle metabolism are still well maintained during the recovery period after exposure to intermittent hypobaric hypoxia.

2. METHODS AND MATERIALS

Nineteen male Wistar rats were randomly divided into 2 groups: an experimental group (HH, n=11) exposed to hypobaric-hypoxia at about 450 Torr for 5 weeks from 21

---

* Michiko Tanaka, Mou Nagasaka, Kiyoshi Sunaga, Miyazaki Prefectural Nursing University, Miyazaki, 880-0924, Japan. Atsuko Masuda, Tokyo Medical and Dental University, Tokyo, 113-8519, Japan, Yoshiyuki Honda, Chiba University, Chiba, 260-8672, Japan.

Frontiers in Modeling and Control of Breathing, edited by Poon and Kazemi.
Kluwer Academic/Plenum Publishers, New York, 2001.

days of age, and a control group (C, n=8) fed for 5 weeks at the sea level the equivalent quantities of diet consumed by the HH group.

Two rats in HH group were exposed to hypobaric-hypoxia in an environmental chamber (45×45×35 cm) at about 450 Torr (~4,300 m high) for 8 hours daily for 5 weeks. The flow rate of air in the chamber was maintained at 18 l/min. The ambient temperature and relative humidity in the animal room with 12-hr light-dark cycle were maintained at approximately 23°C and 55 %, respectively throughout the study. Each rat was weighed and diet intake was measured daily at 10 a.m. before food and water were supplied. All animals were anesthetized with i.p. injection of sodium pentobarbital (50mg / kg body weight) about 16 hr after termination of the last hypoxic or normoxic exposure. Blood was withdrawn into a heparinized syringe from the abdominal aorta. A portion of the whole blood was analyzed for measurement of pH, Po_2, Pco_2, base excess (BE), $[HCO_3^-]$, Sao_2, and P_{50} by Blood gas analyzer (OPTI1CCA, AVL). The remaining blood was analyzed for RBC counts, Hb concentration, hematocrit (Hct) level, mean corpuscular volume (MCV), mean corpuscular Hb (MCH), and mean corpuscular Hb concentraion (MCHC) by a Sysmex autoanalyzer (k-4500, Sysmex, Japan).

Soleus muscle was homogenized in 175 mM KCl buffer (pH 7.4) using a polytron. The activities of succinate dehydrogenase (SDH), lactate dehydrogenase (LDH) and citrate synthase (CS) were spectrophotometrically assayed. The SDH and CS were used to assess the citric acid cycle activity and LDH was used to determine the capacity of lactate production. Protein concentration was measured by the method of Lowry et al[7].

The data between two groups were compared using unpaired Student's t-test. Differences were considered significantly at $p<0.05$.

3. RESULTS

Becasue a marked anorexic state in the HH group was observed in comparison with the normobaric group[8], the rats in the C group were fed restricted amounts of the diet consumed by rats in hypoxic condition on the preceding day. There was no change in body weight between the C and HH group.

Table 1 shows the effects of the recovery from hypobaric hypoxia on blood gases and hematological characteristics. We found no difference in the RBC count, Po_2, and SaO_2 between the C and HH groups. Pco_2 of the HH group was higher than that of the C group, while pH was maintained with no significant difference between the C and HH group. Hb concentration, Hct level, MCV, MCH and MCHC in the HH group was higher than in the C group. In addition, P_{50} of the HH and C groups was 33.8±0.9, and 30.1±1.3 mmHg, respectively.

The weight of the soleus muscle in the C and HH groups was 110±6 and 101±3 mg, respectively. These muscle weights were not affected by the exposure to hypobaric hypoxia. The enzyme activities in soleus muscle are shown in Table 2. The specific activity of SDH in the soleus in the HH group tended to be higher compared with the C group. On the other hand, the activity of LDH of the HH group was less than that of the C group. There was no difference in the activity of CS between the HH and C group. The LDH/CS ratio, which is often used for comparing relative capacities for anaerobic vs. aerobic metabolism in tissues, was significantly lower in the HH group than the C group.

Table 1. Effects of the recovery from hypobaric-hypoxia on blood gases and hematological characteristics

	C group (n=8)	HH group (n=11)	p value
PH	7.48±0.07	7.45±0.02	0.605
Po_2 (Torr)	81.4±5.4	91.2±9.9	0.455
Pco_2 (Torr)	26.6±2.3	36.6±1.7	<0.01
BE	0.94±1.68	1.70±1.66	0.760
$[HCO_3^-]$(mEq/l)	21.7±1.2	24.4±1.3	0.158
Sao_2(%)	94.2±1.5	94.3±1.3	0.967
P_{50} (Torr)	30.1±1.4	34.2±0.8	<0.05
RBC counts ($\times 10^4$ counts/μl)	647±23	678±14	0.249
Hb concentration (g/dl)	13.6±0.4	15.7±0.4	<0.01
Hct level (%)	37.3±1.3	41.4±1.1	<0.05
MCV (fl)	57.7±0.4	61.3±0.6	<0.01
MCH (pg)	21.1±0.2	23.9±0.4	<0.01
MCHC (%)	36.6±0.2	38.2±0.6	0.056

Mean±SEM. C group: control group, HH group: Hypoxic group. BE: base excess, RBC: red blood cell, Hb: Hemoglobin, Hct: Hematocrit, MCV: mean corpuscular volume, MCH: mean corpuscular Hb, MCHC: mean corpuscular hemoglobin concentration.

Table 2. Effects of the recovery from hypobaric hypoxia on enzyme activities of the soleus muscle

	C group (n=8)	HH group (n=11)	p value
LDH			
(μmol/min/g wet wt.)	403±31	289±30	<0.05
(μmol/min/g protein)	3336±209	2628±287	0.066
SDH			
(μmol/min/g wet wt.)	10.1±1.7	11.0±1.2	0.685
(μmol/min/g protein)	68±10	100±12	0.058
CS			
(μmol/min/g wet wt.)	32.7±1.3	32.8±2.5	0.962
(μmol/min/g protein)	274±15	298±24	0.401
LDH/CS	12.3±0.8	8.2±0.7	<0.01

Mean±SEM. C group: control group. HH group: hypoxic group. LDH: Lactate dehydrogenase. SDH: Succinate dehydrogenase. CS: Citrate synthase.

4. DISCUSSION

Many studies have been conducted previously on the adaptations to hypoxic exposure, while studies concerning the changes of the skeletal muscle metabolism from hypoxia to normoxia are rare[5]. In this study, blood gases, hematological characteristics and soleus muscle metabolism in post-hypobaric hypoxia (~ 450 Torr) were studied for 8 hours per day for 5 weeks in rats.

Many investigators have shown that exposure to high altitude induces a loss in body weight due to reduced caloric intake resulting from anorexia, mal-absorption of ingested fat and/or carbohydrates, elevated rate of energy expenditure, and water loss[6]. In this study, the body weight and soleus weight showed no difference between the C and HH group, because rats in both groups were fed the equivalent quantities of diets.

We found that RBC counts of the HH group slightly (but not significantly) increased compared with the C group. However, Hb concentration, Hct level, MCH and MCHC in the HH group were significantly higher than in the C group. Although we observed that Hct level of HH groups was significantly higher than that of the C group, this change of Hct level was within the normal range and thus should not impair O_2 carrying capacity due to increased viscosity[9].

A previous study of capillary O_2 transport during severe hypoxia showed that a decreased P_{50} may be advantageous for tissue oxygenation when the O_2 supply is severely compromised[10]. This observation does not apply to the present study, which was not conducted with an exposure to severe hypoxia. Our study showed that the increased P_{50} in the HH group may also have an important role in O_2 delivery in tissues. The increased P_{50} in HH group, including a shift to the right of the O_2 dissociation curve, might be due to the Bohr effect. These results suggest that the animals in our study may have improved O_2 delivery and O_2 carrying capacity without significant increase in viscosity.

It has been reported that intermittent hypoxia has strong influence on tissue metabolism[11]. It is possible that the intermittent hypoxia for 8 hours per day in this study has the same effect on the skeletal muscle as stimuli in periodically altered environment of hypoxia-normoxia. The specific activity of SDH in the HH group tended to be higher than in the C group; in contrast, activity of LDH in HH group was less than in the C group. Moreover, the LDH/CS activity ratios in soleus muscle were significantly lower in the HH group than in the C group. Our results are in agreement with previous investigations which reported that chronic exposure to altitude induced increased oxidative enzymes activity in tissues[2,12]. We suggest that exposure to hypobaric hypoxia results in depressed anaerobic metabolism, but maintained oxidative metabolism, even during the recovery period after the termination of hypoxia.

Oxygen deliveries to the tissues are determined by the release of O_2 from Hb, the affinity of Hb for O_2, the level of the tissue metabolism, and the amount of blood in the tissue. We conclude that in the post-hypobaric-hypoxia period, O_2 transport capacity may remain elevated as a consequence of increased Hb concentration and increased O_2 delivery to the tissues due to an increased P_{50}, while O_2 utilization capacity of the skeletal muscles may also increase due to an increased mitochondrial enzyme activity.

5. ACKNOWLEDGMENTS

This research was supported by the Miyazaki Prefectural Nursing Foundation.

6. REFERENCES

1. X. Bigard, A. Brunet, C-Y. Guezennec and H. Monod, Effects of chronic hypoxia and endurance training on

muscle capillarity in rats. *Eur. J. Physiol.* **419**, 225-229 (1991)

2. M. Tanaka, K. Mizuta, F. Koba, Y. Ohira, T. Kobayashi and Y. Honda, Effects of exposure to hypobaric-hypoxia on body weight, muscular and hematological characteristics, and work performande in rats. *Jpn. J. Physiol.* **47**, 51-57 (1997)
3. I. Kuwahara, T. Iwamoto, U. Kamiya, Changes in Hb concentration due to splenic contraction in chronic hypoxic rats. In : *Progress in Mountain Medicine and High Altitude Physiology*, edited by Ohno H., Kobayashi Y., Masuyama S. and Nakashima M., Kyoto, Dogura & Co., Ltd, 1998, p.452-453
4. S. Taguchi, A. Ishihara, M. Itoh and K. Itoh, Effects of hypobaric hypoxia on the oxidative capacity of the extensor digitorum longus motor in the rat. *Neurochem. Res.* **15**, 923-926 (1990)
5. R. Favier, H. Spielvogel, D. Desplanches, G. Ferretti, B. Kayser, A. Grunenfelder, M. Leuenberger, L. Tuscher, E. Caceres, and H. Hoppeler. Traning in hypoxia vs. training in normoxia in high-altitude natives. *J. Appl. Physiol.* **78**, 2286-2293 (1995)
6. J. B. West, *Respiratory physiology* (William & Wilkins, Balyimore, 1985)
7. H. Lowry, N. J. Rosebrough, A. L. Farr and R. J. Randall, Protein measurement with the Folin phenol reagent. *J. Biol. Chem.* **193**, 265-272 (1951)
8. M. S. Westertep-Plantenga, K. R. Westeterp, M. Rubbens, C. R. Verwegen, J-P, Richelet and B. Gardette, Appetite at " high altitude" [Operation Everest III (Comex-'97)]: a simulated ascent of Mount Everest. *J. Appl. Physiol.* **87**, 391-399 (1999)
9. R. R. Wolfe and S. M. Horvath, Role hematocrit and blood volume on hypoxia tolerance in the sea level- and altitude- adapted rat. *Aerospace. Med.* **45**, 821-823 (1974)
10. J. C. Stein and M. L. Ellsworth, Capillary oxygen transport during severe hypoxia: role of hemoglobin oxygen affinity. *J. Appl. Physiol.* **75**(4), 1601-1607 (1993)
11. H. Okamoto, S. Yamasaki, S. Ogoh, H. Senge and S. Taguchi. Adaptations to an environmental change of sea level condition to hypobaric hypoxia, and adaptations to reversed change. (in Japanese) *Jpn. J. Mountain Med.* **15**, 97-104 (1995)
12. A. Oshihara, K. Itoh, M. Itoh, C. Hirofuji and H. Hayashi, Hypobaric-hypoxia exposure and histochemical responses of soleus muscle fibers in the rat. *Acta. Histchem.* **96**, 74-80 (1994)

Respiratory Sensation and Behavioral Control

INTERCEPT SHIFT IN THE BREATHLESSNESS/ VENTILATION RELATIONSHIP IN RESPONSE TO PROGRESSIVE INCREASE IN WORKLOAD

Change in exercise breathlessness response

H.R.Ahmad, M.M.F. Subhan, Shabbir Bokhari, Asad Ali and M.N.Khan*

1. INTRODUCTION

The relationship between breathlessness based on the visual analog scale (VAS) and expired pulmonary ventilation (V_E) has been investigated under different experimental conditions. The 'Newcastle effect' revealed for the first time that a repetitive exercise protocol over a period of ten days increased the intercept without change in the slope of VAS/V_E relationship[1]. Whether experiments were conducted daily or weekly, the time course of the intercept plateaued after around 5 to 6 weeks of exercise training[2]. A similar effect has been shown to occur during volitional tracking of an inspiratory resistive breathing pattern[2].

In another study, hypercapnia has been shown to decrease the intercept and increase the slope[3, 4]. Inspiratory resistive loading increased the slope without any change in the intercept[5]. The VAS/V_E relationship, when examined in patients pursuing a COPD-rehabilitation program, showed a decrease in the slope without change in the intercept[6]. The analysis of the intercept and slope dynamics, being the quantitative indices of breathlessness in contrast to VAS scores alone, may open avenues in modeling and understanding the mechanisms of breathlessness.

The aim of this study, firstly, was to see if the 'Newcastle effect' could be reproduced. Secondly, since in clinical practice the recording of heart rate is simpler and easier then that of expired pulmonary ventilation, the dynamics of VAS-heart rate relationship has been additionally investigated. Thus we have a representation of both respiratory and cardiovascular control systems being examined with reference to a change in the degree of breathlessness.

---

* Department of Physiology & Pharmacology, The Aga Khan University, Karachi-74800, Pakistan.

Frontiers in Modeling and Control of Breathing, edited by Poon and Kazemi.
Kluwer Academic/Plenum Publishers, New York, 2001.

2. METHODOLOGY

2.1. Protocol

The effect of repetitive testing on the perceived intensity of breathlessness during bicycle ergometry was examined in a group of healthy subjects. Subjects undertook a progressive exercise test to a symptom-limited maximum during the course of which breathlessness was estimated each minute using a VAS. Subjects exercised weekly on 6 to 10 occasions.

2.2. Subjects

The subjects were recruited from staff of the Aga Khan University (AKU) and were naive as to the aims of the study. All were non-smokers with no history of cardiopulmonary disease. Experiments were performed at approximately the same time of day, at least 2 hours after the last meal. Informed consent was obtained from the subjects before the study commenced as approved by the AKU ethical committee. All had normal lung function with respect to the reference values[7]. The Group had eleven subjects and their mean age was 23.96 ± 5.8 years (Mean ± Standard Deviation). Six subjects completed 10 weeks of testing, the remaining only 6 weeks.

2.3. Exercise Test

The exercise tests were conducted on an ergometer (Tunturi, Finland). The subject breathed through a pneumotachograph, which allowed inspiration and expiration from room air. Once attached to the equipment, the subject was given a few minutes to be accustomed to the apparatus. The heart rate was monitored throughout the experiment using a pulse oximeter (Biox, Ohmeda, USA). Inspired ventilation was measured using a Fleish pneumotachograph (Gould Medical, Holland). The data from the pneumotachograph and pulse oximeter was recorded on a Y-T pen recorder (Figure 1).

The VAS was drawn as a 100 mm linear line on a card, the extremes of which were labeled 'not at all breathless' (0 mm) and 'extremely breathless' (100 mm). Every response of the subject was recorded using a pencil, on a scale of 0-100 mm and for every minute, a separate card was used. The subject was given approximately 5 s to mark the card, which was held before the subject at the end of each minute. Before the start of each experiment the subjects were requested not to confuse breathlessness sensations with other sensations associated with exercise, such as leg fatigue or the awareness of increased breathing. Subjects were also told on subsequent occasions to use the same criteria in estimating their sensations as before.

After 2 minutes at rest, the subject started unloaded pedaling, for one minute, at a frequency of between 50-60 revolutions per minute. Thereafter, the resistance was increased 25 Watts per minute, till a symptom-limited maximum. Subjects were told that if they felt any pains in the arms or legs, or tightness in the chest or dizziness or giddiness in the head then to spit the mouthpiece out, stop pedaling and relax. The test was stopped if the heart rate was within 10 beats/min of the maximum predicted for that individual or where the subjects indicated greater than 90 mm on the VAS. The predicted maximum heart rate was estimated from the equation: $f_{Cmax} = 210-0.65[age]$ in years.[7]

2.4. Analysis

Changes in breathlessness were assessed from the slope and intercept of the linear regression of VAS score and the expired ventilation V_E. A similar analysis was also made with the regression between VAS and heart rate. A paired t-test was used for statistical analysis; all t-tests were two-tailed and the level of probability taken as significant was 5% ($p < 0.05$).

3. RESULTS

3.1. VAS Shows a Threshold Characteristic in Response to Progressive Increase in Workload

The original recording in Figure 1 shows that the increase in tidal volume and heart rate followed a similar time course in response to a progressive increase in work. However, VAS started to increase only after a work rate of 75W. V_T and f_C showed one third of the maximum response at 75W. Therefore, VAS has a typical threshold characteristic. Thereafter, it increased linearly parallel to an increase in V_T and f_C.

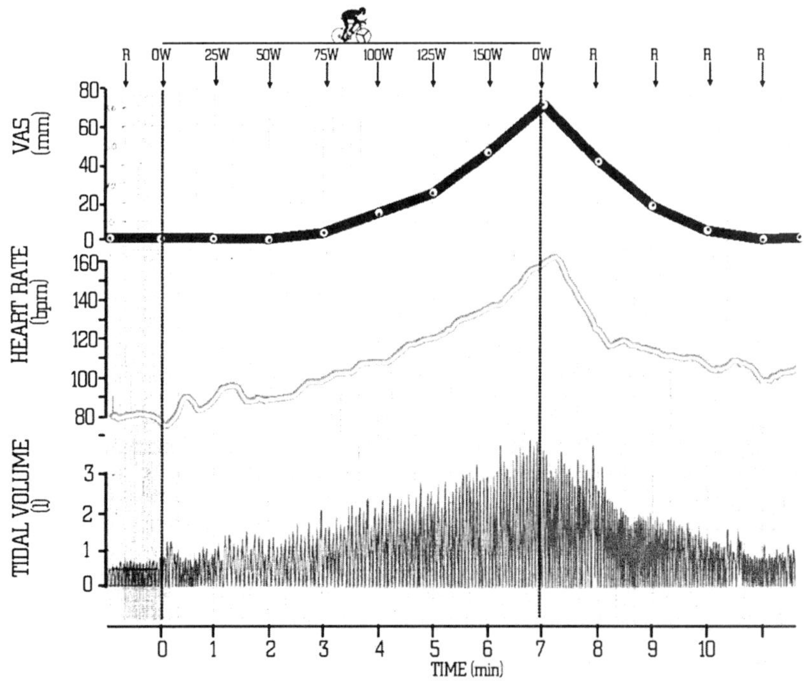

Figure 1. The time course of VAS, heart rate and tidal volume in response to a progressive exercise.

Table 1. Slope and intercept of breathlessness data in relation to ventilation and heart rate and their paired t-test probability values

	VAS – V_E		VAS – f_C	
	Intercept	Slope	Intercept	Slope
Week 1	15 ± 7 (n=11)	1.40 ± 0.50	97 ± 19 (n=10)	0.94 ± 0.40
Week 10	25 ± 9 (n=6)	1.03 ± 0.41	128 ± 21 (n=5)	0.64 ± 0.29
p values	0.0004	0.21	0.0006	0.055

3.2. VAS-V_E Intercept Increases during Repetitive Exercise

The plot of VAS against V_E is effectively linear with a mean r^2 value of 0.93 in Week 1. Table 1 shows the mean slopes of VAS/V_E relationship did not show any significant change (p=0.21) between Weeks 1 and 10. However, the intercept showed a significant increase when compared between Week 1 and 10 (p=0.0004). At Week 6 the increase was not significant (p=0.06).

3.3. VAS-f_C intercept increases during repetitive exercise

The VAS/f_C relationship also showed a similar degree of linearity in Week 1 (r^2=0.91), as that ventilation. The mean slope of VAS/f_C relationship showed a decreasing non-significant trend over a period of 10 weeks (p=0.06). In contrast, the time course of the mean intercept of the VAS/f_C showed a significant increase both at 6 and 10 weeks with a p value of 0.015 and 0.0006 respectively (Table 1).

4. DISCUSSION

This study shows that the slope of the VAS/V_E relationship did not show any significant change over the period of repetitive exercise testing. However, the intercept of the VAS/V_E relationship increased significantly when compared between Week 1 and 10. It may be inferred that these individuals show dyspnoeic pre-conditioning in response to repetitive exercise training. However, this result is partially consistent with the previous findings[1]. It has been demonstrated that the time course of the intercept plateaued between 5-6 weeks[7]. A careful analysis of their data showed a more pronounced plateau in an athletic group[7]. Since individuals in our study were mostly of a sedentary type, this may have contributed to the lack of a plateau in the time course of the VAS/VE relationship. It may be anticipated that they would show the plateau effect beyond 10 weeks of training.

Interestingly the time course of VAS/f_C intercept over a period of 10 weeks did not show a plateau. The conclusion is that neither intercept dynamics of VAS/V_E and VAS/f_C showed the typical 'Newcastle effect'. This discrepancy could be due to difference in athletic and sedentary subjects. This study on normal healthy subjects during repetitive exercise shows that we can assess the intensity of breathlessness relative to the level of expired minute ventilation as well as heart rate.

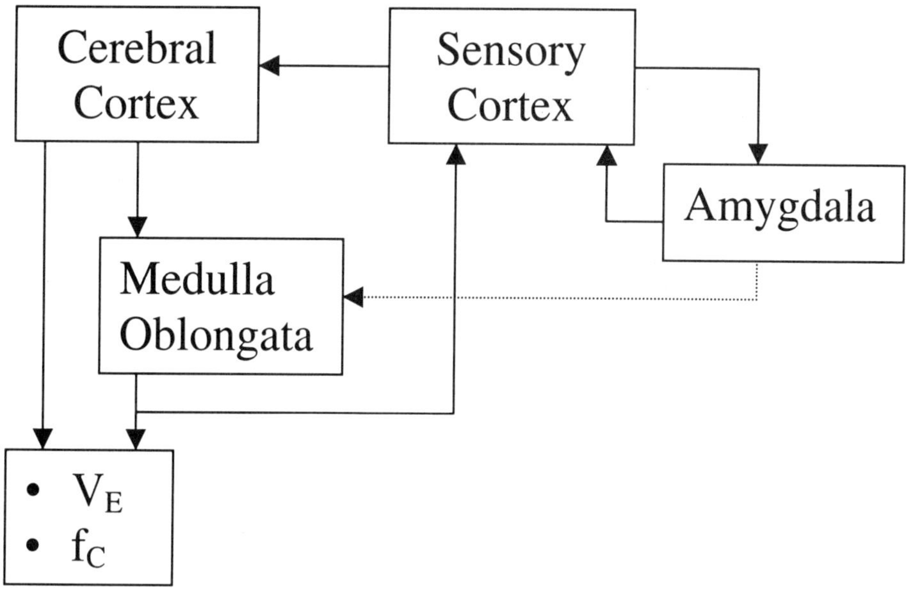

Figure 2. Dyspnea Model showing the interaction of cortical and medullary drives to produce Central Motor Command for V_E and f_C.

The role of peripheral and central chemoreceptors[8-12], intrapulmonary receptors[13-15], chest wall mechanoreceptors[16], and the central motor command[2, 17] in the genesis of breathlessness during exercise have been extensively investigated. Even though the neurophysiological mechanisms of breathlessness and its interpretation are not completely elucidated, the current working hypothesis is that the central motor command from the brainstem is projected not only to the respiratory muscles but also to the sensory cortex[17].

Reed's group has shown that the intercept increases without change in slope in repetitive testing[1] and volitional breathing[2]. The fall in breathlessness with volitional breathing could be due to the cortical drive by-passing the medullary reflex drive. How these two drives interact in exercise is not known. It is postulated in the model presented in Figure 2, that the sensory cortex-amygdala loop may continuously scan the medullary and cortical efferents to modulate the intercept of VAS/V_E and VAS/f_C relationship without change in slope. Thus, this loop may optimize the relationship between the sensation and interpretation of breathlessness and cortical and medullary efferents.

5. ACKNOWLEDGMENTS

Dedicated in fond memory of D.J.C. Cunningham and H.H. Loeschcke.

6. REFERENCES

1. M. M. F. Subhan and J. W. Reed, Effect of repetitive testing on breathlessness in man, *Journal of Physiology* **483.P**, 95-96P (1995).
2. M. M. F. Subhan. Ph.D. Thesis. University of Newcastle upon Tyne, UK (1996).
3. T. Chonan, M. B. Mulholland, J. Leitner, M. D. Altose, and N. S. Cherniack, Sensation of dyspnea during hypercapnia, exercise, and voluntary hyperventilation, *Journal of Applied Physiology*, **68**: 2100-2106 (1990b).
4. J. W. Reed and M. M. F. Subhan, Modelling and control of ventilation ed. Semple, Adams, Whipp: Plenum, New York: 123-127 (1995). (*Adv Exp Med Biol* **393**).
5. R. C. Wilson and P. W. Jones, Influence of prior experience on the estimation of breathlessness during exercise, *Clinical Science*, **78**: 149-153 (1990).
6. A. Ramirez-Venegas, J. L. Ward, E. M. Olmstead, *et al.* The effect of exercise training on dyspnea measures in patients with chronic obstructive pulmonary disease, *J. Cardiopulm Rehab.* **17**:103-09 (1997).
7. J. E. Cotes, Lung Function, assessment and application in medicine, Blackwell Scientific Publications, Oxford (1993).
8. L. Adams, R. Lane, S. A. Shea, A. Cockcroft, and A. Guz, Breathlessness during different forms of ventilatory stimulus: a study of mechanisms in normal subjects and respiratory patients, *Clinical Science*, **69**: 663-672 (1985).
9. S. A. Ward and B. J. Whipp, Effects of peripheral and central chemoreflex activation on the isopneic rating of breathing in exercising humans, *Journal of Physiology*, **411**: 27-43 (1989).
10. T. Chonan, M. B. Mulholland, M. D. Altose, and N. S. Cherniack, Effects of changes in level and pattern of breathing on the sensation of dyspnea, *Journal of Applied Physiology*, **69**: 1290-1295 (1990a).
11. R. Lane, L. Adams, and A. Guz, The effects of hypoxia and hypercapnia on perceived breathlessness during exercise in humans, *Journal of Physiology*, **428**: 579-593 (1990).
12. R. Lane and L. Adams, Metabolic acidosis and breathlessness during exercise and hypercapnia in man, *Journal of Physiology*, **461**: 47-61 (1993).
13. N. R. Banner, M. H. Lloyd, R. D. Hamilton, J. A. Innes, A. Guz, and M. H. Yacoub, Cardiopulmonary response to dynamic exercise after heart and combined heart-lung transplantation, *British Heart Journal*, **61**: 215-223 (1989).
14. R. J. Kimoff, T. H. Cheong, M. G. Cosio, A. Guerraty, and R. D. Levy, Pulmonary denervation in humans, effects on dyspnea and ventilatory pattern during exercise, *American Review of Respiratory Disease*, **142**: 1034-1040 (1990).
15. H. R. Harty, C. J. Mummery, L. Adams, R. B. Banzett, I. G. Wright, N. R. Banner, M. H. Yacoub, and A. Guz, Ventilatory relief of the sensation of the urge of breathe in humans: are pulmonary receptors important? *Journal of Physiology*, **490**: 805-815 (1996).
16. S. C. Gandevia, K. J. Killian, D. K. McKenzie, M. Crawford, G. M. Allen, R. B. Gorman, and J. P. Hales, Respiratory sensations, cardiovascular control, kinaesthesia and transcranial stimulation during paralysis in humans, *Journal of Physiology*, **470**: 85-107 (1993).
17. L. Adams and A. Guz, Dyspnea on Exertion. In: B. J. Whipp, K. Wassermann, ed. Exercise, pulmonary physiology and pathophysiology. New York: Marcel Dekker, 449-494 (1991).

VENTILATORY VS. DYSPNEIC RESPONSES TO AUGMENTING AND SUBSEQUENTLY DECLINING CO_2 STIMULATION IN HUMANS

A. Masuda[*], Y. Sakakibara[†], Y. Ohyabu[‡], C. Yoshino[§], T. Kobayashi,[**] and Y. Honda[††]

1. INTRODUCTION

The PCO_2 - ventilation control system is known to reveal an open-loop condition when the subject rebreathes a high level of CO_2 (such as 7%) according to the Read method.[1] Equilibrium among receptor tissues, mixed venous and arterial blood as well as end-tidal PCO_2 ($P_{ET}CO_2$) is assumed to develop rapidly so that steady-state conditions in the stimulus - response relationship can be achieved promptly. In the present experiment, we first observed the effects of an augmenting $P_{ET}CO_2$ with the conventional Read method (incrementing run, IR). Subsequently, $P_{ET}CO_2$ was lowered at the same rate as that of IR but in an opposite direction (decrementing run, DR). Both the CO_2 - ventilatory as well as dyspneic responses were obtained under iso-oxic condition and the results between IR and DR runs were compared.

2. METHOD

The experimental setup for the IR and DR runs is illustrated in Fig. 1. Nine young college students (five females and four males) participated in the experiments. Their age, height and weight (mean ± SD) were 20.1 ± 3.2 yr, 166 ± 9 cm and 61.7 ± 11.3 kg, respectively. Arterial blood O_2-saturation (SpO_2) as well as respiratory sensation (visual analog scale, VAS) were continuously measured with a pulse oximeter (Ohmeda, Biox III) and with the method according to Adams et al,[2] respectively. Statistical comparisons between two means were made by using Student's paired t-test, whereas multiple comparisons were made by using one-way ANOVA. Significance was accepted at $P < 0.05$.

* Schl. Allied Med. Sci., Tokyo Med. & Den. Univ., Tokyo; †Dept. Human Life Sci. & Environm. System, Kanazawa Inst. Tech., Ishikawa; ‡Dept. Exercise Physiol., Kogakuin Univ., Tokyo; §Schl. Allied Med. Sci., Chiba; **Dept. Hyg. & Prev. Med., Fukushima Med. Univ., Fukushima; ††Dept. Physiol. Schl. Med., Chiba Univ., Japan.

Frontiers in Modeling and Control of Breathing, edited by Poon and Kazemi.
Kluwer Academic/Plenum Publishers, New York, 2001.

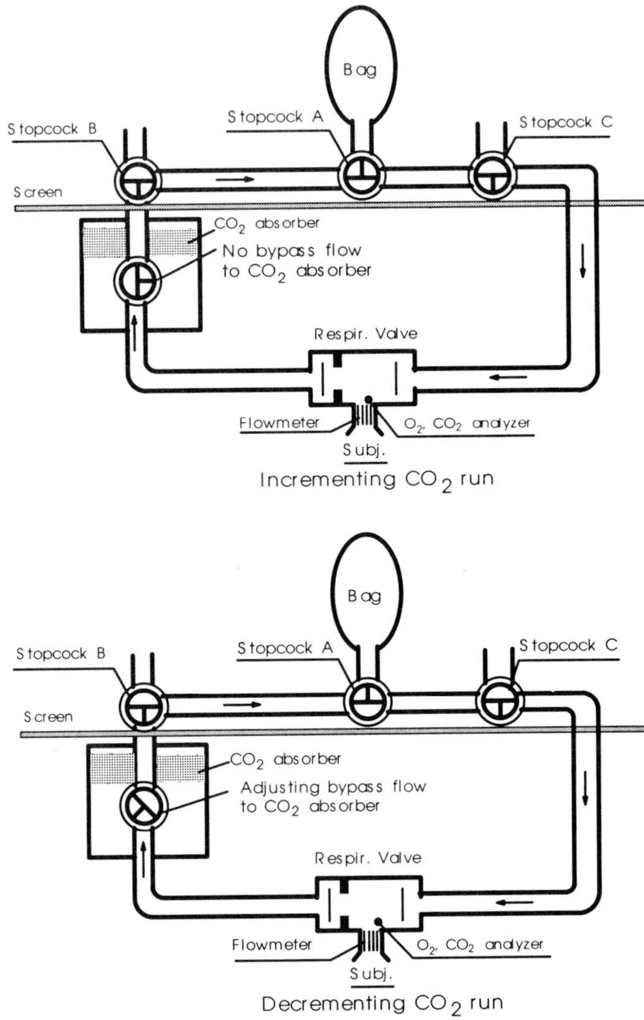

Figure 1. Experimental setup for the IR and DR runs. In the control period, the subject breathes through the respiratory circuit which is open to room air. In the test run, the respiratory circuit is closed and the test gas preliminarily filled in the bag is rebreathed. In the IR run, the bypass flow to CO_2 absorber, which is incorporated in the respiratory circuit, is closed, so that $PETCO_2$ increased with time. On the other hand, in the DR run the stop cock placed at the bypass flow is regulated so as to keep the rate of $PETCO_2$ change in the same with that of IR run, but opposite direction. During the rebreathing test, end-tidal PO_2 ($PETO_2$) was maintained constant by introducing a small amount of O_2 or N_2 in the inspiratory portion of the respiratory valve. Three different iso-PO_2 levels, i.e. 100, 80, and 60 mmHg were chosen and termed normoxic, mild and moderate hypoxic runs, respectively. Each subject conducted the same iso-oxic rebreathing test thrice in random order. In total, therefore, 9 rebreathing tests were performed in each subject. Breath-by-breath tidal volume (V_T) and respiratory frequency (f) were detected by using a hot-wire flow meter (Minato RF-H, Osaka Japan), Minute ventilation was normalized by dividing body surface area. $PETO_2$ and $PETCO_2$ were continuously recorded by a rapid response O_2-CO_2 analyzer (Nihon Denki-Sanei, H-21, Tokyo Japan).

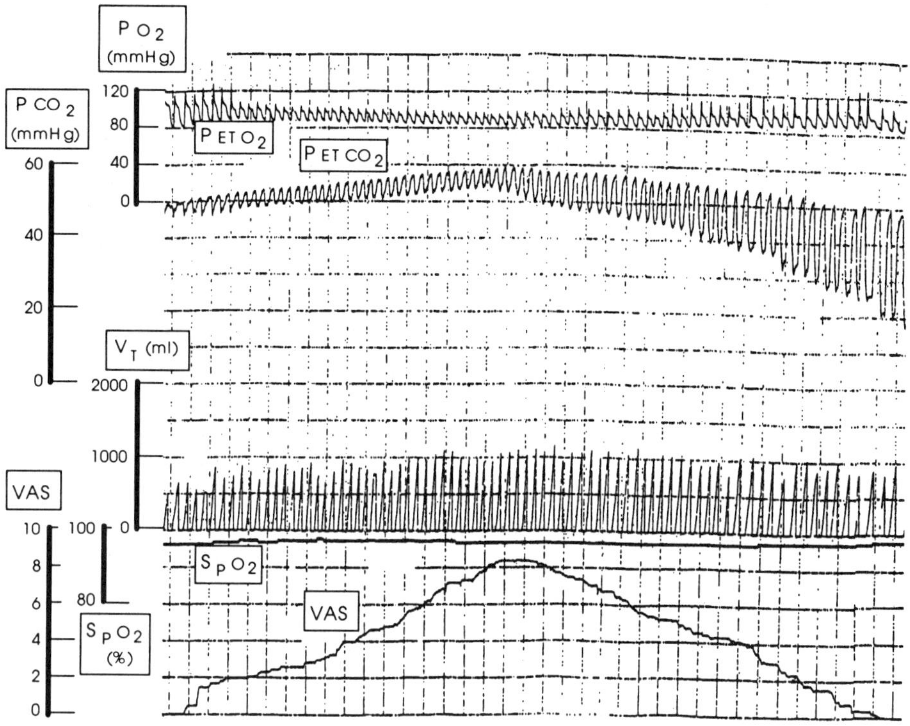

Figure 2. An example of ventilatory and dyspneic responses to incrementing and subsequently decrementing CO_2 stimulation in mild hypoxia. $PETO_2$: end-tidal PO_2; $PETCO_2$: end-tidal PCO_2; V_T: tidal volume; SpO_2: SaO_2 measured by a pulse oximeter; VAS: respiratory sensation measured by a visual analog scale.

3. RESULTS AND DISCUSSION

An example of ventilatory and dyspneic responses to incrementing and subsequently decrementing CO_2 stimulation obtained at mild hypoxia is demonstrated in Fig. 2. The rate of $PETCO_2$ change with time appears to be the same, but in opposite directions in IR vs. DR run. Despite the same $PETCO_2$ depression in the DR period, the degree of ventilatory depression was seen less and VAS suppression greater than those in the IR run.

Figure 3 summarizes the comparison between the IR and DR runs. In panel A, the rates of $PETCO_2$ change with time showed no significant difference between both runs. In panel B, the magnitude of the VAS response in any iso-oxic condition was not significantly different in both runs. Panel C shows that the V_E response to unit $PETCO_2$ change was significantly depressed in the DR run. Panel D shows that the VAS response to unit ventilation was significantly greater in DR than in the IR run, except in the moderate hypoxia condition where the p value was slightly larger than the threshold for statistical significance.

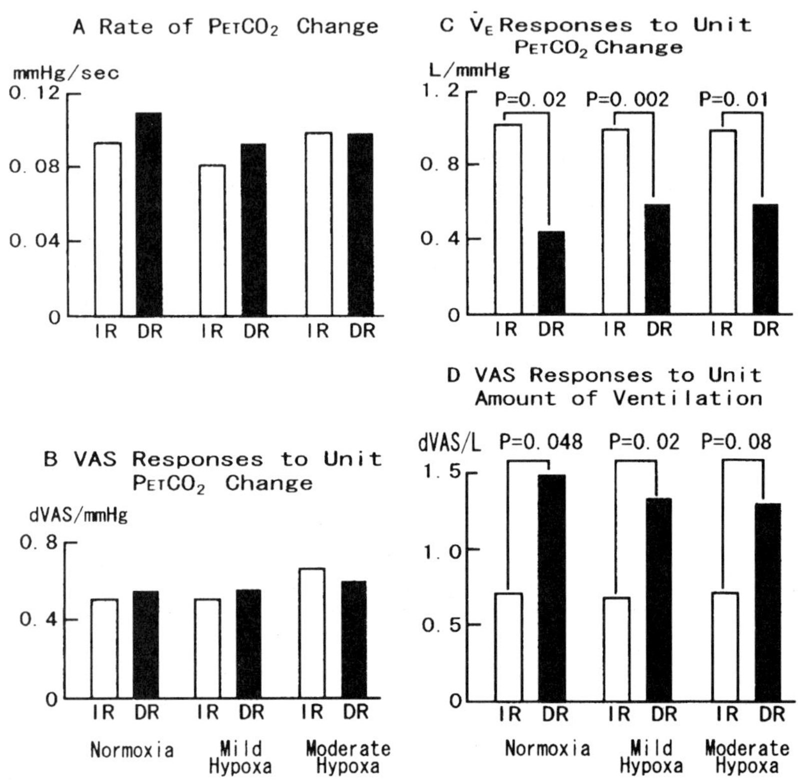

Figure 3. Statistical comparison between incrementing (IR) and decrementing (DR) CO_2 stimulation runs.

In agreement with the above experimental findings, all subjects experienced that their dyspneic sensation was markedly alleviated, so that breathing could easily be performed with less ventilatory effort in DR than in the IR period.

In light of the fact that the VAS response to unit $P_{ET}CO_2$ was maintained almost at the same level, we speculate that the $P_{ET}CO_2$ - VAS axis may represent a significant potential determinant for ventilatory activity in our experimental settings.

4. REFERENCES

1. D. J. C. Read and J. Leigh, Blood-brain tissue PCO_2 relationships and ventilation during rebreathing. *J. Appl. Physiol.*, **23**: 53-70 (1967).
2. L. Adams, N. Chronos, R. Lane and A. Guz, The measurement of breathlessness induced in normal subjects: validity of two scaling techniques. *Clin. Sci.*, **69**: 7-16 (1985).

HYPOXIC-HYPERCAPNIC INTERACTION ON VENTILATORY RESPONSE AND RESPIRATORY SENSATION IN FEMALES

Profile during menstrual cycle

Atsuko Masuda, Yoshio Ohyabu, Toshio Kobayashi, Chikako Yoshino, and Yoshiyuki Honda*

1. INTRODUCTION

Ventilatory control in response to combined CO_2 and hypoxic stimulations has been well documented as inducing positive interaction. Dyspnea sensation or breathlessness is an important determinant for limiting successful achievement of respiratory activities. Although hypoxia and hypercapnia are important chemical factors of respiratory sensation, little investigation of the study of the interaction of these two chemical factors on respiratory sensation has hitherto been reported. On the other hand, previous investigations have demonstrated that progesterone is also a ventilatory stimulant. However, no report has been seen that effect of menstrual cycle on hypoxic-hypercapnic interaction on ventilatory as well as on respiratory sensation. The purpose of this study was to simultaneously examine the effect of menstrual activity on ventilatory sensitivity and respiratory sensation.

2. METHODS

This study was carried out on 16 young healthy females. Their age, height, weight and body surface area were 22.3±3.0 yrs, 159.5±8.2 cm, 50.9±5.6 kg and 1.51±0.12 m^2 (mean±S.D.), respectively. The average menstrual cycle length was 29 days (range 23 to 37). All subjects considered their cycle normal. Half of all subjects tested were in follicular phase and remaining half were in luteal phase.

---

* Atsuko Masuda, Tokyo Medical and Dental University, Tokyo, 113-8519, Japan. Yoshio Ohyabu, Kogakuin University, Tokyo. Toshio Kobayashi, Fukushima Medical University, Fukushima. Chikako Yoshino, Chiba College Allied Medical Sciences, Chiba. Yoshiyuki Honda, Chiba University, Chiba.

Frontiers in Modeling and Control of Breathing, edited by Poon and Kazemi.
Kluwer Academic/Plenum Publishers, New York, 2001.

The subjects were exposed to progressive hypercapnia by the modified Read's rebreathing method[1] with four different conditions. The following gas mixtures with different oxygen content were used: A. 7% CO_2 + 93% O_2 for hyperoxia run, B. 7% CO_2 + 19% O_2 for normoxia run, C. 7% CO_2 + 13% O_2 for mild hypoxia run and D. 7% CO_2 +11% O_2 for moderate hypoxia run. During rebreathing, we maintained the P_{ETO_2} level at >300 for the A run, 100 for the B run, 80 for the C run and 60mmHg for the D run by administering small amounts of O_2 or N_2 gas via a small tubing inserted in the inspiratory portion of the respiratory valve. Breath-by-breath tidal volume (V_T) and respiratory frequency (f) were detected by a hot-wire flow meter (Minato, RF-H, Osaka, JAPAN). Minute ventilation (V_E) was normalized by dividing body surface area. P_{ETO_2} and P_{ETCO_2} were observed with the rapid response O_2-CO_2 analyzer (San-ei, H-21, Tokyo, JAPAN). Respiratory sensation was also continuously recorded by visual analog scale (VAS) [2,3]. The experiment was terminated 3-4 min after the start of rebreathing or when the VAS score reached 10, the ceiling level. Each subject randomly repeated four runs: twice in the morning and afternoon, respectively on the same day.

The results were examined for variance between 4 runs using ANOVA. Statistical differences between luteal and follicular phases were compared by Student unpaired t-test.

3. RESULTS

In the first place, group means of the ventilatory and VAS responses of all 16 subjects were calculated. The slope of the CO_2-ventilation response curve of the A, B, C and D runs were 0.90±0.42, 1.19±0.79, 1.14±0.50 and 1.30±0.78 l^{-1} min^{-1} m^{-2} (mean±S.D.), respectively. There was significant slope augmentation in B, C and D runs compared to the A run. On the other hand, the slope of the CO_2-VAS response curve remained unchanged despite of intensifying hypoxia. The magnitude of the VAS-V_E slopes of the A, B, C and D runs were 0.64±0.36, 0.52±0.23, 0.58±0.34 and 0.53±0.27 l^{-1} min^{-1} m^{-2} (mean±S.D.), respectively. There was significant decreases in B and D runs compared A run, and the amount of VAS elicited from a given ventilation tended to depressed by hypoxia.

Figure 1 shows the comparison of values of 3 response slopes, i.e., CO_2-ventilation, CO_2-VAS and V_E-VAS slopes, between luteal and follicular subjects. V_E/P_{ETCO_2} was lower in follicular than luteal phase on all runs except D run. In contrast, VAS/V_E was higher in follicular than luteal females. No significant difference was noted in VAS/P_{ETCO_2} slope in any of A to D runs.

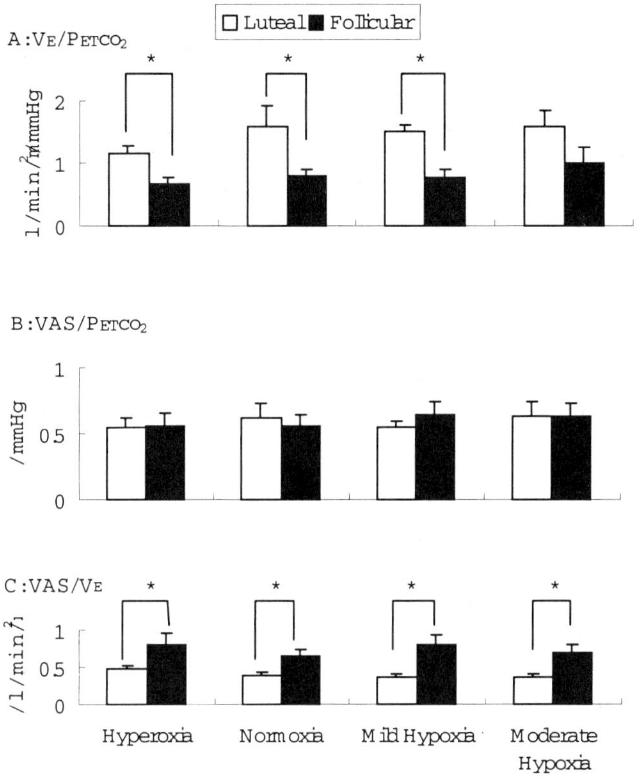

Figure 1. Statistical comparison between luteal and follicular subjects. *: $p<0.05$ Luteal vs. Follicular

4. DISCUSSION

In this study, the slope of the CO_2-ventilation response curve became steeper with decreasing oxygen levels, thus hypoxia induced a synergistic effect on hypercapnic ventilatory response. Several investigators have also reported positive ventilatory interaction between hypercapnia and hypoxia in man[4-6] and animals[7,8]. Recently, Mohan and Duffin[5] also confirmed this phenomenon by using a modified Read's rebreathing method. Although they didn't measure ventilatory response to CO_2 at hyperoxic condition, they conducted rebreathing at severer hypoxia, P_{TEO_2} of 40 mmHg, than us.

Contrary to the ventilatory data, we found no change in the slope of the CO_2-VAS response curve. Hypoxia diminished the slope of the VE-VAS curve and the amount of VAS elicited from a given ventilation was decreased. The depressed slope of VE-VAS might counteract the augmented slope of CO_2-ventilation. Therefore it is possible to infer that the sensitivity of respiratory sensation to CO_2 was unchanged . Underlying mechanisms of depressed respiratory sensation for given amount of ventilation in the present study is considered as follows.

At first, we measured dyspnea sensation by visual analog scale. VAS has been used to evaluate the amount of sensation such as in pain and dyspnea[9]. The two ends of the scale were fixed and designated from "none at all" to "most intense imaginable", respectively. However, because this is primarily a category scale[2], respiratory sensation may be underestimated when approaching the maximal VAS value.

The another possibility is the effect of pulmonary stretch receptor (PSR) activity. It is well known that the respiratory distress induced by breathing with maximal effort is relieved by rebreathing or respiratory movement[10-12] despite worsening blood gas status. It has been suggested that this relief may be due to inhibitory effects from PSR activity[11,12].

We found that ventilatory response to combined hypoxia and hypercapnia was higher in luteal than follicular phase. The difference of menstrual period on ventilatory response has been reported previously by Shoene et al.[13]. They observed a near doubling of both HVR and HCVR from the follicular to the luteal phase. In this study, we further compared the respiratory sensation between both phases and found that the amount of VAS elicited from a given ventilation was lower in luteal than follicular phase. To our knowledge, nothing has been reported on the change of respiratory sensation during menstrual cycle.

Dyspnea is a common feature of pregnancy, but the mechanism of dyspnea is controversial. Initially, breathlessness was attributed to increased mechanical load by chest wall distortion from the gravid uterus[14]. However, dyspnea typically precedes before any upward displacement of the diaphragm[15], suggesting that factors other than mechanical encumbrance may be involved. Physiological dyspnea noted in many pregnant women could be due to an excessive increase in sensitivity to CO_2 or hypoxia[16]. However, if we extend our finding, that VAS for a given amount of ventilation depresses during luteal phase, into the pregnant women, their dyspneic sensation might effectively mitigated despite of excessive hyperventilation due to elevated progesterone levels. From the standpoint of QOL, this can be inferred to be an useful physiological adaptation to maintain pregnancy accompanying with less physical distress.

5. REFERENCES

1. D. J. C. Read, Blood-brain tissue Pco_2 relationships and ventilation during rebreathing, *J. Appl. Physiol.* **23**, 53-70 (1967).
2. S. S. Stevens, On the psychophysical law, *Psychol. Rev.* **64**, 153-181 (1957).
3. L. Adams, N. Chronos, R. Lane, and A. Guz, The measurement of breathlessness induced in normal subjects: validity of two scaling techniques, *Clin. Sci.* **69**, 7-16 (1985).
4. M. Nielsen and H. Smith, Studies on the regulation of respiration acute hypoxia, *Acta. Physiol. Scand.* **24**, 293-313 (1952).
5. A. Dahan, J. DeGoede, A. Berkenbosch, and L. C. Olievier, The influence of oxygen on the ventilatory response to carbon dioxide in man, *J. Physiol. (London).* **428**, 485-499 (1990).
6. R. Mohan and J. Duffin, The effect of hypoxia on the ventilatory response to carbon dioxide in man, *Respir. Physiol.* **108**, 101-115 (1997).
7. T. F. Hornbein, Z. Griffo, and A. Roos, Quantitation of chemoreceptor activity: interaction of hypoxia and hypercapnia, *J. Neurophysiol.* **24**, 561-568 (1961).

8. R. S. Fitzgerald and D. C. Parks, Effect of hypoxia on carotid chemoreceptor response to carbon dioxide in cats. *Respir. Physiol.* **12**, 218-229 (1971).
9. D. D. Price, P. A. McGrath, A. Rafii, and B. Buckingham, The validation of visual analog scales as ratio scale measures for chronic and experimental pain. *Pain* **17**, 45-56 (1983).
10. L. Hill, L and M. Fluck, The effect of excess of carbon dioxide and want of oxygen upon the respiration and the circulation. *J. Physiol. (London)* **37**, 77-111 (1908).
11. W .S. Fowler, Breaking point of breath-holding. *J. Appl. Physiol.* **6**, 539-545 (1954).
12. P. A. Flume, F. L. Eldridge, L. J. Edwards, and L. M. Houser, Role of vagal input in the relief of the distress of breathholding: normal vs. pts with double lung transplants. *Am. Rev. Respir. Dis.* **147**,A550 (1993).
13. R. B. Shoene, H. T. Robertson, D. J. Pierson, and A. P. Peterson, Repiratory drives and exercise in menstrual cycles of athletic and unathletic women. *J. Appl. Physiol.* **50**, 1300-1305 (1981).
14. H. G. Knuttgen and R. K. Emmerson, Physiological response to pregnancy at rest and during exercise. *J. Appl. Physiol.* **36**, 549-553 (1974).
15. R. Gilbert, L. Epifano, and J. H. Auchincloss, Dyspnea of pregnancy: a syndrome of altered respiratory control. *JAMA* **182**, 1073- 1077 (1962).
16. F. Garcia-Rio, J. M.Pino, L. Gomez, R. Alvarez-Sala, C. Villasante, and J.Villamor, Regulation of breathing and perception of dyspnea in healthy pregnant women. *Chest* **110**, 446-453 (1996).

OPIOID EFFECT ON BREATHING FREQUENCY AND THERMOGENESIS IN MICE LACKING EXON 2 OF THE μ-OPIOID RECEPTOR GENE

Elise Sarton, Luc Teppema, Diederik Nieuwenhuijs, Hans WD Matthes, Brigitte Kieffer, and Albert Dahan*

1. INTRODUCTION

Various sites in the peripheral and central nervous system involved in the control of breathing contain cells which express μ-opioid receptors (μOR). For example, it was recently shown that injection of the enkephalin-derived μ-opioid receptor agonist DAMGO ([D-Ala^2, N-Me-Phe^4, Gly-ol^5]-enkephalin) into the preBötzinger complex of rodents decreased endogenous respiratory-related rhythm in an *in vitro* preparation of the ventral medulla.[1] A modulatory effect of the μ-opioid receptor and endogenous opioid peptides on breathing frequency may therefore be expected.

Recently, mice lacking the various opioid receptors have been generated by inactivation of the specific receptor genes by homologous recombination techniques (see for a review reference 2). Studies in these mice enable assessment of the necessity of the opioid receptor gene product (*i.e.* the opioid receptor) for mediating a specific process. Studies in mice lacking the μ-opioid receptor gene and hence the μ-opioid receptor implicate the μ-opioid receptor gene product as molecular site of morphine-analgesia.[3]

In order to study the involvement of μ-opioid receptors in the control of breathing frequency as well as on opioid-induced changes in breathing frequency, we examined breathing frequency in mice lacking exon 2 of the μ-opioid receptor gene (μOR knockout mice or μOR$^{-/-}$) and wild-type mice of the same strain (μOR$^{+/+}$). Studies were performed in the control state (*i.e.* at rest, without any agent administered), after injection of the potent μ-opioid agonists morphine and morphine-6-glucuronide (M6G). We also assessed the effect of morphine and morphine-6-glucuronide on nociception and thermogenesis.

* Elise Sarton, Albert Dahan, Luc Teppema, Diederik Nieuwenhuijs, Departments of Anesthesiology and Physiology, Leiden University Medical Center, 2300 RC Leiden, The Netherlands. Hans Matthes, Brigitte Kieffer, UPR 9050 CNRS, ESBS Université Louis Pasteur, Strasbourg, France.

Frontiers in Modeling and Control of Breathing, edited by Poon and Kazemi.
Kluwer Academic/Plenum Publishers, New York, 2001.

2. METHODS

Mice with disruption of exon 2 of the μ-opioid receptor gene, which contains the main coding region, and wild type animals of the same mouse strain have been reported previously.[3] All animals are 1:1 hybrids from 129/SV and C57BL/6 mouse strains. Studies were performed while the animals were 2 to 6 months of age. Studies were conducted after approval of the protocol by the local animal ethics committee. Respiratory activity was measured using whole body plethysmography, originally described by Drorbaugh and Fenn,[4] with an important adaptation allowing the continuous flow of dry gas through the measurement and reference chamber. The flow and composition of the gas through the chambers was set by three mass flow controllers. The chamber was kept at 24-26°C.

2.1. Control Studies

Breathing frequency was calculated per breath while the animals breathed 21% oxygen in nitrogen (8 $\mu OR^{-/-}$ and 10 $\mu OR^{+/+}$ mice). The data were subsequently averaged over 50 breaths for further analysis. In each animal, measurements were performed on 5 consecutive occasions at least 3 weeks apart.

2.2. Drug Studies

2.2.1. Respiration

The effect of placebo (NaCl 0.9%, 1.2 ml i.p.), morphine (6, 20 and 100 mg/kg, 1.2 ml i.p.) and morphine-6-glucuronide (10 mg/kg, 1.2 ml i.p.) on breathing frequency was assessed (8 $\mu OR^{-/-}$ and 10 $\mu OR^{+/+}$ mice). The different drugs (and drug concentrations) were tested on separate days at least three weeks apart.

2.2.2. Antinociception

The antinociceptive effects of placebo, morphine (6, 20 and 100 mg/kg) and M6G (10 mg/kg) was tested using two nociceptive assays to test for spinal analgesia, using the tail-immersion test (water temperature = 54 °C; cutoff time = 15 s), and supraspinal analgesia, using the hot-plate test (temperature of the plate = 52 °C; cutoff time = 30 s).

2.2.3. Temperature

Body temperature was measured *via* a temperature probe inserted ~1 cm rectally. The effect of placebo, 6, 20 and 100 mg/kg morphine on body temperature was assessed.

2.3. Statistical Analysis

A of the control data was by one-way analysis of variance (factor: genotype). Drug effect was tested by two-way analysis of variance (factors: genotype and treatment) and least significant differences test. *P*-values < 0.05 were considered significant.

3. RESULTS

3.1. Control Studies

The μOR$^{-/-}$ and μOR$^{+/+}$ mice did not differ in weight, morphology and behaviour. Baseline breathing frequencies (*i.e.* at rest before administration of any agent) differed significantly between genotypes with a 15 to 20% higher breathing rate in mice with a disrupted μ-opioid receptor gene. Mean breathing frequencies ± SEM were 188 ± 7 breaths per min in μOR$^{+/+}$ mice and 220 ± 7 breaths/min in μOR$^{-/-}$ mice ($P < 0.0001$).

3.2. Drug Studies

3.2.1. Respiration

3.2.1a. Morphine. Morphine caused a dose-dependent decrease in breathing frequency relative to saline treated animals from 187 ± 13 breaths per min (saline) to 124 ± 11 breaths/min after 100 mg/kg morphine in μOR$^{+/+}$ mice but had no effect in μOR$^{-/-}$ mice (saline: 192 ± 13 breaths/min against 100 mg/kg morphine: 192 ± 12 breaths/min; NS). See legend of figure 1 for *P*-values. Analysis of variance revealed a significant genotype and genotype×treatment effect ($P < 0.01$).

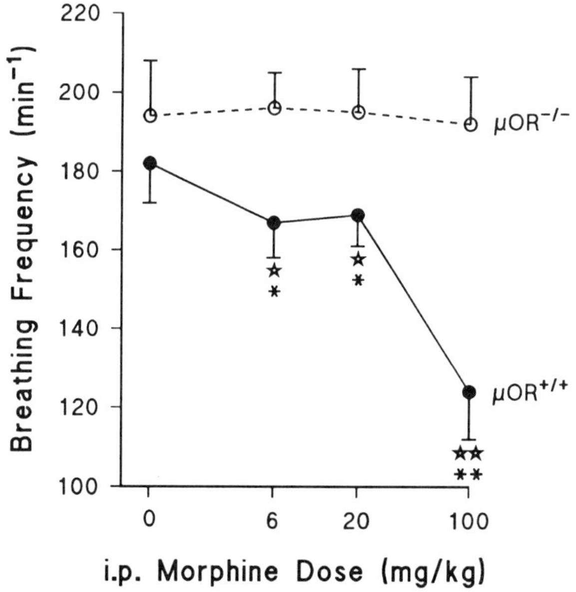

Figure 1. Influence of saline and three morphine concentrations on breathing frequency in μOR$^{-/-}$ and μOR$^{+/+}$ mice. Values are mean ± SEM. μOR$^{-/-}$ *vs.* μOR$^{+/+}$ mice: * $P < 0.05$, ** $P < 0.01$ (two-way ANOVA). Treatment effect: ✶ $P < 0.01$ *vs.* placebo, ✶✶ $P < 0.01$ *vs.* all other treatments (two-way ANOVA).

3.2.1b. Morphine-6-glucuronide. M6G at 10 mg/kg caused a significant decrease in breathing frequency from 214 ± 11 breaths/min (saline) to 140 ± 3 breaths/min ($P < 0.0001$) in $\mu OR^{+/+}$ mice but had no effect in $\mu OR^{-/-}$ mice (saline: 227 ± 10 breaths/min against M6G: 218 ± 8 breaths/min; NS).

3.2.2. Antinociception

Baseline latencies did not differ between genotypes. In both nociceptive assays, $\mu OR^{-/-}$ mice showed absence of morphine and M6G analgesia, with latencies not different from placebo. In contrast, $\mu OR^{+/+}$ mice, morphine (at all doses) and M6G (10 mg/kg) caused a significant increase in latencies. At 100 mg/kg morphine cutoff values were reached in all $\mu OR^{-/-}$ animals.

3.2.3. Temperature

Baseline temperatures did not differ between genotypes. Morphine had a dose-dependent hypothermic effect in wild-type mice, but had no effect on rectal temperature in μ-opioid receptor knockout mice. See figure 2.

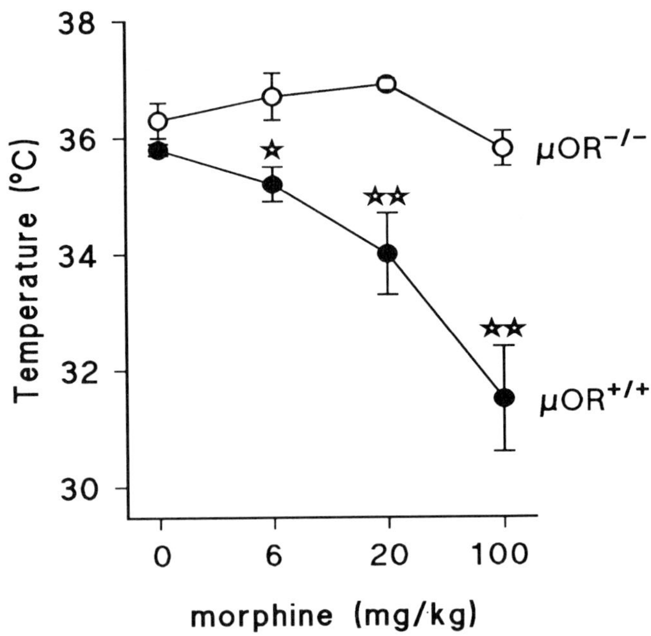

Figure 2. Effect of morphine on rectal temperature in homozygous μ-opioid receptor knockout mice ($\mu OR^{-/-}$) and wild-type animals ($\mu OR^{+/+}$). 0 mg/kg reflects saline treated animals. Values are mean ± SEM. ✶: $P < 0.01$ against $\mu OR^{-/-}$ mice receiving the same dose. ✶✶ $P < 0.01$ against saline treated animals of the same genotype and against $\mu OR^{-/-}$ receiving the same dose.

4. DISCUSSION

Using an *in vivo* knockout mouse model we observed that:

1. μ-opioid receptor deficient mice display higher breathing frequencies compared to mice with an intact μ-opioid receptor gene;
2. Morphine, at doses that cause analgesia, respiratory depression and hypothermia in wild-type mice, do not affect respiratory frequency and rectal temperature in $\mu OR^{-/-}$ mice;
3. Morphine-6-glucuronide, at a dose that produces analgesia and respiratory depression in wild-type mice (10 mg/kg), is ineffective in $\mu OR^{-/-}$ mice.

Ad 1. Our finding of differences in breathing frequencies between mice with a disrupted μ-opioid receptor gene and mice with an intact gene (both genotypes have identical genetic backgrounds) indicates the involvement of the μ-opioid system (receptor and endogenous μ-opioid peptides) in the modulation of breathing frequency. The differences in baseline frequencies between genotypes were small, however (15 to 20%). This may indicate only a modest role for the μ-opioid system in the control of breathing frequency. Alternatively, it may be related to the fact that the knockout mice in our study lacked the μ-opioid gene through development. Functional compensation may have –partly– masked the phenotype (*i.e.* the frequency generated by the respiratory rhythm generator) resulting from the chronic absence of the targeted gene. Furthermore, it may well be that differences in breathing frequencies between genotypes will be accentuated during stressful stimulation of the ventilatory control system such as may occur during exercise, chronic hypoxia, chronic elevations in arterial carbon dioxide concentration, asphyxia and other acute stressful events such as pain and suffering. Interestingly, the differences in breathing frequencies between genotypes remained present during anesthesia with the volatile anesthetic sevoflurane (see reference 4). Note that our current protocol was not designed to assess the modulatory influence of the other opioid-receptors (δ- and κ-receptors) or receptor cross-talk on resting breathing frequency.

Ad. 2 and 3. Despite its many side effects, such as respiratory depression,[5, 6] morphine remains a valuable pain killer. Respiratory depression from morphine can be life-threatening, especially when the balance between respiratory stimulation (*e.g.* from arousal, pain and chemoreflex stimulation) and respiratory depression (from morphine and sedation) is shifted to the latter. The finding that morphine analgesia –morphine's desired effect– , morphine hypothermia and morphine depression of breathing frequency –morphine's side effects– are linked to the same gene product is of importance. It suggests that agents acting at the μ-opioid receptor will invariably cause analgesia, hypothermia and respiratory depression.

In humans (but not mice), morphine-6-glucuronide is an active and potent metabolite of morphine (about 10 to 20% of morphine is metabolized to M6G). Intravenous infusion of M6G causes analgesia (relative to morphine, M6G is 2 to 4 times more potent after intravenous infusion and >100 times after intrathecal injection in humans and rats, respectively) with small changes in arterial PCO_2. Our findings demonstrate that both morphine and M6G act at the μ-opioid receptor causing analgesia and respiratory depression. Our data show further a similar potency with respect to the respiratory depressant effects of M6G and morphine. These findings indicate no advantage of M6G over morphine in providing analgesia, at least in our animal model. If this is also true in humans deserves further study. Furthermore, the importance of the unique receptor for M6G analgesia in M6G-induced respiratory depression needs to be elucidated.[7]

5. CONCLUSIONS

Our data demonstrate that the endogenous μ-opioid system plays a (small) role in the control of breathing frequency, and that the μ-opioid receptor is the essential target for morphine (and morphine-6-glucuronide) induced hypothermia, respiratory depression and analgesia. These findings confirm and extend our earlier observations in the same mouse strain, which showed involvement in morphine spinal and supraspinal analgesia, reward, physical dependence, respiratory depression and immunosuppression.[3, 8, 9]

6. REFERENCES

1. P.A. Gray, J.C. Reckling, C.M. Bocchiaro, and J.L. Feldman. Modulation of respiratory frequency by peptidergic input to rhythmogenic neurons in the preBötzinger complex, *Science* **286,** 1566–1568 (1999).
2. B.L. Kieffer. Opioids: first lessons from knockout mice. *TIPS* **20,** 19–26 (1999).
3. H.W.D. Matthes , R. Maldonado, F. Simonin, O. Valverde, S. Slowe, I. Kitchen, K. Befort, A. Dierich, M. Le Meur, P. Dollé, E. Tzavara, J. Hanoune, B.P. Roques, and B.L. Kieffer. Loss of morphine-induced analgesia, reward effect and withdrawal symptoms in mice lacking the μ-opioid-receptor gene. *Nature* **38,**: 819–823 (1996).
4. A. Dahan, E. Sarton, L. Teppema, C. Olievier, D. Nieuwenhuijs, H.W.D. Matthes, B.L. Kieffer. Respiration, anesthetic potency, and morphine- and anesthesia-effect on respiration in μ-opioid-receptor knockout mice. *Anesthesiology,* accepted for publication.
5. A. Dahan, E. Sarton, L. Teppema, and C. Olievier. Sex-related differences in the influence of morphine on ventilatory control in humans. *Anesthesiology* **88,** 903–913 (1998).
6. E. Sarton, A. Dahan, and L. Teppema. Sex differences in morphine-induced respiratory depression reside within the peripheral chemoreflex loop. *Anesthesiology* **90,** 1329–133813 (1999).
7. A.G.P. Schuller, M.A. King, J. Zhang, E. Bolan, Y. Pan, D.J. Morgan, A. Chang, M.E. Chick, E.M. Unterwald, G.W. Pasternak, and J.E. Pintar. Retention of heroin and morphine–6μ-glucuronide analgesia in a new line of mice lacking exon 1 of MOR–1. *Nat Med* **2,** 151–156 (1999).
8. H.W.D. Matthes, C. Smadja, O. Valverde, J. Vonesch, A.S. Foutz, E. Boudinot, M. Denavit-Saublé, C. Severini, L. Negri, B.P. Roques, R. Maldonado, and B.L. Kieffer: Activity of the μ-opioid receptor is partially reduced, whereas activity of the μ-receptor is maintained in mice lacking the μ-receptor. *J Neurosci* **18,** 7285-7295 (1998).
9. C. Gavériaux-Ruff, H.W. Matthes, J. Peluso, and B.L. Kieffer. Absence of morphine immunosuppression in mice lacking the mu-opioid receptor gene. *Proc Natl Acad Sci USA* **95,** 6326-6330 (1998).

MODULATION OF BREATHING USING IMPERCEPTIBLE UNLOADING

Brett F. BuSha, Martha H. Stella, Scott H. Landry, J. Andrew Daubenspeck, and James C. Leiter*

1. INTRODUCTION

Hering and Breuer first described the reflex effects of volume feedback on respiratory timing in 1868[1]. They found that passive lung inflation above normal tidal volumes either shortened inspiratory time (T_I) or prolonged expiratory time (T_E) depending on the timing of the inflation stimulus within the respiratory cycle. More recently, increased inspiratory airflow (V_T/T_I) also was found to modulate breathing by increasing tidal volume (V_T) and decreasing T_I[2]. Sectioning the vagi abolishes both reflexes. In a previous study,[3] we investigated the effect of unloading the respiratory system using imperceptible amounts of proportional assist ventilation (PAV). A subtle increase in V_T and V_T/T_I elicited by PAV, which was below the perceptual threshold, modified respiratory timing in a way consistent with the Hering-Breuer reflex in awake normal humans: there was a small, but significant, reduction in inspiratory time.

In the present study, we attempted to differentiate the potential contribution of flow and volume feedback in the breath-by-breath control of inspiratory timing during wakefulness. We increased V_T or V_T/T_I in normal awake humans on a breath-by-breath basis following a pseudo-random binary sequence (PRBS) of flow or volume assist ventilation. We assessed the effect of volume and flow related feedback on respiratory timing in the absence of cortical influences by using sub-threshold unloading.

2. METHODS

2.1. Subjects

We recruited 4 normal adult volunteers without prior or current respiratory disease. Except for the participation of one of the authors, the subjects were unaware of the objectives of the study. Informed consent was obtained from all subjects.

* Dartmouth Medical School, Lebanon, New Hampshire, 03756

Frontiers in Modeling and Control of Breathing, edited by Poon and Kazemi.
Kluwer Academic/Plenum Publishers, New York, 2001.

2.2. Instrumentation and Set-up

In each study, the subject sat recumbent in a dental chair, wore a nose clip, and breathed though a 'snorkel' mouthpiece attached to a prototype proportional assist ventilator, the Winnipeg PAV (University of Manitoba, Winnipeg, MB, Canada). Subjects listened to white noise to prevent clues about the activity of the ventilator during the experiment. We designed software to calculate the level of volume or flow assist and commanded the ventilator to give the appropriate assist pressure on a breath by breath basis. The assist pressure was calculated using the instantaneous volume and each subject's E_{RS} or the instantaneous flow and each subject's R_{RS}. Thus, the assist pressure waveform followed either the instantaneous V_T (incrementing ramp) or flow (decrementing wave). For analysis, airway pressure measured at the mouth (P_M) and inspiratory airflow were recorded directly from an analog output supplied by the ventilator. Flow was integrated with a zero crossing reset to calculate V_T. During the measurements of E_{RS} and R_{RS}, airflow at the mouth was occluded with a 2-way shutoff valve (9340 series inflatable balloon controlled with an 8230 series automatic controller, Hans Rudolph, Inc., Kansas City, MO). Programs written with LabVIEW (National Instruments, Austin, TX) software controlled the ventilator and balloon valves and displayed and recorded the data.

2.3. Experimental Protocols

2.3.1. Determination of Respiratory Mechanics

We used a modification of the interrupter method to estimate E_{RS} and R_{RS}[4, 5]. Each subject was mechanically ventilated using controlled mandatory ventilation to achieve passive ventilation. The subjects were invariably hypocapnic during this period. Flow rate, P_M, and V_T were monitored continuously to ensure passive ventilation. The airway was occluded every 5-8 breaths for 400 to 600 ms after delivery of 60-80% of the subject's V_T. Occlusions were judged satisfactory when no respiratory effort was visible during the breath, when the flow and pressure waves were similar to previous breaths, and when the occlusion V_T plateau was constant. Approximately 20 satisfactory occlusions were recorded from each subject. E_{RS} was calculated from the difference between the plateau pressure and the end expiratory pressure divided by V_T of that breath. R_{RS} was calculated from the difference between the P_M measured before the occlusion and the subsequent plateau pressure divided by the flow immediately preceding the occlusion.

2.3.2. Determination of 50% Threshold of Sensation

We used two similar forced choice protocols to determine the threshold of flow and volume assist ventilation[6]. The levels of assistance varied from imperceptible to well above the perceptual threshold for each protocol. For each protocol the subject was 'forced' to decide whether the ventilator 'helped' or 'did not help' on each assisted breath. The ventilator randomly applied five levels of assist ranging from 0% to 40% with three unassisted breaths between each assisted breath. Each 20-breath sequence of assisted breaths was repeated 11 times in a randomized-block design. The first 20-breath sequence from each protocol was discarded from the analysis. The probability of detection of ventilator assistance (P) was calculated at each level of assistance by using logits (1 logit = Ln (P/1-P). We defined the perceptual threshold as the level of assistance that was detected 50% of the time. For each subject, the logits were plotted as a function of percent assistance, and linear regression was used to fit the data. The x-axis crossing in this analysis is equal to the 50% threshold of sensation ($TOS_{50\%}$).

2.3.3. PRBS Assist Protocols

Each breathing trial was comprised of three sequences applied in the following order: 70 unassisted breaths, a 127 PRBS of assisted and unassisted breaths and another 70 unassisted breaths. The 70 unassisted breaths recorded before and after the pseudo-random sequence enhance the ability of the system identification procedure to model the low frequency response characteristics of the system that occur during the initiation and the conclusion of the PRBS sequence. Subjects were tested with both a flow and a volume assist trial. The order of the trials was randomized, and a 10-minute rest period was allowed between trials.

2.4. Determination of 1st Breath Response of Respiratory Variables

Respiratory variables (V_T, T_I, T_E, V_T/T_I), instantaneous respiratory muscle pressure output (P_{MUS}), respiratory muscle output integrated per breath ($\int P_{MUS}$), and P_M were calculated online using programs written in LabVIEW and MATLAB (Math Works, S. Natick, MA). Continuous measurements of flow, V_T and P_M and the values of E_{RS} and R_{RS} determined previously in each subject were used to calculate P_{MUS} as follows:[7]

$$P_{MUS} = E_{RS} \times V_T + R_{RS} \times \text{flow} - P_M$$

for each assist protocol. P_{MUS} was calculated only during inspiration.

Some subjects had augmented breaths (sighs), which would have disproportionately affected the average breath size. Sighs were defined as breaths with a V_T greater than the mean plus three times the standard deviation. If a sigh occurred before or after the PRBS sequence, data for the breath were omitted from analysis. If a sigh occurred during the PRBS sequence, individual values for that breath were replaced with mean values calculated from the PRBS sequence. Breaths were not removed from the PRBS sequence to avoid altering the randomness of the assisted breaths.

We used a general system identification technique known as the prediction-error method[8] to calculate the response to a single assisted breath for each respiratory variable. This method assumes that the output of the system (e.g., V_T) is linearly related to the input function (e.g., application of volume or flow assist) and to random noise. Individual data sets were transformed to remove the mean of the signal. A linear difference equation was constructed to model the data (specifically, the ARX variant) as follows:

$$y(t) = -a_1 \times y(t-1) - \cdots - a_n \times (t - n_a) + b_1 \times u(t-1) + \cdots + b_n \times u(t - n_b) + e(t)$$

where "t" was the sample number, y(t) was the system response (i.e. V_T), u(t) was the system input (the PRBS), and e(t) was the error term. The parameters '$a_{1...n}$' and '$b_{1...n}$' were estimated using a least squares minimization algorithm. Models of different order and with different values of coefficients were constructed. We selected the optimal order of the model and the coefficients with Akaike's information-theoretic criterion (AIC):

$$AIC \approx \log[(1 + 2n/N) \times V]$$

where n is the total number of estimated parameters; N is the length of the data record; and V is the loss function (the difference between the model estimates and real response data). The coefficients producing the model with the lowest AIC value were used to define the impulse-response characteristics[8] of the first through fifth breaths. All system identification procedures were analyzed on a computer using MATLAB and the MATLAB System Identification Toolbox (Math Works, S. Natick, MA).

2.5. Statistical Analyses

Data are expressed as mean ± standard deviation. A one-sample t-test was used to determine if the change in a respiratory variable differed from zero. Individual respiratory variables were compared across trials with paired t-tests or with unpaired t-tests when appropriate. A P value ≤ 0.05 was considered significant.

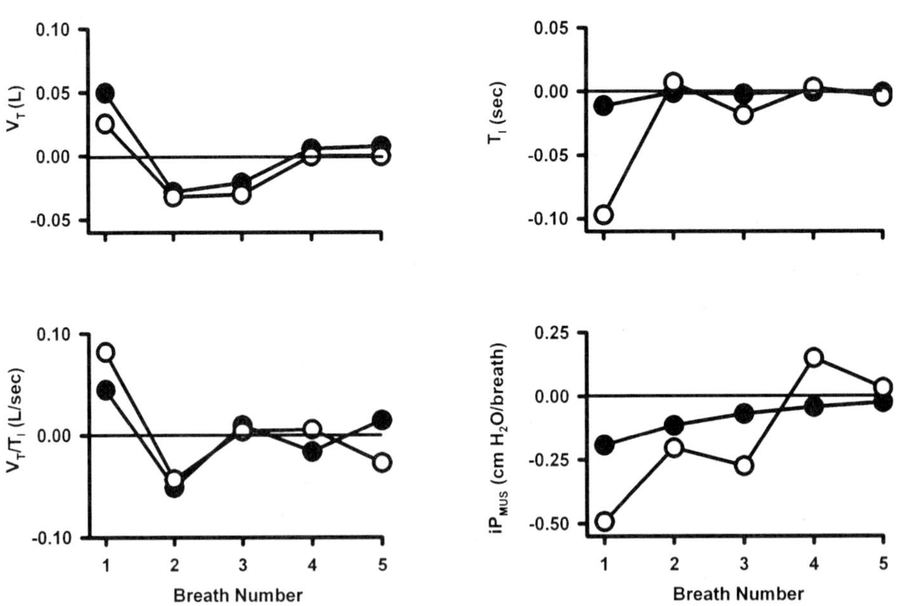

Figure 1. Characteristic impulse responses from one subject during volume (closed circles) and flow (open circles) assist. Assist (delivered at breath one) resulted in an increase (compared to previous non-assisted breaths) in V_T and V_T/T_I, and a decrease in T_I and $\int P_{MUS}$ (iP_{MUS}).

3. RESULTS

3.1. Subject Characteristics

All four subjects (weight, 169 ± 21 lb; height, 69 ± 3 in) completed all of the protocols. The mean value for E_{RS} for all subjects was 13.3 ± 3.2 cm H_2O/L and the mean R_{RS} was 5.0 ± 1.5 cm H_2O/L/s. The mean value for $TOS_{50\%}$ for flow assist was significantly greater than for volume assist (30.7 ± 6.17%; 18.3 ± 4.6%; P = 0.004).

3.2. Responses to Respiratory System Unloading

Although it was not possible to separate volume and flow stimuli completely, we changed the relative amounts of V_T and V_T/T_I during the assisted breaths. The effects of unloading on V_T, V_T/T_I, T_I, and $\int P_{MUS}$ during volume or flow assist from a single subject are shown in Fig. 1. The impulse responses were examined over five breaths: the first assisted breath and the four following breaths. There were significant first breath effects on V_T and V_T/T_I during both volume and flow assist. T_I and $\int P_{MUS}$ were reduced significantly on the first breath only during flow assist. The reduction in T_I and $\int P_{MUS}$ are the expected consequences of vagal feedback related to the elicited changes in V_T and/or V_T/T_I[9, 10, 11].

The average first breath responses for all subjects for volume assist and flow assist unloading are shown in Table 1. The average P_M applied during volume assist was 1.37 ± 0.53 cmH_2O, and during flow assist was 0.96 ± 0.49 cmH_2O. All four subjects had significant individual increases in V_T and V_T/T_I during volume assist. The average increases in V_T and in V_T/T_I were significant. Volume assist elicited no change in T_I or $\int P_{MUS}$. During flow assist unloading, three subjects had individual significant increases in V_T and all four had significant increases in V_T/T_I. The average increases in V_T and V_T/T_I were significant. Flow assist resulted in a significant individual decrease in T_I in two subjects and in $\int P_{MUS}$ in 3 subjects. The average decrease in $\int P_{MUS}$ during flow assist was significant, but the average decrease in T_I was not significant.

Table 1. First breath of impulse responses

	Volume Assist				Flow Assist			
Subject	V_T (L)	V_T/T_I (L/sec)	T_I (sec)	$\int P_{MUS}$ (cmH_2O/ breath)	V_T (L)	V_T/T_I (L/sec)	T_I (sec)	$\int P_{MUS}$ (cmH_2O/ breath)
A	0.050*	0.045*	-0.011	-0.192	0.026*	0.082*	-0.097*	-0.492*
B	0.133*	0.073*	-0.021	-0.185	0.018	0.057*	-0.212*	-0.790*
C	0.083*	0.059*	0.022	0.189	0.047*	0.072*	-0.037	-0.247*
D	0.028*	0.021*	-0.038	-0.172	0.047*	0.028*	-0.007	-0.286
Mean	0.073*	0.049*	-0.012	-0.090	0.035*	0.060*	-0.088	-0.454*
Std. Dev.	0.046	0.022	0.025	0.186	0.015	0.023	0.091	0.249

Values reported are the estimated changes resulting from one breath of assist.

* significantly different from zero.

4. SUMMARY

We investigated the role of V_T and V_T/T_I modulation of breathing in awake human subjects. We applied a PRBS of volume (incrementing ramp) or flow (decrementing wave) assist at levels below the perceptual threshold in order to stimulate respiratory feedback. We modeled the PRBS data with linear difference equations to obtain impulse-response profiles of V_T, V_T/T_I, T_I and $\int P_{MUS}$. We limited cortical responses to our stimuli by applying sub-threshold levels of assist, and we limited humoral effects (O_2 and CO_2) by augmenting mechanical respiratory output intermittently and by small amounts. We found that flow or volume assist elicited similar significant increases in V_T and V_T/T_I. During flow assist there was a significant decrease in $\int P_{MUS}$, and T_I was reduced, albeit not significantly; however, volume assist did not modify T_I or $\int P_{MUS}$. The earlier onset of flow assist, relative to volume assist, may explain the difference between the responses. We conclude that vagally mediated inspiratory flow receptors in the chest wall or lungs may modulate breathing on a breath by breath basis when small, imperceptible increases in airflow occur early during inspiration. Furthermore, lung volume feedback during imperceptible unloading (occurring at the end of inspiration) was less effective. Finally, pseudorandom unloading with imperceptible stimuli provides a useful tool to study reflex regulation of ventilation in awake subjects without confounding cortical influences.

5. ACKNOWLEDGMENTS

This work was support by NIH grants HL 07449 and HL 29068 (JAD). We wish to thank Dr. Magdy Younes for lending us a prototype Winnipeg PAV.

6. REFERENCES

1. J. Breuer, Self-steering of respiration through the Nervus Vagus, in: *Breathing: Hering-Breuer Centenary Symposium*, edited by R. Porter (J. & A. Churchill, London, 1970), pp. 365-394.
2. C. –S. Poon, S. A. Ward, and B. J. Whipp, Influence of inspiratory assistance on ventilatory control during moderate exercise, *J. Appl. Physiol.* **62**, 551-560 (1987).
3. B. F. BuSha, B. J. Judd, H. L. Manning, P. M. Simon, B. C. Searle, J. A. Daubenspeck, and J. C. Leiter, Identification of respiratory vagal feedback in awake normal subjects using pseudo-random unloading, *J. Appl. Physiol. in press*, 2001.
4. P. C. Gay, J. R. Rodarte, M. Tayyab, and R. D. Hubmayr, Evaluation of bronchodilator responsiveness in mechanically ventilated patients, *Am. Rev. Respir. Dis.* **136**, 880-885 (1987).
5. A. Rossi, S. B. Gottfried, B. D. Higgs, L. Zocchi, A. Grassino, and J. Milic-Emili, Respiratory mechanics in mechanically ventilated patients with respiratory failure, *J. Appl. Physiol.* **58**, 1849-1858 (1985).
6. K. J. Killian, E. J. M. Campbell, and J. B. L. Howell, The effect of increased ventilation on resistive load detection, *Am. Rev. Respir. Dis.* **120**, 1233-1238 (1979).
7. M. Younes and W. Riddle, A model for the relation between respiratory neural and mechanical outputs. I. Theory, *J. Appl. Physiol.* **51**, 963-978 (1981).
8. L. Ljung, *System Identification* (Prentice-Hall, New Jersey, 1987).
9. D. Bartlett, Jr. and W. M. St. John, Influence of lung volume on phrenic, hypoglossal and mylohyoid nerve activities, *Respir. Physiol.* **73**, 97-110 (1988).
10. S. T. Kuna, Inhibition of inspiratory upper airway motoneuron activity by phasic volume feedback, *J. Appl. Physiol.* **60**, 1373-1379 (1986).
11. J. Polachek, R. Strong, J. Arens, C. Davies, I. Metcalf, and M. Younes, Phasic vagal influence on inspiratory motor output in anesthetized human subjects, *J. Appl. Physiol.* **49**, 609-619 (1980).

GENERAL MODEL FOR PATIENT–VENTILATOR INTERACTIONS

L. Heyer and P. Baconnier*

1. INTRODUCTION

Understanding and furthermore control Patient–Ventilator interactions is of great importance for partial assisted modes of ventilation. In the classical situation of entrainment of the respiratory rhythm the ventilator imposes a stable stimulation (the "Zeitgeber") on a passive patient, while in partial assisted mode the active patient acts on the ventilator and reciprocally [1-2]. It is the general problem of reciprocal interactions between a complex biological system and a complex physical system. This kind of interactions between two complex systems exhibits complex behaviors. The aims of Patient – Ventilator interactions modeling are to explain the observed interactions, to determine the physical parameters and to characterize the physiological properties affecting interactions. It implies to study both the dynamics and structural characteristics of the Patient and the Ventilator systems.

2. MODELING

For the purpose of this study, simulations of the dynamics of the patient–ventilator interactions are generated by a Boolean motor [2]. The patient's respiratory rhythm is given by sequences of neural inspiratory and expiratory passive or active activity. An active inspiratory onset and inspiratory cut off activity, an active expiratory onset and an active expiratory cut off activity govern transitions between neural states (Figure 1a). The Ventilator's rhythm is given by sequences of inflation and exhalation. A Ventilator cut-off mechanism of inflation and exhalation governs transitions between the two Ventilator states (Figure 1b). Simple linear transformations of the neural activity in muscular output and mobilized volume, a constant flow inflation and the definition of the transition mechanisms according to volume allows to simulate realistic dynamics of patient–ventilator interactions (Matlab program). Even though, this Boolean motor does not take into account the complex structure of the ventilator flow generation system and the patient volume generation system.

* L. Heyer, TIMC/IMAG – UJF, faculté de Médecine de Grenoble, 38700, La Tronche, France.

Frontiers in Modeling and Control of Breathing, edited by Poon and Kazemi.
Kluwer Academic/Plenum Publishers, New York, 2001.

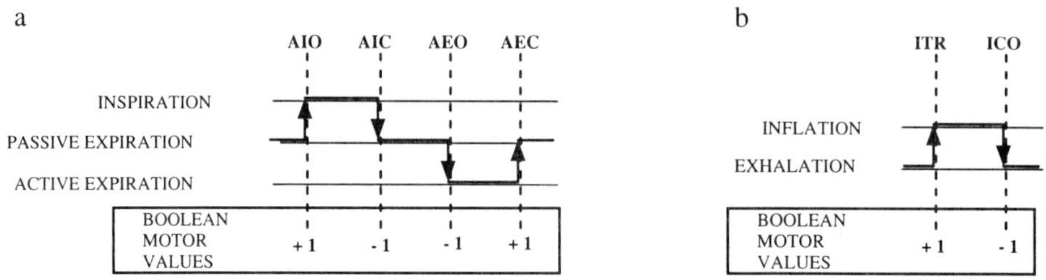

Figure 1. a. Boolean motor of the respiratory neural activity. Boolean motor triggers transitions between neural states. AIO: Active inspiratory onset, AIC: Active Inspiratory cut off, AEO: Active expiratory onset, AEC: Active expiratory cut off. b. Boolean motor of the ventilator. Boolean motor triggers ventilator's transitions between inflation and exhalation states. ITR: Inflation trigger, ICO: Inflation cut off.

To investigate the impact of the model structure on the dynamics of patient–ventilator interactions, we develop a general mathematical model using concepts related to the Object-Oriented programming language. The Object-Oriented modeling describes Objects, each defined by its states (or attributes) and its competencies (or methods) in response to external stimulation (or messages). Thus, it allows modeling the static structure of a system by Objects and theirs relationships. The patient–ventilator interaction process is described as a patient's volume generator and a ventilator flow generator which both act on the passive respiratory system. For this purpose, in this model we define three objects, two active which both act on one passive object (Figure 2).

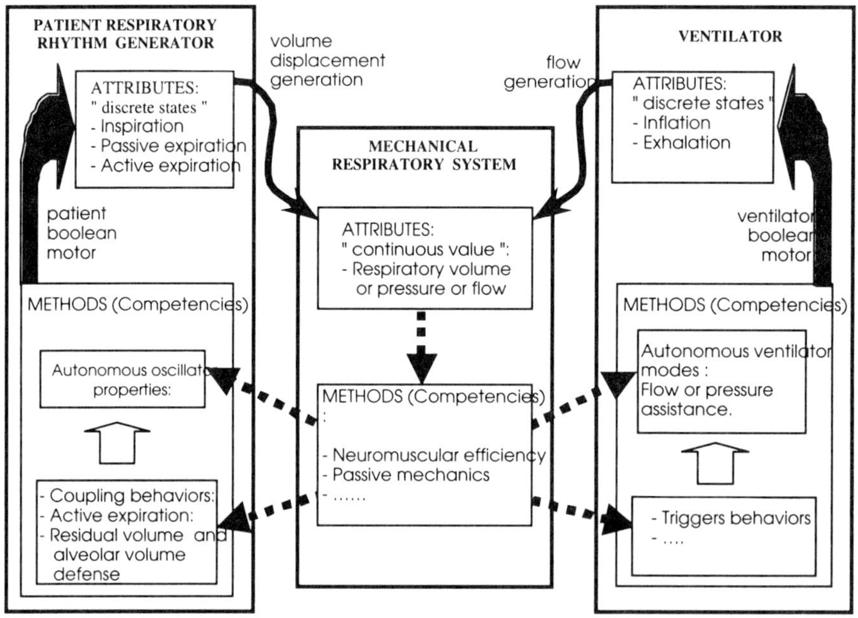

Figure 2. General diagram of the structure of the general model with schematic description of the objects, their attributes and competencies with their relationships. See text for legends and further explanations.

The passive object models the mechanical respiratory system with one state continuously described by the resultant volume, flow or pressure. This respiratory system's descriptor must fulfill two characteristics: One is to continuously summarize mechanical effects of the two other active objects. The other is to be continuously recognized by both generators. The passive respiratory system object's competencies are devoted to mechanical properties, which affects interactions.

The active objects are associated to their own Boolean motor, which determines transitions between object's discrete states accordingly to object's competencies. Two sets of competencies are individualized:

- One set governs transitions between object's states while no external stimulation act on the object. It corresponds to the minimum set of object's competencies required to assure an autonomous function of his corresponding Boolean motor. For the patient's respiratory rhythm, it corresponds to an oscillator in free running configuration and for the ventilator's flow generator to a basic mode able to assure an autonomous flow or pressure generation.
- One second set of competencies is related to events able to affect the free running transitions. For the patient's respiratory rhythm those competencies formalize different type of control strategies and their actions on the transitions, which governs respiratory rhythm generation. For the ventilator's flow generator it corresponds to triggers and cut off mechanisms driven by the resultant effects of patient's activity on the respiratory system state.

3. ANALYSIS

The combination of a Boolean motor, which gives the dynamics, and an Object-Oriented modeling, which characterizes the complex static structure of the two systems, allows the study of the influences of system's structures on the patient-ventilator interaction dynamics.

For this purpose we need taxonomy of the patient–ventilator interactions with the definition of two situations. 1) The ventilator dominates the patient or the patient dominates the ventilator. 2) The patient and the ventilator disclose interactions:

- The first situation describes two extreme clinical situations, one related to an absence of any patient's volume generation and the other to an absence of ventilator flow generation which is qualitatively equivalent to a spontaneous breathing situation. By extension, this situation also corresponds to cases where the patient's or the ventilator's activity don't affect the activity of the other one. In the model, those cases are well characterized by the work of one of the two Boolean motors in his autonomous mode. In those cases, one Boolean motor in his autonomous mode act on the other one but not the reciprocal, thus there is no interactions between the two Boolean motors. When the ventilator's activity is in an autonomous mode, the ventilator dominates the patient whatever his activity. When the patient's activity affect the ventilator's activity and when the patient's respiratory rhythm is in autonomous mode, the patient dominates the ventilator.
- In opposition, when the patient's activity affect the ventilator's activity but when the patient's respiratory rhythm is also affected by some patient's competencies, the patient and the ventilator disclose interactions.

Further clinically relevant classifications of interactions need to define a synchronous functioning situation and to characterize the fact that one generator drives the other. It allows to identify when the patient synchronously drives the ventilator or the reciprocal.

- A first Boolean motor drives the second one when all transitions from one state to another of the second Boolean motor arise in response to transitions of the first Boolean motor.
- Two Boolean motors interact synchronously when each transition of one Boolean motor is followed by a transition in the other. Nevertheless, two Boolean motors could be synchronous while we are unable to determine which Boolean motor drives the other. It is the case when one first Boolean motor drives one type of transition of the second motor and when simultaneously the second motor drives a transition of the first motor.

To have clinical applications, those definitions require the choice of minimal observation duration for the interaction's dynamics and of a hierarchy between different transition's states of the two Boolean motors. In the respiratory physiology field the inspiration or the inflation are highlighted and it's classical to use as a reference duration a respiratory cycle defined by the temporal duration between two successive inspiration onsets. Thus, patient drives the ventilator if an inflation always follows an inspiration whatever the modalities used for the expiratory or exhalation transitions. When each inspiration is followed by an inflation, the patient synchronously drives the ventilator. Consequently, a ventilator drives the patient if the patient never triggers inflation. When each inflation causes a patient's response that affects the ventilator's dynamics, the ventilator synchronously drives the patient. This case supposes that the patient's activity triggers the cut off mechanism of inflation. In case of ventilator's mode with a cut off mechanism insensitive to patient's activity like in volume control mode, the ventilator cannot drive the patient but could only dominate it.

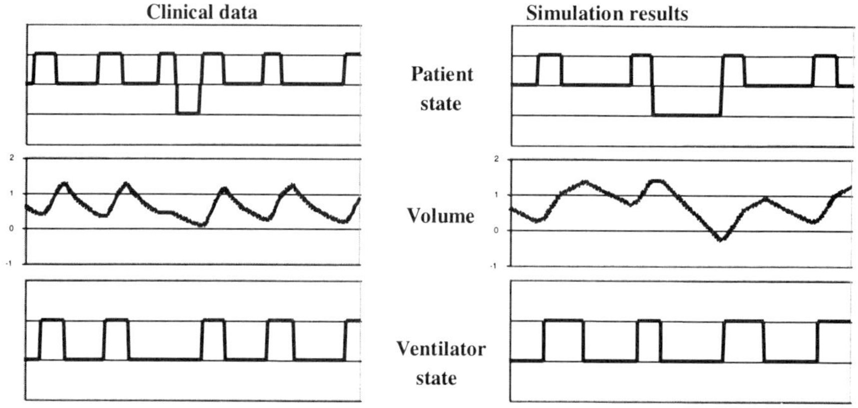

Figure 3. Left : volume signal recorded on a patient undergoing assisted ventilation together with its associated central respiratory (patient) and ventilator states. Right: simulated data obtained with an appropriate choice of model parameters

4. RESULTS

In Figure 3, a clinical example is shown with a qualitatively equivalent result generated with our model by an appropriate choice of its variables. For both clinical example and simulation, the corresponding states of the ventilator and patient's Boolean motor are shown.

5. DISCUSSION

This general model is developed as an investigator tools devoted to the patient–ventilator interactions description and analysis. The first objective of the model is to disclose realistic interactions qualitatively equivalent to the observed patient–ventilator interactions. As other models in that field, this model is based one a classical mechanical ventilation/respiratory system representation composed of a ventilator flow generator and a patient's volume displacement generator which both act on a passive respiratory system [3]. We restrict the structure of the model at only three major objects, two actives and one passive, all of them with limited states but with multiple competencies. In the active object one first set of competencies is introduced only to provide to a Boolean motor an autonomous mode of functioning in absence of any interactions. In opposition, a second set of competencies and their relationships are directly concerned with the problem of patient – ventilator interactions. This hierarchical structure facilitates a characterization of the interactions and provides basis for a classification approach.

With this model, interactions between the two generators result from the fact that each generator affects the other through a passive object. This passive object is only characterized by his mechanical properties while the generators are each characterized by two different functional parts. One part is a motor which gives the dynamics and one other part is concerned by regulatory functions which formalize control procedures that affects the motor's dynamics. This first allows for each generator to dispose of an autonomous mode in absence of any interactions when the generator is not influenced by the action of the other on the passive respiratory system. This particularity gives an easy way to distinguish interactions from domination. It also allows for the generator motor to dispose of different functional states with the introduction of some particular behaviors: This particularity distinguishes our model from others based on a patient generator output that can only be down regulated [4, 5]. By this way, as in the model of Y. Yamada ET col., our model allows changes in the patient's generator behaviors with the introduction of an active expiratory state [6].

In our model we suppose that the nature of the interaction mainly depends on the ability of one generator to continuously detect actions of the other on the passive system. For this purpose, the passive system is characterized by continuous variables that may affect the autonomous mode of each generator. The existence of this continuous variable and its effects are crucial for the interaction dynamic. In the model of E. Giannouli ET col., this variable is not used when the PAV (Pressure Assist ventilation) mode is simulated [5]. As results, the simulations are restricted to only one type of interactions: the patient drives the ventilator.

A definition of the patient's generator output as a volume displacement generator and the description of the passive respiratory system described by a unique variable, the lung volume, allows a detailed analysis of the patient – ventilator interactions. In the model, this volume displacement output is generated by the integration of Boolean neural value. The description of the passive respiratory system by a volume variable is a simple way to formalize regulatory functions of the patient's generator since some of the major chemical or non chemical control's systems loops could be defined with volume variables. Similarly, the trigger or cut off mechanisms of the ventilator may be defined through threshold values obtained as a flow – times duration product.

With this model we are able to describe all type of patient–ventilator interactions given by our taxonomy and thus analyze the determinant factors that can affect each type of interaction. In addition, those patient–ventilator interactions still correspond to relevant clinical situations.

6. REFERENCES

1. S. Muzzin, P. Baconnier and G. Benchetrit, Entrainment of respiratory rhythm by periodic lung inflation: effect of airflow rate and duration, *Am. J. Physiol.*, **263**, R292–R300 (1992).
2. P. Baconnier, J. Demongeot, and G. Benchetrit, Entrainment of respiratory rhythm: a new approach. *J. Theor. Biol.*, **164**, 149 – 162 (1993).
3. P. S. Crooke, J. D. Head, J. J. Marini, J. R. Hotchkiss, Patient–ventilator interaction: A general model for non–passive mechanical ventilation, *IMA. J. Math. Appl. Med. Biol.*, **15**, 321–337 (1998).
4. E. Giannouli, K. Webster, D. Roberts, M. Younes, Response of ventilator-dependent patients to different levels of pressure support and proportional assist, *Am. J. Respir. Crit. Care. Med.*, **159**, 1716–1725 (1999).
5. Y. Yamada, H. Du, Analysis of the mechanisms of expiratory asynchrony in pressure support ventilation: a mathematical approach, *J. Appl. Physiol.*, **88**, 2143–2150 (2000).

PHASE RELATIONS BETWEEN RHYTHMICAL FOREARM MOVEMENTS AND BREATHING IN CEREBELLAR IMPAIRED PATIENTS

Dietrich Ebert* and Harald Hefter†

1. INTRODUCTION

Cerebellar patients often report that they have difficulties in synchronizing speech and breathing movements, especially during additional extremity movements. This clinical observation was the basis for the hypothesis, that the cerebellum may also be involved in the coordination of extremity and breathing movements. The demonstration of a modulation of the cardiovascular and respiratory systems by electrical stimulation of the cerebellar cortex in decerebrated dogs[1] and the cerebellar uvula in decerebrated cats [2] suggests at least some cerebellar influence on breathing. To approach this problem from the clinical end we have investigated the coordination of forearm and breathing movements in cerebellar patients in a tracking paradigm. In normal subjects performing such tracking tasks coupling between limb and breathing movements can be observed at least in a restricted frequency range[3, 4].

In cerebellar patients, breathing pauses of several seconds length may occur during tracking while breathing at rest appears to be normal in the majority of patients[5]. Breathing pauses are known to occur under resting conditions in patients with severe brainstem lesions [6]. This "movement accompanied ataxic breathing" in cerebellar patients became apparent only when the coordination between the extremity and breathing movements was tested specifically and systematically. The results on the grossly disturbed coordination between breathing and rhythmic movement to be reported here show that the analysis of breathing movements in neurological disorders may provide new insights in the pathophysiology of disturbed system coordination.

* Carl-Ludwig-Institute of Physiology, University of Leipzig, Liebigstraße 27, 04103, Leipzig, Germany.
† Department of Neurology, University of Duesseldorf, Moorenstraße 5, D-40225 Duesseldorf, Germany

Frontiers in Modeling and Control of Breathing, edited by Poon and Kazemi.
Kluwer Academic/Plenum Publishers, New York, 2001.

Table 1. Age, sex, diagnosis and the results of clinical scoring of the patient group

patient	age	sex	diagnosis	score sum	dysarthr.	ataxia	gait	stance	refl.	oculo.	tremor	reb.
1	35	m	SAR	3	0	0	1	1	1	0	0	0
2	23	f	ADCA-III	9	1	1	1	1	1	1	2	1
3	46	f	ADCA-III	10	1	1	2	2	1	2	0	1
4	29	m	ADCA-III	12	1	2	2	2	1	1	1	2
5	44	f	ADCA-III	13	2	1	2	2	1	3	0	2
6	52	m	ADCA-III	14	2	1	3	3	1	1	1	2
7	47	m	ADCA-III	16	3	1	2	2	1	3	1	3
8	32	f	ADCA-III	16	3	1	4	3	0	3	0	3
9	63	m	IDCA-C	6	1	1	2	1	0	1	0	0
10	61	f	IDCA-C	14	2	1	3	3	0	2	1	2
11	37	f	SL	9	2	0	1	1	2	2	1	0

The score of a single item ranged between 0 (normal) and 4 (severely impaired). 8 single scores were summed up to yield a total clinical score, cerebellar ataxia was classified according to Currier and Subramony[14] (5th column). SAR denotes subject at risk for ADCA, ADCA-III = autosomal dominant cerebellar ataxia with pure cerebellar symptoms (previously called „Nonne-Marie"), IDCA = idiopathic cerebellar ataxia with pure cerebellar symptoms (previously called "Marie-Foix-Alajouanine"), SL, surgical lesion; m, male; f, female.

2. METHODS

2.1. Patients

In 15 normal subjects and 11 patients with a cerebellar motor disorder phase relations between breathing and forearm tracking were analyzed during a tracking paradigm. Patients underwent careful clinical and neuroradiological investigations including CT and MR imaging. Patients with a clear gaze nystagmus, double vision, gross ataxia or severe action tremor of the upper extremity disabling them to perform the tracking task were excluded from the study as were patients with clinical or neuroradiological signs of a brainstem or pyramidal tract involvement. Age, sex, diagnosis and clinical scoring of symptoms are presented in Table 1. The lowest total clinical score in our population of 3 was observed in an at risk subject whose father suffered from a severe heredoataxia. This patient was diagnosed to be impaired a few months after our investigation and therefore included in this study. Two patients were severely handicapped (pat. 7 and 8; Table 1) and unable to take care for themselves mainly because of their impairment of the lower extremities. They were still able to participate in the study. Patients´ data were compared to those of 15 controls. Normal subjects and patients were not matched because no age effects could be detected in previous studies on coordination [3].

2.2. Experimental Set-up, Paradigm and Data Recording

A sinusoidally moving target was presented as a thick spot on an oscilloscope. Target frequency was increased in steps of 0.1 Hz from 0.1 to 1.0 Hz and in steps of 0.2

Hz between 1.0 and 2.0 Hz. Patients were comfortably seated in a lab-chair. To track the target subjects and patients had to perform 15 degree elbow flexions and extensions which were recorded by a goniometer. The elbow position signal was presented on the oscilloscope as a second thin spot which had to be kept within the larger target spot. By means of a pneumotachograf (PTG) with a FLEISCH-transducer breathing was measured in parallel to the forearm movements. The target, elbow position and breathing signal were digitized with 100 Hz each and stored for off-line analysis.

2.3. Data Analysis

After determination of the onset and end of the inspiration and expiration phase breathing mean breathing rates with pauses ex- and included were determined.

Furthermore, "phase" intervals (PIs, in seconds) were calculated between the breathing and arm position signal quantifying the phase relations between these two oscillating signals. For details of calculation of PIs see [3]. In consecutive phase interval plots stable 1:1 coordination presents as a fairly straight line of points parallel to the x-axis (examples in [3]). Similarly 1:2 coordination presents as two parallel lines of points with little differences between every second phase interval. For quantifacation subsegments of consecutive PIs (PI_n, .., PI_{n+k}) were selected for which the absolute difference between two consecutive PIs (PI_{n+j} - PI_n) was smaller than 10 % of the mean movement cycle duration during the entire recording epoch. The durations of those subsegments fulfilling the 10 % criterion were added and the percentage of total recording length of this sum was determined to yield an estimate how often 1:1 coordination between breathing and arm position was observed within a trial. The same procedure was applied to subsegments where every second interval (PI_{2k}, ,$PI_{2*(k+n)}$) was analysed to quantify 2:1 and 1:2 coordination. 3:1 and 1:3 coordination was quantified similarly. The results are designated as coordination frequency and given in %.

Statistical analysis included the non-parametric u-test and the determination of rank correlation coefficients and linear regression lines.

3. RESULTS

3.1. Dependence of Breathing Rate on Target Frequency

There was no hint that breathing at rest might be disturbed in cerebellar patients. As soon as cerebellar patients start to make additional arm movements their breathing becomes considerably disturbed in contrast to breathing at rest. In the majority of patients breathing was highly irregular and breathing pauses of several seconds length occurred during tracking. Compared to the normals cerebellar patients revealed amplitude fluctuations already at the lowest target frequencies and a reduction of the mean breathing rate.

In Fig. 1 mean breathing rates of normals (upper part) and cerebellar patients (lower part) with (squares) and without pauses excluded (triangels) are compared. In normals with increasing target frequency breathing rate increases with a slight slope. The correlation between breathing rate and target frequency failed to be significant in the normals ($r = 0.5245$).

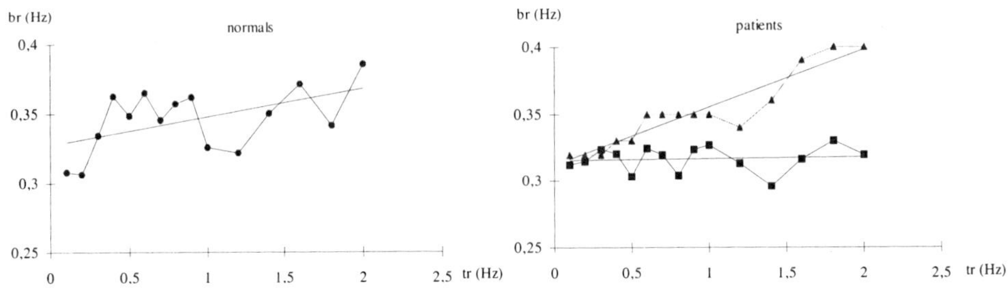

Fig. 1. Mean breathing rates (br) were plotted against target rate (tr) in controls (left panel) and patients (right panel). In the patients breathing rates are determined with pauses included (squares) and with pauses excluded (diamonds). In the controls breathing rates varied considerably with increasing target rate because of the coupling between breathing and limb movements (for details see text). In the patients breathing rates calculated from pause-free segments linearly increased with increasing target rate, whereas no dependence of breathing rate on target rate could be detected when the pauses were included.

In the patients, breathing rates based on complete trials (including breathing pauses) only showed little variation and a mean value of 0.32 Hz and were independent of target frequency (Fig. 1; right panel; squares; $y = 0.0015 * x + 0.3155$; $r = 0.0911$). Breathing rates during segments without pauses significantly increased with target frequency (Fig. 1, right panel; circles; $y = 0.0431 * x + 0.3119$; $r = 0.9395$; $p < 0.001$). Since only a few small pauses occurred at the lowest target frequencies the two regression lines have nearly the same intercept and intersect each other close to zero. Thus in cerebellar patients breathing pauses seem to be used to stabilize the breathing rate.

3.2. Coordination between Breathing and Forearm Movements

Coordination between breathing and forearm movements was analysed by determining phase intervals (PIs) between the breathing and arm position signals. In sequential PI-plots stable 1:1 coordination between both signals becomes visible as constant PIs (c.f. ref. 3). In Fig. 2, PIs between breathing and arm movements are presented for a severely affected cerebellar patient. Only at 0.1 Hz coordination of the 4:1 type seems to be present, in all other trials no clear-cut coordination type can be distinguished. The large peaks at 0.7 and 0.8 Hz reflect breathing pauses.

The coordination frequency was determined for each subject and target rate. Thereafter the mean percentage in the control and patient group was determined for the 1:1 (upper part of Fig. 3), 1:2 and 2:1 (middle part) and 1:3 and 3:1 coordination type (lower part) and plotted against target rate. Compared to normals (light hatched bars) 1:1 coordination is significantly reduced in the patients (dark hatched bars). However, 2:1 and 1:2 coordination could be observed to a similar extent in both groups. 3:1 and 1:3 coordination was less frequently observed in the patient group.

The percentage of all types of coordination decreased in relation to an increasing clinical score (Fig. 4, $r = 0.705$). Correlations between the single items of the clinical score (compare table 1) and the coordination frequencies showed a reciprocal tendency too but being not significant, correlation coefficients were much lower and ranged between 0.001 (for tremor) and 0.67 (for stance), mostly around 0.4.

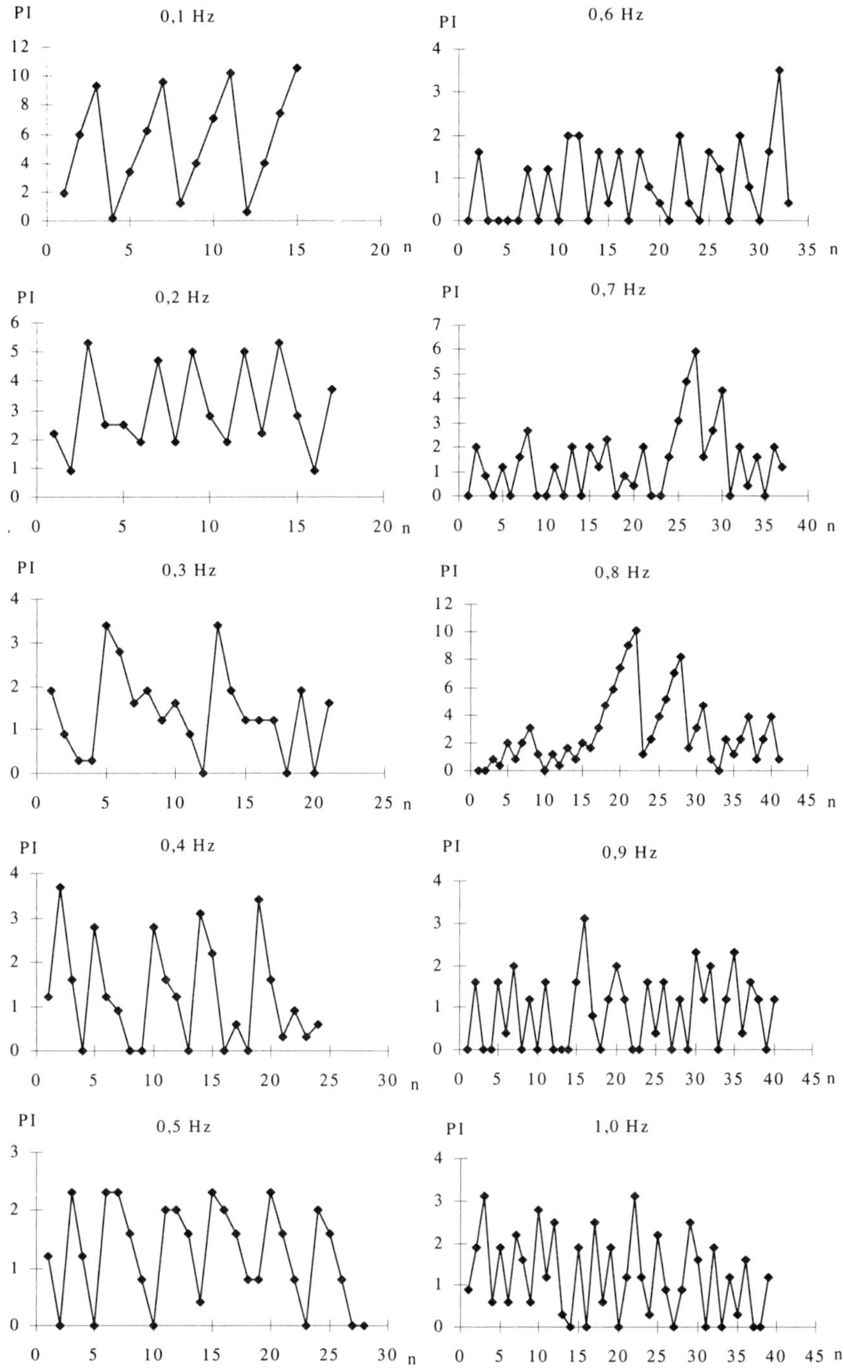

Fig. 2. Sequential phase interval plots for 10 different target rates in a cerebellar patient. 4 : 1 coordination is obvious for tracking at 0.1 Hz. But for higher target frequencies no clear coupling effect can be detected.

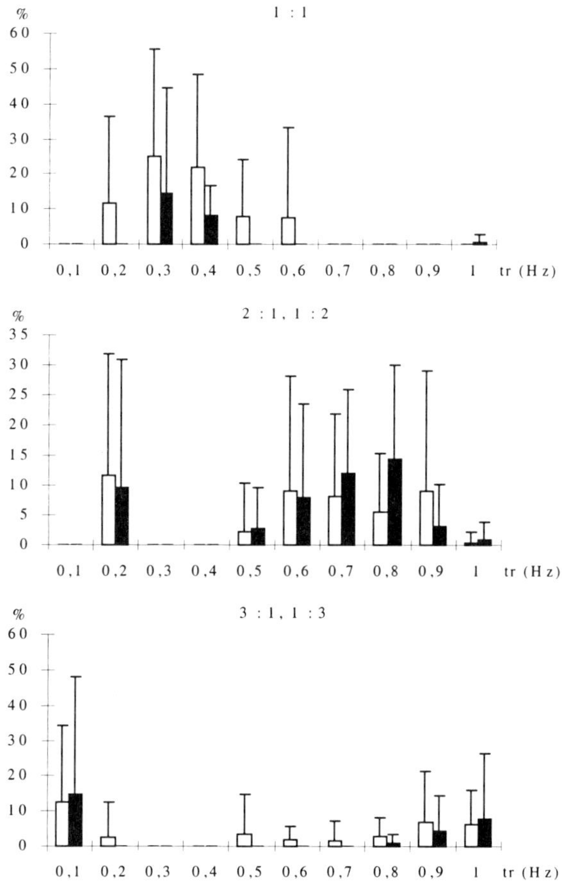

Fig. 3. The mean frequencies of 1:1, 2:1, 1:2, 3:1 and 1:3 coordination as percentage of the total recording duration was plotted in dependence on target rate (tr) for the controls (light hatched bars) and the patients (dark hatched bars). The reduction of 1 : 1 coupling is striking whereas 1 : 2 coordination seems to be preserved in the patients.

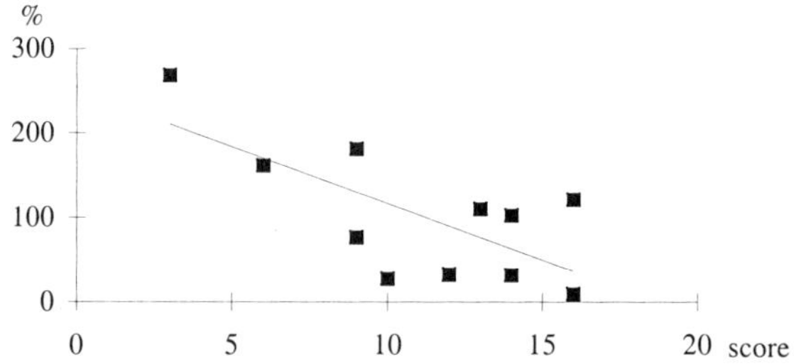

Fig. 4. The summarized percentages of coordination frequencies (ordinates) for all coordination types together were drawn in dependence on the clinical score of Table 1 (abscissa). Note the reciprocal correlation for the pooled data of all coordination types (r = 0,705).

4. DISCUSSION

4.1. Breathing at Rest and Brainstem Involvement

Lesions of the brainstem may lead to severe disturbances of the spontaneous breathing rhythm[6]. In our cerebellar patient group, however, breathing at rest was normal, indicating along with the clinical data that there was no major brainstem involvement. Subclinical brainstem involvement could not completety be ruled out. Because deficits became obvious mainly during tracking we think that these deficits are mainly a coordination disorder and hint more towards a relevant cerebellar deficit than towards a subclinical brainstem involvement

4.2. Interaction of Breathing with Other Rhythmical Activity

In humans coupling of breathing and limb movements during walking [8], running [9], cycling [10], rowing [11, 12], step tracking movements [13] and sinusoidal forearm tracking [3, 4] has been reported. However, interactions of the cerebellum with the respiratory neuronal networks have only been described in a few animal studies [1, 2] and were not subject to clinical investigations.

It cannot be ruled out that lesions of other regions in the brain involved in the sensorimotor system lead to similar failure of coordination. Our assumption that only the cerebellar impairment in question is responsible for the coordination disturbances is supported by the high-resolution MNR imaging, which did not show any structural changes in other brain regions outside the cerebellum. It might also be that coordination failures are serving as a strategy for compensation of disturbed, ataxic pursuit tracking movements. However, we do not know enough about the mutual influences of the different parts of the sensorimotor system to each other for tackling exact tracking behavior. Therefore, it seems to be speculative to conclude details about the final reasons for disturbed coordination behavior based on the descriptive data presented.

In cerebellar patients the physiological entrainment of the breathing rhythm is disturbed. Especially the 1:1 coupling occurs less frequently than in normals whereas the 1:2 coupling type seems to be preserved. More refined analysis will be necessary to see whether more complex coupling types (3:2 or 5:2), which occur during running and forearm tracking movements in normals, are also disturbed in cerebellar patients.

4.3. The Occurrence of Pauses

In normals pauses do not occur during breathing at rest. However, during fast running end-inspiratory pauses can usually be observed. These pauses seem to play a role in the exceptional broad repertoire of different coupling types between respiration and locomotion [9] in humans. Whereas disruptions of the breathing rhythm at rest were not observed in cerebellar patients, their breathing rate (between pauses) increased linearily with target rate during tracking (Fig. 1). There was no clear-cut influence of coupling on breathing rate as observed in normals. However, mean breathing rate calculated with pauses included was independent on target rate.

This result may suggest that in cerebellar patients because of their impairment to coordinate breathing and limb movements pauses were inserted by the respiratory system to stabilize their breathing rate unconsciously for the patient. Whether the occurrence of pauses is the utilization of a physiological mechanisms normally coming into play under extreme conditions such as fast running or during high target frequencies (above 1.6 Hz) in our paradigm, or reflects a pathological deficit, is presently unclear.

4.4. Significance of Coordination Failure and of Pauses in Patients

There was only a weak correlation between the coordination frequency and any single score parameters yet a rather strong reciprocal one between the coordination frequency and the sum score. The lack of coordination described here can not be related to any particular symptoms. Remarkably, the strongest correlation was found for the summarized percentages of all coordination types. This indicates that the disturbance of fixed timing relationships between respiration and extremity movements may act as a sensitive parameter for the severeness of cerebellar impairment in general. Therefore coordination frequency at particular entrainment rates may be considered for cerebellar diagnostics.

5. ACKNOWLEDGMENTS

This study was supported by grants from the Deutsche Forschungsgemeinschaft SFB 194, A5.

6. REFERENCES

1. G. Stella and G. Stevan, Changes in the heart rate from stimulation of the cerebellar cortex in decerebrated dogs, *Arch. Int. Pharmacodyn.* **136**, 1-11 (1962).
2. J.F.G. Paton and M.D. Gilbey, Effect of anaesthetic on sympathetic responses evoked from cerebellar uvula in decerebrated cats, *Am. J. Physio.l (Heart Circ Physiol)* **263**, H1285-H1291 (1992).
3. D. Ebert, B. Raßler and H. Hefter, Coordination between breathing and forearm movements during sinusoidal tracking, *Eur. J. appl. Physiol.* **81**, 288 - 296 (2000).
4. D. Ebert, B. Raßler, and S. Waurick, Phase relations between rhythmical forearm movements and reathing under normacapnic and hypercapnic conditions, *Adv. Expl. Med. Biol.*, **450**, 101 - 110 (1998).
5. D. Ebert, H. Hefter, C. Dohle and H.-J. Freund, Ataxic breathing during alternating forearm movements of various frequencies in cerebellar patients, *Neurosci. Letters.* **193**, 145-148 (1995).
6. F. Plum and J.B. Posner, *The diagnosis of stupor and coma* (F.A. Davis Company, Philadelphia, 1983).
7. S.H. Brown, K.R. Kessler, H. Hefter, J.D. Cooke and H.-J. Freund, Role of the cerebellum in visuomotor coordination, *Exp. Brain Res.* **94**, 478-488 (1993).
8. S.H. Loring, J. Mead and T.B. Waggener, Determinants of breathing frequency during walking, *Resp. Physiol.* **82**, 177-188 (1990).
9. D.M. Bramble and D.R. Carrier, Running and breathing in mammals, *Science* **219**, 251-256 (1983).
10. F. Garlando, J. Kohl, E.A. Koller and P. Pietsch, Effect of coupling the breathing and cycling rhythms on oxygen uptake during bicycle ergometry, *Europ. J. Appl. Physiol.* **54**, 497-501 (1985).
11. D.A. Mahler, C.R. Shuhart, E. Brew and T.A. Stukel, Ventilatory responses and entrainment of breathing during rowing, *Med. Sci. Sport. Exerc.* **23**, 186-192 (1991).
12. J.M. Steinacker, M. Both B.J. Whipp, Pulmonary mechanisms and entrainment of respiration and stroke rate during rowing, *Int. J. Sports. Med.*, **14** (Suppl 1), S15-S19 (1993).
13. B. Rassler, D. Ebert, S. Waurickand R. Junghans, Coordination between breathing and finger tracking in man, *J. Mot. Behav.* **28**, 48-56 (1996).
14. R.D. Currier and S.H. Subramony, Distinguishing between the adult ataxias, in: *Handbook of cerebellar diseases.* edited by R. Lechtenberg (Marcel Dekker, New York, 1993) pp. 337-343.

COMPUTATIONAL MODELING OF INTEGRATION OF VOLUNTARY/BEHAVIORAL AND AUTOMATIC MECHANISMS FOR BREATHING CONTROL

Ilya A. Rybak, Karen A. Moxon, Simon Giszter, and John K. Chapin*

1. INTRODUCTION

The integration of voluntary/behavioral controls with central pattern generators providing low-level control of movement is an important issue of broad relevance to basic neurophysiology, and neuroprosthetic technology. For automatic movements like breathing and locomotion, there is strong evidence that the CNS is organized for their control in a hierarchical fashion. The central pattern generator (CPG) is located in the brainstem (respiration) or spinal cord (locomotion) and generates the motor program for the neuromuscular apparatus. This low-level controller interacts with afferent feedback from the peripheral receptors and in turn is controlled by descending signals from the higher-level controllers (Figure 1A). The CPG defines automatic (involuntary) movements, which may be executed without conscious awareness. At the same time, these movements are subject to voluntary/behavioral regulation (i.e., they are integrated and regulated with the overall movement/system state). Therefore, understanding of complex mechanisms used by the brain for motor control demands investigation of interactions between the automatic and voluntary/behavioral components of this control.

Neural control of respiration is an excellent model object for such studies because the respiratory CPG has been more extensively studied at the neuronal and network levels than other CPG types in mammals. There are two major pathways for voluntary/behavioral control of breathing (see Figure 1B): (i) a direct path, which avoids the brainstem controller and affects motoneurons controlling diaphragm movements and lung inflation[1-5] and (ii) a path which leads to the respiratory center in the brainstem and affects the respiratory motor pattern via the modulation of the respiratory network performance[6-9]. The exact pathways for both above types of descending control are

* Ilya A. Rybak and Karen A. Moxon, School of Biomedical Engineering, Science and Health Systems, Drexel University, Philadelphia, PA 19104. Simon Giszter, Department of Neurobiology and Anatomy, MCP Hahnemann University, 2900 Queen Lane, Philadelphia PA 19129. John K. Chapin, Department of Physiology and Pharmacology, SUNY Health Science Center at Brooklyn, 450 Clarkson Ave., Brooklyn, NY 11203.

unknown, but there are data on their effects on the respiratory motor pattern, as well as on firing patterns of respiratory neurons to warrant and support a theoretical investigation. To this date however, no attempt has been made to model the possible effects of higher-level control systems on the CPG performance. Incorporating the descending control inputs into a CPG model provides an opportunity for computational analysis of voluntary and behavioral regulation of breathing.

2. MODEL

In our simulations, we modified and used a detailed model of the respiratory CPG developed earlier[10, 11]. The model included five types of respiratory neurons (early-I, ramp-I, late-I, post-I, E2 and pre-I) and had pulmonary feedback via the spinal motoneurons, diaphragm, lungs and pulmonary stretch receptors (PSRs) (see Figure 1B). Without influence of the supra-medullary centers, the model generated a normal three-phase respiratory pattern (Figure 2). In order to simulate possible interaction between the automatic control provided by the CPG and the descending control from the supra-medullary centers, we incorporated additional inputs allowing simulation of both supra-medullary control pathways: the direct path to the spinal cord motoneurons and the path to the CPG (Figure 1B).

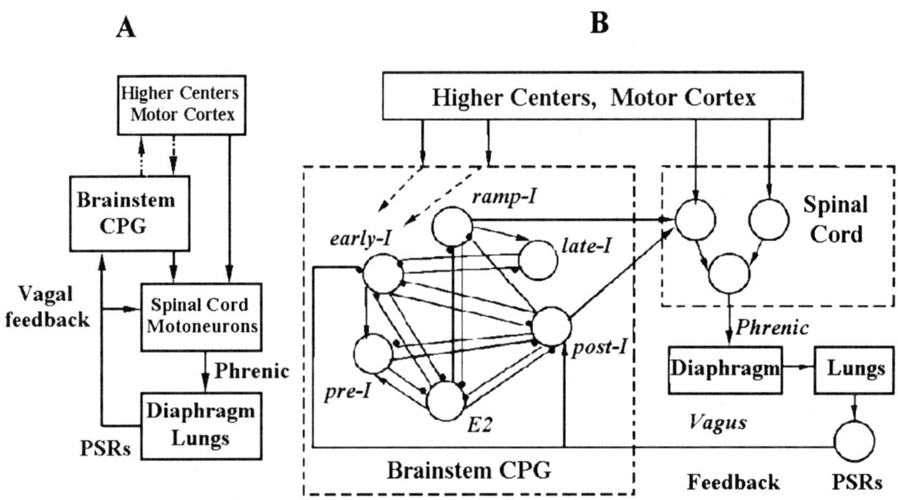

Figure 1. A. Schematic of hierarchical control of movement in context of neural control of breathing. **B.** Schematic of the model of integration of automatic and voluntary/behavioral control of breathing.

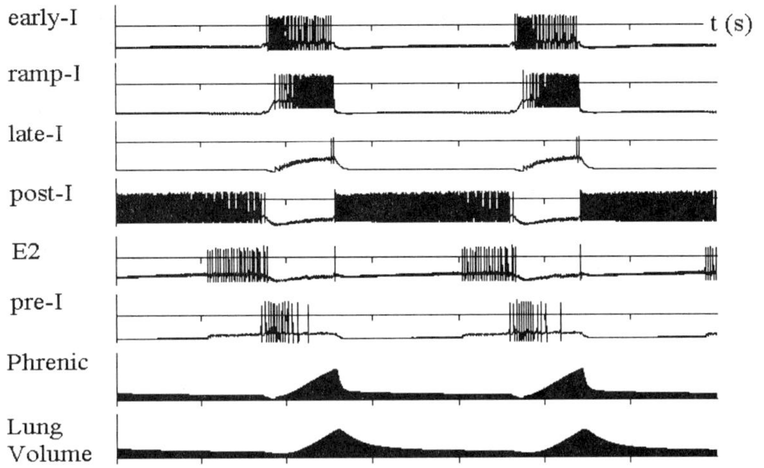

Figure 2. Three-phase respiratory rhythm generated by the CPG without supra-medullary regulation: firing patterns of respiratory neurons, integrated activity of the phrenic nerve, and dynamics of lung volume.

3. RESULTS

Figure 3 shows the results of our simulation of voluntary control of breathing through the direct path from the supra-medullary centers (e.g. the motor cortex) to the motoneurons controlling the diaphragm. The applied control signal produced a "voluntary inspiration" via direct activation of diaphragm and lung inflation. Interestingly, the CPG in turn was affected by vagal feedback (see Figure 1B), which arrested the automatic respiratory pattern (generated by the brainstem CPG) in the postinspiratory phase ("postinspiratory apnea"). This simulation demonstrated a possibility of a *competitive* type of interaction between the voluntary and involuntary control mechanisms when the higher-level voluntary commands overcome the low-level automatic controller (CPG) and provide direct control of breathing on a "movement-to-movement" basis.

The other type of possible high-level behavioral/voluntary control, via modulation of the automatic respiratory CPG, is shown in Figures 4 and 5. In this, *cooperative* type of interaction, the supra-medullary signals may selectively modulate activity of different respiratory neurons in the CPG. In our simulations, these signals provided an additional excitation to the early-I and post-I neurons. A relatively weak modulatory signal produced a "behavioral" increase in the frequency and amplitude of phrenic discharges, i.e. increased the frequency and deepness of breathing (Figure 4). A stronger signal produced a "behavioral" stop of breathing (Figure 5). Interestingly, the exact effect was dependent on the phase in which the signal was applied. While applying during inspiration (first stimulus in Figure 5) it arrested the respiratory pattern in the phase of inspiration ("apneusis") whereas its application during expiration produced postinspiratory apnea (second stimulus in Figure 5).

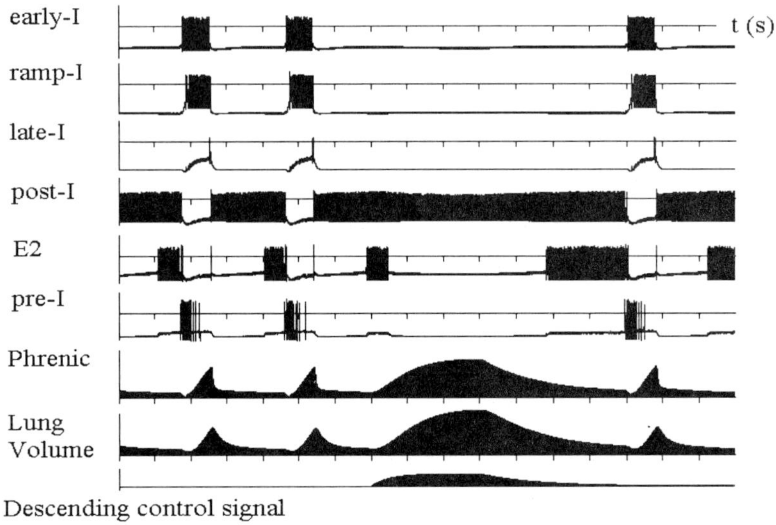

Figure 3. Results of simulation of voluntary control of breathing by supra-medullary centers) through the direct path. The descending control signal (the lowest trace) activated the motoneuron controlling the diaphragm. The applied control signal produced a "voluntary inspiration". Note, that vagal feedback arrested the automatic respiratory rhythm (generated by the CPG) in the postinspiratory phase.

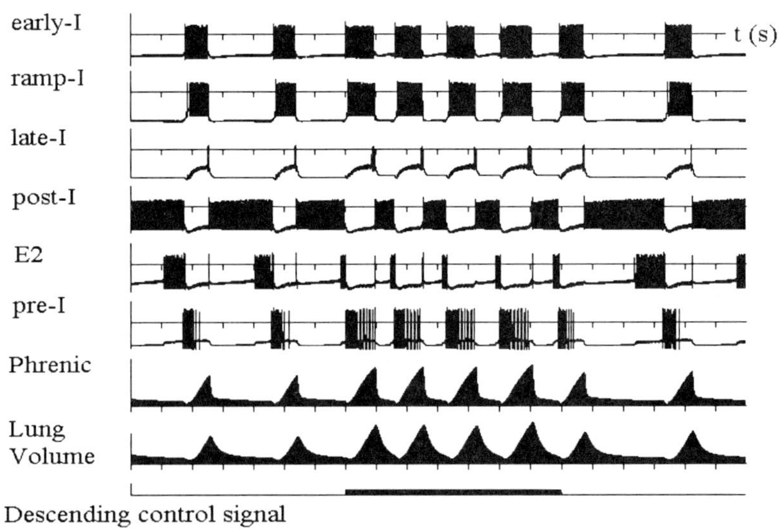

Figure 4. Results of simulation of voluntary/behavior control of breathing by supra-medullary centers through modulation of CPG performance. A weak stimulation (the lowest trace) applied to the early-I and post-I neurons produced a "behavioral" increase in the frequency and amplitude of breathing.

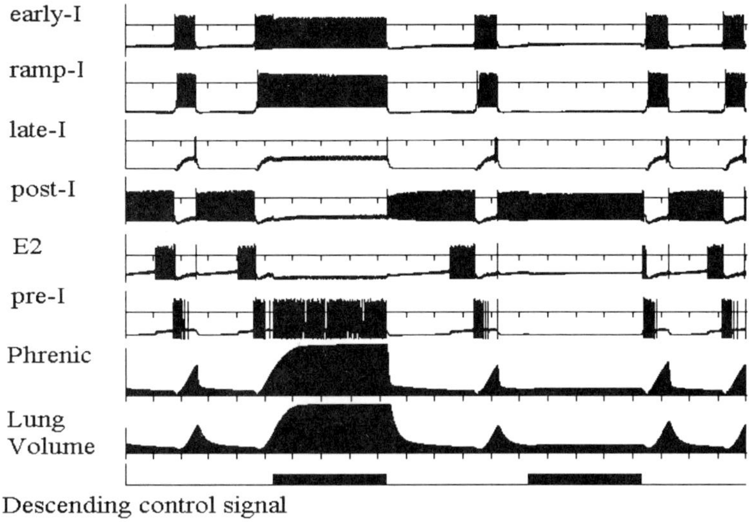

Figure 5. Results of simulation of voluntary/behavior control of breathing by supra-medullary centers through modulation of CPG performance. A relatively strong stimulation (the lowest trace) applied to the early-I and post-I neurons produced a "behavioral" stop of breathing. The effect depended on the phase in which the signal was applied. While applying during inspiration (first stimulus) it arrested the respiratory rhythm in the phase of inspiration whereas its application during expiration (second stimulus) produced "postinspiratory apnea".

4. CONCLUSION

This paper presents a first attempt to model interactions between the automatic low-level and voluntary/behavioral high-level mechanisms for neural control of breathing. Our simulations have demonstrated the possible types of these interactions. However, because of the lack of specific data at the neuronal level, inputs of the supra-medullary control pathway to particular respiratory neurons were set arbitrary. A more careful analysis of all available direct and indirect data, and additional special experimental studies are required for building a more detailed model of interactions between automatic and voluntary/behavioral control of breathing.

5. REFERENCES

1. S. C. Gandevia and D. K. McKenzie, Activation of the human diaphragm during maximal static efforts, *J. Physiol. Lond.* **367**, 45-56 (1985).
2. S. C. Gandevia and D. K. McKenzie, Human diaphragmatic endurance during different maximal respiratory efforts, *J. Physiol. Lond.* **395**, 625-638 (1988).
3. S. C. Gandevia, D. K. McKenzie, and B. L. Plassman, Activation of the human respiratory muscles during different voluntary maneuvers, *J. Physiol. Lond.* **428**, 387-403 (1990).

4. G. Macefield, and S. C. Gandevia, The cortical drive to human respiratory muscles in awake state assessed by premotor cerebral potentials, *J. Physiol. Lond.* **439**, 545-598 (1991).
5. R. A. Mitchell and A. J. Berger, Neural regulation of respiration. *Am. Rev. Respir. Dis.* **111**, 206-224 (1975).
6. K. Budzinska, K. Glowicki, W. A. Karczewski, J. Kulesza, and M. Ryba, The respiratory response to magnetic simulation of the cortex in the unanaesthetized baboon, *Acta Neurobiol. Exp. (Warsz)* **51**, 125-128 (1991).
7. J. Orem, The activity of late inspiratory cells during the behavioral inhibition of inspiration. *Brain Res.* **458**, 224– 230 (1988).
8. J. Orem and A. Netick, Behavioral control of breathing in the cat, *Brain Res.* **366**, 238–253 (1986).
9. J. Orem, and R. H. Trotter, Behavioral control of breathing, *NIPS* **9**, 228-232 (1994).
10. I. A. Rybak, J. F. R. Paton, and J. S. Schwaber, Modeling neural mechanisms for genesis of respiratory rhythm and pattern: II. Network models of the central respiratory pattern generator, *J. Neurophysiol.* **77**, 2007-2026 (1997).
11. I. A. Rybak, J. F. R. Paton, and J. S. Schwaber, Modeling neural mechanisms for genesis of respiratory rhythm and pattern: III. Comparison of model performances during afferent nerve stimulation, *J. Neurophysiol.* **77**, 2027-2039 (1997).

THE CONTROL OF BREATHING AT REST

Safraaz Mahamed*, Arshia F. Ali*, Bernice Wang*, Dominic Ho*, and James Duffin* †

1. INTRODUCTION

Breathing at rest is driven by two possible sources: behavioural, such as "wakefulness," and metabolic, from the respiratory chemoreflexes that automatically control partial pressures of carbon dioxide and pH in the blood[1-3]. However, few measurements of the balance between these two drive sources at rest have been made, and some investigators suggest that behavioural drives may be more important than chemoreflex drives and sufficient to sustain resting breathing[4, 5].

2. METHODS

Each of 11 subjects (6 male, 5 female) was studied on 2 occasions; the first was a familiarisation trial of the protocol and the second was the actual experimental protocol as follows. First, subjects sat at rest for 5 minutes before measurements began. Second, they hypoventilated for 7 minutes at their resting tidal volume, coached to a breathing frequency of 3 breaths min^{-1} by the investigator, after which they rested for 5 minutes. Third, their resting measurements were obtained for 5 minutes, immediately after which they rebreathed until the first equilibration plateau of P_{CO2} was observed. Finally, after a further 5-minute rest, they hyperventilated, coached by the investigator to maintain an end-tidal P_{CO2} of 20 mmHg for 6 minutes before rebreathing at an iso-oxic P_{O2} of 100 mmHg.

3. RESULTS

The mean (SEM) for all subjects of resting end-tidal P_{CO2} (34.3 ± 0.5 mmHg) and the average plateau P_{CO2} of the first rebreathing test was 50.3 ± 0.58 mmHg. The average (SEM) slope and threshold of the second rebreathing test response (iso-oxic P_{O2} of 100

* University of Toronto, Department of Physiology, Toronto, Ontario, Canada M5S 1A8.
† University of Toronto, Department of Anaesthesia, Toronto, Ontario, Canada M5S 1A8.

Frontiers in Modeling and Control of Breathing, edited by Poon and Kazemi.
Kluwer Academic/Plenum Publishers, New York, 2001.

mmHg) was 1.4 ± 0.26 l min^{-1} $mmHg^{-1}$ and 42.0 ± 0.53 mmHg respectively. Average (SEM) basal ventilation (3.1 l min^{-1} ± 0.48) was significantly lower than resting ventilation (9.1 l min^{-1} ± 0.73) and resting end-tidal P_{CO2} was significantly larger than the corrected estimate of chemoreflex threshold P_{CO2} (26.0 ± 0.9 mmHg).

Figure 1 is a plot of ventilation versus P_{CO2} showing the average metabolic hyperbola and ventilatory response to carbon dioxide for all subjects. The latter is also shown left-shifted by the mean resting-plateau difference in P_{CO2} for all subjects (16.0 ± 0.75 mmHg), which estimates the P_{CO2} difference between end-tidal and central chemoreceptor tissue. The intersection of this shifted ventilatory response to carbon dioxide with the metabolic hyperbola estimates the resting state of ventilation and end-tidal P_{CO2}.

However, the shift in threshold also includes the peripheral component, which should not be so shifted. We used a theoretical chemoreceptor threshold (open circle in Figure) at 34.1 mmHg,[6] corrected by the mean resting-plateau P_{CO2} difference for all subjects (filled circle in Figure), as the pivot point about which reductions in sensitivity attributed to the peripheral chemoreflex are factored. A 20% reduction[7] brings the ventilatory response to carbon dioxide close to passing through the measured resting ventilation and end-tidal P_{CO2} point.

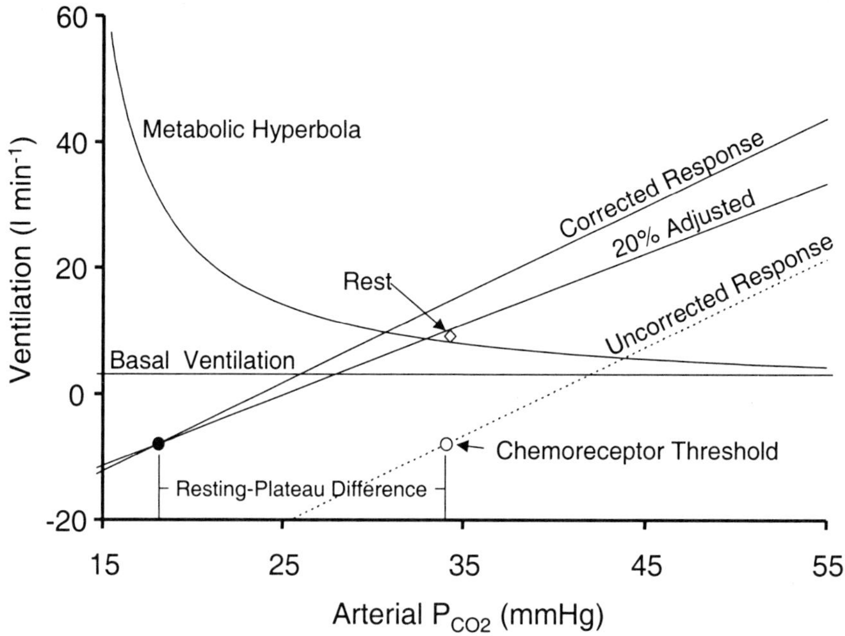

Figure 1. Determinants of resting ventilation. Mean (SEM) ventilation and end-tidal P_{CO2} for all subjects at rest (open diamond), during hyper- and hypoventilation are fitted with a metabolic hyperbola. The mean (SEM) basal ventilation, threshold and sensitivity to carbon dioxide from the rebreathing tests for all subjects determined the uncorrected ventilatory response to carbon dioxide (dashed line) which was corrected by the resting-plateau difference (solid lines) and the peripheral chemoreflex contribution.

4. DISCUSSION

We interpret our findings to show that there is a substantial contribution to ventilation at rest by both behavioural and chemoreflex drives.

5. ACKNOWLEDGMENTS

We acknowledge the expert advice of the Respiratory Research Grouping at the University of Toronto. This research was approved by the University of Toronto Ethical Research Board and supported by a grant from the Ontario Thoracic Society.

6. REFERENCES

1. E. A. Phillipson, and G. Bowes. Control of breathing during sleep, in: *Handbook of Physiology, Section 3, Volume 2,* edited by A. P. Fishman (Bethesda MD: American Physiological Society, 1987), pp. 649-690.
2. J. Orem, and R. H. Trotter, Behavioral control of breathing. *NIPS* **9**, 228-232 (1994).
3. S. A. Shea, Behavioural and arousal-related influences on breathing in humans. *Exp. Physiol.* **81**, 1-26 (1996).
4. D. R. Corfield, C. A. Roberts, A. Guz, K. Murphy, and L. Adams, Modulation of the corticospinal control of ventilation by changes in reflex respiratory drive. *J. Appl. Physiol.* **87**, 1923-30 (1999).
5. C. M. Spengler, C. A. Czeisler, and S. A. Shea, An endogenous circadian rhythm of respiratory control in humans. *J. Physiol. (Lond)* **526**, 683-694 (2000).
6. J. Duffin, R. M. Mohan, P. Vasiliou, R. Stephenson, and S. Mahamed, A model of the chemoreflex control of breathing in humans: model parameters measurement. *Respir. Physiol.* **120**, 13-26 (2000).
7. P. Dejours, Chemoreflexes in breathing. *Physiol. Rev.* **42**, 335-358 (1962).

CENTRAL AND PERIPHERAL CHEMOREFLEX CHARACTERISTICS: PANIC DISORDER PATIENTS VS. HEALTHY VOLUNTEERS

Lukasz Struzik[*†], Martin Katzman[†‡], Nishka Vijay[†], Aimee Coonerty-Femiano[†], Safraaz Mahamed[*], and James Duffin[*]

1. INTRODUCTION

Panic disorder is a psychiatric condition characterised by spontaneous panic attacks. These are discrete episodes of intense anxiety that are perceived to be uncontrollable and are accompanied by unpleasant physical sensations such as palpitations, chest pain, dyspnea, choking, sweating, tremors, faintness and paraesthesia. Although successful treatments for the condition are available, its etiology remains elusive[1]. In 1993 Klein has suggested that panic disorder patients possess a "false suffocation alarm" that may be associated with a lowered threshold for carbon dioxide detection[2]. The theory attributes carbon dioxide hypersensitivity as a causal instigator for panic in these patients, with carbon dioxide as the physiological messenger for detecting potential suffocation. To date there has been no assessment of the respiratory suffocation sensors, namely the central and peripheral chemoreflexes, in response to rising carbon dioxide levels.

2. METHODS

Eleven panic disorder patients (7 male, 4 female) and 9 healthy volunteers (5 male, 4 female) were studied. Iso-oxic, modified rebreathing tests were used to assess differences in peripheral and central chemoreflex characteristics between the two groups[3]. These included the chemoreflex thresholds and sensitivities for carbon dioxide. Each subject underwent a practice test to allow familiarisation with experimental conditions. Following, two modified rebreathing tests were undertaken under a hyperoxic condition (central chemoreflex alone,) and hypoxic condition (combined central and peripheral chemoreflex response) in a randomised fashion.

---

* University of Toronto, Department of Physiology, Toronto, Ontario, Canada M5S 1A8.
† Centre for Addiction and Mental Health, Anxiety Disorders Clinic, Toronto, Ontario, Canada, M5T 1R8.
‡ University of Toronto, Faculty of Medicine, Toronto, Ontario, Canada M5S 1A8.

3. RESULTS

The mean (SEM) for the chemoreflex threshold under hyperoxia (150 mmHg) for patients and controls was 43.6 ± 1.5 and 43.2 ± 1.0 (mmHg) respectively. The central chemoreflex sensitivity (slope) of response to carbon dioxide in patients and controls was 1.60 ± 0.19 and 1.63 ± 0.23 (L/min/mmHg) respectively. Similarly, under hypoxia (50 mmHg) the mean chemoreflex threshold for patients and controls was 41.1 ± 1.1and 39.9 ± 1.2 (mmHg) respectively. The average peripheral chemoreflex sensitivity to carbon dioxide in panic patients and controls was 2.85 ± 0.32 and 2.93 ± 0.55 (L/min/mmHg) respectively. No significant differences between the 2 groups were observed in any of the variables under study.

Figure 1 is a plot of ventilation versus P_{CO2} showing the average hyperoxic (150 mmHg) response (central chemoreflex alone) to carbon dioxide for panic disorder patients (solid line) and healthy controls (dashed line.) Figure 2 is a similar plot of the average hypoxic (50 mmHg) response (combined central and peripheral chemoreflex) to carbon dioxide for the two test groups. Chemoreflex thresholds and sensitivities are indicated with associated standard errors of the mean.

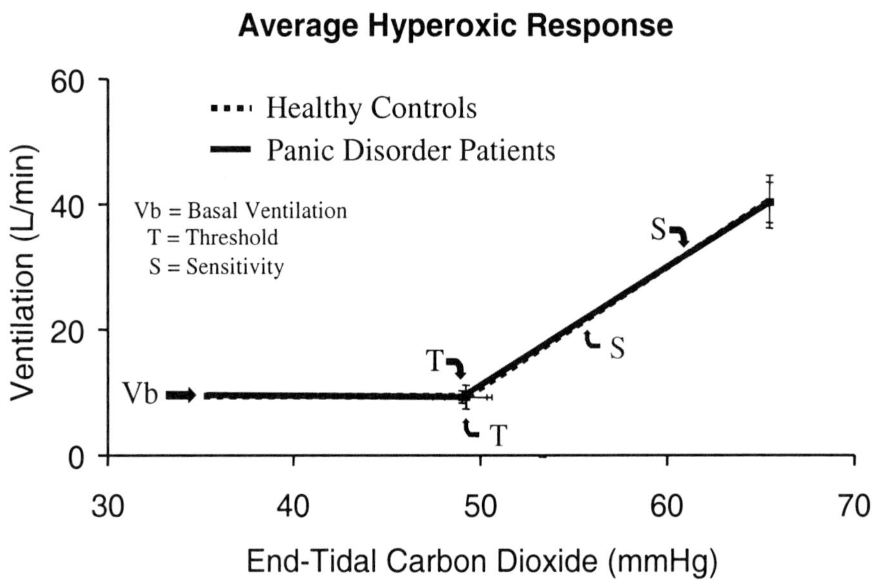

Figure 1. Average hyperoxic (150 mmHg) ventilatory response (central chemoreflex alone) to carbon dioxide. Mean (SEM) threshold (first brake point) and sensitivity (slope) of response to carbon dioxide for panic disorder patients (solid gray) and healthy volunteers (dotted black) are indicated.

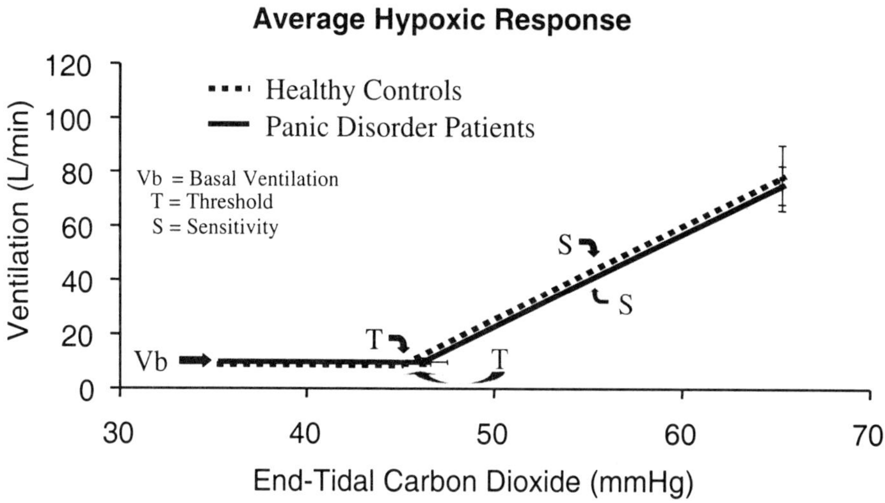

Figure 2. Average hypoxic (50 mmHg) ventilatory response (combined central and peripheral chemoreflex) to carbon dioxide. Mean (SEM) threshold (first brake point) and sensitivity (slope) of response to carbon dioxide for panic disorder patients (solid gray) and healthy volunteers (dotted black) are indicated.

4. DISCUSSION

Our results show no hypersensitivity of the chemoreflexes in panic disorder. These findings challenge Klein's association of a "false suffocation alarm" with heightened chemoreflex responsiveness as an explanation for panic initiation. The central and peripheral chemoreflexes, the sole respiratory sensors of suffocation, are not hypersensitive to carbon dioxide in panic disorder patients. The respiratory chemoreflexes therefore may participate in the genesis of a panic attack, but the triggering event lies in the interpretation of the chemoreflex signals. Alternatively, pathology elsewhere in the central nervous system may act on the chemoreflex arc to generate panic.

5. ACKNOWLEDGMENTS

We acknowledge the expert advice of the Respiratory Research Grouping at the University of Toronto. This research was approved by the University of Toronto Ethical Research Board.

6. REFERENCES

1. L. A. Papp, D. F. Klein, J. M. Gorman, CO_2 hypersensitivity, hyperventilation, and panic disorder. *Am. J. Psychiatry* **150**, 1149-1157 (1993).
2. D. F. Klein, False suffocation alarms, spontaneous panics, and related conditions. *Arch. Gen. Psychiatry* **50**, 306-317 (1993).
3. J. Duffin, R. M. Mohan, P. Vasiliou, R. Stephenson, and S. Mahamed, A model of the chemoreflex control of breathing in humans: model parameters measurement. *Respir. Physiol.* **120**, 13-26 (2000).

A PHYSIOLOGICAL AND PSYCHOLOGICAL MODEL OF IDIOPATHIC HYPERVENTILATION

Sandy Jack and Christopher J. Warburton*

1. INTRODUCTION

Hyperventilation was first described in the middle of the 19th century, with soldiers in the American Civil war complaining of symptoms originally thought to be cardiac in nature[1]. However, these symptoms were later discovered to be identical to idiopathic hyperventilation. Since this time there has been limited published data and little consensus as to existence and cause of this condition. The symptomatology is varied; however the principle symptoms include shortness of breath, chest tightness and parasthesia. The physiological variables associated with idiopathic hyperventilation are disproportionate breathlessness and evidence of hyperventilation at rest and during exercise. A disorganised breathing pattern has also been described, with no relationship between the ventilatory response and the level of exercise. This altered physiological response has been suggested to be due to a neuromuscular disturbance. Patients with idiopathic hyperventilation (IH) have also been described to have an altered psychological profile. Soley and Shock[2] in 1938 reported a relationship between symptomatology consistent with IH and psychological factors. Lowry in 1967 suggested that IH was identical to hysteria, patients consider themselves physically inferior and have perfectionist traits[3]. Lum[4] described the most comprehensive explanation of the physiological and behavioural aspects of this condition. He hypothesised that IH may be initiated by anxiety. This emotional state creates symptoms, which are perceived and interpreted by the patient. A vicious circle then ensues whereby patients interpret the symptoms as life threatening, this creates further anxiety with increased hyperventilation and symptomatology. Plum[5] proposed a model of interaction of the motor behavioural pathway and metabolic pathway. The motor behavioural pathway functions in addition to the metabolic pathway of ventilatory control. It is thought that copies of motor signals from respiratory muscles (corollary discharge) are sent to areas of the cortex where they are perceived as sensations of effort. Patients with IH may have an inappropriate perception of work, which makes them feel that their exertion is extreme and they are

* Sandy Jack, Research Scientist and Christopher J. Warburton, Consultant Physician, Aintree Chest Centre, University Hospital Aintree Liverpool United Kingdom.

Frontiers in Modeling and Control of Breathing, edited by Poon and Kazemi.
Kluwer Academic/Plenum Publishers, New York, 2001.

required to ventilate at a higher level. This feeling self-perpetuates. Howell[6] in 1997 further developed this model and suggested that emotional factors may play a large part in influencing the sensorimotor cortex and the behavioural pathway . This may result in a disturbance in normal ventilatory control particularly at times of emotional distress, which may become a conditioned response. It is known that psychological factors, in particular chronic anxiety, are associated with irregularities of breathing pattern. Ventilatory inefficiency as reflected by an increase in V_E/VO_2 ratio has also been observed in high-trait-anxiety groups and we have previously demonstrated that there is a negative correlation between $P_{ET}CO_2$ and anxiety[7].

Voluntary and involuntary forebrain influences can interact with and override the automatic control of breathing in awake man. Behavioural influences diminish with increased metabolic demands therefore these influences may be more pronounced during relaxed wakefulness[8]. It has been suggested that emotional disturbance caused by mental stress can alter breathing pattern and ventilatory control in man. It has been shown that increased anxiety causes an increase in minute ventilation (V_E l/min) which is achieved predominately by increased breathing frequency and not by tidal volume[9]. Dudley and Pitts-Poarch[10] in 1980 reported similar findings in their group of normal patients. They demonstrated that anger and anxiety caused an increase in breathing frequency with decreased PCO_2 .

There is also evidence to suggest that sensory stimuli and mental activity alter breathing patterns and that this may become a conditioned response[11]. Patients with chronic anxiety have been reported to have irregularities in rhythmicity of breathing pattern. Several reviews in the literature have investigated the role of the amygdala in the control of emotions and the regulation of breathing pattern. The central nucleus of the amygdala has major projections to the brainstem respiratory areas implicated in respiratory timing. Studies have shown that anxiety affects both T_I and T_E and that there is a negative correlation between anxiety and decrease T_E [12].

Our study aims to investigate the physiological response and psychological profile of a group of patients with idiopathic hyperventilation. This will include an assessment of the ventilatory response during a cardiorespiratory exercise test including variability in breathing pattern. These data in conjunction with the psychological data may elucidate the influence of behavioural and emotional factors on ventilatory control mechanisms.

2. METHODS

We performed a controlled physiological study to investigate the altered physiological response and psychological profile in patients with IH. We studied 24 patients 10 males, 14 females mean age (SD) 56.3(10.7) years with IH. Patients had normal lung function, no cardiac, neural or endocrine abnormalities and normal bronchial reactivity. Hyperventilation was defined as $PetCO_2$ of less than 30mmHg for greater than one minute at rest. The study protocol included a symptom limited cardiorespiratory exercise test on an electronically braked cycle ergometer using a ramped incremental protocol. To study a normal physiological response to exercise we included an age matched control group of 24 normal subjects, 6 males mean age (SD) 48(13) years. Normality was defined as normal cardiorespiratory exercise test with no other evident pathologies. The patients with IH also completed Hospital Anxiety and Depression scores (HADs) and Nijmegen questionnaires. The HAD scores were used to assess the psychological profile of the patients, determining their anxiety and depression levels. The Nijmegen questionnaire[13] consists of 16 complaints, the frequency of incidence of these can be indicated on a five-point ordinal scale, and is used to indicate the presence of hyperventilation.

The exercise tests were performed using a Medgraphics CPX breath by breath computer analysis machine. The subjects performed a flow volume loop prior to exercise. This was to establish a baseline FEV_1 for comparison with post testing values to confirm that exercise induced asthma was not present. The workload was set to a ramped incremental protocol at a predetermined level of subjective fitness to achieve maximal exercise in which the subject reached volitional termination in 8 to 12 mins to avoid fatigue. The cycle ergometer workload increased on a continuous ramp so that the increase was between 15 to 20 watts per minute. Resting data was recorded for 2 mins and then the subjects freewheeled for 1 minute, they then began the pre-set protocol. The subjects were encouraged to exercise to maximum to keeping their pedalling rpm at 60. At the end of exercise, subjects performed a flow volume loop manoeuvre at 5, 10 and 15 mins post exercise. A fall of 15 % in FEV1 (suggestive of asthma) led to exclusion from the study.

The data was analysed using non-parametric Mann-Whitney U tests to compare each variable at different work rates on both patient groups. The variability in breathing pattern was analysed during the first half of exercise (below anaerobic threshold), excluding the first 30 seconds of pedalling to allow for stability. The data was analysed using the mean, standard deviation, mean x SD (which we have published as the exercise breathing index EBI) and the coefficient of variation of the V_E/VO_2. This data was then analysed using non-parametric tests to compare the mean, SD, EBI and Coefficient of Variation of patients with idiopathic hyperventilation compared to the control group.

3. RESULTS

Table 1 shows that there is a significantly high Nijmegen score suggestive of hyperventilation. There was also a significantly increased anxiety level reported and evidence of depression in some subjects.

There was a significantly augmented ventilatory response to low work rate exercise in patients with IH compared to the control group and significant hypocapnia throughout. Ventilation (V_E l/min) and ventilatory equivalent for oxygen (V_E /VO_2), breathing frequency (B_f br/min) tidal volume (V_T litres) and $P_{ET}CO_2$ (mmHg) were significantly different in the two groups at rest and at 40 watts exercise.

Table 1. Questionnaire data on patients with Idiopathic Hyperventilation

QUESTIONNAIRES	IH PATIENTS	CLINICAL SIGNIFICANCE
Anxiety	10.5 (4.6)	Score >7
Depression	6.6 (2.6)	Score >7
Nijmegen	46.4. (14.0)	Score >23

Table 2. Resting data on patients with idiopathic hyperventilation compared to age match controls

REST	IH PATIENTS	CONTROLS	SIGNIFICANCE
V_E (l/min)	16.7 (6.6)	10.5 (3.1)	p<0.0001
V_E/VO_2	54.2 (14.9)	34.6 (8.9)	p<0.001
$P_{ET}CO_2$ (mmHg)	27.8 (5.3)	33.9 (4.5)	p<0.0001
B_f (br/min)	24.9 (7.2)	17.8 (5.6)	p<0.0001
V_T (litres)	0.7 (0.3)	0.6 (0.2)	p =NS

Table 3. Exercise data at 40 watts on patients with idiopathic hyperventilation and age match controls

EX (40 W)	IH PATIENTS	CONTROLS	SIGNIFICANCE
V_E (l/min)	38.6(11.5)	20.7(5.6)	p<0.0001
V_E/VO_2	52(16.2)	29.8(3.7)	p<0.001
$P_{ET}CO_2$ (mmHg)	27.3(5.7)	37.0 (4.6)	p<0.0001
B_f (br/min)	36.8 (8.2)	22.6 (6.0)	p<0.0001
V_T (litres)	1.11 (0.3)	1.24 (0.4)	p =NS

The disorganised breathing pattern was assessed using the mean, standard deviation (SD), coefficient of variation (CV) of V_E/VO_2 and the (mean x SD) of V_E/VO_2 EBI)[14]. These variables were measured at rest and during the first half of exercise (below anaerobic threshold) excluding the first 30 seconds of pedalling to allow for stability. There was significantly increased variability (SD) and EBI in V_E/VO_2 at rest in the IH patients as compared to normals (p<0.001), but no difference in CV. During exercise once more SD and EBI were significantly higher in the IH patients (p<0.001), with no difference in CV in patients with IH compared to the control group (21 (10) Vs 19 (9)). These results are shown below. EBI takes into account both the level and variability of ventilation and identifies abnormal breathing patterns in individuals.

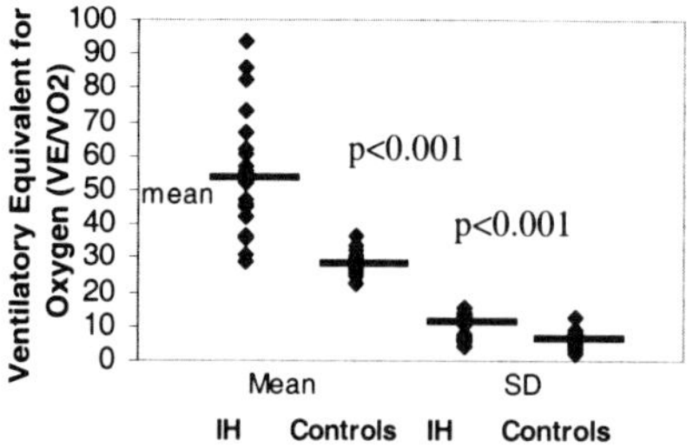

Fig. 1. Mean and SD of VE/VO2 During Exercise in Patients with IH and Controls

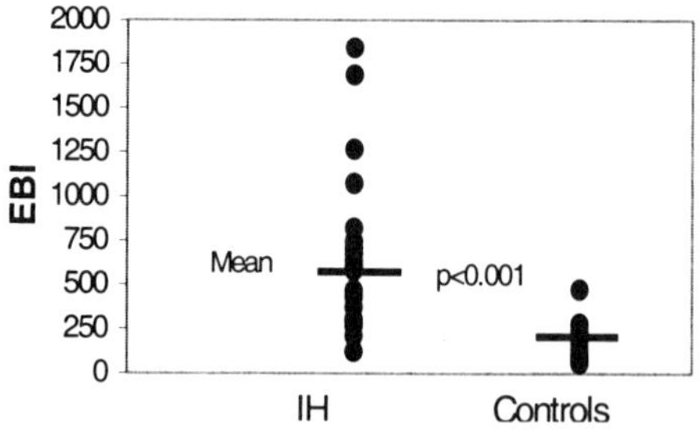

Fig. 2. EBI (Mean x SD VE/VO2) During Exercise in Patients with IH and Controls

4. DISCUSSION

In this study comparing patients with IH and normal subjects, we have demonstrated both psychological and physiological differences between the two groups. The patients with IH have high levels of anxiety some have depression and all have an augmented and abnormal ventilatory response to exercise.

One must consider the cause and effect of the psychological and physiological abnormalities demonstrated. It is our view that the abnormal physiological response to exercise is caused by the presence of anxiety (rather than the abnormal physiological response causing the subject to become anxious), for the following reasons.

It has been previously suggested that behavioural factors, particularly anxiety and anger may influence the physiological response to exercise and alter breathing patterns. Anxiety has been shown to cause an increase in minute ventilation mediated by increased breathing frequency. Our data demonstrate that patients with IH have an augmented ventilatory response to exercise with significant hypocapnia and a disorganised breathing pattern. The increased ventilatory response was mediated mainly by increased breathing frequency. This pattern or response would be consistent therefore with IH being an anxiety-related condition.

The disorganised and inappropriate breathing pattern demonstrated during exercise was unrelated to metabolic demands. This suggests that the abnormal ventilatory response during early exercise may be mediated via the behavioural pathway. Behavioural factors have been suggested to be more pronounced during relaxed wakefulness and diminish with increased metabolic demands. We have previously shown that breathing pattern becomes more physiological as work rate increased[15]. Anecdotally we recognise that there are several patterns of abnormal ventilatory response in IH. These include a consistently high V_E/VO_2 during exercise, increased variability (SD) with only marginally raised V_E/VO_2 and both a high level and variability of V_E/VO_2. We believe that the EBI is the best value to capture and quantify all of these patterns.

The amygdala may contribute to the instability in ventilation. It is involved in the control of rate and depth of breathing and regulation of breathing pattern via the parabrachalis nucleus medialis (NPBM). This centre is known to be influenced by emotions i.e. fear and anxiety via the sympathetic autonomic nervous system. Previous studies have shown that there is a correlation between respiratory timing and anxiety.

The interaction between chemical (metabolic) and non-chemical (behavioural) drives at rest and during early exercise, may cause random variations in ventilation from breath to breath. Therefore the physiological and psychological findings of this study strongly support the model suggested by Howell. As a consequence of a persistently abnormal ventilatory response to exercise, altered chemoresponsiveness to CO_2 may develop and this in turn may lead to perpetuation of the condition. We feel therefore that a complex interaction between physiological and psychological factors initiates and subsequently sustains inappropriate hyperventilation in these subjects.

5. REFERENCES

1. DaCosta J.M. On irritable heart: a clinical study of a form of functional cardiac disorder and its consequences. *Am. J Med Sci* 1871; **61:**17–52.
2. Soley M, Shock N. The aetiology of effort syndrome. *Am J Med Sci* 1938; **196**: 840 – 51.

3. Lowry T. P. Hyperventilation and hysteria. 1967; *Springfield, III, Charles C. Thomas*
4. Lum C. The syndrome of habitual chronic hyperventilation. In: Modern trends in psychosomatic medicine. O. W. Hill ed. Bitterwoth publ, London. 1976; **11:**196-230.
5. Plum F. Neurological integration of behavioural and metabolic control of breathing In: Porter R, ed. Breathing: *Hering – Breuer Centenary Symposium. London: Churchill* 1970;159 – 81.
6. Howell J.B.L. The hyperventilation syndrome: a syndrome under threat? *Thorax* 1997; **52**(3) s: 30-34.
7. Owen M.J, Jack S, Oates K.E, Pearson M.G and Warburton C. The effect of psychological factors on physiological variables in patients with idiopathic hyperventilation. *European Respir J* 1998; **12**(28): 412S.
8. Morgan N. and Cameron I. R. Respiratory response to behavioural and metabolic stimulation. *Clin Sci* 1984; **67:** (Suppl 9): 61 p.
9. Masoaka Y and Homma I Anxiety and respiratory patterns: their relationship during mental stress and physical load. *Int J Psychophysiol.* 1998; **27:** 153 – 159.
10. Dudley D. L and Pitts-Poarch A.R Psychological aspects of respiratory control. *Clinics in Chest Medicine* 1980; **1**(1): 131- 143.
11. Mador M. J., and Tobin M.J. Effects of alteration in mental activity on the breathing in healthy subjects. *Am Rev Respir Dis* 1991; **144:** 481 – 487.
12. Bechbache R.R, Chow H.H.K, Duffin J and Orsini E.C. 1979 The effects of hypercapnia, hypoxia, exercise and anxiety on the pattern of breathing in man. *J Physiol* 1979; **293:** 285-300.
13. Dixhoorn Van J. and Duivendoorn H.J. Efficacy of Nijmegen questionnaire in recognition of the hyperventilation syndrome. *J. Psychosom* 1984; 29: 199 – 206.
14. Jack S. Pearson M.G. and Warburton C.J. A clinical index to assess breathing pattern in early exercise: The Aintree Breathing Pattern Index (ABPI) *Am Rev Resp Crit Care Med* 2000.
15. Jack S. Wilkinson M. and Warburton C.J. Behavioural breathing pattern during hypocapnic and normocapnic exercise in idiopathic hyperventilators. *Am Rev Resp Crit Care Med* 2000; **161**(3): A431

BEHAVIORAL STATE CONTROL AND AIRWAY INSTABILITY

M. A. Haxhiu[*†], S.O. Mack[*], R. J. Martin[†], D.G. Ferguson[†], G. Longobardo[‡], and N.S. Cherniack[‡]

1. INTRODUCTION

Sleep is an actively generated state induced by excitation of specific networks that alter the activity of different cell populations along the neuraxis, including serotonin-containing neurons. The main region identified in slow wave sleep (SWS) generation is located within the ventrolateral preoptic area (VLPO), which contains neurons that are specifically activated during sleep[20]. These cells innervate cell bodies and dendrites of neurons that participate in arousal, such as histamine containing cells[13]. In addition, the lateral preoptic area projects to midline neurons (Haxhiu, unpublished data), indicating possible pathways by which sleep influences activities of upper airway dilating muscles and cholinergic outflow to the tracheobronchial system.

While the main CNS sites involved in the generation of SWS lie outside of the brainstem, all major events of rapid eye movement (REM) sleep are controlled by brainstem structures[10, 15]. It is well established that microinjection of cholinergic agonists or acetylcholine esterase inhibitors such as physostigmine into the medial pontine reticular formation (mPRF) causes a REM sleep-like state[15,21]. During sleep, the discharge of serotonin-containing neurons is inhibited via activation of GABAergic inputs[19] and the activity of upper airway dilating muscles is decreased[2,3,9,12,22]. However, cholinergic outflow to the tracheobronchial tree is enhanced during both SWS and REM sleep[16,23].

It has been shown that GABA is the principal inhibitory neurotransmitter of the circadian system[18], and sleep is associated with activation of GABA containing neurons of the VLPO region[20]. We postulate that excitation of VLPO neurons or pontine REM 'on cells' leads to activation of GABAergic mechanisms which inhibit serotonin containing neurons, causing a decrease in serotonin at targeted sites, a decrease in the discharge of upper-airway dilating muscles, and an increase in cholinergic outflow to the tracheobronchial system.

* Howard University, Washington, District of Columbia, 20059; † Case Western Reserve University, Cleveland, Ohio, 44106; ‡ UMDNJ-NJMS, Newark, New Jersey, 07103

Frontiers in Modeling and Control of Breathing, edited by Poon and Kazemi.
Kluwer Academic/Plenum Publishers, New York, 2001.

2. NEUROANATOMICAL STUDIES

To test our hypothesis, in the first series of experiments, we used the pseudorabies virus (PRV) transneuronal labeling method to identify the neuronal circuits that control the activity of airway-related vagal preganglionic neurons, and of upper airway dilating muscles[8,24].

The neuroanatomical basis of coupling between midline neurons and motor cells innervating phrenic and airway-related preganglionic cells was also studied[7]. In rats, the retrograde tracer technique was employed to identify bulbospinal cells projecting to phrenic motoneurons. This transneural labeling method was used to localize second-order medullary neurons that innervate airway parasympathetic preganglionic cells. Specifically, fluorogold was injected into the phrenic nucleus, and PRV was administered into extrathoracic trachea, following sympathetic denervation of the airways. After four days of survival, double-labeled neurons were seen in different regions, including the ventromedial medulla oblongata and midline and parapyramidal serotonin-containing neurons. An example of a midline neuron that projects to airway-related preganglionic cells and phrenic motoneurons is shown in Fig. 1. This finding indicates that a subset of midline neurons that projects to phrenic motoneurons also innervates airway-related preganglionic cells and/or hypoglossal motoneurons (not shown in the figure), providing parallel control of inspiratory pumping muscles, upper airway dilating muscles, and parasympathetic outflow to the airways.

Most recently, we examined the expression of $GABA_AR$ on PRV infected neurons of caudal raphe nuclei, following the injection of PRV into the phrenic nerve. Figure 2 (left) shows the expression of $GABA_AR$ on midline neurons that project to phrenic motoneurons. This suggests that GABA, if released within midline neurons following excitation of VLPO or pontine REM 'on cells', will inhibit caudal raphe neurons via the activation of $GABA_AR$.

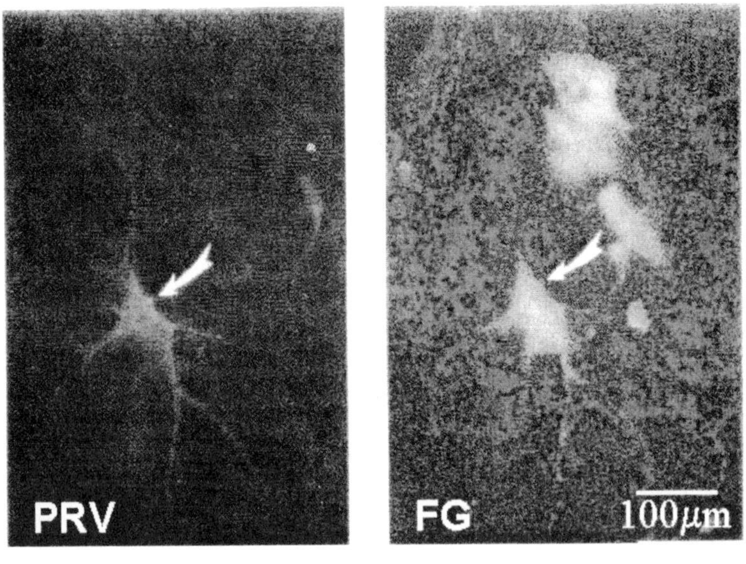

Figure 1. An example of a double-labeled neuron (arrow) in the rat ventromedial region of the medulla oblongata following fluorogold (FG) injection into phrenic motoneurons and pseudorabies virus (PRV) into the tracheal wall. For details see Ref. 7. For a color representation of this figure, see color insert facing this page.

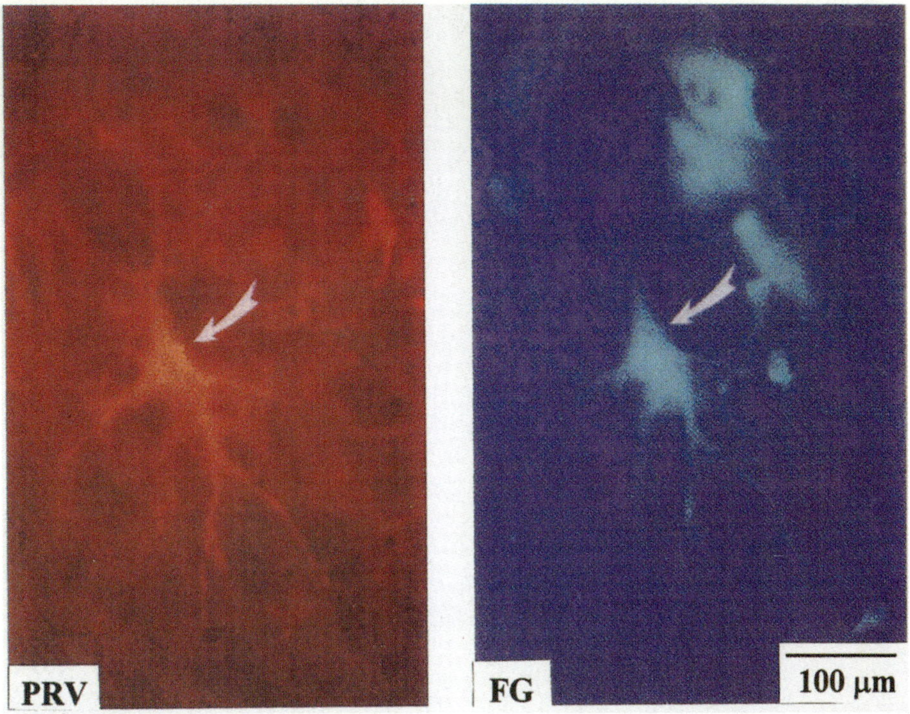

Figure 1. An example of a double-labeled neuron (arrow) in the rat ventromedial region of the medulla oblongata following fluorogold (FG) injection into phrenic motoneurons and pseudorabies virus (PRV) into the tracheal wall. For details see Ref. 7.

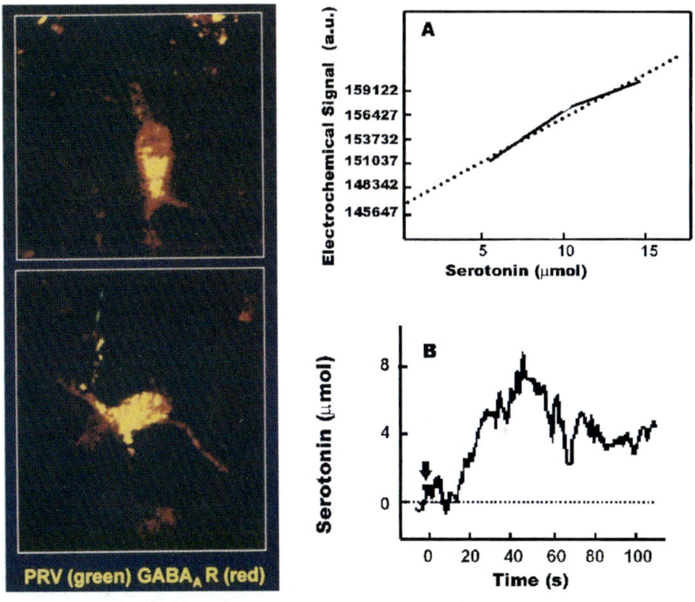

Figure 2. Left panel: Confocal microscopic images showing GABA$_A$ receptors visualized with rhodamine (red) on transneurally labeled inspiratory bulbospinal neurons in the raphe obscurus. PRV infected neurons are visualized with fluorescein (green). In these double-labeled cells the yellow signal is the result of superimposed red on green signals. Right panel: An example of the effect of raphe neurons stimulation (L-glutamate, 4 nmol/80 nl) on serotonin release in the rostral ventrolateral medulla in an anesthetized cat. A) Calibration curve for serotonin. B) Release of serotonin following stimulation of raphe nuclei. Serotonin release was associated with a decrease in airway smooth muscle tone (not shown). For details see Ref. 8.

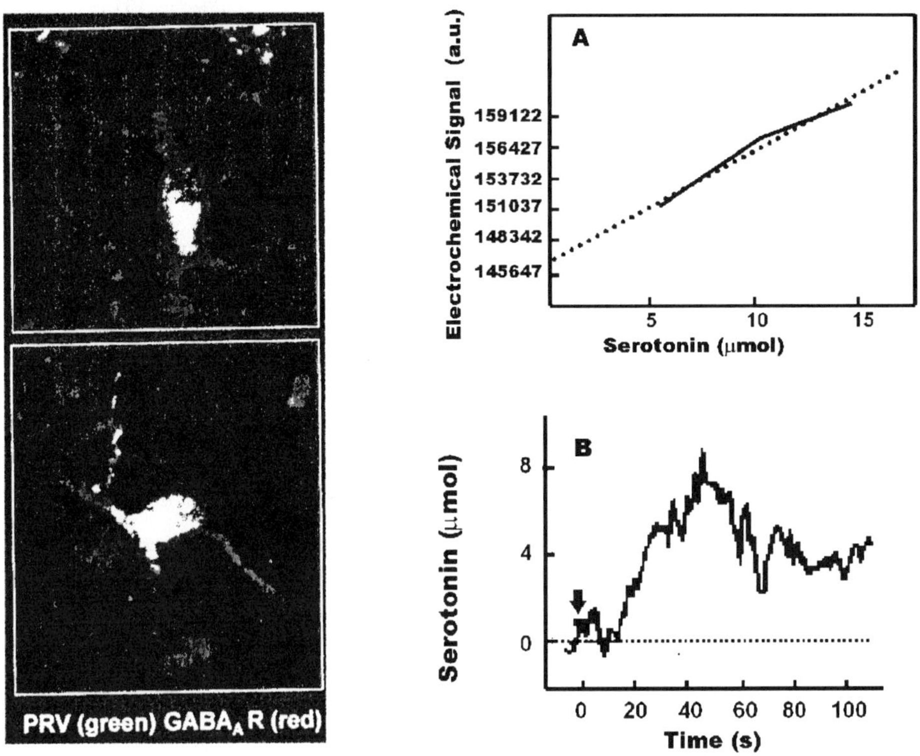

Figure 2. Left panel: Confocal microscopic images showing GABA$_A$ receptors visualized with rhodamine (red) on transneurally labeled inspiratory bulbospinal neurons in the raphe obscurus. PRV infected neurons are visualized with fluorescein (green). In these double-labeled cells the yellow signal is the result of superimposed red on green signals. Right panel: An example of the effect of raphe neurons stimulation (L-glutamate, 4 nmol/80 nl) on serotonin release in the rostral ventrolateral medulla in an anesthetized cat. A) Calibration curve for serotonin. B) Release of serotonin following stimulation of raphe nuclei. Serotonin release was associated with a decrease in airway smooth muscle tone (not shown). For details see Ref. 8. For a color representation of this figure, see color insert facing page 446.

3. PHYSIOLOGICAL STUDIES

We postulated that sleep related changes in respiratory drive could be due to inhibitory influences of sleep-activated cells on midline neurons (caudal raphe). We performed physiologic experiments in cats, and comparative studies in ferrets and dogs. In addition, the release of 5-HT from the rostral ventrolateral medulla was measured by an electrochemical method using microelectrodes, and an IVEC-10 hardware and software system. This system runs on a DOS compatible computer that can rapidly detect and discriminate monoamine neurotransmitters such as serotonin. The results showed that microinjection of L-glutamate, but not buffered saline, into the caudal raphe, caused a significant increase in 5-HT (Fig. 2, right), which was associated with a withdrawal of cholinergic outflow to the airways[6]. In addition, we examined the effects of 5-HT antagonists on raphe stimulation-induced responses. Bilateral topical applications or microinjections of methylsergide in the rostral ventrolateral medulla blocked airway smooth muscle relaxation induced by raphe stimulation

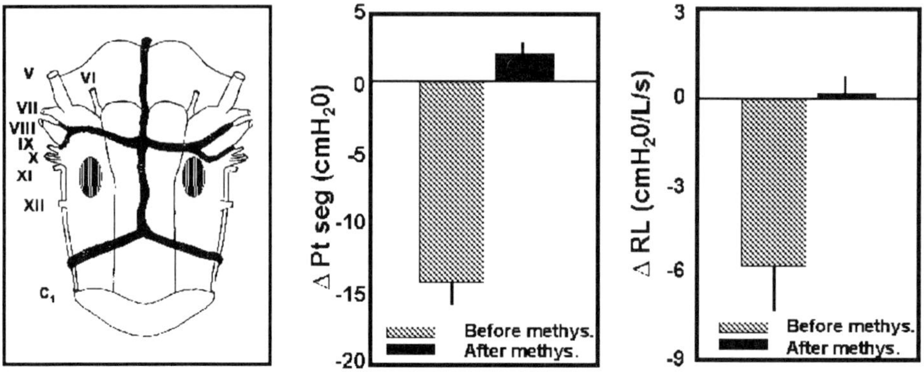

Figure 3. The average decreases in tracheal tone and lung resistance induced by raphe neurons stimulation, before and after blockade of 5-HT receptors within the ventrolateral medulla oblongata. For details see Ref. 8.

(Fig. 3), indicating that serotoninergic inputs are involved in central withdrawal of cholinergic outflow to the airways. However, activation of midline neurons preferentially increased the activity of the genioglossus muscle (not shown).

4. DISCUSSION

In the last 30 years, enormous progress has been made in the neurobiology of sleep, including sleep-related neurochemical substrates in the brain[10,15]. However, little is known about CNS cell groups that connect sleep activated neuronal pools with cranial nerve motoneurons, such as hypoglossal and airway-related vagal preganglionic cells. One reason could be that retrograde transport studies using conventional retrograde markers provide limited information.

One way to avoid this type of problem is to use a highly specific transneuronal marker such as Bartha pseudorabies virus (PRV), which is an ideal tool for determining the cells of origin that project to functionally specific neurons[8,14]. We used this tracing method as a guide for physiological investigations, which showed that serotonin-containing neurons project to airway related-vagal preganglionic cells[5,8], motoneurons of cranial nerves that innervate upper airway dilating muscles[24], phrenic motoneurons[11], and they express $GABA_AR$. Furthermore, using different retrograde tracers, we found that a single midline neuron projects to phrenic motoneurons and to airway-related vagal preganglionic cells[7], and/or hypoglossal motoneurons (Haxhiu, unpublished data). It has been shown that the activity of serotonin-containing neurons is decreased during sleep. Upon entering SWS, neuronal activity slows to approximately 50% of the quiet waking level and loses its regularity. Finally, during REM sleep, the activity of most serotonin neurons is suppressed completely via activation of GABAergic input[19]. Furthermore, serotonin levels in the hypoglossal nucleus region and hypoglossal nerve activity are reduced during carbachol induced REM sleep[12]. Stimulation of serotonin-containing neurons is associated with inhibition of airway-related vagal preganglionic cells and excitation of hypoglossal motoneurons. With in situ voltametry, we observed a release of serotonin and a decrease in airway smooth muscle tone following stimulation of midline neurons[6].

Serotonin exerts modulatory effects in many brain regions involved in regulation of motor, sensory, and behavioral state responses. Diverse actions of serotonin are due to the existence of at least 15 different 5-HT receptor types. For example, activation of 5-HT2_A receptors induces cell depolarization, while activation of 5-HT1_A has an opposite effect[1]. Hypoglossal motoneurons and motor cells innervating pharyngeal and laryngeal muscles express 5-HT2_A receptors [4]. During sleep, particularly entering REM sleep, the activity of serotoninergic receptors reaches its nadir. Microinjection of 5-HT into the hypoglossal nucleus can significantly attenuate the REM sleep-like suppression of XII nerve activity, in part, by substituting for the decreased endogenous amount of 5-HT in the XII nucleus[12].

The 5-HT_{1A} receptors are expressed throughout the CNS[1], including the airway-related vagal preganglionic neurons [6]. In most of these nuclei the receptor functions postsynaptically, responding to serotonin release from projections of raphe nuclei. The most obvious effect of serotonin, which is mediated by 5-HT_{1A} receptors, is membrane hyperpolarization. This membrane-delimited signaling mechanism involves a G protein of the Gi/Go family, the opening of inwardly rectifying potassium channels, and inhibition of calcium currents[1,17].

In summary, the data presented here support the hypothesis that coordinated behavioral changes in the activity of upper airway dilating muscles, chest wall pumping muscles, and cholinergic outflow to the airways may occur via modulation of serotoninergic neurons, in addition to changes induced by the rhythm generating network (Fig. 4). Stimulation of raphe neurons preferentially increases the respiratory-related drive to hypoglossal motoneurons, but diminishes cholinergic outflow to the airways. Hence, inhibition of serotonin-containing neurons during sleep may lead to a decrease in the activity of upper airway dilating muscles and to an elevation of cholinergic outflow to the airways, which will predispose individuals to upper airway occlusion and bronchoconstriction. Qualitatively different responses of hypoglossal and airway-related vagal preganglionic neurons to endogenously released serotonin could be related to its simultaneous action on different serotonin receptor subtypes expressed by these motoneurons, eliciting functionally appropriate physiological responses of the organism to its surroundings.

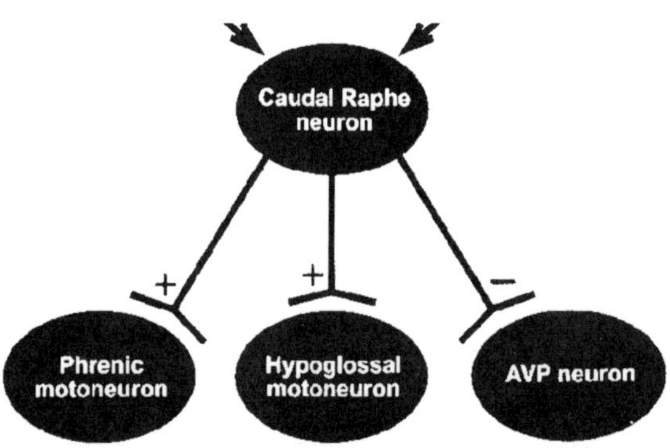

Figure 4. Summary diagram of the differential effects of raphe neurons stimulation on phrenic, hypoglossus, airway-related vagal preganglionic (AVP) neurons.

5. ACKNOWLEDGMENTS

This work was supported by NIH grants HL-50527 and I U 54 NS 39407.

6. REFERENCES

1. Andrade, R. Regulation of membrane excitability in the central nervous system by serotonin receptor subtypes. *Ann. N. Y. Acad. Sci.* **861**:190-203, 1998.
2. Cherniack, N.S, Respiratory dysrhrythmias during sleep. *N.Engl. J. Med.* **305**:325-330, 1981.
3. Dempsey J.A, Smith C.A, Harms C.A, Chow C.M, Saupe K.W., Sleep-induced breathing instability. *Sleep.* **19**: 236-247, 1996.
4. Fay, R, Kubin L, Pontomedullary distribution of 5HT2A receptor-like protein in the rat. *J. Comp. Neurol.* **418**:323-345, 2000.
5. Hadziefendic S and Haxhiu M.A., CNS innervation of vagal preganglionic neurons controlling peripheral airways: a transneuronal labeling study using pseudorabies virus. *J. Auton. Nerv. Syst.* **76**:135-145, 1999.
6. Haxhiu, MA. Erokwu,BO, Bhardwaj, V, Dreshaj, IA., The role of the medullary raphe nuclei in regulation of cholinergic outflow to the airways. *J. Auton. Nerv. Syst.* **69**: 64-71, 1998.
7. Haxhiu MA, Erokwu B, Cherniack N.S., The brainstem network involved in coordination of inspiratory activity and cholinergic outflow to the airways. *J. Auton. Nerv. Syst.* **61**: 155-161, 1996.
8. Haxhiu MA, Jansen A.S.P, Cherniack NS, Loewy A.D. CNS innervation of airway-related parasympathetic preganglionic neurons: a transneuronal labeling study using pseudorabies virus. *Brain Res.*, **618**:115-134, 1993.
9. Haxhiu MA, van Lunteren E, Mitra J, Cherniack N.S. Comparison of the response of diaphragm and upper airway dilating muscle activity in sleep cats. *Respir. Physiol.* **70**: 183-193, 1987.
10. Hobson JA, Lydic R, Baghdoyan HA. Evolving concepts of sleep cycle generation: from brain centers to neuronal populations. *Behav. Brain Res.* **9**: 371-448, 1986
11. Holtman, JR Jr, Dick TE, Berger AJ. Involvement of serotonin in the excitation of phrenic motoneurons evoked by stimulation of the raphe obscurus. *J. Neurosci.* **6**:1185-1193, 1986.
12. Kubin L, Reignier C, Tojima H, Taguchi O, Pack A.I, Davies R.O. Changes in serotonin level in the hypoglossal nucleus region during carbachol-induced atonia. *Brain Res.* **645**: 291-302, 1994.
13. Lin J-S, Hou Y, Sakai K, Jouvet M. Histaminergic descending inputs to the mesopontine tegmentum and their role in the control of cortical activation and wakefulness in the cat. *J. Neurosci.* **16**: 1523-1537, 1996.
14. Loewy, A.D., Franklin, M.F., Haxhiu, M.A. CNS monamine cell groups projecting to pancreatic vagal motor neurons: a transneuronal labeling study using pseudorabies virus. *Brain Res.* **638**: 248-260, 1994.
15. Lydic R, Baghdoyan H.A, Zwillich C.W. State-dependent hypotonia in posterior cricoarytenoid muscles of the larynx caused by cholinergic reticular mechanisms. *FASEB J.* **3**: 1625-1631, 1989.
16. Martin R.J, Cicutto L.C, Ballard R.D. Factors related to nocturnal worsening of asthma. *Am. Rev. Respir. Dis.* **141**: 33-38, 1990.
17. Marszalec W.R., Scroggs R.S, Anderson E.G. Serotonin induced reduction of the calcium dependent plateau in frog dorsal root ganglion cells is blocked by serotoninergic agents acting at 5-hydroxytryptamine 1A site. *J. Pharmacol. Exp. Ther.* **247**:399-404,1988.
18. Moore R.Y and Speh J.C. GABA is the principal neurotransmitter of the circadian system. *Neurosci. Lett.* **150**: 112-116, 1993.
19. Nitz D, Siegel J. GABA release in the dorsal raphe nucleus: Role in the control of REM sleep. *Am.J. Physiol.* **273**:R451-R455, 1977.
20. Sherin JE, Shiromaniu P.J, McCarley R.W, Saper C.B. Activation of ventrolateral preoptic neurons during sleep. *Science.* **271**: 216-219, 1996.
21. Shuman SL, Capece M.L, Baghdoyan H.A, Lydic R. Pertussis toxin-sensitive G proteins mediate carbachol-induced REM sleep and respiratory depression. *Am. J. Physiol.* **269**: R308-R317, 1995.
22. Strohl K.P, Cherniack N.S, Gothe B. Physiological basis for therapy of sleep apnea. *Am. Rev. Respir. Dis.* **134**: 791-802, 1986.
23. Turner-Warwick M. Epidemiology of nocturnal asthma. *Am. J. Med.* **85**(1B): 6-8, 1988.
24. Waldbaum S, Hadziefendic S, Erokwu B, Zaidi SI, Haxhiu M.A. CNS innervation of posterior cricoarytenoid muscles: a transneuronal labeling study. *Respir. Physiol.* (2000; submitted).

Stability and Variability of Breathing

NEURAL DRIVES AND BREATHING STABILITY

Guy Longobardo, Carlo Evangelisti, Nanduri Prabhakar, and Neil S. Cherniack*

1. INTRODUCTION

Cessation of breathing and cyclic variations in tidal volume and frequency are far more likely to be precipitated when hypoxia or transient hyperventilation produces subnormal levels of PCO_2 in NREM sleep and at sleep onset, than in awake subjects. The ability of the awake subject to breathe with only rare apneas and oscillations in ventilation at levels of PCO_2 considerably below resting seems to depend on the interplay of chemical and neural drives.

Two excitatory neural mechanisms, seem to be important in ventilation in conscious humans. The first arises from disturbances in the environment like noise and light, and from mental activity, which we shall call the alertness drive, and includes the "wakefulness drive" described by Fink.[1] The studies of Nielsen and Smith[2] and by Mohan and Duffin[3] support the presence of such an alertness drive. However, Nielsen and Smith found that hypoxia altered the magnitude of this drive while Mohan and Duffin did not. The second mechanism, post-synaptic potentiation results in an excitatory response that persists for a minute or so after the termination of a period of respiratory stimulation.[4] This stimulatory effect disappears with hypoxia if it is sustained for more than five minutes. Theoretically both post-synaptic potentiation and alertness drives can reduce the amplitude and duration of oscillations in breathing and diminish tendencies for the occurrence of apneas.

Our previous mathematical models of respiration, which were used to simulate Cheyne-Stokes breathing or sleep apnea,[5, 6] had no need to accommodate neurally stimulated ventilation. They have controllers which necessarily allow apneas to occur at sub-normal levels of PCO_2. The ventilation/carbon dioxide relations for sub-normal values of CO_2 in these models are extrapolations of the quasi-linear relationships found experimentally at normal and higher values of PCO_2. Although these models are successful in duplicating the qualitative characteristics of the highly oscillatory breathing patterns of patients with Cheyne-Stokes breathing and that also occur in sleep, they are

*Guy Longobardo, Bronxville, New York 10708. Carlo Evangelisti, Jefferson Valley, New York 10535. Nanduri Prabhakar, Case Western Reserve University, Cleveland, Ohio 44106. Neil Cherniack, University of Medicine and Dentistry of New Jersey, Newark, New Jersey 07103.

Frontiers in Modeling and Control of Breathing, edited by Poon and Kazemi.
Kluwer Academic/Plenum Publishers, New York, 2001.

not adequate however to simulate the more stable breathing of awake humans. In conscious humans, subnormal levels of PCO_2 caused by increased breathing because of either hypoxia or hyperventilation produce apneas and ventilatory oscillations but only occasionally.

Our current model focuses on breathing in conscious man. It is based on our earlier models of the chemical regulation of breathing but has several new features. Like the earlier models, it includes a multi-compartment concept of the body gas stores of O_2 and CO_2 and a chemical controller that includes the effects of interaction between central and peripheral chemoreceptors. However, unlike the controller in the previous models, the current model has no discrete CO_2 thresholds. The CO_2 responses of the central and peripheral chemoreceptors are considered to be non-linear with activity even at very low levels of CO_2. While the evidence for this is far from conclusive, there is experimental support for the idea that chemoreceptors can respond to PCO_2 in the subnormal range albeit with a much diminished sensitivity.[7, 8] In addition, in this model we have added the effects of hypoxic depression, post-synaptic potentiation, and an alertness drive roughly similar to the wakefulness drive proposed by Fink and used in the Khoo model.[1, 9] Both post-synaptic potentiation and hypoxic depression are considered to be present during sleep and the awake state. Excitatory neural drives add to chemoreceptor responses.

The underlying concept is that some minimum number of afferent impulses is required to give rise to ventilation. These impulses come from both neural and chemical sources. During deep sleep, chemical drives during hypocapnia may be too low to produce ventilation. But as sleep lightens additional impulses (increasing "wakefulness-like" drives) act on respiratory neurons until, with arousal they are sufficient to produce ventilation. (see Figure 1)

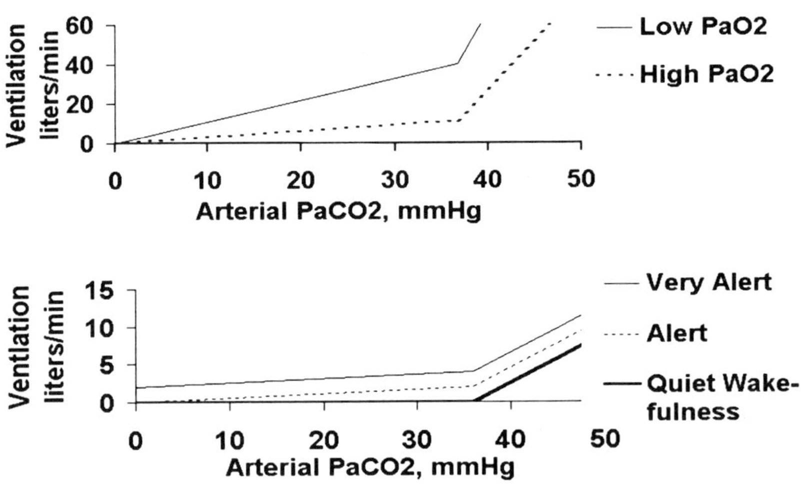

Figure 1. Top: Ventilation vs. Arterial $PaCO_2$ at Constant PaO_2. Bottom: The Effect of Alertness on Ventilatory Response to CO_2

2. FORMULATION OF THE MODEL

The mathematics of a model for the chemical control of ventilation (Cheyne-Stokes) model have been previously described by us.[4, 5] Others have similar models.[9] This model contains a controller with input from the central and peripheral sensors connected to gas stores which consist of a brain compartment, an "all other" compartments (mostly muscle) and a lung compartment where O_2 and CO_2 transfer occur between the blood and the alveolar air. The same formulation of these gas stores is used in the current model.

The inputs to the chemical controller in the Cheyne-Stokes models are arterial PCO_2 and arterial PO_2 measured at the peripheral sensor, and brain PCO_2, the tension of the venous blood leaving the brain. The gas tension values at the chemical controller are delayed about 2 breaths in time from when they appeared at the lung, about 6 seconds, during normal operation. The output of the chemical controller is ventilation, in liters per minute. The constants defining the controller were chosen so that at $PaCO_2$= approximately to 39mmHg, PaO_2=105mmHg, and brain venous PCO_2=48mmHg, ventilation is 6.27 liters/min. Also, the controller CO_2 gain is about 2 liters/minute/mmHg arterial PCO_2 and O_2 gain is 1 liter/min per % decrease in saturation. In this Cheyne-Stokes model these values place the thresholds for both the central receptor and the peripheral receptor at normoxia at 36.25mmHg $PaCO_2$.

In the current model we assume that the chemical controller actions can be defined in two different regions, one above a transition PCO_2, ($TPCO_2$), and the other below. $TPCO_2$ defines a change in slope in the chemical controller from the value in hypercapnia to one of much lower slope at subnormal PCO_2 values. Below $TPCO_2$ the peripheral ventilation and central ventilation are considered to diminish linearly down to zero PCO_2 (Figure 1).

Also, in the current model, the excitatory neural drives described earlier, and hypoxic depression participate in ventilatory regulation. Hypoxic ventilatory depression is assumed to act centrally. It is saturation and time dependent. The data of Khamnei and Robbins and others were used with the general principles that beginning after 5 minutes of hypoxia ventilation gradually decreases and reaches about 50% of its peak value in about 30 minutes.[10] Ventilation remains depressed after 7-15 minutes of normoxia. To shape the function a third order response to hypoxia was used with three equal time constants of 2.5 minutes with a steady state depression of about 10 liters/minute at a saturation of 84.8% (50mmHg arterial PO_2).

Post-synaptic potentiation (PSP) is simulated by a module that acts as a 'respiratory memory'.[4] As a disturbance increases ventilation, the module charges to a potential equal to the incremental ventilation over normal caused by the disturbance. This charging occurs exponentially in time, defined by a first order differential equation. During charging there is no effect on ventilation. The module is gated however so that if the controller asks for a sudden drop in ventilation, the filter "discharges", again exponentially, allowing ventilation to fall gradually instead. In the model the PSP is made to disappear after about five minutes of hypoxia by introducing a threshold, described by a high order differential response to hypoxia. The threshold is designed to increase slowly during the first 2.5 minutes of hypoxia, and then rise rapidly during the next several minutes so that the PSP disappears by 5 minutes.

The alertness drive is assumed to vary with excitement and is likely to be affected by stress and personality. At zero alertness drive, ventilation disappears only at zero PCO_2.

With sufficient decrease in the alertness drive the model reverts to the Cheyne-Stokes model.

3. RESULTS

3.1. Hyperventilation

Increased neural drive reduces apneas and oscillations in ventilatory response. Figure 2 shows the response to hyperventilation followed by normal breathing while breathing 37% oxygen at several levels of neural drive. As the neural drive is increased the off-transient response is less oscillatory and apneas disappear. The response to an alertness drive of -2 liters/minute is in accord with the observations of Meah and Gardner; that apneic periods rarely occur immediately after voluntary over-breathing, but gradually increase in number and length, the largest pauses occurring 2 minutes after the end of the hyperventilation.[11] Apneas in their studies appeared with mean times between 0.8 and 5.6 minutes after the end of hyperventilation, and the end of the apneic period was associated with an end tidal PCO_2 of about 36mmHg. In the model, as in the experiment, the first apnea starts 2 minutes after hyperventilation, which reflects post-synaptic potentiation, and the period during which apneas occur ends after about 5 minutes. The arterial $PaCO_2$, which the controller sees, is about 37 mm Hg compared to Meah and Gardner's 36mmHg end tidal PCO_2. We have called this level of neural drive "quiet wakefulness".

At the start of the post-hyperventilation period arterial $PaCO_2$ is subnormal, and PaO_2 is high. As a result both the central and peripheral drives are driven to low levels. When alertness drive is less than zero, ventilation can be oscillatory, both because apneas occur and because ventilation is controlled mainly by input from the peripheral sensor whose CO_2 gain is highly O_2 sensitive. When the alertness drive is higher the possibility for apneas lessens, because the higher ventilations cause oxygen levels to rise more quickly, reducing the tendency to oscillate.

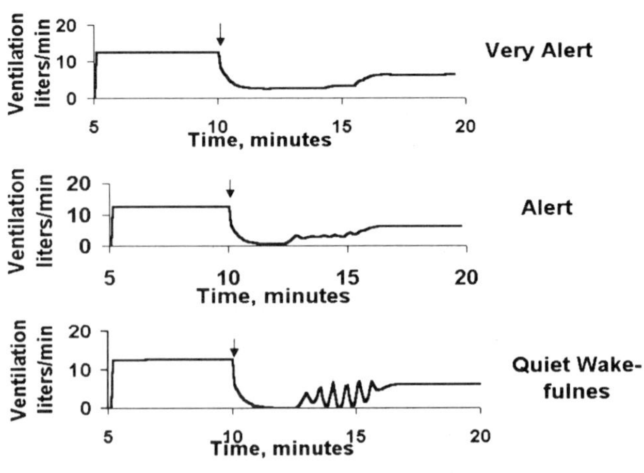

Figure 2. The effect of alertness on breathing after voluntary hyperventilation (which ends at arrow). At quiet wakefulness alertness drive = -2 l/min.

$TPCO_2$ as well as neural drive can influence responses at subnormal levels of PCO_2. There is very little data on its actual value although others have reported non-linear ventilation response at low values of chemical stimulation.[2, 3] When we tested responses for varying values of $TPCO_2$, we found that higher values of $TPCO_2$ provided a more stable response. Post hyper-ventilation, CO_2 drive is significantly greater when $TPCO_2$ is higher allowing oxygen levels to recover more rapidly.

3.2. Hypoxia

Figure 3 shows the ventilation responses to two minutes of breathing 8% O_2, followed by breathing 37% O_2, for neural drives of +2 liters/minute (very alert) and -2 liters/minute. The greater value again results in a more stable response.

Since hypoxic duration is only 2 minutes, post-synaptic potentiation is included in Figure 3. When post-synaptic potentiation is eliminated from the simulation, the stability characteristics of the responses are essentially unchanged; low neural drive produces larger swings in ventilation compared to higher neural drive. Post-synaptic potentiation by itself does not eliminate the oscillations in ventilation.

When hypoxia is prolonged, hypoxic ventilatory depression has an important role in producing instability. In experiments, which simulate breathing during hypoxia for several minutes, periodic breathing occurs at higher levels of oxygen, because of the inclusion of hypoxic depression.

3.3. Other Results

Decreasing progressively alertness drive allows the model to simulate breathing during various stages of sleep. Also, sufficient lengthening of circulation time even during waking conditions produces Cheyne-Stokes breathing.

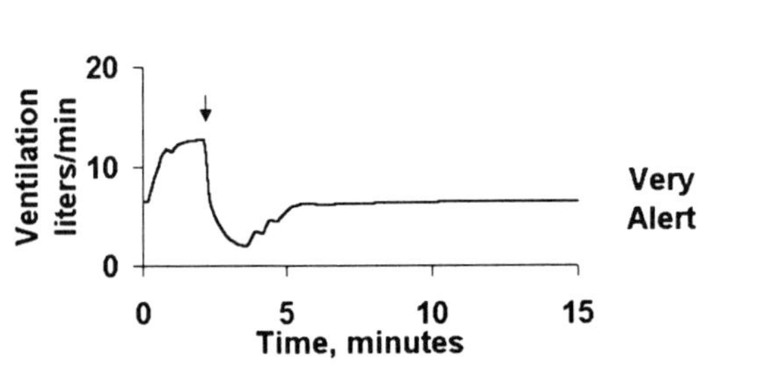

Figure 3. The effect of alertness on breathing after short-term hypoxia. Hypoxia ends at arrow.

4. DISCUSSION

The present study uses mathematical models to evaluate potential mechanisms, which might account for the scarcity of apneas and periodic breathing in the awake state. We found that the simplest model that allowed the essential features of breathing to be reproduced in both the awake and sleeping states had a non-linear chemical controller, which was active over a very wide range of PCO_2 and PO_2 but resulted in ventilation only when there was a sufficiently large neural component.

In this model, stability was greater, with increasing excitatory neural drives and when the PCO_2 at which the high and low slope segments of the chemical controller met ($TPCO_2$), was higher. The variability in the occurrence of apneas in the awake state in different individuals probably arises from the difference in $TPCO_2$ levels and neural drive intensity. Post-synaptic potentiation also helped stabilize breathing but could be overridden by sufficient hypocapnia or prolonged hypoxia and by itself did not adequately explain the rarity of repetitive apneas in conscious breathing responses. Hypoxic depression had a destabilizing effect both awake and asleep.

5. REFERENCES

1. Fink, B. R., Hanks, E. C., Ngar, S. H., and Papper, E. M., 1963, Central regulation of respiration during anesthesia and wakefulness, *Ann. NY Acad. Sci.* **109**:892-899.
2. Nielsen, M., and Smith, H., 1952, Studies on the regulation of respiration in acute hypoxia, *Acta Physiol. Scand.* **24**:293-313.
3. Mohan, R., and Duffin, J., 1997, The effect of hypoxia on the ventilatory response to carbon dioxide in man, *Resp. Physiol.* **108**(2);101-115.
4. Georgopoulos, D., Bshouty, Z., Younes, M., and Anthonisen, N. R., 1990, Hypoxic exposure and activation of the afterdischarge mechanism in conscious humans, *J. Appl. Physiol.* **69**(3);1159-1164.
5. Longobardo, G. S., Cherniack, N. S., Fishman, A. P., 1966, Cheynes-Stokes breathing produced by a model of the human respiratory system, *J. Appl. Physiol.* **21**(6), 1839-1846.
6. Cherniack, N. S., and Longobardo, G. S., 1994, Periodic breathing during sleep, in: *Sleep and Breathing*, 2nd ed, N. A. Saunders, and C. E. Sullivan, eds, Marcel Dekker, Inc., New York, pp. 157-190.
7. Cummin, A. R., Sidhu, V. S., Telford, R. J., and Saunders, K. B., 1992, Ventilatory responsiveness to carbon dioxide below the normal control point in conscious normoxic humans, *Eur. Resp. J.* **5**(5):512-516.
8. Patrick, W., Webster, K., Puddy, A., Sanii, R., and Younes, M., 1995, Respiratory response to CO_2 in the hypocapnic range in awake humans, *J. Appl. Physiol,* **18**:1053-1056.
9. Khoo, M. C. K., 1999, Periodic breathing and central apnea, in: *Control of breathing in health and disease*, M. D. Altose, and Y. Kawakami, eds, Marcel Dekker, Inc., New York, pp. 203-250.
10. Khamnei, S., Robbins, P. A., 1990, Hypoxic depression of ventilation in humans: alternative models for the chemoreflexes, *Resp. Physiol.* **81**(1):117-134.
11. Meah, M. S., Gardner, W. N., 1994, Post-hyperventilation apnoea in conscious humans, *J. Physiol.* **477**(3):527-538.

VARIABILITY OF CO_2-SENSITIVITY DURING SLEEP

Thorsten Schäfer and Marianne E. Schläfke*

1. INTRODUCTION

The hypercapnic ventilatory response (HCVR) in man varies markedly during sleep. Rebreathing tests during well-defined sleep/wake states in adults showed that the HCVR was significantly reduced in all stages of sleep compared with wakefulness, falling to less than 50 % of wakefulness during NREM sleep and less than 30 % during REM sleep[1].

Furthermore a circadian rhythm of the HCVR in awake men with an amplitude of 30 % of the mean HCVR and lowest slopes of the HCVR in the early morning hours were described by Raschke and Möller[2] and were recently confirmed by Spengler et al.[3], who found a variation of 12.1 % of the 24 h mean. The combination with sleep state-related variability of the HCVR might result in a severe reduction of CO_2-related respiratory drive in the early morning.

Recent findings of chemosensitive neurons in or near nuclei involved in sleep/wake regulation (medullary raphé) suggest that chemoreceptors at different locations within the brain may differ in inherent sensitivity and in their responsiveness in different arousal states[4]. These experiments in rats as well as differentiation of pH/PCO_2 effects of superfusion or inner perfusion of isolated brain stem preparations of guinea pigs[5] demonstrated different CO_2 responsiveness of tidal volume and respiratory frequency, respectively.

In order to examine the changes in the HCVR during night directly, we continuously repeated CO_2-challenges by moderate increases of the inspired fraction of CO_2 to 2.5 and 4.0 % throughout the whole night[6] and focused the analysis on the circadian variability and the contribution of tidal volume and frequency responses to CO_2 during different sleep states.

* Deptartment of Applied Physiology, Ruhr-Universität Bochum, D-44780 Bochum, Germany

Frontiers in Modeling and Control of Breathing, edited by Poon and Kazemi.
Kluwer Academic/Plenum Publishers, New York, 2001.

2. METHODS

Experiments were performed in ten healthy men with a mean age of 25.0 ± 2.6 yrs, range 23-32 yrs. They were non-obese, without a history of cardiac, pulmonary or neurological disorders. They were regular nocturnal sleepers and had refrained from stimulants like coffee since the early morning of the study day.

Each subject gave informed consent. The study was approved by the Ethics Committee of the Medical Faculty at the Ruhr-Universität Bochum.

The subjects breathed through a tightly fitting nose mask from and into a background flow of 60 l/min. The gas composition was changed every 6-7 min so that end-tidal PCO_2 rose by 3 and 6 Torr, followed by room air. Respiration was measured by means of a pneumotachograph (Jaeger, Würzburg), end-tidal and inspiratory PCO_2 by capnography (Datex, Bremen), simultaneously with a complete sleep EEG (Schwarzer, München).

Sleep was scored visually according to the rules of Rechtschaffen and Kaleș. Respiration and PCO_2 were analyzed breath-by-breath using a laboratory software on the basis of the Spike2 script language (CED, Cambridge, UK) including careful elimination of artifacts. As NREM 1 episodes were only transient, NREM 1- and NREM 2-sleep epochs were merged as "light sleep". NREM 3- and NREM 4-sleep tended to intermingle, therefore these two states were merged as "deep sleep".

Data capture started when the subjects were still awake. Continuous application of the protocol resulted in approximately 60 3-point steady-state CO_2 response curves (baseline, +3, +6 Torr) in each subject.

3. RESULTS

We obtained 567 steady-state CO_2 response curves in our 10 subjects. Figure 1 shows individual data of the time course of slopes, apnea points, resting PCO_2 and sleep states in one subject. In this case the slopes of the HCVR varied between 0.26 and 1.24 l/min/Torr. In the whole group the flattest slopes of the HCVR with a mean CO_2-induced increase in ventilation of 0.31 ± 0.13 (SD) l/min/Torr occurred at 4:45 ± 1:11 a.m., in one subject during wakefulness, in one during light, in four during deep and in three during REM sleep (fig. 2). Mean highest slopes of the HCVR were 1.45 ± 0.35 l/min/Torr.

The CO_2 responses of tidal volume (V_T, fig. 3) did not differ between wakefulness (60 ± 28 ml/Torr), light (60 ± 38 ml/Torr) and deep sleep (53 ± 22 ml/Torr). During REM sleep, however, the V_T response to CO_2 was significantly lower (37 ± 23 ml/Torr) than during all other states ($P < 0.05$; t-test for paired samples).

The effect of CO_2 on respiratory frequency was much smaller. During wakefulness there was a tendency to lower frequencies (-0.20 ± 0.55 breaths per min/Torr). During sleep, however, respiratory frequency tended to increase during CO_2-breathing, but the differences failed to reach the level of significance.

Despite the marked variability of the slopes of the HCVR, baseline ventilation and baseline PCO_2 were not different between sleep states. This was due to offsetting shifts of the apnea points (fig. 1) which stabilized ventilation at and slightly above baseline PCO_2.

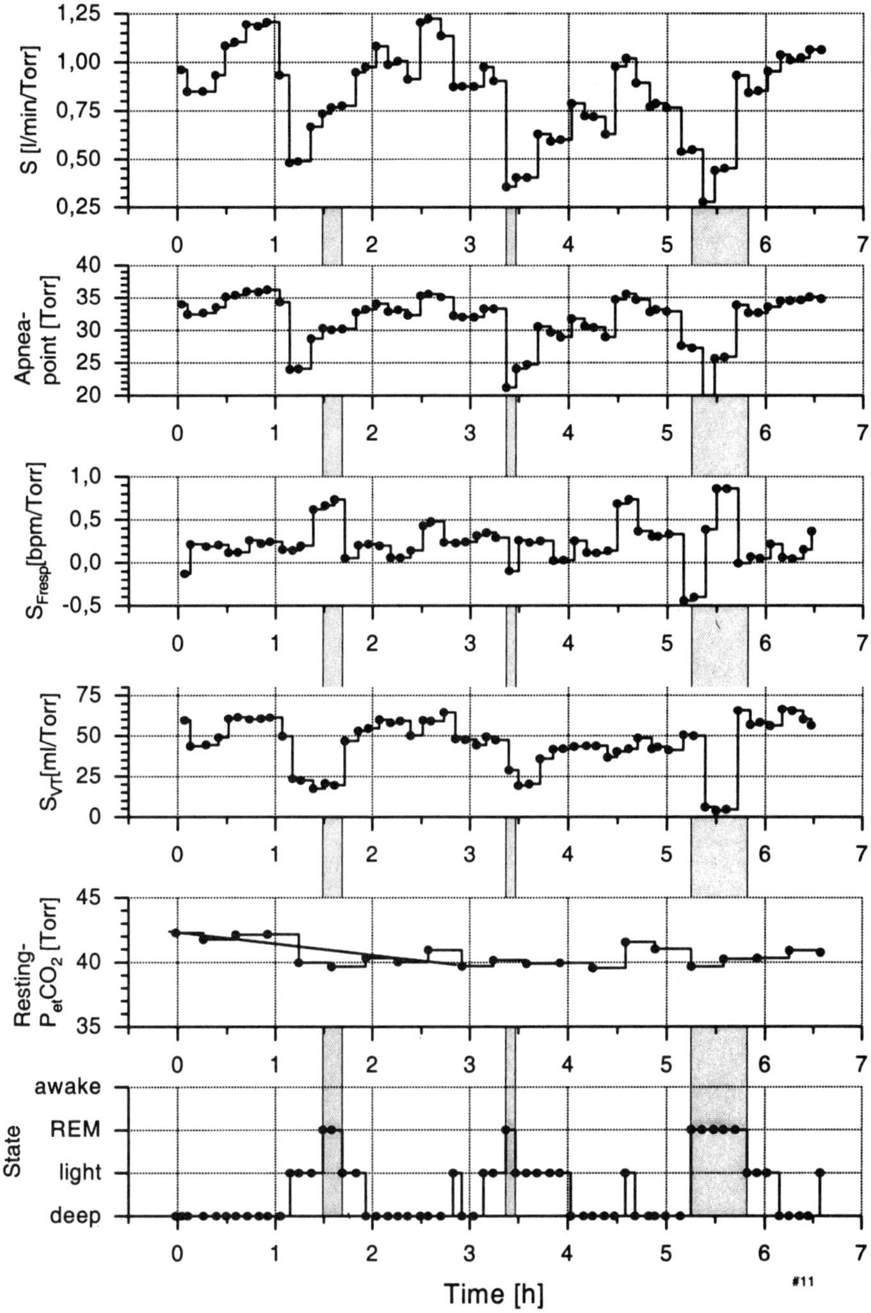

Figure 1. Profile of the slopes (S) and apnea points (intercept with the x-axis) of the ventilatory CO_2 response curves, slopes of the frequency response (S_{Fresp}) and tidal volume response (S_{VT}), resting end-tidal PCO_2, and the sleep states ("hypnogram"), calculated from the 61 three-point steady-state hypercapnic ventilatory responses in a single subject. REM sleep periods are indicated by the shaded bars.

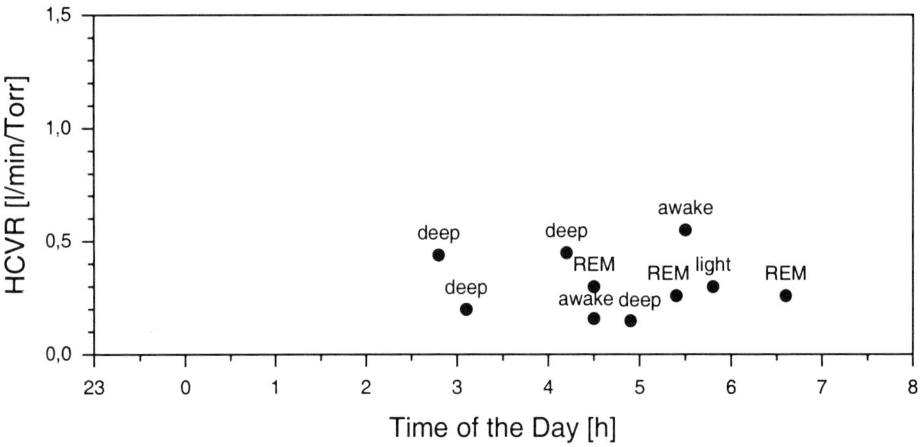

Figure 2. Occurrence of the flattest CO_2 response curves in the course of the night. The dots represent each subject's lowest HCVR in l/min/Torr. The states of sleep or wakefulness, in which these response curves occurred, are indicated (REM = REM sleep, light = light sleep/NREM 1- and NREM 2-sleep, deep = deep sleep/NREM 3- and NREM 4-sleep).

4. DISCUSSION

Repeated HCVR (n = 60) during night showed marked intraindividual variability between 0.31 and 1.45 l/min/Torr with lowest slopes in the early morning hours and with a sleep state-dependent pattern of the tidal volume response. The reduced tidal volume response during REM sleep, which was not completely offset by the increased frequency response is one of the reasons for the lower overall HCVR during REM sleep[1, 6].

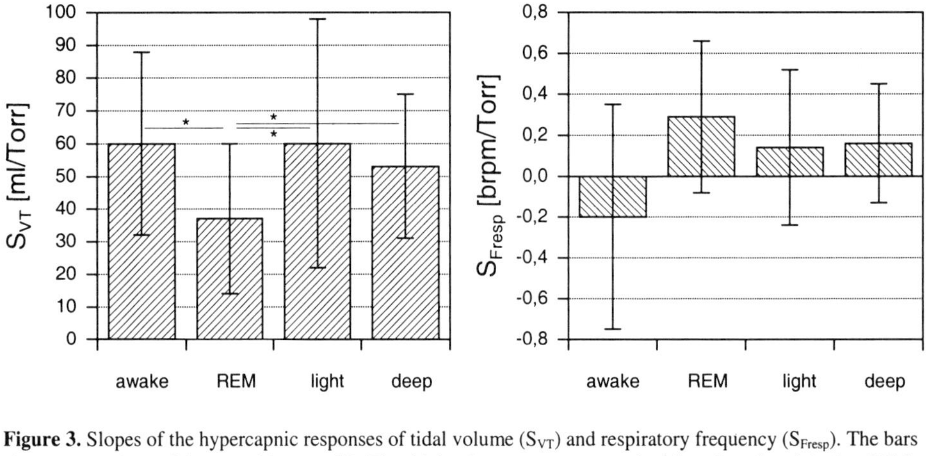

Figure 3. Slopes of the hypercapnic responses of tidal volume (S_{VT}) and respiratory frequency (S_{Fresp}). The bars show mean values of the ten subjects ± SD. The tidal volume response was significantly reduced during REM sleep (* $P < 0.05$; t-test for paired samples), frequency responses did not differ significantly.

Similar HCVRs during wakefulness and NREM sleep in this study may result from the fact that the periods of wakefulness also occurred after sleep onset in the middle of the night. In this case the level of arousal and the respiratory drive might have been lower than before sleep onset[1].

The fact that the lowest HCVRs did not always occur during REM sleep, and the analysis of the individual time course of the HCVR show that there may be other factors that cause variability of the HCVR during night, as for example circadian rhythmicity[2, 3] and variations in brain blood flow[7]. Whether or not the observed differences in response patterns with increases in tidal volume and decreases or increases in respiratory frequency are due to different chemoreceptor sites or to changes in respiratory mechanics and/or efferent motor systems (e.g. muscle atonia and phasic phenomena during REM sleep[8]) must remain open.

Despite the large scatter of the slopes of the HCVR, baseline ventilation and resting PCO_2 remained constant due to offsetting shifts of the apnea points. We, therefore, propose the explanation that variations of the HCVR during sleep may indicate adapting mechanisms to changes in non-chemical respiratory drives caused by sleep/waking regulation in order to stabilize ventilation at and slightly above baseline PCO_2 values[6].

5. REFERENCES

1. N. J. Douglas, D. P. White, J. V. Weil, C. K. Pickett, and C. W. Zwillich, Hypercapnic ventilatory response in sleeping adults. *Am. Rev. Respir. Dis.* **126**, 758-762 (1982).
2. F. Raschke and K. H. Möller, Untersuchungen zur Tagesrhythmik der Chemosensitivität und deren Beitrag zu nächtlichen Atmungsregulationsstörungen. *Pneumologie* **43**, 568-571 (1989).
3. C. M. Spengler, C. A. Czeisler, and S. A. Shea, An endogenous circadian rhythm of respiratory control in humans. *J. Physiol. (Lond.)* **526**, 683-694 (2000).
4. E. Nattie, Multiple sites for central chemoreception: their roles in response sensitivity and in sleep and wakefulness. *Respir. Physiol.* **122**, 223-236 (2000).
5. M. P. Morin-Surun, E. Boudinot, T. Schäfer, M. Denavit-Saubié, Localization of chemosensitive structures in the isolated brainstem of adult guinea-pig. *J. Physiol. (Lond.)* **485**, 203-212 (1995).
6. T. Schäfer, Variability of vigilance and ventilation: studies on the control of respiration during sleep. *Respir. Physiol.* **114**, 37-48 (1998).
7. R. A. Parisi, J. A. Neubauer, M. M. Frank, T. V. Santiago, and N. H. Edelman, Linkage between brain blood flow and respiratory drive during sleep. *J. Appl. Physiol.* **64**, 1457-1465 (1988).
8. T. Schäfer and M. E. Schläfke, Respiratory changes associated with rapid eye movements in normo- and hypercapnia during sleep. *J. Appl. Physiol.* **85**, 2213-2219 (1998).

DYNAMICS OF HEART RATE CONTROL IN OBSTRUCTIVE SLEEP APNEA

A minimal model

Michael C.K. Khoo, Vasily Belozeroff, and Richard B. Berry*

1. INTRODUCTION

Recent studies indicate that cardiovascular regulation is impaired and sympathetic activity is abnormally elevated in patients with obstructive sleep apnea (OSA) during both wakefulness and sleep [1, 2]. Heart rate variability (HRV) has been found to be substantially reduced in OSA[3]. Spectral analysis shows an increase in the proportion of low-frequency (0.04 to 0.15 Hz) content relative to high-frequency (0.15 to 0.4 Hz) content of heart rate oscillations in awake OSA subjects, suggesting increased sympathetic modulation along with decreased vagal cardiac activity[3]. Although statistical and spectral analyses of HRV offer a simple and non-intrusive means of assessing cardiovascular autonomic function in these subjects, the measures derived can be confounded significantly by intra- and inter-subject differences in ventilation level and breathing pattern[4, 5]. To circumvent this limitation, we have introduced a minimal modeling approach that allows a clear delineation of the respiratory-related influences on HRV from the effects that are intrinsically circulatory in nature.

2. METHODS

2.1. Experimental Procedures and Data Processing

Thirteen patients recently diagnosed with moderate-to-severe OSA and 9 normal controls participated in the study after giving informed consent. All the subjects were male except for 1 female in the control group. There was no significant difference between the ages of the OSA and normal groups (mean ± SE: 46.2 ± 2.8 vs. 44.6 ± 2.6 years, respectively). However, the OSA subjects were significantly heavier than the

* M. C. K. Khoo and V. Belozeroff, University of Southern California, Los Angeles, California 90089. R. B. Berry, University of Florida & V. A. Medical Center, Gainesville, Florida 32610.

Frontiers in Modeling and Control of Breathing, edited by Poon and Kazemi.
Kluwer Academic/Plenum Publishers, New York, 2001.

normals (223.8 ± 8.9 vs. 168.7 ± 7.3 lbs, P<0.001). Apnea-hypopnea index for the OSA subjects ranged from 26 to 154. None of the patients had begun treatment for OSA. Five of the OSA subjects were hypertensive. For reasons of safety, these subjects were asked to continue taking their medication for hypertension prior to the study.

Breathing was monitored using calibrated respiratory inductive plethysmography (Respitrace, Ambulatory Monitoring, Inc., Ardsley, NY). Since the experimental protocol required the measurement of ventilation in the supine and standing postures, calibration of the respiratory inductive plethysmograph was performed separately in both positions. The electrocardiogram (ECG) was measured using a 3-lead system and amplified using a Model BMA-831 bioamplifier (CWE Inc., Ardmore, PA).

At the start of each study, the subject lay supine while ECG and spontaneous respiration were monitored for approximately 5 minutes. Subsequently, the subject was asked to breathe in a manner so as to track the respiratory waveform measured in the previous 5 minutes. Both target and tracking waveforms were displayed on a computer monitor. This procedure allowed the subject to become familiarized with the task of tracking the displayed breathing pattern. Finally, the subject was asked to track a waveform with respiratory durations that varied randomly from breath to breath. The tidal volumes of the target breath pattern were selected so that the average minute ventilation could be maintained at an approximately constant level equal to that deduced from the subject's previously monitored spontaneous breathing pattern. This ensured that chemical drive would remain relatively unchanged over the course of the procedure. The purpose of employing a randomized breathing pattern was to improve the accuracy with which the model parameters could be estimated, since such a pattern effectively broadens the bandwidth of the "input" (i.e., respiration)[7]. The entire randomized breathing protocol lasted 5 minutes. Subsequently, the subject changed his/her posture to standing, following which the respiratory inductive plethysmograph was recalibrated. The entire set of procedures carried out in the supine posture were repeated, including the initial 5 minutes of spontaneous breathing. During each procedure, the respiratory signal was digitized at 10 Hz, while ECG and BP were sampled at 200 Hz; all signals were recorded and stored in an IBM-compatible computer using custom-designed Matlab (Mathworks, Natick, MA) software.

Data collected in each study were processed in the following way. First, the time-locations of the R-waves in the ECG tracing were detected using a computer algorithm. The results of this procedure were reviewed manually and edited for detection errors. Subsequently, the intervals between successive R waves were computed. The sequence of R-R intervals (RRI) was converted into an equivalent uniformly spaced time-series (sampling rate: 2 Hz) using a resampling algorithm similar to that of Berger et al.[7] The breathing waveform was also resampled at 2 Hz so that each respiratory value would be synchronized with the corresponding resampled RRI values.

2.2. Minimal Model

The minimal model used in this study is displayed in Fig.1. The model assumes that the circulatory control system is continually perturbed by respiration (labelled "$\Delta r(\tau)$" in Fig.1) and influences from other extraneous sources. Thus, current fluctuations in RRI result from a combination of respiratory sinus arrhythmia (RSA), stochastic noise

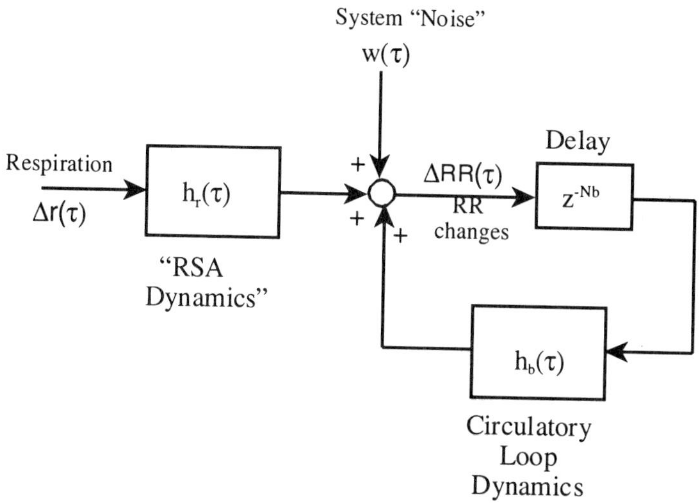

Figure 1: The minimal model of heartrate control. See text for definition of terms.

("w(τ)"), and contributions mediated by the baroreflexes. These changes in RRI, in turn, lead to variations in cardiac output and arterial pressure which subsequently (ie. after some delay, N_b) act on the baroreflexes to produce future RRI fluctuations. This feedback mechanism is labelled as "circulatory loop dynamics" in Fig.1 and is characterized mathematically by the impulse response, $h_b(\tau)$. The impulse response $h_r(\tau)$ characterizes the dynamics of the RSA mechanism.

Given the measurements of respiration and RRI, the unknown impulse responses, $h_b(\tau)$ and $h_r(\tau)$, and loop delay were estimated by reducing the problem to one of least-squares minimization[8]. Substantial reduction in the parametrization of $h_b(\tau)$ and $h_r(\tau)$ was achieved by representing these responses as Laguerre expansions[9]. However, due to the closed-loop nature of the problem, it was necessary to constrain the minimization procedure so that only physiologically feasible candidate solutions were accepted. We imposed the restriction that the sum of values in the first 1.5 s of $h_r(\tau)$ had to be negative, reflecting the fact that initial phase in RSA is an acceleration of heartrate.

2.3. Model Validation Using Simulations

The validity of the minimal model and estimation algorithm was first tested using "data" generated by a computer model of cardiovascular control based on that of Saul et al.[10] The latter was extended to include simple circulatory mechanics and sympathetic control of peripheral resistance. Gaussian white noise was added to the model to simulate the effects of system and measurement noise. Simulation of the model was implemented using SIMULINK software (Mathworks, Natick).

3. RESULTS

3.1. Results Based on Simulated Data

Application of the minimal model to the simulated data showed that the estimation procedure was feasible in principle over a wide range of conditions. The RSA impulse response, $h_r(\tau)$ could be estimated accurately over the frequency range spanned by the breathing pattern (~0.15-0.5 Hz). Reasonably good estimates of the circulatory loop impulse response, $h_b(\tau)$, were also obtained, although these were not as accurate as $h_r(\tau)$, especially in the higher frequency range. A comparison of the estimated and "true" impulse responses is displayed in Figure 2.

3.2. Results from Experimental Data

Figure 3 shows the group-averaged estimates (± SE) of $h_r(\tau)$ for normals (left panel) and OSA (right panel) in both the supine (filled circles) and standing (open circles) postures. In all cases, we found that heartrate started to increase approximately 1 s prior to the start of inspiration, consistent with previous observations[11]. The peak (negative) values of $h_r(\tau)$ were substantially lower in magnitude in OSA versus normals (14.7 ± 1.9 vs. 58.6 ± 19.5, P<0.003, supine posture). Peak $h_r(\tau)$ magnitude decreased by 45.6 ± 11.5 % (P<0.005) during postural change from supine to standing in normals, but remained relatively unchanged in OSA.

The corresponding estimates for $h_b(\tau)$ are displayed in Figure 4. Estimates for the circulatory loop delay ("z^{-Nb}" in Fig.1) have been incorporated into the plots of $h_b(\tau)$. In all cases, there was a delay of 0.5 to 1 s prior to the start of any circulatory-related changes in RRI. Peak $h_b(\tau)$ magnitude was significantly lower in OSA relative to normals (0.54 ± 0.06 vs 0.72 ± 0.06, P<0.05, supine posture). In normals, peak $h_b(\tau)$ magnitude increased by 35.2 ± 13.8 % (P<0.04) from supine to standing. The corresponding posture-induced change in OSA was not significantly different from zero.

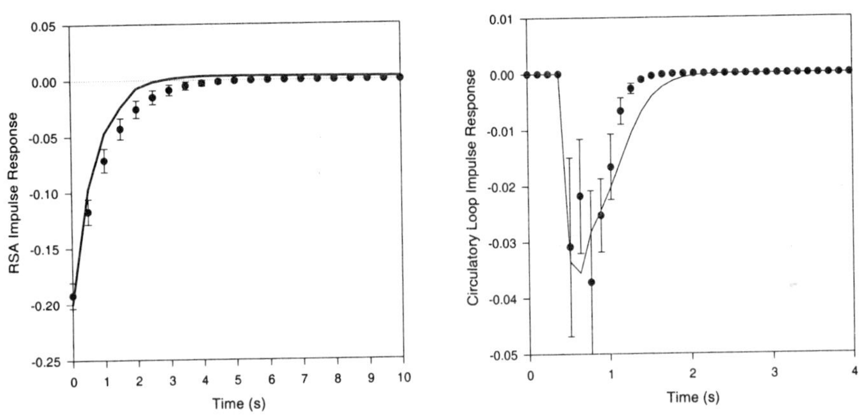

Figure 2: "True" $h_r(\tau)$ and $h_b(\tau)$ compared to corresponding impulse responses estimated from simulated data. See text for details.

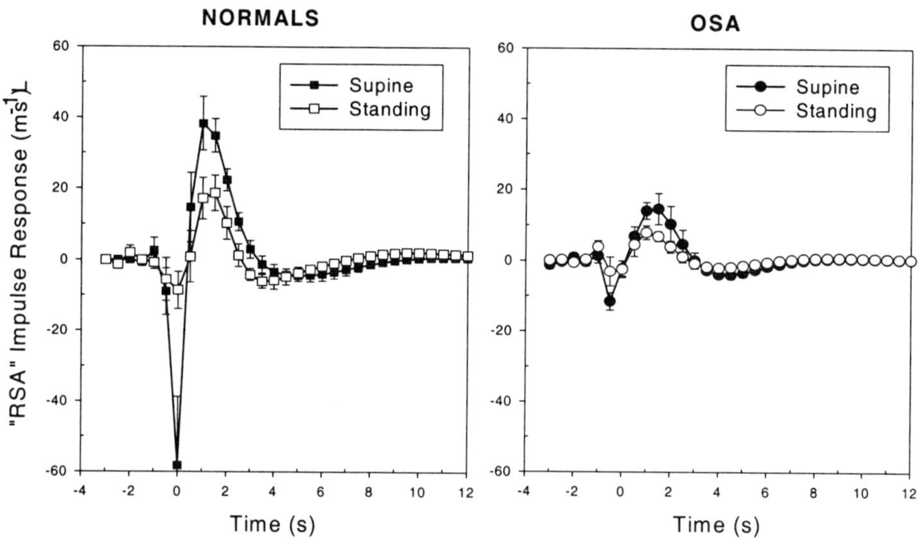

Figure 3: Group-averaged estimates of $h_r(\tau)$ (RSA dynamics) in normals and OSA in both supine and standing postures. Error bars represent standard errors.

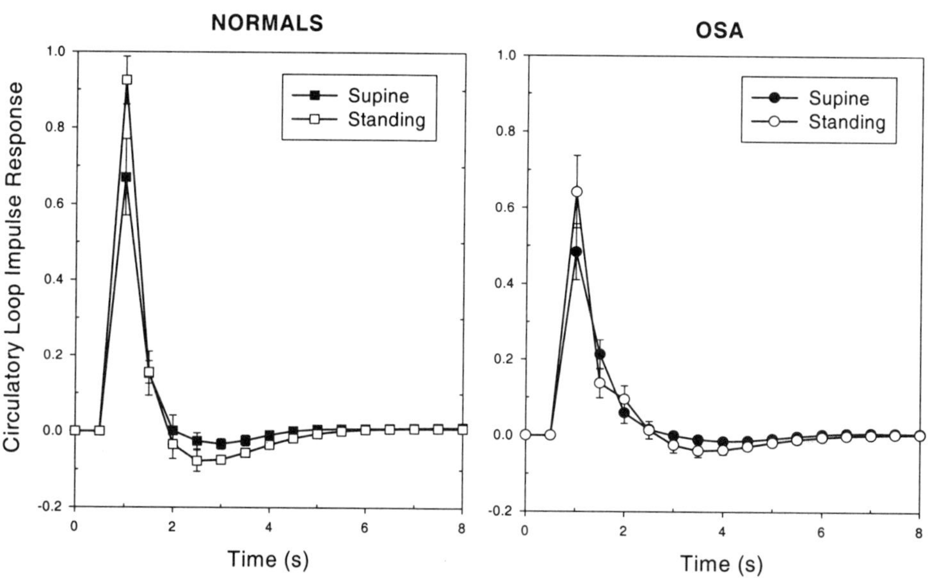

Figure 4: Group-averaged estimates of $h_b(\tau)$ (circulatory loop dynamics) in normals and OSA in both supine and standing postures. Error bars represent standard errors.

4. DISCUSSION AND CONCLUSIONS

We have introduced a new approach for quantifying cardiovascular autonomic control in OSA. The advantage of the minimal modeling technique over spectral analysis of HRV is that it allows a clear separation of the respiratory-related contributions from those that are mediated by the baroreflexes, as well as stochastic influences. Similar minimal models have been proposed previously[12, 13], but these are more comprehensive and require the simultaneous monitoring of continuous arterial pressure. In contrast, our approach is truly "minimal", requiring only measurements of heartrate and respiration, and therefore can be easily applied in the clinical setting. Our findings suggest that decreased HRV in OSA is likely due to reductions in both RSA gain and baroreflex sensitivity. There may also be accompanying changes in circulatory mechanics. Orthostatic stress induced by postural change produced significant changes in RSA and circulatory loop dynamics in normals, but not in OSA. This may be due to the pre-existing high levels of sympathetic tone in OSA even under supine, resting conditions. This differential response to postural change suggests a simple clinical means of assessing the impairment of autonomic function in OSA before and during treatment.

5. ACKNOWLEDGMENTS

We thank Edwin Valladares for technical assistance in the experiments. This work was supported in part by NIH Grants HL-58725 and RR-01861.

6. REFERENCES

1. J. R. Shepard Jr., Hypertension, cardiac arrhythmias, myocardial infarction, and stroke in relation to obstructive sleep apnea, *Clin. Chest Med.* **12**, 437-458 (1992).
2. V. K. Somers, M. E. Dyken, M. P. Clary, and F. M. Abboud, Sympathetic neural mechanisms in obstructive sleep apnea, *J. Clin. Invest.* **96**,1897-1904 (1995).
3. K. Narkiewicz, N. Montano, C. Cogliati, P. J. H. van de Borne, M. E. Dyken, and V. K. Somers, Altered cardiovascular variability in obstructive sleep apnea, *Circulation* **98**, 1071-1077 (1998).
4. T. E. Brown, L. A. Beightol, J. Koh, and D. L. Eckberg, Important influence of respiration on human R-R interval power spectra is largely ignored, *J. Appl. Physiol.* **75**, 2310-2317 (1993).
5. Khoo MCK, Kim T-S, Berry RB. Spectral indices of cardiac autonomic function in obstructive sleep apnea. *Sleep* **22**, 443-451 (1999).
6. T. S. Kim, and M.C.K. Khoo, 1997, Estimation of cardiorespiratory transfer under spontaneous breathing conditions: A theoretical study, *Am. J. Physiol.* **273**, H1012-1023 (1997).
7. R. D. Berger, S. Akselrod, D. Gordon, and R. J. Cohen, An efficient algorithm for spectral analysis of heart rate variability, *IEEE Trans. Biomed. Eng.* **33**, 900-904 (1986).
8. M. C. K. Khoo, *Physiological Control Systems: Analysis, Simulation and Estimation* (IEEE Press, Piscataway, NJ, 2000).
9. V. Z. Marmarelis, Identification of nonlinear biological systems using Laguerre expansion of kernels. *Ann Biomed. Eng.* **21**, 573 – 589 (1993).
10. J.P. Saul, R.D. Berger, P. Albrecht, S. P. Stein, M. H. Chen, and R. J. Cohen, A transfer function analysis of the circulation: unique insights into cardiovascular regulation, *Am. J. Physiol.* **261**, .H1231-H1245 (1991).
11. K. Yana, J. P. Saul, R. D. Berger, M. H. Perrott, and R. J. Cohen, A time domain approach for the fluctuation analysis of heart rate related to instantaneous lung volume, *IEEE Trans. BME* **40**,74-81 (1993).
12. T. J. Mullen, M. L. Appel, R. Mukkamala, J. M. Mathias, R. J. Cohen, System identification of closed-loop cardiovascular control: effects of posture and autonomic blockade, *Am. J. Physiol.* **41**, H448-461 (1997).
13. V. Belozeroff, R. B. Berry, and M. C. K. Khoo, Minimal model of circulatory control in obstructive sleep apnea, *Sleep* **23**, A14-A15 (2000).

COMPUTATIONAL MODEL OF THE EFFECTS OF BREATHING ON CARDIAC FUNCTION

Binh Q. Tran* and Eric A. Hoffman†

1. INTRODUCTION

Previous work by our group and others have shown that manipulations in intrathoracic pressure and lung inflation have the effect of altering cardiac size and function.[1,2] The mechanisms by which these changes affect the circulatory system are not clearly understood. Possibilities include alterations in venous return[3], stroke volume[4,5], ventricular preload and afterload[6], pulmonary vascular resistance[7], and other mechanisms. Previous work by Cournand et al. have shown that mechanical ventilation (i.e. positive pressure) has deleterious effects on cardiac performance.[8]

Using imaging-based techniques, we observed a strong relationship between positive pressure static lung inflation and a reduction in cardiac size, i.e. all four chambers of a canine heart.[2] Simultaneous to these observations, we observed a decrease in stroke volume and cardiac output with positive pressure lung inflation from 0 to 25 cm H_2O airway pressure, while ejection fractions remained essentially unchanged. Removal of the external pressure load resulted in a return of cardiac performance to pre-loaded levels.

From these observations, we hypothesized that if positive pressure lung inflation and deflation have the effect on decreasing and increasing cardiac size, respectively, then, targeted timing of these ventilation events to the cardiac cycle may assist the heart during ejection and during filling, thus increasing cardiac efficiency. We refer to this special mode of timing ventilation to specific cardiac events as Cardiac-Gated Ventilation (CGV; systolic-gated ventilation-SGV and diastolic-gated ventilation-DGV). Using a computational circulation model developed by us[9,10], we performed simulations to study the effects of SGV and DGV on cardiac function. The model was used for optimal parameter estimation. Model predictions were compared to experimental data obtained by us previously in canine models using imaging-based methodologies.

* Binh Q. Tran, Ph.D., Department of Biomedical Engineering, The Catholic University of America, Cardinal Station, Washington DC, 20064.
† Eric A. Hoffman, Ph.D., Department of Radiology, University of Iowa College of Medicine, Iowa City, IA 52242.

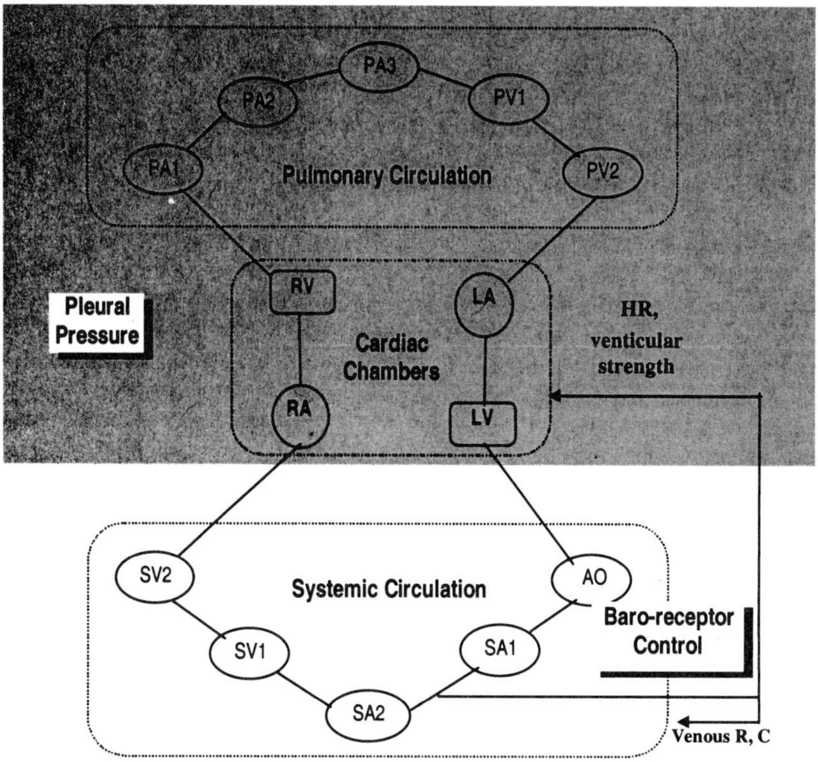

Figure 1. Closed-loop, hydrodynamic model of the circulation system. PA= pulmonary artery, PV=pulmonary veins, AO=aorta, SA=systemic artery, SV=systemic veins; RA=right atrium RV=right ventricle, LA=left atrium, LV=left atrium.

2. MODEL AND SIMULATION

A schematic of the mathematical, closed-loop model of the circulatory system is shown in Figure 1. A detailed description of the model and model parameters has been published elsewhere.[10] Using basic principles of hydrodynamics, a 14-compartment model of the circulatory system was developed. The venous compartments (systemic and pulmonary) was modeled using a simple Windkessel compartment model. To account for inertial effects, a Westkessel model was used as the basic functional unit for the arterial (systemic and pulmonary) compartments. Pressure, flow and volume relationships for the Windkessel model are as given by Eqs. (1)-(3) below:

$$\text{Flow} = Q(i) = (P_i - P_o) / R \tag{1}$$

$$\text{Volume} = V(i) = \int Q\, dt \tag{2}$$

$$\text{Pressure} = P(i) = (V - V_u) / C \tag{3}$$

where P_i, P_o are the input and output pressure, V_u is the unstressed volume, R is the resistance, and C is the compliance of the respective compartments.

Similarly, for the Westkessel model, pressure and volume are calculated using Eqs. (2) and (3), however, the flow equation is modified to account for inertial effects (L=inertance) and is given by Eq. (4):

$$Q = \int [(P_i - P_o) - R^*Q(i\text{-}1)] / L \tag{4}$$

The compartments comprising the four cardiac chamber, two atria and two ventricles, consisted of Windkessel models. However, to simulate pulsatility, ventricular compartments were modified to incorporate time-varying compliance features, thereby mimicking the contractile function of the heart. Time-varying compliance parameters were selected so as to achieve similar characteristics of the underlying physiology (i.e. stroke volume, ventricular pressures, cardiac output, etc.). Using a typical pressure-volume curve, Figure 2a shows how ventricular stiffness parameters for end-systole and end-diastole were chosen. Figure 2b shows a schematic of how these parameters were incorporated into a time-varying stiffness (inverse of compliance) model. As shown in Fig. 2b, the contractile phase (i.e. systole) of the cardiac cycle was modeled as a half-sinusoid for simplicity, while the diastolic or relaxation phase was modeled as a period of constant compliance.

Inherent to this circulation model is a control mechanism similar to the Frank-Starling mechanism. As more volume is returned to the heart, more pressure is generated following the relationship in Eq. (3), and hence, more blood will be ejected on the next contractile phase. Additional features of the circulation model include an algorithm to incorporate baroreceptor control. The computational model tracks arterial pressure in the carotids (i.e. small artery compartment, SA2) comparing it against a set target pressure (i.e. 72 mm Hg). As arterial pressure deviates from this target pressure , the difference between actual and target pressure is proportionally fed back into the model to alter the strength of ventricular contraction, heart rate, and venous return (by influencing systemic venous resistance). The goal of this proportional control system is to maintain carotid pressure at or near the target settings.

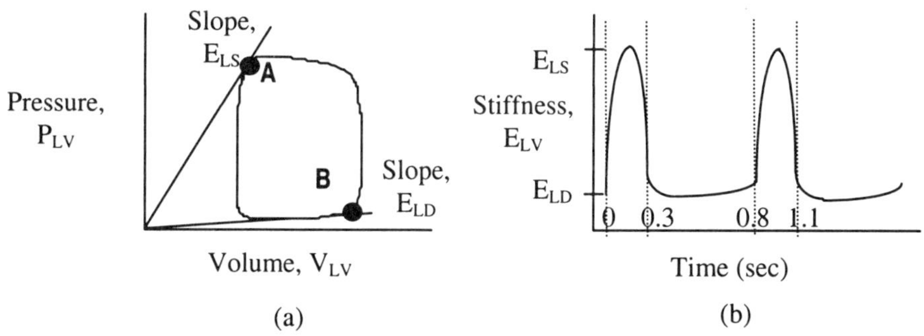

Figure 2. Time-varying compliance parameters for the left ventricle. (a) Normal LV pressure-volume loop. (b) Time-varying stiffness of the LV over two cardiac cycles. Systolic duration, TS, is 300 ms. Diastolic duration is 500 ms. Time for one cardiac cycle is 800 ms or 75 beats per minute.

Upon verification that the model pressure, flow, and volume characteristics are comparable to normal physiologic values described elsewhere,[10] simulations were performed to evaluate the following test cases on cardiac function:

i) static increases in pleural pressure (+15 cm H_2O; i.e. Valsalva maneuver)
ii) static decreases in pleural pressure (-15 cm H_2O; i.e. Mueller maneuver)
iii) cardiac-gated ventilation (CGV)-optimal parameter estimation

3. RESULTS

Results from the computational model of the circulation system were compared to experimental findings obtained from imaging studies with magnetic resonance imaging (MRI) and electron-beam computed tomography (EBCT).

Static increase in pleural pressure (+15 cm H_2O) showed decreases in cardiac function. Right ventricular (RV) and left ventricular (LV) stroke volumes (and cardiac output) decreased by 19% and 23%, respectively. However, ejection fraction did not change appreciably with the maneuver (-1%). The decreased stroke volume resulted in decreased pulmonary artery and aortic blood flow. Additionally, the model predicts a shift in thoracic blood volume to systemic compartments, as both pulmonary (-2%) and cardiac (-12%) volumes decreased. Decreased thoracic blood volumes (in compartments comprising the lungs and heart) coincided with decreased venous return from the systemic venous compartment.

Negative pleural pressure (-15 cm H_2O relative) produced the opposite affect on cardiac performance. Stroke volume and cardiac output increased by 25% compared to baseline, apneic states. From the mathematical model, the simulated negative pressure maneuver resulted in a predicted increase in vena cava blood flow returning to the thorax. Additionally, blood volume in the lungs and heart increased by 2% and 12% respectively. These findings are similar to those by Brecher and colleagues.[3]

Applying the same circulation model to incorporate changes in intrathoracic pressure due to ventilation showed phasic changes in hemodynamics with changes in pleural pressure. Synchronizing these changes in intrathoracic pressures to cardiac-cycle specific events (i.e. CGV) resulted in several qualitative observations. First, the timing of ventilation relative to cardiac phases could appreciably alter cardiac hemodynamics and function. SGV produced maximal LV end-diastolic volume, stroke volume, and cardiac output. Whereas, when mechanical ventilation was timed to diastolic events (DGV), stroke volume and cardiac output was minimized on a beat-by-beat basis. The model predicted that differences in cardiac function during SGV vs. DGV were by as much as 30%.

4. DISCUSSION

Experimental studies performed with MRI techniques in which the human subjects performed Mueller (generating negative pleural pressures) and Valsalva maneuvers (generating a positive pleural pressure environment) showed comparable results to that predicted by the model.[11] During systolic-timed image acquisitions, no significant effects on LV chamber were observed in healthy human subjects, indicating minimal effects of intrathoracic pressures on end-systolic volumes.

However, during imaging timed to diastolic events, sustained Valsalva maneuvers resulted in decreased LV size as predicted by the circulation model proposed. Additionally, stroke volume and cardiac output were also observed to decrease with the Valsalva maneuver experimentally, as predicted by the model.

Sustained experimental results in humans while performing the Mueller maneuver revealed an initial significant ($p<0.05$) decrease in LV chamber size upon performing the maneuver, but showed recovery after several heartbeats to pre-maneuver levels. This observation was predicted by the model and may indicate additional features not accounted for by the model (i.e. effects of pericardial sac).

Using EBCT and cardiac cycle-specific synchronized ventilation in canine models,[9,12] we observed that SGV resulted in increased stroke volume and cardiac output of 22% ($p<0.05$) compared to a non-synchronized protocol. These experimental findings support the prediction by the model that optimal timing of ventilation to specific events of the cardiac cycle can result in improved cardiac performance (i.e. stroke volume and cardiac output) by up to 30%.

5. CONCLUSION

Using a computational, hydrodynamic circulation model, we were able to study the interaction between the heart and lungs while performing various ventilation maneuvers. The major finding of this study was that heart-lung interactions may be harnessed for augmenting cardiac function if ventilation is properly timed to the cardiac cycle (i.e. CGV). This notion was demonstrated previously in canine models by our laboratory. However, using a mathematical model of the circulation system, we were able to study the mechanisms whereby this augmentation occurs. Analysis of model results showed that increased cardiac stroke volume and cardiac output was achieved due to increased ventricular filling during diastole and due to assisted ejection during systolic events. Computational and experimental findings suggest that the primary mechanism affecting heart-lung interaction during breathing maneuvers is dominated by pleural pressure changes during ventilation.

Thus, by using a hydrodynamic model of the circulatory system incorporating the effects of ventilation, we were able to extract the mechanisms by which (positive and negative) pressure breathing affect cardiac function. Additionally, we were able to specifically study the underlying parameters that may serve to augment cardiac performance in the special case of cardiac-gated ventilation. The computational model was also used for optimal parameter estimation and to aid in developing experimental protocols to further examine the benefits of CGV.

6. REFERENCES

1. Hoffman, E. A. & Ritman, E. L. Shape and dimensions of cardiac chambers via computed tomography: Role of imaged slice thickness and orientation. *Radiology* **155**, 739-744 (1985).
2. Hoffman, E. A. & Ritman, E. L. Heart-lung interaction: effect on regional lung air content and total heart volume. *An. of Biomed. Eng.* **15**, 241-257 (1987).
3. Brecher, G. A. & Hubay, C. A. Pulmonary blood flow and venous return during spontaneous inspiration. *Circ. Res.* **3**, 210-214 (1955).

4. Franklin, D. L., van Citters, R. L. & Rushmer, R. Balance between right and left ventricular output. *Circ. Res.* **10**, 17-26 (1962).
5. Hoffman, J. I. E., Guz, A., Charlier, A. A. & Wilken, D. E. I. Stroke volume in conscious dogs: effect of respiration, posture, and vascular occlusion. *J. Appl. Physiol.* **20**, 865-877 (1965).
6. Robotham, J. L. & Mitzner, W. A model of the effects of respiration on left ventricular performance. *J. Appl. Physiol.* **46**, 411-418 (1979).
7. Scharf, S. M., Brown, R., Saunders, N. & Green, L. H. Changes in canine left ventricular size and configuration with positive end-expiratory pressure. *Circ. Res.* **44**, 672-678 (1979).
8. Cournand, A., Motley, H. L., Werko, L. & Richards Jr., D. W. Physiological studies of the effects of intermittent positive pressure breathing on cardiac output in man. *Am. J. Physiol.* **47(2)**, 453-461 (1948).
9. Tran, B. Q. Cardio-pulmonary interaction during cardiac-gated ventilation: Combined imaging and model-based study, Ph.D. Thesis in Biomedical Engineering, University of Iowa, Iowa City, 1998).
10. Tran, B. Q. & Hoffman, E. A. Role of breathing on cardiac performance: experimental and mathematical models. *SPIE Medical Imaging* **3660**, 150-158 (1999).
11. Tran, B. Q. & Hoffman, E. A. Role of pleural pressure on LV volume assessed via high-speed MR imaging. *Circulation* **96 (8)** Supp, I514 (1997).
12. Hanson, W. C. & Hoffman, E. A. in *SPIE Medical Imaging* **2433**: 303-308 (1995).

INSPIRATORY TIMING DIFFERENCES AND REGULATION OF *GRIA2* GENE VARIATION: A CANDIDATE GENE HYPOTHESIS

Clarke G. Tankersley, Heather Kulaga, and Michael M. Wang*

1. INTRODUCTION

Individual variation in eupneic breathing pattern has been shown in human subjects and laboratory animal models. Genetic predisposition is one of several factors proposed as the basis for the variation among individuals. It is unclear, however, which genes may determine diversity in breathing pattern and by what mechanism(s) genetic factors play a role. Our laboratory has addressed these questions using variation among inbred mouse strains as genetic models of differential respiratory control mechanisms. This report summarizes a series of studies that considers the role of the *Gria2* gene, which encodes an ionotropic glutamate receptor subunit, in determining strain variation in eupneic breathing pattern.

Strain differences in inspiratory timing (T_I) between C3H/HeJ (C3) and C57BL/6J (B6) mice have been described during eupneic, hypoxic and hypercapnic breathing.[1-3] Inheritance studies, including first and second generation offspring of C3 and B6 progenitors, suggested that the rapid breathing pattern of B6 mice is dominant relative to the slow pattern of C3 mice.[1] Genotype-phenotype association studies (i.e. linkage analysis) demonstrated that a candidate region on mouse chromosome 3 determines a substantial proportion of the variance in eupneic T_I between C3 and B6 mice.[2] The *Gria2* gene is located adjacent to the linkage site, and represents a functional candidate in regulating differential T_I.

The *Gria2* gene encodes a subunit (i.e. GluR2) of ionotropic glutamate receptors responsive to α-amino-3-hydroxy-5-methyl-4-isoxazolepropionate (AMPA). AMPA glutamate receptors have an excitatory role in mammalian central nervous system function, and are known to influence respiratory control mechanisms. Feldman and colleagues[4] have suggested that AMPA receptor activation is critical for respiratory rhythm generation arising from the preBötzinger complex. In addition, Whitney et al.[5] showed that systemic administration of AMPA receptor antagonist significantly altered the pattern of breathing during eupneic, hypoxic and hypercapnic conditions. These investigators suggested that AMPA receptor antagonism reduced eupneic breathing frequency by specifically prolonging

* Clarke G. Tankersley, Division of Physiology, Johns Hopkins University, School of Public Health.

T_I. As shown by an age-dependent increase in GluR2 expression in the nucleus tractus solitari, AMPA receptor modulation of breathing pattern may be most influential during the maturation of respiratory control mechanisms.

Collectively, these observations suggest that variation in DNA sequence in the *Gria2* gene may contribute to the strain differences in eupneic T_I between C3 and B6 mice. We tested this hypothesis in two ways: 1) searching for single nucleotide polymorphisms in the regulatory and coding regions of the *Gria2* gene, and 2) examining differential gene expression in brainstem tissue of C3 and B6 mice.

2. METHODS

2.1. Animals

Male C3 and B6 inbred mice were purchased from Jackson Laboratories (Bar Harbor, ME) at 10-12 wks of age. Initial studies were performed using tissue samples obtained from animals at ~25 wks of age. Animals were provided water and mouse chow ad libitum, and all procedures were reviewed and approved by the Animal Care and Use Committee at the Johns Hopkins School of Public Health.

Strain-specific differences in T_I during eupneic, hypoxic and hypercapnic conditions were determined by whole-body plethysmography. The method and results have been described in detail and reported elsewhere.[1] Among ten standard inbred mouse strains, C3 and B6 mice represent the extremes of several trait-specific strain distribution patterns. During eupneic ventilation, for example, the strain difference between C3 and B6 mice is ~80 msec, which represents the magnitude of the T_I response that is genetically regulated. A putative quantitative trait loci (QTL) for eupneic T_I was assigned the symbol *Itbq1*, and mapped to mouse chromosome 3 by The Jackson Laboratory. The results of the present study consider whether or not *Itbq1* and *Gria2* are the same gene.

2.2. Gria2 DNA Sequence Analysis

While the sequence of cDNA provides information concerning the coding regions of the *Gria2* gene, the sequence of genomic DNA permits a search for single nucleotide polymorphisms (SNPs) in the regulatory regions of the gene including promoter and intronic regulatory sites. In addition, splice variants of the *Gria2* gene involving exons 14 and 15 can be investigated using genomic DNA.

To generate cDNA, reverse transcription was performed following the protocol of Life Technologies (http://www.lifetech.com, 1999). Total RNA was extracted from brainstem and isolated using a Trizol protocol (Gibco BRL, MD). Strain-specific genomic DNA was purchased from the Jackson Laboratory. Genomic DNA and cDNA were used as templates during polymerase chain reaction (PCR) amplification using standard conditions. The PCR products were purified using a QIA kit (Qiagen, CA), mixed with specifically-designed oligonucleotide primers (see Table 1), and submitted to Genetic Resource Core Facility at the Johns Hopkins School of Medicine.

2.3. Differential Gria2 Gene Expression

Northern analysis was performed using total RNA obtained from brainstem (and other brain regions including forebrain and cerebellum) and run in a 0.9% agarose gel. The northern gel was transferred to a Nytran membrane and hybridized to oligonucleotide probes (Table 1), which were produced using standard methods. To evaluate differential gene expression of two alternative splice variants encoded by either exon 14 or 15 of the *Gria2* gene (i.e. designated as flip or flop, respectively), northern analysis was accomplished using oligonucleotide probes specifically designed based on DNA sequencing.

Northern gels were quantified using standard phosphorimaging techniques. Data were expressed as means (± SD), and comparisons between strains were evaluated with a Wilcoxin signed rank test. Statistical significance was established at an α-level of 0.05.

Table 1. Oligonucleotide PCR primers for C3 and B6 *Gria2* gene variation

Sequence	Start Site	Purpose
DNA Sequencing		
5' ATGTTTTATCGTTTCAAGA 3'	2163	Intronic primers for exon 14
5' ATTCCAAGGTCAGCCCCCGA 3'	2343	(ref 6)
5' TATTTTCCACGTGAAGAACC 3'	2127	Intronic primers for exon 15
5' GTAAGGTCAGTTGCTGCAGG 3'	2286	(ref 6)
5' TGCAGAGAGACTTCCACGCA 3'	2	Primers for SNP in promoter
5' CTCTGAGCCGCGCTTTCTCG 3'	951	(ref 7)
Differential expression		
5' AGGAAATGCGGTTAACCTCGC 3'	2491	Exon 14
5' TATATGGCATCGAGAGTGTT 3'	2826	(ref 8)
5' ACCCCAGTAAATCTTGCAGT 3'	2226	Exon 15 (ref 6)
5' TATATGGCATCGAGAGTGTT 3'	2826	(ref 8)

3. RESULTS

3.1. Gria2 DNA Sequence Variation between C3 and B6 Mice

The mRNA derived from the mouse *Gria2* gene is a 938 amino acid sequence fragment. There are two splice variants that encode either exon 14 or 15. There were no detectable variations between C3 and B6 mice in the cDNA sequence, including flip and flop splice variants.

The organization of the promoter region of the mouse *Gria2* gene incorporates a sequence of >7,000 nucleotides.[9] At least three significant regulatory sites are proposed as described by Myers et al.[10] Our sequencing efforts uncovered a SNP between C3 and B6 at nucleotide 147.[7] The 16 bp sequence that contains this strain-specific (i.e. B6) DNA

B6 ➔ CAATCCAGATACCCAA ➔➔ INTERACTS WITH **eHand**
5' 3'
C3 ➔ CAGTCCAGATACCCAA
⇑ ⇑
145 bp 160 bp

Figure 1. Schematic of the promoter region[7] depicting the sequence variation between C3 and B6 strains of mice. The proposed sequence variation is detected in a 16 bp fragment known to interact with a recently described basic helix-loop-helix transcription factor, eHand.

variation appears capable of interacting with a basic helix-loop-helix transcription factor called eHand. The B6 variant, but not the C3 variant, is likely to interact with the transcription factor, eHand. Figure 1 summarizes the sequence variation between in C3 and B6 mice located in the regulatory region of the *Gria2* gene.

3.2. Gria2 Expression Differences between C3 and B6 Mice

As shown in figure 2, *Gria2* gene expression in brainstem was significantly ($P < 0.05$) greater in C3 compared with B6 mice. The strain difference was detected using oligonucleotide probes specific to flip and flop splice variants. Furthermore, *Gria2* gene expression differences between strains appeared to be specific to brainstem, since the expression levels were similar between strains in cerebellum and forebrain sections obtained from the same animals.

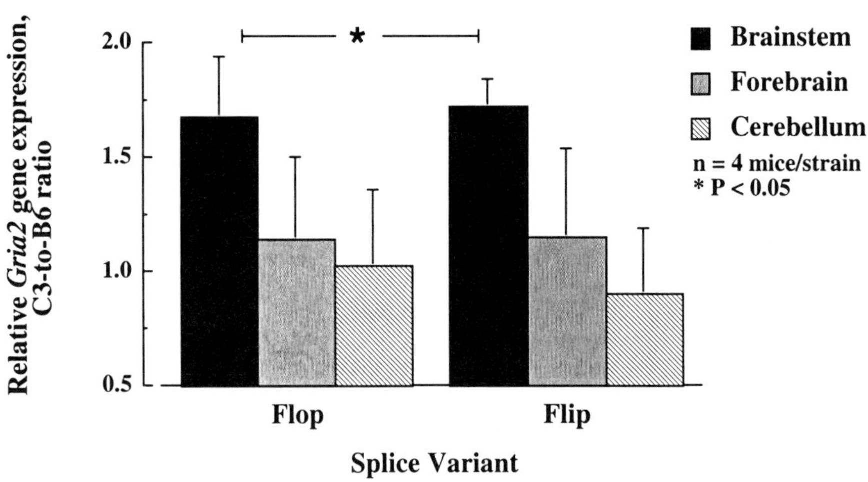

Figure 2. Strain differences in *Gria2* gene expression in various brain sections of B6 and C3 mice. The significantly ($P < 0.05$) greater *Gria2* gene expression in C3 mice is specific to the brainstem. Also, there were no within-strain differences in the regulation between flip and flop splice variants.

4. DISCUSSION

The observations from the present study examine the role of differential *Gria2* gene regulation in determining strain differences in eupneic T_I between C3 and B6 mice. Two important findings are demonstrated in this regard. First, there appears to be a significant single nucleotide polymorphism (SNP) between C3 and B6 mice at a regulatory site in the promoter region of the *Gria2* gene. Secondly, there is a significantly greater *Gria2* gene expression unique to the brainstem of C3 mice compared with B6 mice. However, it is premature to conclude that the sequence variation in the promoter region is coupled to differential *Gria2* gene expression. In addition, it is unclear whether or not the SNP or differential expression in the *Gria2* gene is linked to the strain variation in eupneic T_I. Both of these uncertainties will require future experiments.

Specific glutamate receptors subunits are integral to synaptic excitation involved in neural pattern generation and long-term potentiation. The protein product of the *Gria2* gene is a 115 bp receptor subunit, GluR2, which can alter cationic channel function including Ca^{2+} permeability. These neural processes are fundamental mechanisms in the control of breathing. Although little information exists concerning breathing responses in GluR2 knockout or transgenic mice, other phenotypes have been reported. For example, GluR2 knockout mice show impaired motor coordination, enhanced long-term potentiation, and a ninefold increase in Ca^{2+} permeability.[11] Alternatively, GluR2 flip-overexpressing mice experience enhanced neuronal death following focal ischemia.[12] It is unknown whether or not phenotypic variation between C3 and B6 mice show similar effects.

Kazemi and colleagues have pioneered the role glutamate plays as an amino acid neurotransmitter in the central regulation of eupneic, hypoxic, and hypercapnic breathing.[eg.13] In addition, Feldman and associates have recognized the importance of glutamate receptors in central respiratory rhythm generation.[eg.4] The most recent evidence has emphasized the significance of AMPA glutamate receptors in timing aspects of hypoxic ventilation.[5] Collectively, these data and the results from our genetics studies compelled us to consider the *Gria2* gene as a functional candidate. While our studies rejected the possibility of *Gria2* gene diversity in the coding region, our search revealed a polymorphism in the regulatory region of the gene. More specifically, the data suggest that the SNP is contained within a 16 bp fragment capable of binding to a basic helix-loop-helix transcription factor known as eHand (or Thing1). Although eHand is known to influence neural crest development, no evidence is available relating eHand regulation to the *Gria2* gene or the control of breathing.

The hypothesis that *Gria2* gene regulation is different between C3 and B6 mice is supported by strain variation in *Gria2* gene expression. This strain difference is specific to brainstem regions, where central control of breathing and respiratory pattern generation occur. The expression levels of both flip and flop splice variants are greater in C3 relative to B6 mice, but are not different within C3 or B6 mice. Sommer et al.[12] demonstrated the flip and flop versions were heterogeneously distributed throughout different regions of the brain suggesting that the splice variants provide functional diversity to the GluR2 receptor subunit. Finally, our findings may be age-dependent, since differential *Gria2* gene expression between C3 and B6 mice did not occur in younger mice (i.e. 10 vs. 25 wks of age; data not shown).

In summary, this study demonstrates a SNP between C3 and B6 mice in the regulatory region of the *Gria2* gene. In addition, the findings suggest that *Gria2* gene expression differs between C3 and B6 strains. Future studies using second-generation offspring of C3 and B6 will test whether or not the polymorphism between C3 and B6 mice is linked to eupneic T_I variation by using the SNP as a marker to examine *Gria2* gene diversity.

5. REFERENCES

1. C.G. Tankersley, R.S. Fitzgerald, R.C. Levitt, W.A. Mitzner and S.R. Kleeberger. Genetic control of differential baseline breathing pattern. *J. Appl. Physiol.* **82**, 874-881 (1997).
2. C.G. Tankersley, D.A. DiSilvestre, A.E. Jedlicka, H.M. Wilkins and L. Zhang. Differential inspiratory timing is genetically linked to mouse chromosome 3. *J. Appl.Physiol.* **85**, 360-365 (1998).
3. C.G. Tankersley, R.C. Elston and A.H. Schnell. Genetic determinants of acute hypoxic ventilation: Patterns of inheritance in mice. *J. Appl. Physiol.* **88**, 2310-2318 (2000).
4. Q. Ge and J.L. Feldman. AMPA receptor activation and phosphatase inhibition affect neonatal rat respiratory rhythm generation. *J. Physiol.* **509**, 255-266 (1998).
5. G.M. Whitney, P.J. Ohtake, N. Simakajornboon, Y. Xue and D. Gozal. AMPA glutamate receptors and respiratory control in the developing rat: anatomic and pharmacological aspects. *Am. J. Physiol. Regulatory Integrative Comp. Physiol.* **278**, R520-R528 (2000).
6. B. Sommer, K. Keinanen, T.A. Verdoorn, W. Wisden, N. Burnashev, A. Herb, M. Kohler, T. Takagi, B. Sakmann and P.H. Seeburg. Flip and flop: cell-specific functional switch in glutamate-operated channels of the CNS. *Science.* **249**, 1580-1585 (1990).
7. M. Kohler, H.C. Kornau and P.H. Seeburg. The organization of the gene for the functionally dominant alpha-amino-3-hydroxy-5-methylisoxazole-4-propionic acid receptor subunit GluR-B. *J. Biol. Chem.* **269**, 17367-17370, (1994).
8. K. Sakimura, H. Bujo, E. Kushiya, K. Araki, M. Yamazaki, M. Yamazaki, H. Meguro, A. Warashina and M. Mishina. Functional expression from cloned cDNAs of glutamate receptor species responsive to kainate and quisqualate. *FEBS Lett.* **272,** 73-80 (1993).
9. S. Brené, C. Messer, H. Okado, M. Hartley, S.F. Heinemann and E.J. Nestler. Regulation of GluR2 promoter activity by neurotrophic factors via a neuron-resistive silencer element. *Eur. J. Neurosci.* **12**, 1525-1533 (2000).
10. S.J. Myers, J. Peters, Y. Huang, M.B. Comer, F. Barthel and R. Dingledine. Transcriptional regulation of the GluR2 gene: Neural-specific expression, multiple promoters, and regulatory elements. *J. Neurosci.* **18**, 6723-6739 (1998).
11. Z. Jia, N. Agopyan, P. Miu, Z. Xiong, J. Henderson, R. Gerlai, F.A. Taverna, A. Velumian, J. MacDonald, P. Carlen, W. Abramow-Newerly and J. Roder. Enhanced LTP in mice deficient in the AMPA receptor GluR2. *Neuron.* **17**, 945-956 (1996).
12. D. Le, S. Das, Y.F. Wang, T. Yoshizawa, Y.F. Sasaki, M. Takasu, A. Nemes, M. Mendelsohn, P. Dikkes, S.A. Lipton and N. Nakanishi. Enhanced neuronal death from focal ischemia in AMPA-receptor transgenic mice. *Mol. Brain Res.* **52**, 235-241 (1997).
13. M.D. Burton and H. Kazemi. Neurotransmitters in central respiratory control. *Respirat. Physiol.* **122**, 111-121 (2000).

EXPLAINED AND UNEXPLAINED VARIABILITY OF CO_2-SENSITIVITY IN HUMANS

Christina M. Spengler and Steven A. Shea*

1. INTRODUCTION

When modeling the respiratory control system, it is useful to know the inherent noise in the variables of interest, such as the arterial P_{CO_2} and the sensitivity of the hypercapnic ventilatory response (HCVR). These variables can be greatly affected by behavioral and environmental conditions including fluctuations in arousal level, emotion, cognitive activity, subject-experimenter interaction, light, noise and temperature. For instance, even when in an ostensible steady-state, merely opening one's eyes results in systematic increases in breathing frequency, ventilation and HCVR slope (reviewed by Shea[1]). In addition, challenges to the respiratory control system (e.g., added loads, hypoxia or hypercapnia) can result in adaptation such that subsequent measurements are affected by prior measurements (e.g., Powell et al[2]). Furthermore there can be non-physiological variability that is intrinsic to the measurement technique itself (e.g., instability of sensors). Overall, influences from behaviors, arousal, the environment, adaptation and the measurement technique contribute to the 'noise' in the measured data.

Numerous studies in humans have assessed the variability of HCVR during repeated measurements. These studies report a large within-day coefficient of variation for HCVR slope with group averages ranging from 14% to 20%[3-6]. The between-day variability is much larger than the within-day variability (e.g., Sahn et al[3]). There have been few formal assessments of the within- or between-day variability of resting arterial or end-tidal P_{CO_2} under steady-state conditions, although arterial P_{CO_2} is thought to be a tightly regulated variable[1,7]. The aim of this paper was to assess the stability of the resting eucapnic level and the slope of the HCVR in humans when measured under the most stable of behavioral and environmental conditions. This analysis utilizes a unique data set originally collected to provide estimates of the effects of circadian rhythms on respiration[8] and the effects of sleep deprivation on respiration[9]. The residual

* Christina M. Spengler, Exercise Physiology, Institute for Human Movement Sciences, Swiss Federal Institute of Technology and Institute of Physiology, University of Zurich, 8057 Zurich, Switzerland. Steven A. Shea, Harvard Medical School and Sleep Disorders Program, Brigham & Women's Hospital, 221 Longwood Avenue, Boston, MA 02115, USA.

Frontiers in Modeling and Control of Breathing, edited by Poon and Kazemi.
Kluwer Academic/Plenum Publishers, New York, 2001.

(unexplained) variation, after accounting for the systematic variability related to time (circadian rhythm and duration of wakefulness) provides our best estimate of the inherent noise in our measurement of the ventilatory control system under steady-state conditions.

2. METHODS

The study was approved by the Internal Review Board at Brigham and Women's Hospital. Each subject provided consent to participate. The study was conducted on 10 healthy males (mean age = 23.7 years; having normal BMI and pulmonary function). The methods are described briefly below; for further detail see Spengler et al[8].

To ensure a stable baseline before entry into the laboratory, subjects kept a regular sleep-wake schedule for >2 weeks. They were then studied in an individual laboratory suite isolated from sunlight and external time cues (e.g., clocks, radios, visitors, and mail). Laboratory temperature was maintained at ~23 ^{O}C. The laboratory protocol included 2 baseline days that enabled the subjects to acclimate to the environment, become familiarized with the respiratory tests and establish a steady-state. Subjects woke up at their usual time on the morning of the 3rd day in the laboratory and remained in bed and awake in a semi-recumbent position for 41 h in a 'constant routine' protocol[10]. This protocol has been designed to assess the underlying circadian rhythms as well as the effects of duration of wakefulness while excluding the 'masking' effects of changes in environment and behavior (e.g., sleep). Small identical snacks were given every 2 h. Experimenters were present throughout to ensure that the subjects remained awake (verified from EEG recordings). A rectal temperature sensor was used to measure core body temperature (CBT), providing a marker of the endogenous circadian pacemaker.

Measurements of resting end-tidal P_{CO_2} (P_{ET,CO_2}) and the slope of the HCVR were performed every 2 h while subjects breathed through a mouthpiece with a nose-clip in place. To ensure a steady state of relaxed wakefulness, subjects had their eyes open and focused upon a picture for 10 min, and only data from the 5th to 9th min were analyzed. Breath-by-breath values for P_{ET,CO_2} were recorded. The magnitude of HCVR was measured using the hyperoxic Read re-breathing method[11]. After 2 min basal breathing, subjects breathed via a low resistance spirometer circuit initially filled with a mixture of 7% CO_2 and 93% O_2 to a volume equal to the subject's vital capacity plus 1 liter. Thus, each test was started at a P_{CO_2} that approximated mixed-venous P_{CO_2}, and hyperoxia diminished peripheral chemoreceptor stimulation for testing central chemoreceptor sensitivity. Subjects re-breathed from the spirometer until P_{ET,CO_2} reached 65 mmHg. Tidal gases were analyzed by an instrument that compensated for alterations in barometric pressure. Spirometer volume and P_{ET,CO_2} signals were analyzed using an objective technique for determining the slope of the HCVR[12].

To further ensure measurements were made in basal conditions, the first 5 h of constant routine data were excluded from all analyses to eliminate any residual effects of sleep and postural change upon respiration. 2-h averages of all variables were calculated, centered around the minimum of CBT for every subject (this aligns the data according to circadian phase). The circadian phase and period of the CBT rhythm were estimated for each subject by a 2-harmonic regression analysis of the temperature data using a non-linear least-squares method[13]. The period was used as a constant in the subsequent analysis of the circadian rhythmicity of the other variables, which was performed on both individuals' data and group mean data. The group mean period of subjects' CBT-rhythms was 24.08 h. The circadian phases and amplitudes of the respiratory variables as well as any systematic linear trend were estimated by cosinor analysis[14], using the equation:

$$y = M + A \cdot \cos(\omega \cdot t + \varphi) + s \cdot t$$

(where y = variable; M = mesor; A = amplitude of circadian rhythm [peak-mean]; ω = 2π/period of rhythm; φ = phase of rhythm, t = time [h]; s = underlying linear trend [h^{-1}]). The fraction of the variability in y explained by the model (i.e., r^2) was calculated, and the significance of the model was determined from analyses of variance (F ratio = mean square error from the model [3 df] divided by the residual mean square error [12 df]). To test whether the cosinor models accounted for all of the systematic changes, autocorrelation analyses were performed on the residuals around the models. A lack of autocorrelation (no remaining serial dependence) in the residuals would suggest that the residuals are random and that the model is quite suitable.

In addition, the coefficient of variation (CV) of original data and of corrected data (accounting for systematic effects of time) were compared using paired t-tests to assess whether or not application of the model significantly reduced the inherent variability in the data. The CV relative to the model fit provides an estimate of the inherent noise in the data after accounting for the systematic effects of time (circadian and prior wakefulness). In addition, to compare the current data with published short-term variability data, we calculated the CV from all combinations of measurements over 2 h within each subject.

3. RESULTS

3.1. Group Results

The magnitude of the circadian variation was larger for HCVR slope (amplitude = ± 12.2% of mesor) than for P_{ET,CO_2} (± 1.4%). The effects of duration of wakefulness were small for HCVR and P_{ET,CO_2} (see equations below). Circadian rhythms and effects of prior wakefulness explained approximately two-thirds of the variability in the group mean data (HCVR, $r^2 = 0.72$; P_{ET,CO_2}, $r^2 = 0.66$). These statistics suggest that the cosinor model may be appropriate for analysis of group data. The derived group mean equations were:

$$HCVR\ slope\ (l \cdot min^{-1} \cdot mmHg^{-1}) = 2.9 + 0.36 \cdot \cos(0.26 \cdot t - 0.75) + 0.02 \cdot t$$

$$P_{ET,CO_2}\ (mmHg) = 42 + 0.59 \cdot \cos(0.26 \cdot t - 0.17) - 0.01 \cdot t$$

Applying the cosinor model to the group mean data significantly reduced the sum of the squared errors (SSE) compared to the SSE around the mean (HCVR, $F_{3/12} = 11.3$, $p<0.001$; P_{ET,CO_2}, $F_{3/12} = 8.4$, $p<0.01$), and it resulted in a reduction in unexplained variability, with the CVs decreasing from 11.7 to 6.1% for HCVR and from 1.2 to 0.7% for P_{ET,CO_2}. There was no significant autocorrelation of residuals around the model (i.e., no remaining serial dependence), suggesting that the unexplained variability is random.

3.2. Individuals' Results

As with the group data, for individuals the circadian variation was larger for HCVR (mean amplitude = ± 18.8% of mesor) than for P_{ET,CO_2} (± 2.2%). Again, there was little change related to the duration of prior wakefulness (on average, s = 0.021 $l \cdot min^{-1} \cdot mmHg^{-1} \cdot h^{-1}$ for HCVR and s = −0.010 $mmHg \cdot h^{-1}$ for P_{ET,CO_2}). Applying the model to individual's data reduced SSE in almost all subjects, being significant in 6 subjects for P_{ET,CO_2} but only 3 subjects for HCVR and the unexplained variability was significantly reduced, with the average CVs decreasing from 25.2 to 19.5% for HCVR and from 3.2 to 2.3% for P_{ET,CO_2}. In all subjects, the CV of HCVR was far greater than the CV of P_{ET,CO_2}. Overall, circadian rhythms and effects of duration of wakefulness usually explained less than half of the individuals' variability (HCVR, mean r^2 = 0.38 [range 0.05 to 0.77]; P_{ET,CO_2}, mean r^2 = 0.42 [0.03 to 0.74]). The residuals around the model had no remaining serial dependence in most subjects (7 of 10 subjects for HCVR, and 6 of 10 subjects for P_{ET,CO_2}), suggesting that the unexplained variability is random. Representative HCVR data from one subject are shown in Figure 1. As for the group, in this subject, the CV of HCVR was reduced by application of the model (from 31.3 to 18.7% [and from 3.7 to 2.7% for P_{ET,CO_2}]).

For comparison with published short-term variability data, the uncorrected CV from consecutive paired measurements of HCVR over 2 h averaged 15.1% for the group (range, 8.8 – 20.8%, from 17 consecutive measurements in each subject). For P_{ET,CO_2} the group average short term CV was 2.1% (range 1.3 – 3.0%). There was no systematic change in this short term CV throughout the constant routine for either variable.

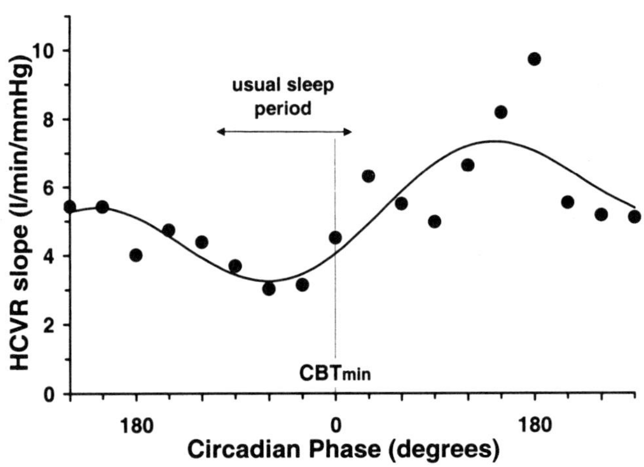

Figure 1. HCVR slope for one subject during 41 h constant routine. Shown are data (points) and cosinor model (line) *vs*. circadian phase (360° ~ 24 h). Vertical line represents minimum core body temperature (assigned 0° phase). Subject's usual sleep period is indicated (although data collected during wakefulness).

4. DISCUSSION

Previous studies in humans have reported large within-day CVs for HCVR slope (CVs among studies range from 14 to 20%[3-6]) and even greater between-day variability[3]. We attempted to assess the stability of the slope of the HCVR in humans when measured under the most stable of behavioral and environmental conditions. The residual (unexplained) variation, after accounting for the systematic variability related to time (circadian rhythm and duration of wakefulness) provides our best estimate of the inherent noise in the ventilatory control system (including measurement error). CVs calculated from the current study over the same time-period as previous studies (i.e., $\leq$ 2.5 hours) resulted in an average of 15.1 $\pm$ 3.9%. This amount of variability is at the low end of the range seen by other authors who presumably had less stringent control of behavioral and environmental conditions. From analysis of the full HCVR data set, our average CV was 25.2% (from both within- and between-day measurements). Applying the model to the full data set (i.e., accounting for serial dependence related to underlying circadian rhythms and the duration of wakefulness) reduced the CV in HCVR to 19.5% and left only 30% unexplained variability in the group mean data. However, more than 50% of the variability remained unexplained for individuals.

Sources of this unexplained variation could include random fluctuations in measurement error, inadequacy of the model or as yet uncharacterized physiological processes. The cosinor analyses relied on the assumptions that any circadian effects on respiratory variables would be sinusoidal (without detectable harmonics), and that any effects of sleep deprivation would be linear. Since the residuals were generally random, these assumptions and the model appear to be valid. In practice, to predict the effects of underlying circadian rhythms and the duration of wakefulness for group analysis, it would be necessary to estimate the circadian phase of the subjects before averaging. In subjects with a regular nocturnal sleep schedule, the circadian period can be assumed to be ~24 h, and the time of minimum core body temperature (assigned 0° circadian phase in our equations) as ~2 h before habitual wake time. Nonetheless, this model for group data has yet to be validated in less stable behavioral and environmental conditions, which could either dampen or accentuate such effects. Prediction of the effects of underlying circadian rhythms and the duration of wakefulness for given individuals is currently hampered by the high degree of unexplained variability in the respiratory variables as well as a large degree of variability among subjects in the phase and amplitude of the circadian rhythms[8].

The CV is independent of units and can permit comparison among variables (assuming those variables have a ratio scale). In all subjects, the CV of HCVR was almost an order of magnitude greater than the CV of P_{ET,CO_2}. In one regard, this may be expected if we consider resting eucapnic P_{ET,CO_2} to be actively regulated by negative feedback, whereas HCVR slope is subject to influences from numerous physiological, behavioral and environmental conditions. While this may be true, the comparison of CVs between HCVR and P_{ET,CO_2} is not strictly valid because P_{ET,CO_2} does not have a true zero (due to ongoing metabolism), HCVR and basal P_{CO_2} may not be independent variables, and the measurement error of HCVR slope would be expected to be larger than the measurement error of P_{ET,CO_2} (because HCVR incorporates measurement error in both P_{ET,CO_2} and ventilation, as well as in the linear regression estimate of slope).

In summary, approximately one-third of the group average variability in P_{ET,CO_2} and HCVR slope remains unexplained, and more than half of the within-subject variability in P_{ET,CO_2} and HCVR slope remains unexplained, even when measurements were made under extremely stable behavioral and environmental conditions and after accounting for the systematic changes in the variables related to the passage of time (circadian and sleep deprivation). This 'inherent noise' is related to as yet uncharacterized physiological processes and/or random fluctuations in measurement error. This high degree of unexplained variability, particularly for HCVR slope, is an important consideration in the design of experiments and in interpretation and the modeling of respiratory control data.

5. ACKNOWLEDGMENTS

Funding NIH HL62149 and Swiss Foundation for Medical & Biological Research.

6. REFERENCES

1. S.A. Shea, Behavioural and arousal-related influences on breathing in humans, *Exp Physiol* **81**: 1-26 (1996).
2. F.L. Powell, W.K. Milsom and G.S. Mitchell, Time domains of the hypoxic ventilatory response, *Respir Physiol* **112**:123-134 (1998).
3. S.A. Shan, C.W. Zwillich, N. Dick, R.E. McCullough, S. Lakshminarayan and J.V. Weil, Variability of ventilatory responses to hypoxia and hypercania, *J Appl Physiol* **43**:1019-1025 (1977).
4. J.R.A. Rigg, E.M. Inman, N.A. Saunders, S.R. Leeder and N.L. Jones, Interaction of mental factors with hypercapnic ventilatory drive in man, *Clin Sci Mol Med* **52**:269-175 (1977).
5. T.Y. Sullivan and P.L. Yu, Reproducibility of CO_2 response curves with ten minutes separating each rebreathing test, *Am Rev Respir Dis* **128**:23-26 (1984).
6. M. Nishimura, M. Yamamoto, A. Yshioka, Y. Akiyama, F. Kishi and Y. Kawakami, Longitudinal analyses of respiratory chemosensitivity in normal subjects, Am Rev Respir Dis **143**:1278-1281 (1991).
7. D.J.C. Cunningham, P.A. Robbins and C.B. Wolff, Integration of respiratory responses to changes in alveolar partial pressures of CO_2 and O_2 and in arterial pH, in: *Handbook of Physiology*, section 3, *The Respiratory System*, vol. II, *Control of Breathing*, part 2, edited by N.S. Cherniack and J.G. Widdicombe (American Physiological Society, Baltimore, 1986) pp. 475-528.
8. C.M. Spengler, C.A. Czeisler and S.A. Shea, An endogenous circadian rhythm of respiratory control in humans, *J Physiol* **526**:683-694 (2000).
9. C.M. Spengler and S.A. Shea, Sleep deprivation per se does not decrease the hypercapnic ventilatory response in humans, *Am J Respir Crit Care Med* **161**:1124-1128 (2000).
10. C.A. Czeisler, R.E. Kronauer, J.S. Allan, J.F. Duffy, M.E. Jewett, E.N. Brown and J.M. Ronda, Bright light induction of strong resetting of the human cirdadian pacemaker, *Science* **244**:1328-1333 (1989).
11. D.J. Read, Clinical method for assessing the ventilatory response to CO_2, *Australas Ann Med* **16**:20-32 (1967).
12. Z. Lorinc, J. Derr, M. Snider and R. Lydic, Defining origin of positive slope in hypercapnic ventilatory response curve, *Am J Physiol* **285**:455-470 (1991).
13. E.N. Brown and C.A. Czeisler, The statistical analysis of circadian phase and amplitude in constant-routine core-temperature data, *J Biol Rhythm* **7**:177-202 (1992).
14. W. Nelson, Y. Liang Tong, J-K. Lee and F. Halberg, Methods for cosinor-rhythmometry, *Chronobiologia* **6**:305-323 (1979).

COMPARISON BETWEEN THE RESPIRATORY INDUCTANCE PLETHYSMOGRAPHY SIGNAL DERIVATIVE AND THE AIRFLOW SIGNAL

André Eberhard, Pascale Calabrese, Pierre Baconnier, and Gila Benchetrit*

1. INTRODUCTION

The use of respiratory inductance plethysmography (RIP) for measurement of breathing is appealing not only because of its noninvasive nature but also because it provides rib cage and abdomen cross sectional area changes[1].

Assessment of the RIP signal and the RIP derivative signal has been undertaken mostly during sleep where the use of invasive methods may perturb the course of sleep. However, adequacy of the use of RIP and accuracy of results obtained in studies on patients with sleep associated breathing disturbances are not unanimously accepted. Indeed, Clark et al.[2] comparing estimates of flow rate from differentiated sum RIP to nasal pressure, found that the latter provides the most sensitive measurement of flow rate and flow limitation and recommended it for the assessment of flow limitation in sleeping subjects. Whereas, Bloch et al.[3] found that breathing patterns recorded by RIP identify characteristics of respiration that differentiate sleep disruptive from non-disruptive snoring and Kaplan et al.[4] concluded that inspiratory flow limitation may be assessed by computer assisted analysis of respiratory inductive plethysmography derived breathing patterns with clinically acceptable accuracy.

There are very few comparison of between airflow and RIP derivative signals in physiological conditions. Sackner et al.[5] reported comparisons of time-volume and flow volume components at forced vital capacity. It would be necessary to have comparisons in controlled conditions in order to provide a quantitative basis of pertinence of RIP sum derivative signal and therefore to possibly detect the limits of the use of RIP derivative as a flow signal.

* André Eberhard, Laboratoire de Modélisation et Calcul, Pascale Calabrese, Pierre Baconnier and Gila Benchetrit, Physiologie Respiratoire Expérimentale, Théorique et Appliquée, Université Joseph Fourier, Grenoble, France. Corresponding author : A. Eberhard, Laboratoire de Modélisation et Calcul, Université Joseph Fourier, BP 53, 38041 Grenoble Cedex 9, France., E-mail : andre.eberhard@imag.fr

Frontiers in Modeling and Control of Breathing, edited by Poon and Kazemi.
Kluwer Academic/Plenum Publishers, New York, 2001.

The aim of the present work has been to evaluate the consistency between airflow and RIP derivative. This was done by recording both airflow (pneumotachography) and RIP in healthy subject in three different postures (seated, lateral and dorsal supine) and also in control and in two resistive loaded conditions and by (i) comparing the volume obtained by the two methods, (ii) calculating the goodness of fit of RIP derivative to airflow and (iii) comparing the values of the resistance calculated with airflow and RIP derivative in control and loaded conditions.

2. METHODS

2.1. Recordings

Ten (4 female) healthy volunteers recruited from amongst graduate students participated in the study. Informed consent was obtained from all subjects. The experimental protocol was examined and approved by the Institutional Ethics Review Board. Subjects were seated or supine and wore a face mask on which was mounted a flowmeter (Fleish head No.1) and a differential pressure transducer (163PC01D36, Micro Switch). Mouth pressure was measured with another differential pressure transducer (142PC01D, Micro Switch). End tidal CO_2 was recorded continuously using an infra-red CO_2 analyzer (Engström Eliza/Eliza MC). Rib cage and abdomen surface area changes were recorded with a computer-assisted respiratory inductance plethysmography (Visuresp, RBI).

Nine recordings were performed for each subject : in seated and supine (dorsal and lateral) postures, without (control) and with two (R1 and R2) different levels of resistive load added throughout the entire breath.

2.2. Signal Processing

All signals were digitized at a rate of 64 Hz. For each recording (of 70 breaths on average) performed on a subject at a given posture and a given condition (control or loaded), a least-squares method was used over one breath to obtain a respiratory inductance plethysmography volume (VRIP) signal by a combination of rib cage and abdomen signals compared to the integrated flow signal (VPNT). We applied the following formula to get an approximation of the volume for all other breaths (k = 2, 3,...) in the recording:

$$\mathrm{VRIP}_k = \tau.\mathrm{RCRIP}_k + \alpha.\mathrm{ABDRIP}_k$$

The derivative of VRIP was then calculated by using centered divided differences:

$$\mathrm{FRIP}_k^* = \left(\mathrm{VRIP}_{k+1} - \mathrm{VRIP}_{k-1}\right)/2\Delta t$$

The resulting signal was smoothed by using low-pass filter:

$$\mathrm{FRIP}_k = \sum_{r=-p}^{+p} \omega_r \, \mathrm{FRIP}^*_{k-r}$$

2.3. RIP and PNT Comparisons

Mean value and SD of tidal volume were calculated for each recording (on average 70 breaths) both with RIP and integrated PNT. The goodness of fit of the respiratory inductance plethysmography derivative signal (FRIP) to the airflow signal (FPNT) was evaluated by calculating:

$$\rho = 1 - \sum_{i=1}^{N} \left(y_i^* - y_i\right)^2 \Big/ \sum_{i=1}^{N} \left(y_i - \bar{y}\right)^2$$

where y_i and y_i^* are respectively the FPNT and FRIP at each instant and $\bar{y}$ is the mean value of y_i's over the whole recording.

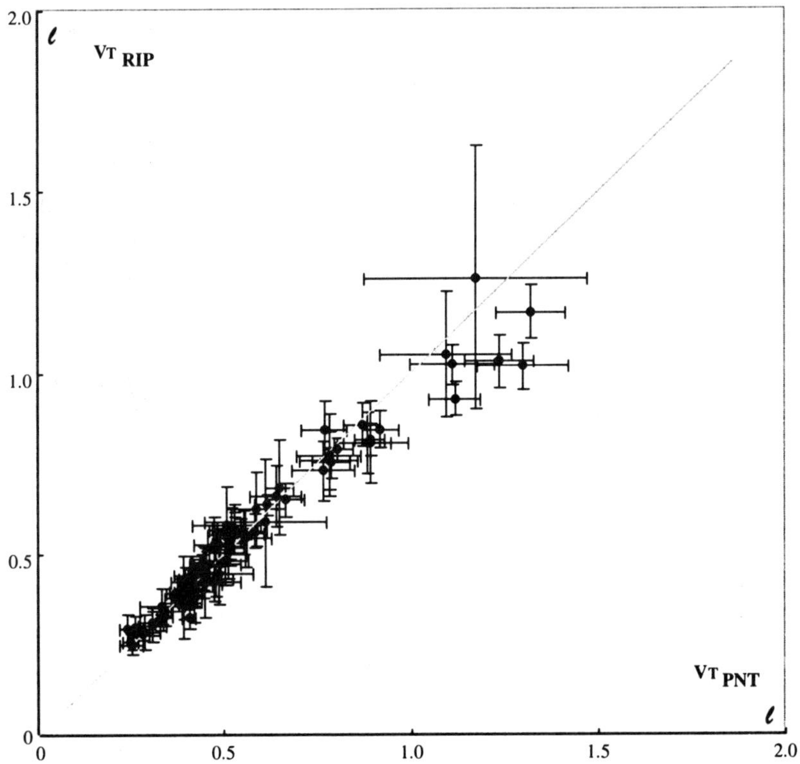

Figure 1. Tidal volume obtained by integration of flow signal versus tidal volume obtained with the sum RIP signal for all subjects in the nine conditions. Each plot is the mean value for a recording represented with its SD.

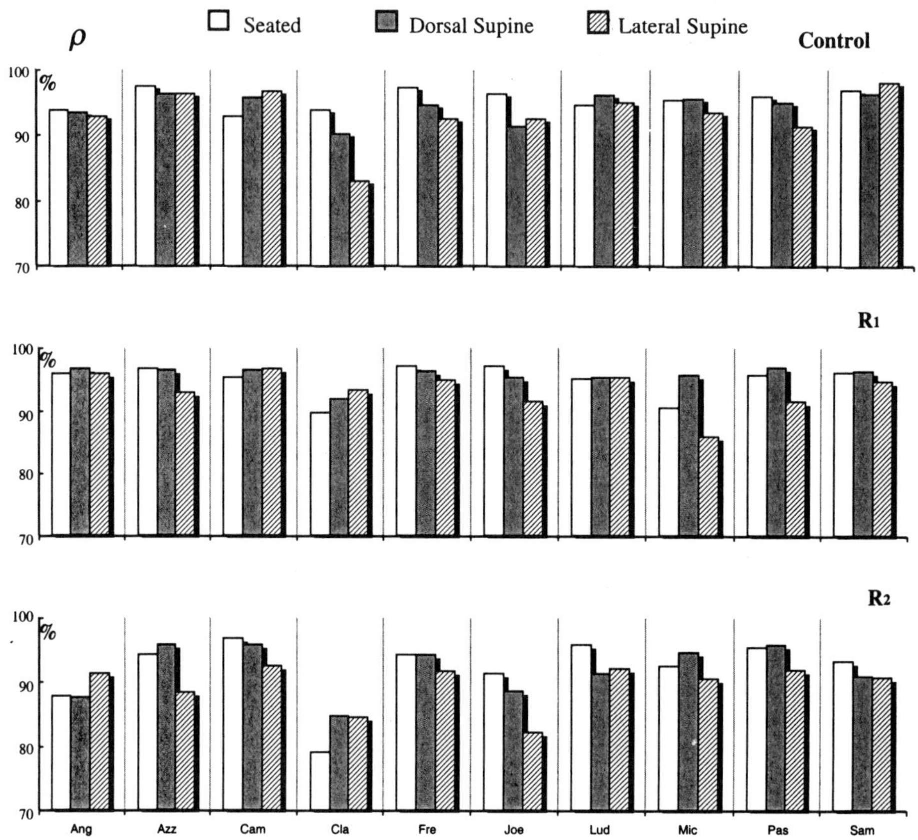

Figure 2. Goodness of fit of the RIP sum derivative to the airflow signal.

Resistances were calculated breath-by-breath both from mouth pressure-F_{PNT} and mouth pressure-F_{RIP} plots according to $P = -(R_{CT} + R_F.|F|).F$, with R_{CT} and R_F resulting from a least squares approximation on a breath. For a breath #k, we have:

$$R_M^k = R_{CT} + R_F \bullet \underset{i \in breath}{mean} |F_i|$$

A mean value was then calculated for the whole recording both with F_{PNT} and F_{RIP}:

$$R_M = \underset{k}{mean} \cdot R_M^k$$

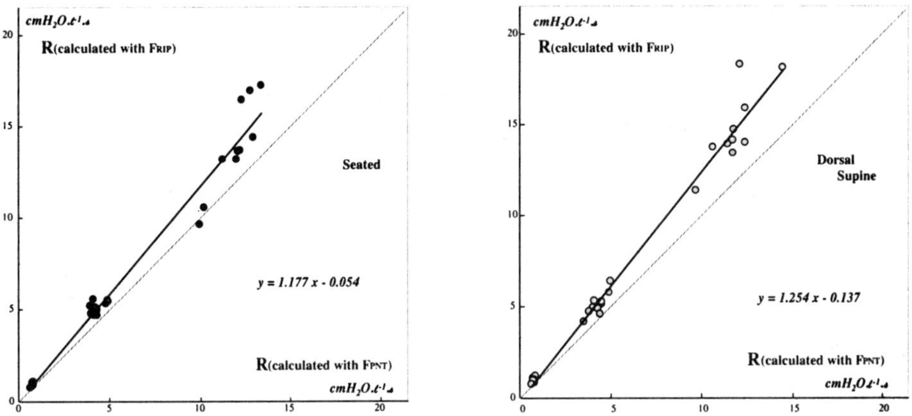

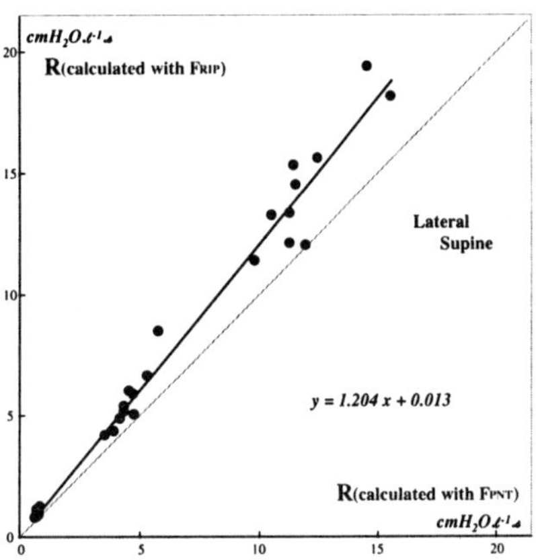

Figure 3. Comparison of the resistances calculated with airflow and with RIP sum derivative in control and loaded conditions.

3. RESULTS

Figure 1 shows VRIP versus VPNT for all of the recordings performed for this study. Calibration was performed following the method of Banzett et al.[6].

Figure 2 shows an estimate of the differences between the flow signal obtained from the derivative of RIP and the standard flow signal.

Figure 3 shows the value of the resistances calculated with these two signals using the same pressure values.

4. DISCUSSION

4.1. Volume Comparison

VRIP versus VPNT appears to be satisfactory. Discrepancies appear for those values of tidal volume higher than 1 liter. This may be explained by the fact that for higher values of tidal volume, the cross sectional area changes of rib cage and abdomen are not the sole components of volume changes.

4.2. Flow Comparison

Figure 2 shows the goodness of fit of the RIP sum derivative to the airflow signal for all subjects and different conditions. This figure shows that the best fit is obtained in the control condition. At the resistance R_1 (~ 5cm $H_2O/l/s$) the fit remains well as ρ in most cases is higher than 90%. At R_2 (~ 14cm $H_2O/l/s$) the fit is clearly lesser for all subjects. Whereas posture does not appear to modify these results, it is noteworthy that there are differences among subjects. Indeed, for subject Cla the value of ρ is lesser than for other subjects in all conditions and for subjects Ang and Joe ρ decreases more than for the other subjects at the highest load condition.

4.3. Resistance Comparison

These comparisons are the most important as far as the measurements obtained with RIP sum and RIP sum derivative are intended to be used for calculation of mechanical parameters of the respiratory system.

These plots show that the resistance calculated with RIP sum derivative compared to the value calculated with the flow is, on the whole, overestimated by 20% suggesting that RIP sum derivative underestimates the flow values. However, it should be noticed that these are external added resistances and results may be different with respiratory resistances.

In conclusion, these results suggest that further study deserve to be carried out to predict the conditions (posture, load, subjects or the airflow shape) for which RIP sum derivative may provide reasonable estimation of ventilatory parameters.

5. ACKNOWLEDGMENT

This work was supported by RBI Company.

6. REFERENCES

1. P. Martinot-Lagarde, R. Sartene, M. Mathieu and G. Durand. What does inductance plethysmographie really measure ? *J. Appl. Physiol.* **64**(4), 1749-1756 (1988).
2. S. A. Clark, C. R. Wilson, M. Satoh, D. Pegelow, and J. A. Dempsey. Assessment of inspiratory flow limitation invasively and noninvasively during sleep. *Am. J. Respir. Crit. Care Med.* **158**,713-722 (1998).
3. K. E. Bloch, Y. Li, M. A. Sackner and E. W. Russi. Breathing pattern during sleep disruptive snoring, *Eur. Respir. J.* **10**, 576-586 (1997).
4. V. Kaplan, J. N. Zhang, E. W. Russi and K. E. Bloch. Detection of inspiratory flow limitation during sleep by computer assisted respiratory inductive plethysmography. *Eur. Respir. J.* **15**, 570-578, (2000).
5. M. A. Sackner, A. S. V. Rao, S. Birsh, N. Atkins, L. Gibbs, and B. Davis. Assessment of time-volume and flow-volume components of forced vital capacity, *Chest* **3**, 272-278 (1982).
6. R. B. Banzett, S. T; Mahan, D.M. Garner, A. Brughera and S. H. Loring. A simple and reliable method to calibrate respiratory magnetometers and respitrace. *J. Appl. Physiol.* **79**(6), 2169-2176 (1995).

RESPIRATORY INSTABILITY AND VARIABILITY: WHY IS RESPIRATION SUCH A MESS?

A workshop summary

Workshop Chair: Michael C.K. Khoo*

1. INTRODUCTION

Although the terms "instability" and "variability" are often used synonymously, the respiratory control system does not have to be unstable for it to produce variable ventilatory patterns. A wide range of physiological mechanisms could lead to respiratory variability, such as fluctuations in state or behavioral inputs, cardiovascular and cerebrovascular influences, time-varying changes in the chemoreflexes, and nonlinear behavior of the central pattern generator. The panel of experts we have assembled here will discuss their latest views on these mechanisms, and comment on how these views have influenced and continue to influence the modeling of the control of breathing.

2. BREATH-TO-BREATH AND LONG-TERM VARIABILITY (P.A. Robbins*)

It has long been recognised that breath-by-breath pulmonary ventilation has some autocorrelative structure. Two questions that naturally arise from this observation are: (a) where does the autocorrelative structure come from, and (b) what form does the structure take. In relation to the first question, two alternatives are that the autocorrelative structure arises from the feedback associated with the chemoreflex control of breathing or that the structure arises within the central neural mechanisms controlling breathing. In our previous study (Liang et al., *J. Appl. Physiol.* 81:2274-2286, 1996), in which respiration was monitored at rest and during exercise, we found that the autocorrelative structure persisted under conditions where end-tidal P_{CO2} was regulated via end-tidal forcing (59 out of 60 datasets were non-white), suggesting that a considerable component of the

* M.C.K. Khoo (Chair), University of Southern California, Los Angeles, CA 90089. G. Benchetrit, Faculte de Medicine de Grenoble, France; N.S. Cherniack, UMDNJ-New Jersey Medical School, Newark, NJ; R. Hughson, University of Waterloo, Canada; G. Longobardo, Bronxville, NY; D. Paydarfar, University of Massachusetts, Worcester, MA; P.A. Robbins, Oxford University, UK; S. Shea, Brigham & Women's Hospital, Boston, MA.

structure arises through central neural mechanisms. In order to exclude the possibility that the residual errors in end-tidal P_{CO2} were responsible for the autocorrelation, we employed a multivariate approach to model the effects of the remaining variations in end-tidal P_{CO2} on ventilation. We found that incorporating these variations into a multivariate model only had modest (~10%) effects on the autoregressive coefficients for ventilation. In relation to the form of the structure, white residuals could normally be obtained after fitting a simple autoregressive (AR_1MA_1) model, or an analogous simple state-space model. We also employed evolutionary spectral analysis in order to investigate whether the data were stationary. Only 13 of the 60 data sets had a constant power spectrum throughout, but 50 out of the 60 had a uniformly modulated power spectrum. The question then arises, what effect does this form of nonstationarity have on the estimated coefficients of the models. We used an autoregressive estimate of the variance to demodulate the data, and then refitted the models to these demodulated data. We found that the coefficients of the models were not significantly affected by demodulating the data prior to fitting the models. A theoretical basis to this result is given through a heteroscedastic form of state-space model. In this form, the variance of the white noise sources is modulated via some general state- and time-dependent function. Despite this added complexity, it can be shown that such a model results in the same Kalman filter equations as the simpler case when the white noise sources have a constant variance. In other words, a simple state-space model is adequate to describe the noise processes even though the data may show a degree of nonstationarity and the noise sources in the model may not have constant variance.

Turning to the issues of long-term variability, we have recently demonstrated that the between-day variability in the ventilatory response to hypoxia is substantially greater than the within-day variability (Zhang and Robbins, *J. Appl. Physiol.* 88:1924-1932, 2000). Possible causes clearly include differences in underlying metabolic acid-base status or metabolic rate, but we were not able to detect differences which were adequate to explain the magnitude of the between-day variability. Thus, the question of what does underlie this variability remains unanswered. Physiologically, however, one speculation of some importance is whether we should view the chemoreflex sensitivities as much more plastic than we have to date. If so, then understanding the source of the between-day variability in these sensitivities may be the same problem as understanding the factors that regulate the chemoreflex sensitivities themselves.

3. STATE-RELATED AND CIRCADIAN VARIABILITY (S. Shea*)

My view is that breathing, in fact, is not such a mess – at least when we are in deep non-rapid-eye movement (NREM) sleep. During that state, breathing is at an essentially steady state and we know the characteristics of the control system – namely chemical control system as classically described, mechanoreceptive reflexes and damping properties of the neural network (e.g., short-term potentiation). Breathing is only a mess when we consider breathing during consciousness (or during REM sleep). In addition to those pathways involved in respiratory control during NREM sleep, during wakefulness, respiratory control is subject to influences from a myriad of other sources – sometimes competing for use of the respiratory muscles (e.g., changing thoracic position requires activation of respiratory muscles for non-respiratory purposes). These variable influences include voluntary control of breathing (probably cortical); emotional inputs (probably limbic); arousal related inputs (probably reticular activating system); and direct effects of afferent stimulation (such as visual or auditory afferents, in addition to mechanoreceptive afferents from working limbs). Furthermore, metabolism can change from moment-to-moment – which can affect breathing pattern from breath-to-breath (likely via metaboloreceptors in the working limbs and chemoreceptors, i.e. classical central and peripheral chemoreceptors plus distributed chemoreceptors throughout the CNS and potential receptors sensitive to flow of carbon dioxide to the lungs). Some of these respiratory influences associated with wakefulness do not involve classical negative feedback, but rather involve feedforward (anticipatory) control, or may be considered non-homeostatic by-products of altered postures, behavior or levels of arousal. Many of my studies have carefully documented the numerous behavioral changes in breathing in relatively steady-state behavioral conditions (Shea, *Exp. Physiol.* 81:1-26, 1996). This conglomeration of inputs clearly contribute to the variability of breathing patterns on a breath-to-breath basis in conscious humans. A striking example of such behavioral inputs to breathing comes from some personal observations on Cheyne-Stokes breathing in awake patients with congestive heart failure. When ventilation was measured with noninvasive techniques – such as respiratory inductance plethysmography, we observed classic Cheyne-Stokes cycling. However, breathing became essentially regular from breath to breath when the subject was asked to breathe through a mouthpiece or to perform mental arithmetic. I have found that merely opening the eyes to look at a blank white screen results in a systematic increase in breathing of 6% (range among 36 subjects was –15% to +45%). As a field, if we are to understand the variability of breathing we need to appreciate that all of these behavioral inputs to breathing are changing on a moment-to-moment basis, as we close and open our eyes or think of different things. Furthermore, the situation is further complicated by the fact that with these ostensibly modest behavioral changes, there can be concomitant moment-to-moment changes in metabolism, which can subsequently affect breathing. For instance, I have measured systematic increases in metabolism when subjects open their eyes or read, or listen to a story, or perform mental arithmetic. However, I have only measured the overall changes over relatively long periods (4-5 minutes). If we want to understand what affects breathing on a breath-by-breath basis – then we have to develop the tools to measure changes in voluntary, emotional, arousal, afferent and metabolic effects on breathing all on a breath-to-breath basis. Currently these tools do not exist. Perhaps the most promising data in this area relates the change in breathing to moment-to-moment changes in arousal level detected in the EEG during sleep onset. As one oscillates between wakefulness and sleep, it has been shown that there is a very good correlation between the EEG alpha power and ventilation – even in healthy subjects (who do not snore or have sleep apnea). It may be that breathing is equally well correlated with changes in EEG characteristics even during different stages of wakefulness (encompassing drowsiness to focused attention). However, to my knowledge, this type of analysis has not been performed.

We have also tried to establish how regular breathing can be in humans when measured under the most stable of behavioral and environmental conditions over long periods of time. Working with Dr. Christina Spengler, who presents data in a separate chapter of this book, we found that by keeping subjects awake in constant environmental and behavioral conditions after appropriate acclimatization, there emerged a systematic change in breathing related to the output of the underlying circadian pacemaker, as well as a small effect of sleep deprivation. However, even when we account for effects of circadian rhythms and of the duration of wakefulness, more than half of the variability in the individuals' data remained unexplained for the resting eucapnic end-tidal P_{CO2} level and the slope of the ventilatory response to hypercapnia (HCVR). The coefficient of variation that remained was about 20% for HCVR. Presumably this unexplained variability in the resting eucapnic level and the slope of the HCVR is related to random fluctuations in behavioral inputs to breathe (even under these stable conditions, we cannot control what subjects are thinking about).

4. MODELING OF RESPIRATORY VARIABILITY (G. Benchetrit*)

In my opinion, the breathing pattern is not a mess since we can predict part of its variability and the rest is about the same amount that the variability of other biological variables. Analyzing time-series derived from various breathing pattern variables (T, Vt, Ti, Te, etc.), several groups have shown that the breath-to-breath fluctuations in these variables were not purely random. A simple autoregressive model was proposed : $(X_n - \underline{\mathbf{X}}) = \alpha\,(X_{n-1} - \underline{\mathbf{X}}) + e_n$ where $\underline{\mathbf{X}}$ is the mean value of the variable over the series, α is a constant and generally found to be positive (~ 0.5), and e_n is the disturbance. Such a model implies that the series are stationary with a mean value and that there are values of X which are above and below this mean value. But it also means that $(X_n-\underline{\mathbf{X}})$ and $(X_{n-1}-\underline{\mathbf{X}})$ have the same sign, that the more X_{n-1} is far from $\underline{\mathbf{X}}$, the more is the difference between X_n and X_{n-1}. It must be noted that the series we are taking into consideration arise from recordings at rest on subjects which are seated quietly in a calm environment. There is no voluntary or behavioral component in the control of the observed pattern and automatic control solely is in play. This is an assumption, because we do not know or at least we have not measured the effect of 'thinking' on breathing. In addition, it has not been shown that in animals with isolated respiratory centers or in humans at Stage 4 sleep breathing is less variable. It remains that we do not know why these disturbances arise and what is their effect. On the other hand, with the autoregressive model, the unexplained variability is less than 10% of the mean value. This is in the range of the biological variability.

5. IS THE VARIABILITY IN BREATHING FRACTAL AND/OR CHAOTIC? (R. Hughson*)

Some preliminary observations from my laboratory have indicated that this variability might be fractal in nature. It has been observed that the slope of the log spectral power versus log frequency relationship ($1/f^{\alpha}$) can be modified by changing the input to the peripheral chemoreceptors (Tuck et al., *Adv. Exp. Med. Biol.* 393:297-302, 1995). That is, the slope became steeper with hyperoxia and flatter, approaching white noise, in hypoxia. These results suggested that increased input from the peripheral chemoreceptors during hypoxia regulated the respiratory control system within much tighter limits.

Several investigations have attempted to determine if the pattern of breathing can be described as chaotic. The Lyapunov exponent has been used as a tool to investigate the deterministic nature of the pattern of breathing. Although positive Lyapunov exponents have been measured, there is need for considerable caution in attributing this to a deterministic control system because surrogate data constructed from the original data sets also yielded a positive Lyapunov exponent (Hughson et al., *Adv. Exp. Med. Biol.* 393:15-19, 1995). Still, the excellent work of Sammon et al. (*J. Appl. Physiol.* 75:887-901, 1993) has shown that a chaotic attractor can be described in the pattern of lung volume in rats. Perhaps it is the continuous nature of the data sets examined by these investigators versus the breath-by-breath data processed in other studies that could account for the different findings. It is appropriate to be cautious when attempting to describe fractal and/or chaotic properties in the pattern of breathing in both health and disease.

6. NONLINEAR DYNAMICS OF THE RESPIRATORY OSCILLATOR (D. Paydarfar*)

A conspicuous feature of biological oscillators is their ability to exhibit abrupt and lasting transformation of activity from an oscillatory to an arrhythmic state. Conversely, arrhythmic behaviors can switch to highly regular oscillations. The mechanisms that initiate and maintain such leaps in activity, between arrhythmic and oscillatory states, are poorly understood but observations in some experimental systems suggest that very small stimuli that impinge upon or arise within the oscillator may play a critical role (Paydarfar and Buerkel, *Chaos* 5:18-29, 1995). We have analyzed the Bonhoeffer-van der Pol and the Hodgkin-Huxley models of neural excitability. Numerical analysis provides a comprehensive map of the effects on oscillatory state of continuous aperiodic stimulation (noise), and point to several principles of temporal organization: (a) Noise can force abrupt transitions between oscillatory and arrhythmic behaviors, by perturbing the oscillator's trajectory between a limit cycle and a singular point. (b) The duration of arrhythmic lapses in oscillatory behavior is critically dependent upon the attributes of stimulation (e.g., intensity, directional effects, frequency spectrum). The duration is maximized when the noise drives the oscillator into the singular point's basin of attraction but the noise is not so intense that the trajectories are displaced away from the singular point and back to the limit cycle. (c) Stimuli that mildly inhibit the rising action potential are among the most potently pro-rhythmic because such perturbations can widen the distance between the singular point and oscillatory trajectories (Paydarfar and Buerkel, *Med. Hypoth.* 49:19-26, 1997). These studies provide a framework for understanding how a variety of biological oscillations are regulated by low-level stimuli, and point to novel methods of elucidating physiological mechanisms of excitability. We are currently applying this approach to the respiratory oscillator, and disturbances in control mechanisms that sporadically initiate and terminate central apnea.

7. CHEMOREFLEX INSTABILITY (G. Longobardo* and N.S. Cherniack*)

While variability in breathing can occur as a result of perturbation from other physiological systems, environmental stimuli, and other sensory inputs, the response of the chemoreflex control system to disturbances during sleep can still be oscillatory. When disturbed, ventilation can be damped or can be patently unstable. Current chemoreflex models tell us some of the variables that determine just how oscillatory the ventilation will be when disturbed. A few among these are controller ventilatory responses and brain dynamics, which includes metabolism, mass and cerebral blood flow. Although there is generally agreement on the controller characteristic for eucapnic and hypercapnic operation, there is still some disagreement about the structure of the oxygen behavior of the controller. However at sub-normal CO_2 levels there are still some issues to be resolved. For example, the location of the apex of the oxygen fan--is it on the zero ventilation axis or is it below it? The former possibility results in less stable control, and also implies that one can hyperventilate oneself to dangerously low oxygen levels. And at these subnormal CO_2 levels, how should ventilatory activity be described? In the very early models that we used to simulate Cheyne-Stokes breathing, controllers had, necessarily, provision for apnea at subnormal carbon dioxide levels. The carbon-dioxide threshold for breathing varied with oxygen, lower for lower oxygen. The model can simulate apneas even during sleep, but not the responses of normal awake people, where apneas do occur, but only rarely, as in voluntary hyperventilation. Nielsen and Smith in experiments on awake subjects found that at subnormal P_{CO2}, ventilation was independent of P_{CO2} and varied only with oxygen, the so-called "dogleg". They suggested that the P_{CO2} "threshold", where the ventilation starting rising with increasing P_{CO2}, was the common threshold for both the "central" receptor and the peripheral receptor at normoxia. Duffin has proposed constant ventilation at subnormal P_{aCO2}, independent of both CO_2 and O_2 tension with the P_{CO2} threshold varying with oxygen tension. The threshold is the intersection between the constant "basal" ventilation and the controller fan. And like the Nielsen and Smith model the chemical controller takes over at that point where ventilation starts rising with P_{CO2}. For the several models, comparing the responses to various disturbances shows that the Cheyne-Stokes model and the Duffin implementation with its "basal" ventilation yield similar ventilatory responses to disturbances like hyperventilation and short and long-term hypoxia. They are both highly underdamped, and highly oscillatory. The Nielsen and Smith model is well damped and shows few oscillations after hyperventilation or hypoxia, only oscillating for brief periods when P_{aCO2} is going through the "dogleg" threshold. Of course, none of the models can withstand high circulation delays, and all showed continuous oscillations when circulation time was 30 seconds.

We are presenting at this conference a different controller model. It consists of a chemical controller that recognizes the experimental evidence for chemoreceptor activity to low levels of P_{CO2}. The controller is the usual chemical controller at normal and above normal P_{CO2} levels but at lower P_{CO2} levels there is a transition to a much lower slope for a given oxygen level, a quasi-curvilinear controller. Added to the impulses from the chemoreceptors is a sensory input, a wakefulness drive, similar to what Dr. Khoo proposed several years ago, but here it is always present regardless of P_{CO2} level and does not disappear at some P_{CO2} threshold, as the others do. The impulses of the neural input are added to those of the chemical input. If there are sufficient impulses to exceed the ventilation threshold, there is breathing. If not, there is apnea. The sensory input can vary with alertness, emotion, behavior, and is in effect a variable factor added in an attempt to simulate awake breathing. This model is more stable than those previously discussed, but with variations in the wakefulness drive it is able to mimic variations seen in awake breathing. We have also included in the model another sensory input, post-synaptic potentiation. It was not included in the results discussed previously. Our preliminary results show that while PSP shortens apneic periods, after hyperventilation for example, it does not change the essential character of the oscillations in the response.

INDEX